Fertilization

FERTILIZATION

Edited by

DANIEL M. HARDY

Department of Cell Biology and Biochemistry
Texas Tech University Health Sciences Center
Lubbock, Texas

ACADEMIC PRESS

A Division of Harcourt, Inc.

San Diego San Francisco New York Boston London Sydney Tokyo

Cover credits:
Pig spermatozoa labeled by immunofluorescence to detect the protein zonadhesion on the apical head. Phase contrast and epifluorescence double exposure. Photograph by Daniel Hardy.

Inset shows mouse fertilization *in vitro*. Differential interference contrast image of a mouse egg with some cumulus cells and interacting spermatozoa. Photograph by Nathaly Cormier, laboratory of Gail Cornwall.

This book is printed on acid-free paper. ♾

Academic Press
A Division of Harcourt, Inc.
525 B Street, Suite 1900, San Diego, California 92101-4495, USA
http://www.academicpress.com

Academic Press
Harcourt Place, 32 Jamestown Road, London NW1 7BY, UK
http://www.academicpress.com

Library of Congress Catalog Card Number: 2001091530

International Standard Book Number: 0-12-311629-5

PRINTED IN THE UNITED STATES OF AMERICA
01 02 03 04 05 06 EB 9 8 7 6 5 4 3 2 1

Contents

3

CAPACITATION

BIJAY S. JAISWAL AND MICHAEL EISENBACH

4

FUNCTION OF THE EGG'S EXTRACELLULAR MATRIX

RICHARD A. CARDULLO AND CATHERINE D. THALER

5

SPERM ADHESION TO THE EXTRACELLULAR MATRIX OF THE EGG

MING BI, MICHAEL J. WASSLER, AND DANIEL M. HARDY

6

SIGNAL TRANSDUCTION MECHANISMS REGULATING SPERM ACROSOMAL EXOCYTOSIS

GREGORY S. KOPF

7

REGULATION OF SPERM ION CURRENTS

ALBERTO DARSZON, FELIPE ESPINOSA, BLANCA GALINDO, DANIEL SÁNCHEZ, AND CARMEN BELTRÁN

8

FUNCTION OF THE SPERM ACROSOME

GEORGE L. GERTON

9

GAMETE FUSION IN MAMMALS

PAUL PRIMAKOFF AND DIANA G. MYLES

10

MEMBRANE EVENTS OF EGG ACTIVATION

KARL SWANN AND KEITH T. JONES

SECTION II

UNIQUE PROBLEMS AND APPLICATIONS

11

MOLECULAR GENETICS OF FERTILIZATION

PATRICIA OLDS-CLARKE AND STEPHEN H. PILDER

12

GAMETE IMMUNOBIOLOGY

M. G. O'RAND AND I. A. LEA

13

FERTILIZATION BIOPHYSICS

D. P. L. GREEN

14

APPLICATIONS OF FERTILITY REGULATION FOR THE MANAGEMENT OF WILD AND DOMESTIC SPECIES

MARK P. BRADLEY AND PETER BIRD

CONTRIBUTORS

Numbers in parentheses indicate the pages on which the authors' contributions begin.

Carmen Beltrán (225), Departamento de Genética y Fisiología Molecular, Instituto de Biotecnología, Universidad Nacional Autónoma de México, Cuernavaca, Morelos 62271, México

Ming Bi (159), Department of Cell Biology and Biochemistry, Texas Tech University Health Sciences Center, Lubbock, Texas 79430

Peter Bird (401), Therapeutics Goods Administration (Bacterial Vaccine Stream), Woden, Australian Capital Territory 2606, Australia

Mark P. Bradley (401), Xcelerator Ltd., North Ryde, New South Wales 2113, Australia

Richard A. Cardullo (119), Department of Biology, University of California, Riverside, California 92521

Alberto Darszon (225), Departamento de Genética y Fisiología Molecular, Instituto de Biotecnología, Universidad Nacional Autónoma de México, Cuernavaca, Morelos 62271, México

Michael Eisenbach (57), Department of Biological Chemistry, The Weizmann Institute of Science, 76100 Rehovot, Israel

Felipe Espinosa (225), Departamento de Genética y Fisiología Molecular, Instituto de Biotecnología, Universidad Nacional Autónoma de México, Cuernavaca, Morelos 62271, México

Blanca Galindo (225), Departamento de Genética y Fisiología Molecular, Instituto de Biotecnología, Universidad Nacional Autónoma de México, Cuernavaca, Morelos 62271, México

David L. Garbers (29), Cecil H. and Ida Green Center for Reproductive Biology Sciences, University of Texas Southwestern Medical Center, Dallas, Texas 75390

George L. Gerton (265), Center for Research on Reproduction and Women's Health, University of Pennsylvania Medical Center, Philadelphia, Pennsylvania 19104

D. P. L. Green (387), Department of Anatomy and Structural Biology, University of Otago Medical School, Dunedin, New Zealand

Daniel M. Hardy (159), Department of Cell Biology and Biochemistry, Texas Tech University Health Sciences Center, Lubbock, Texas 79430

Bijay S. Jaiswal (57), Department of Biological Chemistry, The Weizmann Institute of Science, 76100 Rehovot, Israel

Keith T. Jones (319), Department of Physiological Sciences, University of Newcastle, Newcastle NE2 4HH, United Kingdom

Gregory S. Kopf (181), Center for Research on Reproduction and Women's Health, University of Pennsylvania School of Medicine, Philadelphia, Pennsylvania 19104

I. A. Lea (367), Department of Cell and Developmental Biology, University of North Carolina, Chapel Hill, North Carolina 27599

Diana G. Myles (303), Section of Molecular and Cell Biology, School of Medicine, University of California, Davis, California 95616

Patricia Olds-Clarke (349), Department of Anatomy and Cell Biology, Temple University School of Medicine, Philadelphia, Pennsylvania 19140

Michael G. O'Rand (367), Department of Cell and Developmental Biology, University of North Carolina, Chapel Hill, North Carolina 27599

Stephen H. Pilder (349), Department of Anatomy and Cell Biology, Temple University School of Medicine, Philadelphia, Pennsylvania 19140

Paul Primakoff (303), Department of Cell Biology, School of Medicine, University of California, Davis, California 95616

Daniel Sánchez (225), Departamento de Genética y Fisiología Molecular, Instituto de Biotecnología, Universidad Nacional Autónoma de México, Cuernavaca, Morelos 62271, México

Susan S. Suarez (3), Department of Biomedical Sciences, College of Veterinary Medicine, Cornell University, Ithaca, New York 14853

Karl Swann (319), Department of Anatomy and Developmental Biology, University College, London WC1E 6BT, United Kingdom

Catherine D. Thaler (119), Department of Biology, University of Central Florida, Orlando, Florida 32816

Timothy A. Quill (29), Cecil H. and Ida Green Center for Reproductive Biology Sciences, University of Texas Southwestern Medical Center, Dallas, Texas 75390

Michael J. Wassler (159), Department of Cell Biology and Biochemistry, Texas Tech University Health Sciences Center, Lubbock, Texas 79430

PREFACE

Fertilization is the bridge between generations. It is an amazing process that might appear vulnerable to failure for its complexity, but in fact is noteworthy for its robustness. Its product is a single cell with the extraordinary capacity to develop into a new, genetically unique member of its species. This book relates the saga of egg and sperm, each in its own right a fascinating cell, as they unite to form that pluripotent zygote. The fourteen chapters are written by scientists actively conducting research on their respective topics. Collectively, the chapters tell how we think the bridge between generations is built.

In the ten chapters of Section I, events constituting the fertilization process are reviewed in approximate temporal order, exceptions being events that occur simultaneously or that need not occur sequentially for fertilization to succeed. The emphasis is on fertilization in animals, particularly mammals, reflecting the current predominance of research on these species. In Section II, four chapters review unique problems and applications of fertilization research. For example, sterility cannot "run in the family," so how can we do genetics? Also, what are the biophysical implications of a sperm cell's unusual shape? The chapters in both sections answer many questions, but of course leave many more unanswered.

What is fertilization research? Is it cell biology? Leeuwenhoek's 300-year-old sketches of spermatozoa and countless micrographs published since then argue yes. Is it animal science? Yes; consider the numerous fertilization studies done in swine, cattle, horses, and chickens. Is it marine biology? Again the answer is yes. Many of the great discoveries in the field were made at marine labs, their faculty studying sea urchins, starfish, abalones, and others. Biochemistry? Yes. The first members of several gene families, including the receptor guanylyl cyclases ADAMs, and CRISPs were discovered by fertilization biochemists. Medicine? Of course. Think of the thousands who would not be among us were it not for assisted reproduction procedures.

Is there a scientific concept that unites these disciplines as applied to the study of fertilization? Here is one possibility. To clarify the distinction between a then new field and its predecessor, Sydney Brenner explained that traditional biochemistry focused on the flow of matter and energy, whereas the new "molecular biology" focused on the flow of *information*. Much of molecular biology research aims to characterize how information flows from mother cell to daughter cell, and from gene to protein. Fertilization research has a similar aim. It seeks to determine how the information in two halves of an *entire genome* is assembled to create that extraordinary cell with the capacity to become a new individual. In this sense, regardless of experimental approach or animal model, fertilization research is molecular biology on a grand scale. In fact, with recent progress in understanding relationships between the species specificity of fertilization and the actual formation of new species, we are learning not just how information flows from generation to generation within species, but also how it has flowed from ancestral species to their descendants throughout all of animal life.

Life scientists are quick to point out the practical benefits of their work. We do it in an attempt to communicate the need for the work (and funding) to continue. Fertilization researchers are no exception. Our field has enjoyed some spectacular successes that are already making the world a better place. Many previously infertile couples can now have children thanks to treatments such as *in vitro* fertilization and intracytoplasmic sperm injection. The proven ability to make transgenic animals holds out hope that germ line gene therapy will one day be possible. Discoveries by fertilization researchers also established the feasibility of technologies such as human cloning, production of embryos for stem cell therapies, and manipulations to dictate the sex of a child. Regardless of one's position on the ethics of such applications, it is clear that fertilization research is raising and defining key issues that challenge humanity's values. Ultimately the discoveries are forcing us to decide how we will cope with our own power to shape the future. In short, fertilization is where the action is.

It is easy to use practical successes to justify further research on fertilization. But ends cannot and need not justify means. We do not need examples of beneficial applications to make strong arguments for more research. The plain truth is that we must understand fertilization to appreciate fully who we are and how we came to be. The importance of the question is enough to justify research that is conducted simply to satisfy our curiosity.

It appears that many book prefaces are written with a couple of intentions. Some seem intended to make potential readers want to own the book. Others suggest ways for readers to use the book, thereby increasing its effectiveness as a source of information. If this preface has done either, so much the better; those were not my objectives. More than anything else, I hope to convey the significance of fertilization studies and to provide a glimpse of the fascination researchers in the field have for the topic.

This book effectively delineates the major puzzles that remain in our field. Perhaps it will also help sustain and promote interest in solving them.

Daniel M. Hardy

Acknowledgments

A book like this one depends heavily on the hard work of the chapter authors. My thanks go to all of them for their fine efforts. I also thank the current and past members of my lab, especially Ming Bi, Tony Cheung, John Hickox, Steve Tardif, and Michael Wassler. Their dedication and great attitudes make it fun to come to work every day. Finally, I thank Amelia and Kenneth, who are living proof of Nature's greatness, and Diana for her inspiration and support.

SECTION

I

Fertilization Events

1

Gamete Transport

Susan S. Suarez

Department of Biomedical Sciences, Cornell University, Ithaca, New York

I. Introduction

How do spermatozoa and oocytes reach each other? The detailed answer for each species is unique and the variety of answers, even among the vertebrates, is astonishing. In fishes, for example, the sperm cell may reach the egg externally or within the female—or the eggs may be deposited in a compartment in the male. Because there is substantial variation in the details of the transport of gametes, this chapter is principally focused on mechanisms for transporting gametes in eutherian mammals.

Bringing gametes together presents challenges. In mammals, oocytes are usually fertilized within a few hours of ovulation. Spermatozoa, however, may have to survive for months after the completion of spermatogenesis and release from the testis—and not just within the body of the individual that produced them. They must also survive, sometimes for long periods, in the female. In mice, fertilization takes place within a few hours of mating, but in some bats, mating takes place during winter and fertilization occurs in the spring (Hosken *et al.*, 1996; Bernard *et al.,* 1997; Bernard and Cumming, 1997). Spermatozoa are terminally differentiated cells deprived of an active nucleus and a synthetic apparatus; therefore, they must survive without benefit of the renewal mechanisms available to other cells. In addition to surviving the ravages of time, sperm cells must endure or avoid the defenses of the female immune system. After all, these cells are allogeneic to females and contain antigens that can elicit immune responses (Menge and Edwards, 1993). For spermatozoa of many species, however, survival is not enough to ensure fertilization. They may be forced to compete with the sperm cells from other males to pass on their genetic complement [reviewed by Roldan *et al.* (1992), Birkhead and Moeller (1993), and Hosken (1997)]. They may also compete with other spermatozoa from the same ejaculate, because postmeiotic haploid gene expression places them in competition with their siblings for passing certain genotypes to the next generation (Manning and Chamberlain, 1994). Spermatozoa must not just get to the oocyte, but they must get there first. Thus, survival and competition are perhaps the two main problems to be considered in the sperm cell's strategy for fertilization success.

For the unfertilized oocyte, problems of survival and competition are probably minor. Fertilizaton occurs soon after an oocyte is released from the ovary. The major consideration for oocyte transport is to get this enormous cell directly and quickly from the ovary into the oviduct, thereby avoiding ectopic fertilization and the dangerous consequences of ectopic implantation.

II. SPERM TRANSPORT

A. SPERM TRANSPORT AND OVULATION

Spermatozoa may be required to survive for quite a long time in species with long estrous periods or long periods between estrus and ovulation. For example, some species of bats mate when they gather to hibernate or even during hibernation, and sperm cells are stored until spring, when food is plentiful for the newborns (Birkhead and Moeller, 1993). It is likely that some mammals have long estrous periods to increase chances of encountering the best males. As a result, mating early in estrus necessitates storage of spermatozoa. Mares ovulate about 5 days after the onset of estrus and therefore must be able to store sperm cells for that long (Daels *et al.,* 1991). An alternate solution to long-term sperm storage is induced ovulation. Rabbit does, for example, remain in estrus for long periods of time until they mate, which triggers the hormonal induction of ovulation.

B. SITE OF SPERM DEPOSITION

The semen of humans and other primates [reviewed by Harper (1994)], and of cattle and other ruminants [reviewed by Hawk (1987) and Harper (1994)], is deposited in the cranial vagina at the external os of the cervix. Besides being quite a distance from the ovary, this part of the tract is closest to the exterior and so is the region in which female defenses against microbial agents are most often expressed. In the rabbit, deposition of semen results in a rapid invasion of neutrophils into the vagina. Numerous leukocytes were observed in the vaginae of rabbits 3–24 hours postcoitus, many containing ingested spermatozoa (Phillips and Mahler, 1977a,b). Probably to avoid immune defenses, some species bypass the vagina and deposit spermatozoa directly into the uterine cavity, where they may quickly gain access to the oviduct. Boars and stallions deposit semen directly into the uterine cavity (Harper, 1994). Rodent semen is deposited in the anterior cervix but is rapidly moved en masse into the uterus, causing visible distention (Zambori, 1972; Carballada and Esponda, 1997; Bedford and Yanagimachi, 1992). In uterine deposition, a large volume of seminal plasma accompanies the spermatozoa. The seminal plasma inhibits immune responses (Dostal *et al.*, 1997) and it would seem to carry sperm cells rapidly to the uterotubal junction by distending the uterine lumen.

In species that use vaginal semen deposition, less semen is deposited, but it is placed right at the cervical os. Spermatozoa enter the cervical mucus within minutes [for studies in humans, see Sobrero and McLeod (1962); for studies in rabbits, see Bedford (1971)]. Large volumes of cervical mucus are produced during estrus (late follicular phase in primates) and it is highly hydrated, often exceeding 96% water in women (Katz *et al.*, 1997). Cervical mucus serves as a barrier to abnormal spermatozoa that cannot swim properly or that present a poor hydrodynamic profile; however, morphologically normal, vigorously motile cells can swim through it rapidly (Hanson and Overstreet, 1981; Barros *et al.*, 1984; Katz *et al.*, 1990, 1997).

In murine rodents, most semen is deposited or rapidly transported into the uterine cavity, but some of the seminal fluid enters the vagina, where it coagulates to form a copulatory plug. The plug appears to form a cervical cap that promotes sperm transport into the uterus (Blandau, 1969; Carballada and Esponda, 1992).

C. TRANSPORT THROUGH THE CERVIX

The cervix is immunologically competent. In rabbits and humans, insemination into the vagina stimulates leukocyte migration into the cervix as well as into the vagina, the leukocytes being primarily neutrophils and macrophages (Tyler, 1977; Pandya and Cohen, 1985). Neutrophils migrate readily through midcycle human cervical mucus (Parkhurst and Saltzman, 1994). Neutrophils were found to infiltrate rabbit cervices heavily within half an hour of mating or artificial insemination; however, the invasion was confined to the posterior cervix at the portio vaginalis (Tyler, 1977). Interestingly, it was discovered that if female rabbits were mated to a second male during the neutrophilic infiltration induced by an earlier mating, spermatozoa from the second male were still able to fertilize (Taylor, 1982); therefore, neutrophils may not present a significant barrier to spermatozoa. It has been demonstrated that neutrophils will bind to human sperm cells and ingest them only if serum that contains both complement and complement-fixing antisperm antibodies is present (D'Cruz *et al.,* 1992). This can happen *in vivo* if the female somehow becomes immunized against sperm antigens. Perhaps the leukocytic invasion normally functions to protect against microbes that accompany spermatozoa and does not present a barrier to normal, motile cells.

An elegant three-dimensional reconstruction of serial sections of the bovine cervix produced by Mullins and Saacke (1989) led them to conclude that mucosal folds in the cervix form channels leading to the uterine cavity. Furthermore, based on histochemical staining characteristics of the mucus in the sections of the cervix, they concluded that mucus deep in the channels is different in composition and less dense than that in the central portion of the cervix during the follicular phase (Figure 1.1). They proposed that spermatozoa may readily reach the uterine cavity by entering the deep channels at the external os and following them through to the uterus, avoiding, in the center of the cervical lumen, the more viscous mucus that carries out retrograde discharge of uterine contents. Mattner (1968) found that, when he flushed the cervices of goats and cows 19–24 hours after mating at the onset of estrus, he recovered approximately 90% of the mucus, more than 90% of the lumenal leukocytes, and only about half of the spermatozoa. The remaining half of the cells were found deep in the mucosal grooves, presumably protected somewhat from the leukocytes. These observations also indicate that the cervix supports the passage of normal, motile spermatozoa while discouraging passage of microbes and sperm cells with abnormal form or motility. Normal, fresh, motile spermatozoa can avoid the area most populated by neutrophils and they appear to be resistant to phagocytosis.

Immunoglobulins IgG and IgA have been detected in cervical mucus. Secretory IgA is produced locally by plasma cells in human cervical mucosa. The concentration increases in the follicular phase and decreases at about the time of ovulation (Kutteh *et al.,* 1996). The immunoglobulins would afford greater protection from microbes at the time when the cervical mucus is highly hydrated and offers the least resistance to penetration. However, when there are antibodies directed against antigens accessible on the surface of ejaculated spermatozoa, infertility can result (Menge and Edwards, 1993). Complement proteins are also present in cervical mucus (Matthur *et al.,* 1988), along with regulators of complement activity (Jensen *et al.,* 1995). Thus, the potential exists in the mucus for antibody-mediated destruction of spermatozoa in the cervix, as long as there are antisperm antibodies. Some antisperm antibodies are not of the type that activate complement; however, they can still interfere with movement of sperm cells through cervical mucus by physical obstruction (Menge and Edwards, 1993; Ulcova-Gallova, 1997).

D. TRANSPORT THROUGH THE UTERUS

Sperm transport through the uterus of vaginal depositors is thought to depend significantly on uterine contractions [reviewed by Hawk (1987), Hunter (1988), and Harper (1994)]. In uterine semen depositors, spermatozoa may leave the uterus

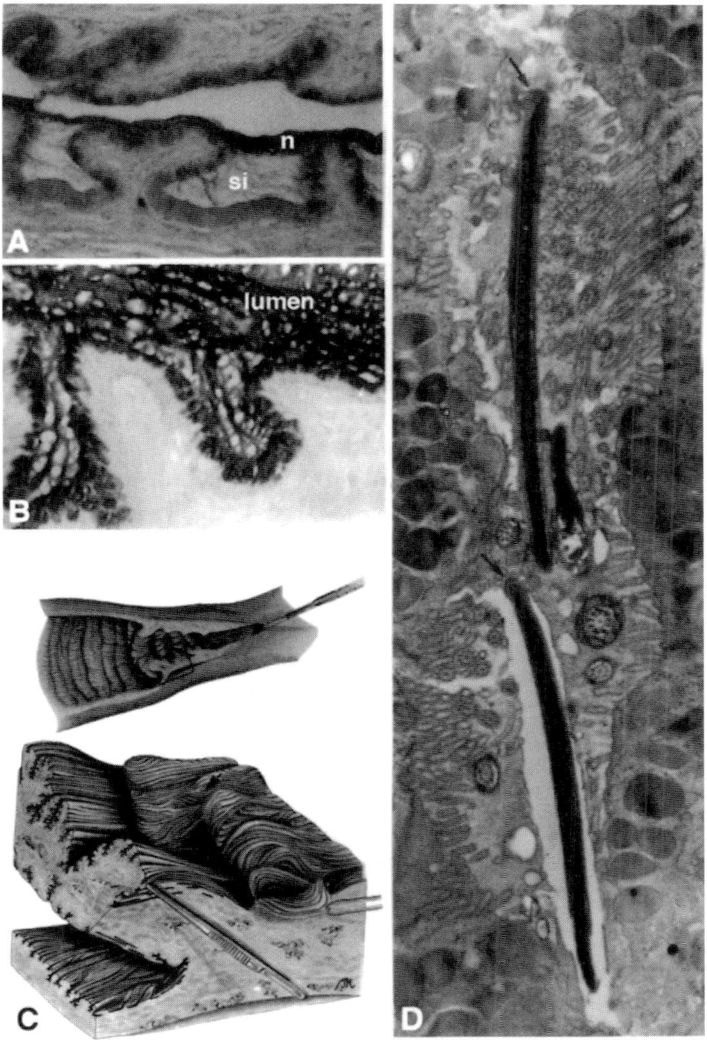

FIGURE 1.1 (A) A section of a primary fold of cervical mucosa taken from a cow (*Bos taurus*) in the follicular phase of the estrous cycle. The tissue was fixed in Bouin's solution, dehydrated, embedded in paraffin, sectioned, and stained with Alcian blue and high-iron diamine (HID). Alcian blue-positive sialomucins (si) appear to be confined to basal areas of minor grooves in the mucosa by denser-staining HID-positive neutral (n) and sulfomucin layers. This staining pattern indicates that spermatozoa encounter types of mucus in the basal area of grooves that are different from those they encounter in the central lumen of the cervix (170× magnification). (B) In a similar section, taken from a cow in the luteal phase, the Alcian blue and HID staining reveal a loss in the layered organization seen in the estrous cow (170× magnification). (C) An illustration by K. J. Mullins of the three-dimensional structure of the folds of the cervical mucosa, derived from stereomicroscopic examination of tissue stained on its mucosal surface and from three-dimensional reconstruction of serial sections. (D) Transmission electron micrograph of cervical tissue showing spermatozoa within grooves of the cervical mucosa. The rostral tips of the heads of sperm cells are indicated by arrows (7850× magnification). (*continues*)

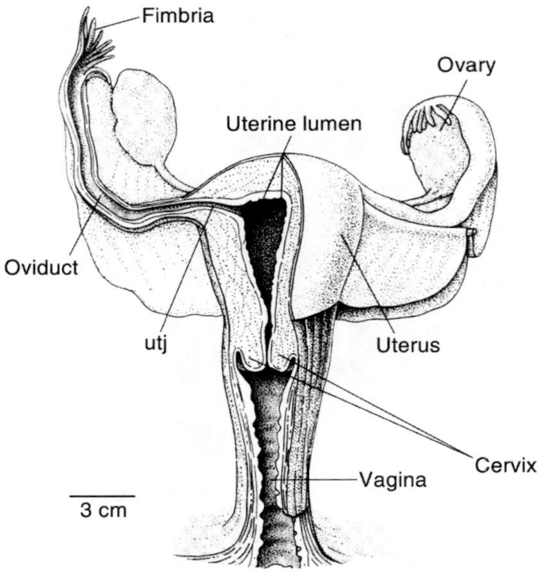

FIGURE 1.2 A drawing of the human female reproductive tract. The actual size of the cervical lumen is about 3 cm in length, and the distance from the internal os of the cervix to the entrance to the uterotubal junction (utj) is about 4.5 cm. The luminal space has been exaggerated for illustrative purposes; actually, the lumen is much narrower throughout the tract. Modified after Solomon *et al.* (1990).

rapidly to enter the uterotubal junction. In pigs, for example, sufficient numbers of cells reach the oviduct to support fertilization within half an hour of insemination (Hunter, 1981). In the case of vaginal semen deposition, spermatozoa must pass through the entire length of the uterus. The human uterine cavity is relatively small, only a few centimeters, and could be passed through rather quickly (Figure 1.2). In contrast, bovine spermatozoa have to pass through a uterine body 2.5–4 cm long and uterine horns that are 20–40 cm long before reaching the oviducts (Figure 1.3) (Roberts, 1986). About 9 hours are required for bovine spermatozoa to reach sufficient numbers in the oviduct to support fertilization (Hunter and Wilmut, 1982, 1984).

Ultrasonography has been used to reveal cranially directed waves of uterine muscle contractions that increase in the late follicular phase of humans (Kunz *et al.,* 1996). Electromyography has indicated that similar strong contractile activity occurs during estrus in cows and ewes, whereas contractions are weak and localized during the luteal phase (Hawk, 1983). Kunz and collaborators (1996) used 5- to 40-μm albumen macrospheres radioactively tagged with technetium to deter-

FIGURE 1.1 (*Continued*) Parts A–D are from Mullin and Saacke, Study of the functional anatomy of bovine cervical mucosa with special reference to mucus secretion and sperm transport. *Anat. Rec.* **226,** 106–117. Copyright © 1989. Reprinted by permission of Wiley-Liss, Inc., a subsidiary of John Wiley & Sons, Inc.

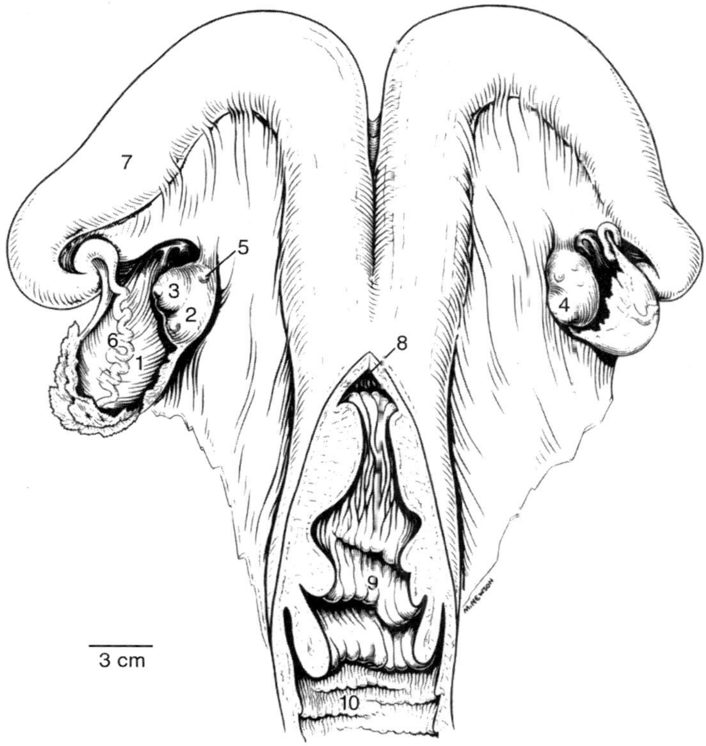

FIGURE 1.3 The dorsal aspect of the female reproductive tract of a cow (*Bos taurus*). 1, Ovarian bursa; 2, ovary; 3, corpus luteum; 4, follicle; 5, corpus albicans; 6, oviduct; 7, uterine horn; 8, uterine body; 9, cervix; 10, vagina. From Roberts (1986) by permission.

mine how such contractions might transport spermatozoa in humans. They found that spheres were rapidly and maximally transported into the uterine cavity and even into the oviductal isthmus during the late follicular phase. Interestingly, transport of the spheres was greater to the isthmus ipsilateral to the dominant follicle than to the contralateral isthmus. Contractile activity might also propel sperm cells and watery midcycle cervical mucus into the uterus and allow the mucus to aid sperm movement through the uterine cavity. Fukuda and Fukuda (1994) interpreted ultrasound images of the uteri of women in the late follicular phase to indicate that the uterine cavity was filled with mucus. This interpretation is reasonable because the volume of uterine fluid in midcycle women is only about 100 μl (Casslen, 1986) and cervical mucus is plentiful.

Rapid transport through the uterus by contraction of the myometrium may be required to enhance sperm survival. As is the case with the cervix, coitus also induces a leukocytic infiltration of the uterine cavity, which reaches a peak several hours afterward in mice (Austin, 1957). The leukocytes are primarily neutrophils and have been observed phagocytizing uterine spermatozoa in mice, rats, and rabbits (Austin, 1957; Bedford, 1965). Nevertheless, this phagocytosis was observed several hours after insemination and might be directed only against damaged cells. Also, by the time spermatozoa have entered the uterus, they may have lost some protective coating provided by seminal plasma. Seminal plasma inhibits immune responses (Suarez and Oliphant, 1982; Dostal *et al.,* 1997). Timing may be the key: shortly after coitus, sperm cells outnumber the leukocytes in the uterus and may have some protective coating. As time goes on, the leukocytes begin to outnumber the spermatozoa, which may have lost some coating, thereby exposing binding sites for leukocytes. To ensure fertilization, sperm cells may have to pass through the uterine cavity before significant numbers of leukocytes arrive.

E. TRANSPORT THROUGH THE UTEROTUBAL JUNCTION

The uterotubal junction presents a barrier to spermatozoa in most eutherian mammals. Anatomically, the lumen in species as disparate as dairy cattle and mice is particularly tortuous and narrow (Figures 1.4 and 1.5) (see Hook and Hafez, 1968; Hafez and Black, 1969; Beck and Boots, 1974; Wrobel *et al.,* 1993; Suarez *et al.,* 1997). There are large and small folds in the mucosa, some of which create grooves that end blindly. A physiological valve may be created by a vascular plexus in the wall that resembles erectile tissue and could serve to reduce the lumen: this plexus has been well described in cattle (Wrobel *et al.,* 1993). The walls of the junction and adjacent tubal isthmus contain a thick muscular layer that could further compress the lumen. Compression of the lumen may be accentuated in some species by muscular ligaments, which could act to increase the flexure of the sigmoidal passageway (Hook and Hafez, 1968; Hafez and Black, 1969). The narrowness of the lumen is especially apparent in living tissue (Suarez, 1987) and in frozen sections (Figure 1.4), wherein tissue does not shrink as it does during standard preparation of paraffin-embedded sections (Suarez *et al.,* 1997). A biochemical barrier may be present in the form of a viscous mucus, which has been described in rabbits (Jansen, 1978; Jansen and Bajpai, 1982), pigs (Suarez *et al.,* 1991), dairy cattle (Suarez *et al.,* 1990, 1997), and humans (Jansen, 1980).

For uterine semen depositors, the uterotubal junction may serve the filtration function provided by the cervix in vaginal depositors. That is, it may filter out spermatozoa with abnormal morphology or motility. In pigs (Baker and Degen, 1972), rats (Gaddum-Rosse, 1981), and hamsters (Smith *et al.,* 1988) motile spermatozoa pass through the uterotubal junction much more successfully than do immotile cells. Spermatozoa demonstrating activated (progressive) motility are more successful at passing through the uterotubal junction than are hyperactivated cells (Gaddum-Rosse, 1981; Shalgi *et al.,* 1992). In addition to removing abnormal sperm cells, the uterotubal junction may filter out seminal plasma. Seminal plasma components are left behind in the uterus and are not detected in the oviducts of rats (Carballado and Esponda, 1997).

Spermatozoa have been recovered in the cranial reaches of the oviductal ampulla only minutes after mating or insemination in several species of mammals

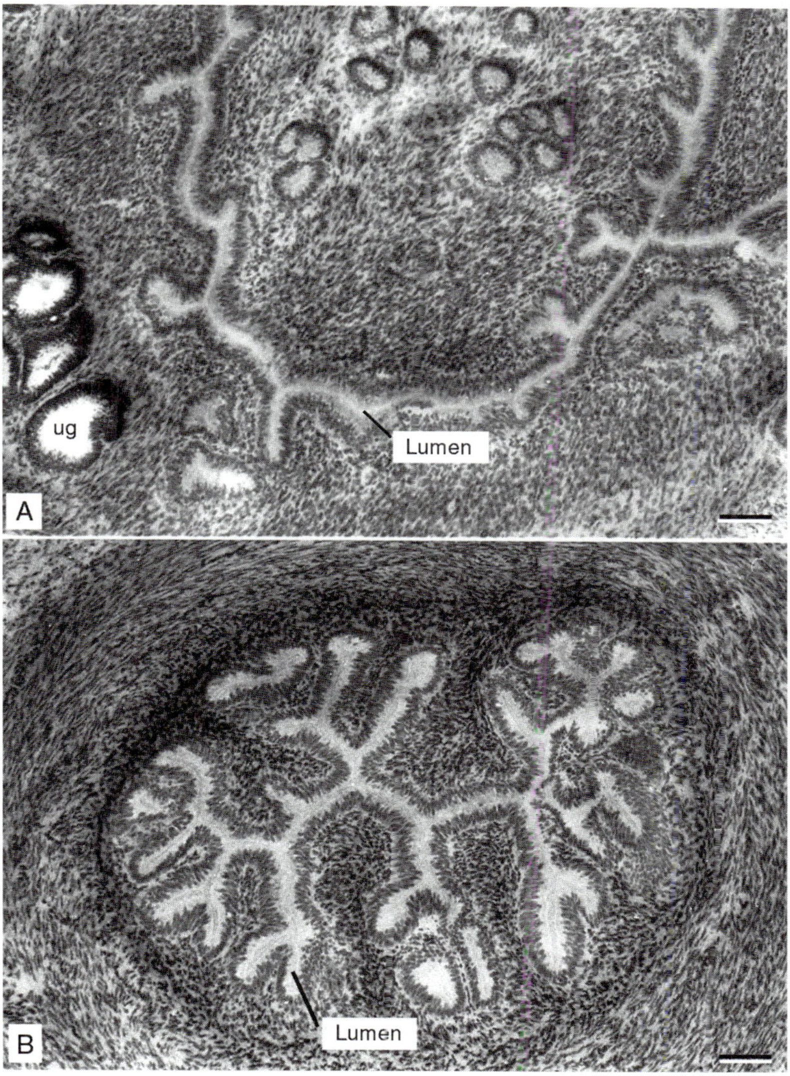

FIGURE 1.4 Frozen sections of the bovine uterotubal junction and caudal isthmus, stained with periodic-acid Schiff and counterstained with hematoxylin according to methods published in Suarez *et al.* (1997). (A) A section of the uterotubal junction near the uterine lumen. The lumen of the junction is a narrow space between the lightly stained cytoplasm of the mucosal epithelial cells. It is only a few micrometers across in most regions. Uterine glands (ug) can be seen in the submucosa (bar = 300 μm). (B) A section of caudal isthmus, near the uterotubal junction. The mucus-filled lumen may be seen as a slightly darker region bounded by the lightly stained cytoplasm of the mucosal epithelial cells (bar = 300 μm). Photomicrographs by S. Suarez.

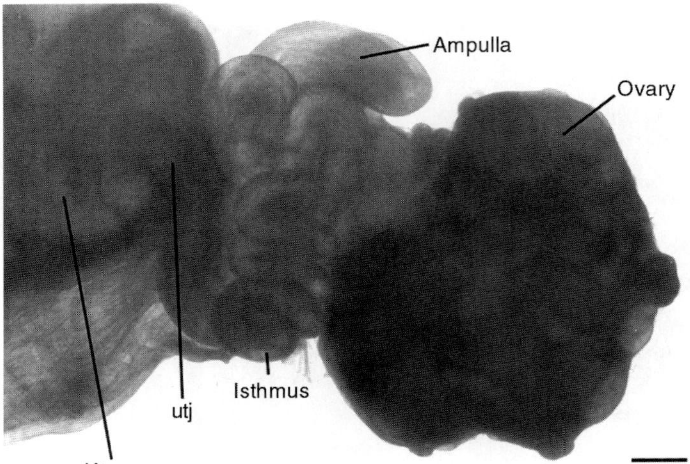

FIGURE 1.5 Photograph of a transilluminated, freshly dissected oviduct of the mouse, illustrating the long, sigmoidal uterotubal junction (utj) and the highly coiled isthmus and ampulla. The uterotubal junction projects into the uterine lumen, presenting a narrow opening for spermatozoa (bar = 300 μm). Photograph by S. Suarez.

(Overstreet and Cooper, 1978; Hawk, 1983). Rapid transport of spermatozoa into the oviduct would seem to counter the proposed model of cells swimming one-by-one through the uterotubal junction. However, when Overstreet and Cooper (1978) evaluated the condition of rabbit spermatozoa recovered from the cranial ampulla shortly after mating, they found that most of these cells were immotile and damaged. They proposed that waves of contractions stimulated by insemination transport spermatozoa rapidly to the site of fertilization, but these cells are mortally damaged by the associated shear stress and do not fertilize. Later, motile cells gradually pass through the uterotubal junction to establish a population capable of fertilizing the egg.

F. OVIDUCTAL SPERM RESERVOIR

On entering the oviduct, spermatozoa become trapped and form a reservoir. The sperm reservoir may have been first discovered in hamsters by Yanagimachi and Chang (1963) and has since been reported to exist in a variety of eutherian mammals [hamsters (Smith *et al.,* 1987), rabbits (Harper, 1973; Overstreet *et al.,* 1978), cows (Hunter and Wilmut, 1984), pigs (Hunter, 1981), and sheep (Hunter and Nichol, 1983)].

The oviductal reservoir of spermatozoa may serve three functions. First, it may prevent polyspermic fertilization by allowing only a few spermatozoa at a time to reach the oocyte in the ampulla. Sperm numbers have been artificially increased at the site of fertilization in the pig by surgical insemination directly into the oviduct (Polge *et al.*, 1970; Hunter, 1973), by resecting the oviduct to bypass the reservoir (Hunter and Leglise, 1971), and by administering progesterone into the muscularis to inhibit contractions (Day and Polge, 1968; Huner, 1972). In all of these experiments, the incidence of polyspermy increased. Second, the oviductal reservoir may maintain the fertility of spermatozoa between the onset of estrus and the time of ovulation. Sperm fertility and motility are maintained longer *in vitro* when they are incubated with oviductal epithelium [bovine (Pollard *et al.*, 1991; Chian and Sirard, 1994), porcine (Suarez *et al.*, 1990), equine (Ellington *et al.*, 1993), canine (Pacey *et al.*, 2000), and human (Kervancioglu *et al.*, 1994)]. Third, capacitation and hyperactivation of spermatozoa may be regulated within the reservoir. Capacitation is defined herein as a set of changes in the sperm plasma membrane that enables a cell to undergo the acrosome reaction. Hyperactivation involves an increase in flagellar bend amplitude and asymmetry that is observed in spermatozoa recovered from the ampulla of the oviduct near the time of ovulation. Capacitation of bull spermatozoa is enhanced by incubation in medium conditioned by oviductal epithelium (Chian *et al.*, 1995) or in oviduct fluid (Mahmoud and Parrish, 1996).

Sperm storage structures have developed in other groups of vertebrates. Spermatozoa are stored in folds of ovarian tissue in several species of viviparous fishes (Koya *et al.*, 1997). In several families of turtles, there exist sperm storage tubules in the region of the oviduct homologous to the mammalian isthmus (Gist and Jones, 1989). Evidently, these storage tubules allow females to fertilize multiple clutches of eggs, sometimes years after mating. In several species of snakes and lizards, sperm storage structures have been described in the anterior vagina and infundibulum of the oviduct (Gist and Jones, 1987; Srinivas *et al.*, 1995; Perkins and Palmer, 1996; Murphy-Walker and Haley, 1996). Neither of these two sites is homologous to the mammalian isthmus, however. Sperm storage tubules have been discovered at the uterovaginal junction in several species of birds (Bakst, 1987, 1994; Birkhead *et al.*, 1993; Birkhead and Moeller, 1993), which also allows them to lay multiple clutches of eggs after a single mating.

In most eutherian mammals, the oviductal reservoir is created by binding of spermatozoa to oviductal epithelium. Motile sperm cells have been observed to bind to the apical surface of the oviductal epithelium in cattle (Figure 1.6) (Suarez *et al.*, 1990), mice (Suarez, 1987), hamsters (Smith and Yanagimachi, 1991), pigs (Suarez *et al.*, 1991), and horses (Thomas *et al.*, 1994). The narrow, sometimes mucus-filled lumen of the uterotubal junction and isthmus would slow the progress of spermatozoa and increase their contact with the mucosal surface, until they were trapped by adhesive molecules.

Sperm binding to oviductal epithelium involves carbohydrate recognition. Fetuin and its terminal sugar, sialic acid, were found to inhibit binding of hamster spermatozoa to the epithelium (DeMott *et all,* 1995). Colloidal gold-labeled fetuin bound to the heads of hamster sperm cells. Fetuin also bound to certain glycoprotein bands on Western blots of membrane extracts from hamster spermatozoa (DeMott *et al.*, 1995). These data indicate that there is a lectinlike molecule on the heads of hamster sperm cells that binds sialic acid and is responsible for attach-

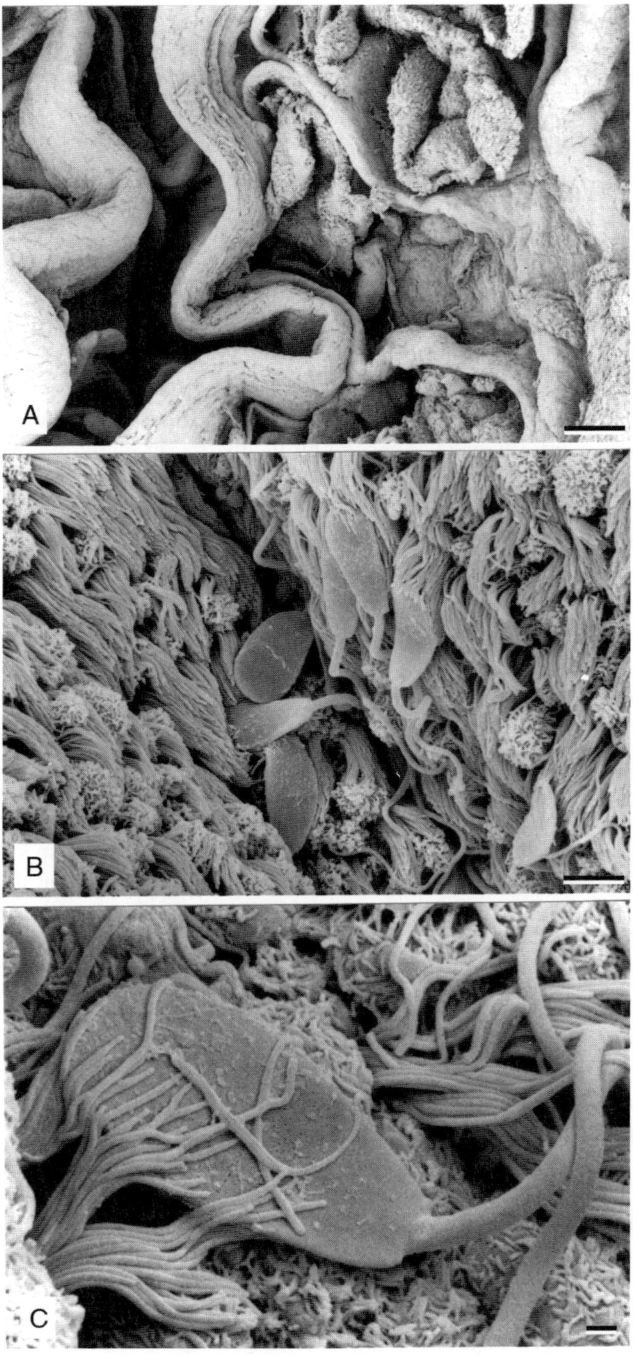
A
B
C

ment of the cells to the epithelium. Attachment of stallion spermatozoa to explants of oviductal epithelium was inhibited by asialofetuin and its terminal sugar, galactose (Lefebvre *et al.,* 1995b; Dobrinski *et al.,* 1996). Bull sperm attachment to oviductal epithelium was determined to be specifically blocked by fucoidan and its component fucose (Lefebvre *et al.,* 1997). Fucose in an α(1-4) linkage to *N*-acetylglucosamine, as in the trisaccharide Lewis-A, inhibited attachment more efficiently than fucose alone. Furthermore, Lewis-A tagged by conjugation to fluorescein-labeled polyacrylamide labeled the heads of live bull spermatozoa (Suarez *et al.,* 1998). Pretreatment of bovine oviductal epithelium with fucosidase, but not galactosidase, reduced sperm attachment (Lefebvre *et al.,* 1997). The protein on bull sperm that binds to fucose has been identified as PDC-109, also known as BSP-A1/A2. This protein inhibits bull sperm binding to oviductal epithelium and restores the ability of capacitated bull sperm to bind fucose (Ignotz *et al.,* 2001). It is produced by the seminal vesicles and is thought to associate with sperm via choline phospholipids in the sperm plasma membrane (Manjunath *et al.,* 1994). In conclusion, carbohydrate involvement in sperm attachment to epithelium appears to be a widespread phenomenon, although the particular carbohydrate comprising the binding site varies according to species. In each of the three species studied so far, a different sugar inhibited binding *in vitro.* However, rapid evolution of carbohydrate specificity is possible because changing a single amino acid residue can alter the carbohydrate specificity of a lectin (Kogan *et al.,* 1995; Revelle *et al.,* 1996), and closely related lectins are known to have different carbohydrate specificities (Weiss, 1994).

Binding between other types of cells involves carbohydrate recognition. Examples are the selectins, which mediate leukocyte adhesion to endothelium (Varki, 1992), and glycolipid ligands on ciliated respiratory cells, which are recognized by mycoplasmas (Zhang *et al.,* 1994). Selectins mediate temporary adhesion between the two cell types, just as interaction between spermatozoa and oviductal epithelium is temporary. Carbohydrate recognition is also implicated in sperm–zona adhesion [reviewed by Yanagimachi (1994) and Sinowatz *et al.* (1997)] and sperm–Sertoli cell adhesion (Raychoudhury and Millette, 1995).

The oviductal mucosa protects spermatozoa against aging damage during storage. Sperm cells incubated with oviductal epithelium remain viable longer *in vitro* than when they are incubated in medium alone [as seen in porcine (Suarez *et al.,* 1990), equine (Ellington *et al.,* 1993), and human (Kervancioglu *et al.,* 1994) studies] or with tracheal epithelium [bovine sperm (Pollard *et al.,* 1991)]. Viability can be extended by incubating spermatozoa with vesicles prepared from the

FIGURE 1.6 Scanning electron micrographs of bovine spermatozoa and the mucosal epithelium of the oviductal isthmus. The sperm cells are located in grooves created by mucosal folds. They appear to be stuck to cilia, as observed with living tissue *in vitro.* (A) A low-magnification view of the isthmus (bar = 75 μm). (B) A higher magnification of a mucosal groove (bar = 5 μm). (C) A high-magnification view of a sperm cell associated with the cilia of the epithelium (bar = 1 μm). From Lefebvre *et al.* (1995a).

apical membranes of oviductal epithelium [in the rabbit (Smith and Nothnick, 1997), equine species (Dobrinski *et al.,* 1997), and humans (Murray and Smith, 1997)], indicating that the epithelium produces the effect by direct contact. It was reported that attachment of equine sperm cells to epithelium or membrane vesicles maintained low levels of cytoplasmic Ca^{2+} compared to free-swimming cells or to cells incubated with vesicles made from kidney membranes (Dobrinski *et al.,* 1996, 1997). Equine and human spermatozoa incubated with oviduct membrane vesicles also capacitated more slowly than did cells incubated in capacitating medium alone (Dobrinski *et al.,* 1997; Murray and Smith, 1997). Possibly, viability is maintained by preventing capacitation and its concomitant rise in cytoplasmic Ca^{2+}. The mechanism for preventing rises of cytoplasmic Ca^{2+} in spermatozoa is not known, but one suggestion is that catalase, which is present in the oviduct, serves to protect against peroxidative damage to the sperm membranes (Lapointe *et al.,* 1998).

Little is known about how spermatozoa are released from the epithelium so that they may fertilize oocytes. Changes in the hormonal state of the oviductal epithelium related to impending ovulation apparently do not affect the density of binding sites for spermatozoa (Lefebvre *et al.,* 1995a; Suarez *et al.,* 1991; Thomas *et al.,* 1994); therefore, it appears that the epithelium does not release spermatozoa by enzymatically destroying or failing to replenish binding sites. Instead, a change in the surface of sperm cells brings about their release. Capacitation involves changes in the plasma membrane over the sperm head and, therefore, may lead to sperm release by eliminating or modifying binding molecules on the head. Hyperactivation may provide the force necessary for overcoming the attraction between spermatozoa and the oviductal epithelium. Smith and Yanagimachi (1991) reported that hamster sperm cells that had undergone both capacitation and hyperactivation *in vitro* did not bind to epithelium when infused into hamster oviducts. While using transillumination to study motile cells within oviducts removed from mated mice, DeMott and Suarez (1992) noted that only hyperactivated spermatozoa detached from the epithelium. Attachment of bull and boar spermatozoa is significantly reduced by capacitation *in vitro,* even in the absence of hyperactivation (Lefebvre and Suarez, 1996; Fazeli *et al.,* 1999). Therefore, it is evident that changes in the sperm head surface are responsible for loss of binding affinity, although the pull produced by hyperactivation may enhance the ability of spermatozoa to release themselves. Although the binding sites present on the epithelium may not be reduced in number or affinity, epithelial secretions initiated by signals of impending ovulation could enhance sperm capacitation, thereby bringing about sperm release. Soluble oviductal factors do enhance capacitation of bull spermatozoa (Chian *et al.,* 1995; Mauhmoud and Parrish, 1996). During early estrus, the oviduct may delay capacitation. As the time of ovulation approaches, the oviduct may respond to hormonal messages to secrete substances that initiate or enhance sperm capacitation.

The lectinlike molecule on spermatozoa that appears to be responsible for binding to the epithelium is lost or loses its specific carbohydrate binding affinity in capacitated cells. Capacitated hamster spermatozoa were no longer labeled by fetuin over the acrosomal region, indicating a loss of binding affinity for sialic acid (DeMott *et al.,* 1995). Fetuin bound to certain protein bands on Western blots of electrophoretically separated membrane proteins extracted from fresh, epididymal hamster spermatozoa, but binding was reduced on proteins extracted from cells that were hyperactivated and partially capacitated (DeMott *et al.,* 1995). When bull spermatozoa were capacitated *in vitro,* they were no longer labeled with fluorescein-labeled fucosylated bovine serum albumin (Revah *et al.* 2000; Ignotz *et al.,* 2001).

In summary of what is known about sperm attachment in the oviduct to date, the following picture emerges. The sperm reservoir forms in the uterotubal junction and/or isthmus by binding of a lectinlike molecule on sperm cells to a glycoconjugate on the surface of the oviductal mucosa. The narrowness of the lumen, and perhaps the mucus within the lumen, enhance sperm attachment by slowing their progress and increasing contact with the epithelial surface. Direct contact with mucosal epithelium prolongs sperm survival and delays capacitation. Capacitation may be initiated by secretions as the time of ovulation approaches. The lectin on the surface of spermatozoa is lost or modified during the complex process of capacitation, thereby allowing the cells to be released. Hyperactivaton may provide the force to pull spermatozoa away from their attachment sites.

In marsupial mammals (Bedford, 1991; Taggart, 1994) and birds (Bakst, 1992; Bakst *et al.,* 1994), spermatozoa are stored in mucosal crypts (i.e., tubules) in the oviduct. However, the sperm heads do not attach to the epithelium in the crypts. Many of the sperm cells in the crypts of the marsupial *Sminthopsis crassicuadata* were observed to be immotile (Bedford and Breed, 1994) and it is thought that the motility of avian spermatozoa is suppressed in the crypts (Bakst *et al.,* 1994). Thus motility suppression may serve to keep spermatozoa in the crypts until ovulation. In the primitive eutherian mammals, the shrews, some species have been reported to possess distinctive bubblelike outpocketings of the oviduct wall in the caudal ampulla. Spermatozoa enter these structures and do not adhere to the epithelium (Bedford *et al.,* 1997a,b). In more advanced eutherian mammals, the storage structures are less tubular and less distinctive, being organized as grooves created by folds of the mucosa. Adhesion may be more effective at trapping spermatozoa in these structures. Motility suppression has been observed in the isthmus of rabbits and has been proposed as a mechanism of storage (Overstreet *et al.,* 1980; Overstreet and Cooper, 1975; Burkman *et al.,* 1984). In hamsters (Smith and Yanagimachi, 1990) and mice (Suarez, 1987), immotile spermatozoa have been observed in the central part of the isthmic lumen; however, in this case, it is thought that these cells are damaged and may not fertilize (Smith and Yanagimachi, 1990). It is curious that distinctive storage structures would be lost and sperm binding would evolve to replace them.

So far, there has been no conclusive evidence for a distinct oviductal sperm reservoir in humans (Williams *et al.,* 1993). Human spermatozoa have not, for the most part, been observed to adhere tightly to oviductal epithelium *in vitro* (Yeung *et al.,* 1994; Murray and Smith, 1997), although some spermatozoa have been observed to stick under certain conditions (Pacey *et al.,* 1995). Nevertheless, human sperm viability is maintained by incubation with oviductal epithelium (Murray and Smith, 1997), as it is in species in which there is strong attachment of spermatozoa to epithelium (Pollard *et al.,* 1991; Chian and Sirard, 1994). As an alternative to oviductal storage, the human cervix may serve as the site of a sperm reservoir. The lumen of the human cervix is 3 cm in length (Insler *et al.,* 1980). The human uterus is rather small in proportion to body size, compared with those of ruminants, for example, and human spermatozoa must travel only a few centimeters through the lumen to reach the uterotubal junction (Figure 10.2). The entrance to the uterotubal junction in humans is shaped rather like a funnel (Hafez and Black, 1969; Beck and Boots, 1974). In comparison, the uterotubal junctions of rodents, pigs, dogs, and ruminants present an elaborate entrance surrounded by mucosal folds. So, human spermatozoa may be guided right into the uterotubal junction, but sperm of other species may be presented with more of a barrier. The evidence that could be used to argue against a cervical reservoir is that very few sperm cells have been recovered from human or primate uteri 24 hours after coitus (Rubenstein *et al.,* 1951; Moyer *et al.,* 1970). Furthermore, the leukocytic infiltration of the uterus, which becomes significant several hours after coitus (Harper, 1994), could present a barrier to passage of spermatozoa that had been stored in the cervix. Leukocytes appear to outnumber human spermatozoa in the uterus at 4 hours after coitus (Williams *et al.,* 1993a). Unless sperm cells are protected from phagocytosis (and they might be!), it is unlikely that they could travel from the cervical reservoir to the oviduct several hours postcoitus.

Alternatively, human spermatozoa could be stored for long periods of time in the oviduct, but not in a distinct reservoir and not by adhering tightly to the mucosal surface. The mucosal folds of the human oviductal lumen, which are quite small in the isthmus, increase in size and complexity toward the ovary, thus offering increasingly greater barriers to the advancement of spermatozoa. Sperm progress could be slowed by the mucus in the lumen (Jansen, 1980) and by sticking lightly to the mucosa (Pacey *et al.,* 1995). So, rather than having a distinct reservoir, human sperm advancement to the site of fertilization could be slowed in such a manner so as to increase the likelihood that a few will be present at the site of fertilization when ovulation occurs. Muscular contractions and secretions at the time of ovulation could move or activate spermatozoa and increase chances of encountering the oocyte. It has been proposed that human sperm cells are chemotactically attracted to the oocyte by follicular fluid introduced into the oviduct by the cumulus mass at ovulation (Ralt *et al.,* 1991). Data on sperm distribution in the tubes of women have not provided a clear picture of the events of sperm transport. Spermatozoa recovered at various times in different regions of the human oviduct have varied so much in numbers that the data do not permit the construction of a model for the pattern of sperm transport (Williams *et al.,* 1993b). Perhaps fertilization is a relatively inefficient and unregulated process in humans, because evolutionary pressures have worked to support long-term pair bonding in addition to fertilization success, providing another important function for coitus.

After fertilization, mammalian spermatozoa may be phagocytosed by isthmic epithelial cells (Chakraborty and Nelson, 1975) or may be eliminated, passing into the peritoneal cavity (Mortimer and Templeton, 1982) and then being phagocytosed. Phagocytosis within the oviduct may be employed by species such as mice, which have an extensive ovarian bursa that would limit passage of spermatozoa into the peritoneal cavity.

III. OOCYTE TRANSPORT

There are two issues involved in oocyte transport. The first is the capture of the oocyte from the surface of the ovary or from the ovarian bursa. The second is transport of the oocyte through the ampulla.

Richard Blandau's films of ovulation and oocyte pickup were pioneering and gave us an appreciation of the process. He developed a system for filming these events *in situ* in the rabbit. The films revealed that the mesosalpinx contracts rhythmically during ovulation, causing the fimbria to slide over the surface of the ovary (Blandau, 1969). The mesosalpinx also moves the oviduct, mesovarium, and ovary to aid in the positioning of the fimbria over the ovary. In addition, contractions of the muscularis of the wall of the fimbria contribute to moving the fimbria over the surface of the ovary. At ovulation, the follicular contents are extruded as a long, sticky strand of cumulus, matrix, and oocyte. The strand soon makes contact with the cilia on the surface of the fimbria, is drawn away from the surface of the ovary, and is rapidly transported into the ampulla. In the rabbit and hamster, if the cumulus and its matrix are removed, the naked oocyte is not picked up by the fimbria (Blandau, 1969; Mahi-Brown and Yanagimachi, 1983).

In some species, such as rats and hamsters, the fimbria are small and cannot sweep over the surface of the ovary. Nevertheless, the space between the ovary and ostium is nearly completely enclosed by the mesovarium and mesosalpinx, which form a bursa. The cumulus mass is actually ovulated into the bursa, where it is jostled by movement of the ovary and oviduct until it comes into contact with the fimbrial surface and is picked up (Blandau, 1969; Mahi-Brown and Yanagimachi, 1983).

Very little is known of the nature of the interaction between cumulus and the fimbria. The site of interaction on the surface of the fimbria is the tips of its cilia (Norwood and Anderson, 1980). Polycationic molecules block ovum pickup in the rabbit and hamster (Norwood and Anderson, 1980; Mahi-Brown and Yanagimachi, 1983). Neuraminidase pretreatment of the fimbria prevents ovum pickup (Mahi-Brown and Yanagimachi, 1983), indicating the involvement of a sialylated molecule on the surface of the fimbria. Although the cumulus is involved in the pickup, the hyaluronic acid of the extracellular cumulus matrix is unlikely to be the molecule primarily responsible for sticking the cumulus to the fimbria. This is based on experiments conducted by Mahi-Brown and Yanagimachi (1983), in which hyaluronate gel was not picked up by hamster fimbria, a solution of hyaluronate did not block pickup, and pretreatment of the cumulus with hyaluronidase did not prevent pickup. It remains to be determined whether there is a specific adhesive interaction between the cumulus mass and the fimbrial surface.

Although the cumulus appears to be important for ovum pickup in most eutherian mammals, there is no cumulus oophorus in marsupial mammals. The granulosa cells do not accompany the oocyte at ovulation (Bedford, 1991, 1996; Breed, 1994). In some shrews, which are considered primitive eutherian mammals, the cumulus does not have a visible matrix at the time of ovulation, although a matrix may be produced after fertilization, when the oocyte is in the oviduct (Bedford *et al.,* 1994, 1997a). If shrews actually accomplish oocyte pickup without a cumulus matrix, and if hyaluronic acid is not responsible for oocyte pickup in more advanced eutherian mammals, then perhaps the cumulus cells are responsible for attachment to the fimbria in all eutherian mammals.

Bedford noted that the diameter of the central lumen of the oviduct matches the size of the ovulatory products (Bedford, 1996). In most eutherian mammals, there is a large, expanded cumulus mass surrounding the oocytes, which fills the relatively large central space of the ampulla. Cumulus expansion is accomplished by the secretion of hyaluronic acid and other matrix materials, followed by hydration. In marsupial mammals, there is no cumulus surrounding the ovulated egg, but the ampulla is narrow (Breed, 1994). In the shrews, an intermediate situation exists: there is a cumulus around the oocytes, but it is not expanded (Bedford *et al.,* 1994, 1997a,b). Correspondingly, the ampullar lumen is intermediate in diameter and fits closely around the compact cumulus mass (Figure 1.7). So, in all cases, the oocyte and its vestments fit snugly in the central ampullar lumen. This raises interesting questions: During the course of evolution, did the ampulla or the oocyte vestments

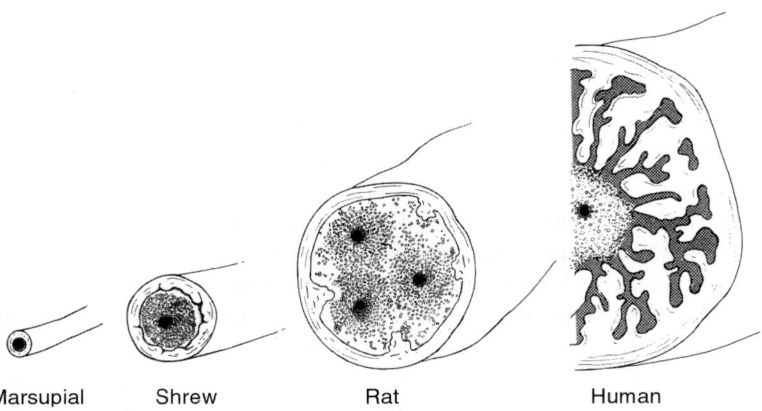

FIGURE 1.7 Diagram of the spatial relationship between the egg or egg–cumulus complex and the site of fertilization in the oviduct of various representative mammals. When the oviduct is much larger than the egg, the ability of the cumulus to fill the space is maximized by a variable degree of cumulus expansion. From Bedford (1996).

expand first? What is the function of these developments? Is the fit of the oocyte and vestments in the lumen important for oocyte transport, or to trap spermatozoa, or both?

Once the cumulus mass containing one or more oocytes enters the ampulla in eutherian mammals, it moves rapidly to the ampullary–isthmic junction. Potential effectors of this movement are the oviductal musculature and the cilia. When smooth muscle contractions were blocked by isoproterenol in the oviducts of rabbit does, the net rate of transport of cumulus–oocyte masses down the ampulla (about 0.12 mm/second) was not affected (Halbert *et al.*, 1976). This indicates that the cilia alone can move the cumulus–oocyte mass to its destination. When muscular contraction was allowed, back-and-forth motion of the mass was observed, but when it was blocked, the mass moved smoothly down the ampulla. So, although the rate of transport is not affected by inhibiting muscular action, the pattern of transport is changed. The back-and-forth movement could serve to enhance infiltration of the cumulus matrix with ovarian secretions or to initiate the process of cumulus removal.

There have been no studies conducted to demonstrate an absolute requirement for ciliary activity for oocyte transport. There is some circumstantial evidence, however, that normal ciliary activity is required. The fact that some female patients diagnosed as having Kartagener's syndrome (immotile cilia syndrome) are infertile (McComb *et al.*, 1986; Halbert *et al.*, 1997), whereas others are fertile (Bleau *et al.*, 1978), indicates that cilia are not absolutely necessary for oocyte transport. Some women with Kartagener's syndrome have some motile cilia (McComb *et al.*, 1986; Halbert *et al.*, 1997) and this could explain why some with the syndrome are fertile. Epidemiological data have revealed a correlation between smoking and ectopic pregnancy in women. The soluble components of mainstream cigarette smoke inhibit ciliary activity in the oviduct and oocyte pickup in hamsters (Knoll *et al.*, 1995; Knoll and Talbot, 1998; Talbot *et al.*, 1998). These data indicate that cilia play an important role in oocyte transport, yet still an absolute requirement for cilia remains to be determined.

IV. A MODEL FOR GAMETE TRANSPORT

Although much remains to be discovered, and some issues remain to be settled, a general model for gamete transport in eutherian mammals can be derived from what is currently known. Spermatozoa are deposited at coitus into the vagina or uterus. Those deposited in the vagina swim through the cervix. Muscular contractions move the sperm cells through the uterine cavity. Eventually a few thousand cells swim through the uterotubal junction. In the junction or the caudal isthmus, they face a narrow lumen filled with mucus that can slow their progress. Prolonged contact with the wall of the junction or isthmus results in specific attachment of spermatozoa to the mucosal epithelium. This serves to create a distinct reservoir in most species. As the time of ovulation approaches, sperm cells become capacitated and hyperactivated and they release from the epithelium. Meanwhile, the oocyte or oocytes, invested in a cumulus mass, are released from the ovary. The mass is picked up by cilia on the mucosal surface of the fimbria and is transported rapidly into the ampulla and down to the ampullary–isthmic junction. During this time, a few spermatozoa reach the cumulus mass. They can hardly avoid it, because the mass nearly fills the ampullar lumen, and because a chemotactic activity likely draws the sperm cells toward it. Fertilization occurs soon thereafter, as spermatozoa penetrate the cumulus oophorus, contact and penetrate the zona pellucida, and finally fuse with the oocyte plasma membrane.

REFERENCES

Austin, C. R. (1957). Fate of spermatozoa in the uterus of the mouse and rat. *J. Endocrinol.* **14,** 335–342.

Baker, R. D., and Degen, A. A. (1972). Transport of live and dead boar spermatozoa within the reproductive tract of gilts. *J. Reprod. Fertil.* **28,** 369–377.

Bakst, M. R. (1987). Anatomical basis of sperm-storage in the avian oviduct. *Scanning Microsc.* **1,** 1257–1266.

Baskt, M. R. (1992). Observations on the turkey oviductal sperm-storage tubule using differential interference contrast microscopy. *J. Reprod. Fertil.* **95,** 877–883.

Baskt, M. R., Wishart, G., and Brillard, J.-P. (1994). Oviductal sperm selection, transport, and storage in poultry. *Poultry Sci. Rev.* **5,** 117–143.

Barros, C., Vigil, P., Herrera, E., Arguello, B., and Walker, R. (1984). Selection of morphologically abnormal sperm by human cervical mucus. *Arch. Androl.* **12** (Suppl.), 95–107.

Beck, L. R., and Boots, L. R. (1974). The comparative anatomy, histology and morphology of the mammalian oviduct. *In* "The Oviduct and Its Funcitons" (A. D. Johnson and C. W. Foley, eds.), pp. 2–51. Academic Press, New York.

Bedford, J. M. (1965). Effect of environment on phagocytosis of rabbit spermatozoa. *J. Reprod. Fertil.* **9,** 249–256.

Bedford, J. M. (1971). The rate of sperm passage into the cervix after coitus in the rabbit. *J. Reprod. Fertil.* **25,** 211–218.

Bedford, J. M. (1991). The coevolution of mammalian gametes. *In* "A Comparative Overview of Mammalian Fertilization" (B. S. Dunbar and M. G. O'Rand, eds.), pp. 3–35. Plenum Press, New York.

Bedford, J. M. (1996). What marsupial gametes disclose about gamete function in eutherian mammals. *Reprod. Fertil. Dev.* **8,** 569–580.

Bedford, J. M. (1997). Unusual nature and possible evolutionary implications of the male vesicular gland secretion in the tree shrew, *Tupaia glis. Anat. Rec.* **247,** 199–205.

Bedford, J. M., and Breed, W. G. (1994). Regulated storage and subsequent transformation of spermatozoa in the fallopian tubes of an Australian marsupial, *Sminthopsis crassicaudata. Biol. Reprod.* **50,** 845–854.

Bedford, J. M., and Yanagimachi, R. (1992). Initiation of sperm motility after mating in the rat. *J. Androl.* **13,** 444–449.

Bedford, J. M., Cooper, G. W., Phillips, D. M., and Dryden, G. L. (1994). Distinctive features of the gametes and reproductive tracts of the Asian musk shrew, *Suncus murinus. Biol. Reprod.* **50,** 820–834.

Bedford, J. M., Mock, O. B., and Phillips, D. M. (1997a). Unusual ampullary sperm crypts, and behavior and role of the cumulus oophorus, in the oviduct of the least shrew, *Cryptotis parva. Biol. Reprod.* **56,** 1255–1267.

Bedford, J. M., Phillips, D. M., and Mover-Lev, H. (1997b). Novel sperm crypts and behavior of gametes in the fallopian tube of the white-toothed shrew, *Crocidura russula* Monacha. *J. Exp. Zool.* **277,** 262–273.

Bernard, R. T. F., and Cumming, G. S. (1997). African bats: Evolution of reproductive patterns and delays. *Q. Rev. Biol.* **72,** 253–274.

Bernard, R. T. F., Happold, D. C. D., and Happold, M. (1997). Sperm storage in a seasonally reproducing African vespertilionid, the banana bat (*Pipistrellus nanus*) from Malawi. *J. Zool. (London)* **241,** 161–174.

Birkhead, T. R., and Moeller, A. P. (1993). Sexual selection and the temporal separation of reproductive events: Sperm storage data from reptiles, birds, and mammals. *Biol. J. Linnean Soc.* **50,** 295–311.

Birkhead, T. R., Pellatt, E. J., and Fletcher, F. (1993). Selection and utilization of spermatozoa in the reproductive tract of the female zebra finch *Taeniopygia guttata*. *J. Reprod. Fertil.* **99,** 593–600.

Blandau, R. J. (1969). Gamete transport—comparative aspects. *In* "The Mammalian Oviduct," pp. 129–162. The University of Chicago Press, Chicago.

Bleau, G., Richer, C.-L., and Bousquet, D. (1978). Absence of dynein arms in cilia of endocervical cells in a fertile woman. *Fertil. Steril.* **30,** 362–363.

Breed, W. G. (1994). How does sperm meet egg?—in a marsupial. *Reprod. Fertil. Dev.* **6,** 485–506.

Burkman, L. J., Overstreet, J. W., and Katz, D. F. (1984). A possible role for potassium and pyruvate in the modulation of sperm motility in the rabbit oviductal isthmus. *J. Reprod. Fertil.* **71.** 367–376.

Carballada, R., and Esponda, P. (1992). Role of fluid from seminal vesicles and coagulating glands in sperm transport into the uterus and fertility in rats. *J. Reprod. Fertil.* **95,** 639–648.

Carballada, R., and Esponda, P. (1997). Fate and distribution of seminal plasma proteins in the genital tract of the female rat after natural mating. *J. Reprod. Fertil.* **109,** 325–335.

Casslen, B. (1986). Uterine fluid volume: Cyclic variations and possible extrauterine contributions. *J. Reprod. Med.* **31,** 506–510.

Chakraborty, J., and Nelson, L. (1975). Fate of surplus sperm in the fallopian tube of the white mouse. *Biol. Reprod.* **12,** 455–463.

Chian, R.-C., and Sirard, M.-A. (1994). Fertilizing ability of bovine spermatozoa cocultured with oviduct epithelial cells. *Biol. Reprod.* **52,** 156–162.

Chian, R. I.-C., LaPointe, S., and Sirard, M. A. (1995). Capacitation *in vitro* of bovine spermatozoa by oviduct cell monolayer conditioned medium. *Mol. Reprod. Dev.* **42,** 318–324.

Daels, P. F., Hughes, J. P., and Stabenfeldt, G. H. (1991). Reproduction in horses. *In* "Reproduction in Domestic Animals" (P. T. Cupps, ed.), 4th Ed., pp. 414–444. Academic Press, San Diego.

Day, B. N., and Polge, C. (1968). Effects of progesterone on fertilization and egg transport in the pig. *J. Reprod. Fertil.* **17,** 227–230.

D'Cruz, O. J., Wang, B.-L., and Haas, G. G., Jr. (1992). Phagocytosis of immunoglobulin G and C3-bound heman sperm by human polymorphonuclear leukocytes is not associated with the release of oxidative radicals. *Biol. Reprod.* **46,** 721–732.

DeMott, R. P., and Suarez, S. S. (1992). Hyperactivated sperm progress in the mouse oviduct. *Biol. Reprod.* **46,** 779–785.

DeMott, R. P., Lefebvre, R., and Suarez, S. S. (1995). Carbohydrates mediate the adherence of hamster sperm to oviductal epithelium. *Biol. Reprod.* **52,** 1395–1403.

Dobrinski, I., Ignotz, G. G., Thomas, P. G. A., and Ball, B. A. (1996). Role of carbohydrates in the attachment of equine spermatozoa to uterine tubal (oviductal) epithelial cells *in vitro*. *Am. J. Vet. Res.* **57,** 1635–1639.

Dobrinski, I., Smith, T. T., Suarez, S. S., and Ball, B. A. (1997). Membrane contact with oviductal epithelium modulates the intracellular calcium concentration in equine spermatozoa *in vitro*. *Biol. Reprod.* **56,** 861–869.

Dobrinski, I., Suarez, S. S., and Ball, B. A. (1996). Intracellular calcium concentration in equine spermatozoa attached to oviductal epithelial cells *in vitro*. *Biol. Reprod.* **54,** 783–788.

Dostal, J., Veselsky, L., Marounek, M., Zelezna, B., and Jonakova, V. (1997). Inhibition of bacterial and boar epididymal sperm immunogenicity by boar seminal immunosuppressive component in mice. *J. Reprod. Fertil.* **111,** 135–141.

Ellington, J. E., Ignotz, G. G., Varner, D. D., Marcucio, R. S., Mathison, P., and Ball, B. A. (1993). *In vitro* interaction between oviduct epithelia and equine sperm. *Arch. Androl.* **31,** 79–86.

Fazeli, A., Duncan, A. E., Watson, P. F., and Holt, W. V. (1999). Sperm–oviduct interaction: Induction of capacitation and preferential binding of uncapacitated spermatozoa to oviductal epithelial cells in porcine species. *Biol. Reprod.* **60,** 879–886.

Fukuda, M., and Fukuda, K. (1994). Uterine endometrial cavity movement and cervical mucus. *Hum. Reprod.* **9,** 1013–1016.

Gaddum-Rosse, P. (1981). Some observations on sperm transport through the uterotubal junction of the rat. *Am. J. Anat.* **160,** 333–341.

Gist, D. H., and Jones, J. M. (1987). Storage of sperm in the reptilian oviduct. *Scanning Microsc.* **1,** 1839–1849.

Gist, D. H., and Jones, J. M. (1989). Sperm storage within the oviducts of turtles. *J. Morphol.* **199,** 379–384.

Hafez, E. S. E., and Black, D. L. (1969). The mammalian uterotubal junction. *In* "The Mammalian Oviduct: Comparative Biology and Methodology" (E. S. E. Hafez and R. J. Blandau, eds.), pp. 85–128. The University of Chicago Press, Chicago.

Halbert, S. A., Patton, D. F. L., Zarutskie, P. W., and Soules, M. R. (1997). Function and structure of cilia in the Fallopian tube of an infertile woman with Kartegener's syndrome. *Hum. Reprod.* **12,** 55–58.

Halbert, S. A., Tam, P. Y., and Blandau, R. J. (1976). Egg transport in the rabbit oviduct: The roles of cilia and muscle. *Science* **191,** 1052–1053.

Hanson, F. W., and Overstreet, J. W. (1981). The interaction of human spermatozoa with cervical mucus *in vivo. Am. J. Obstet. Gynecol.* **140,** 173–178.

Harper, M. J. K. (1973). Relationship between sperm transport and penetration of eggs in the rabbit oviduct. *Biol. Reprod.* **8,** 441–450.

Harper, M. J. K. (1994). Gamete and zygote transport. *In* "The Physiology of Reproduction" (E. Knobil and J. D. Neill, eds.), 2nd Ed., pp. 123–187. Raven Press, New York.

Hawk, H. W. (1983). Transport and fate of spermatozoa after insemination of cattle. *J. Dairy Sci.* **70,** 1487–1503.

Hawk, H. W. (1987). Sperm survival and transport in the female reproductive tract. *J. Dairy Sci.* **66,** 2645–2660.

Hook, S. J., and Hafez, E. S. E. (1968). A comparative study of the mammalian uterotubal junction. *J. Morphol.* **125,** 159–184.

Hosken, D. J. (1997). Sperm competition in bats. *Proc. R. Soc. Lond. B.* **264,** 385–392.

Hosken, D. J., O'Shea, J. E., and Blackberry, M. A. (1996). Blood plasma concentrations of progesterone, sperm storage and sperm viability and fertility in Gould's wattled bat (*Chalinolobus gouldii*). *J. Reprod. Fertil.* **108,** 171–177.

Hunter, R. H. F. (1972). Local action of progesterone leading to polyspermic fertilization in pigs. *J. Reprod. Fertil.* **31,** 433–444.

Hunter, R. H. F. (1973). Polyspermic fertilization in pigs after tubal deposition of excessive numbers of spermatozoa. *J. Exp. Zool.* **183,** 57–64.

Hunter, R. H. F. (1981). Sperm transport and reservoirs in the pig oviduct in relation to the time of ovulation. *J. Reprod. Fertil.* **63,** 109–117.

Hunter, R. H. F. (1988). "The Fallopian Tubes. Their Role in Fertility and Infertility," pp. 53–80. Springer-Verlag, Berlin.

Hunter, R. H. F., and Leglise, P. C. (1971). Polyspermic fertilization following tubal surgery in pigs, with particular reference to the role of the isthmus. *J. Reprod. Fertil.* **24,** 233–246.

Hunter, R. H. F., and Nichol, R. (1983). Transport of spermatozoa in the sheep oviduct: Preovulatory sequestering of cells in the caudal isthmus. *J. Exp. Zool.* **228,** 121–128.

Hunter, R. H. F., and Wilmut, I. (1982). The rate of functional sperm transport into the oviducts of mated cows. *Ann. Reprod. Sci.* **5,** 167–173.

Hunter, R. H. F., and Wilmut, I. (1984). Sperm transport in the cow: Periovulatory redistribution of viable cells within the oviduct. *Reprod. Nutr. Dev.* **24,** 597–608.

Ignotz, G. G., Lo, M. C., Perez, C. L., Gwathmey, T. M., and Suarez, S. S. (2001). Characterization of a fucose-binding protein from bull sperm and seminal plasma that may be responsible for formation of the oviductal sperm reservoir. *Biol. Reprod* **64,** 1806–1811.

Insler, V., Glezerman, M., Zeidel, L., Bernstein, D., and Misgav, N. (1980). Sperm storage in the human cervix: A quantitative study. *Fertil. Steril.* **33,** 288–294.

Jansen, R. P. S. (1978). Fallopian tube isthmic mucus and ovum transport. *Science* **201,** 349–351.

Jansen, R. P. S. (1980). Cyclic changes on the human fallopian tubes isthmus and their functional importance. *Am. J. Obstet. Gynecol.* **136,** 292–308.

Jansen, R. P. S., and Bajpai, V. K. (1982). Oviduct acid mucus glycoproteins in the estrous rabbit: Ultrastructure and histochemistry. *Biol. Reprod.* **26,** 155–168.

Jensen, T. S., Bjorge, L., Wollen, A.-L., and Ulstein, M. (1995). Identification of the complement regulatory proteins CD46, CD55, and CD59 in human fallopian tube, endometrium, and cervical mucosa and secretion. *Am. J. Reprod. Immunol.* **34,** 1–9.

Katz, D. F., Morales, P., Samuels, S. J., and Overstreet, J. W. (1990). Mechanisms of filtration of morphologically abnormal human sperm by cervical mucus. *Fertil. Steril.* **54,** 513–516.

Katz, D. F., Slade, D. A., and Nakajima, S. T. (1997). Analysis of preovulatory changes in cervical mucus hydration and sperm penetrability. *Adv. Contracept.* **13,** 143–151.

Kervancioglu, M. E., Djahanbakhch, O., and Aitken, R. J. (1994). Epithelial cell coculture and the induction of sperm capacitation. *Fertil. Steril.* **61,** 1103–1108.

Knoll, M., and Talbot, P. (1998). Cigarette smoke inhibits oocyte cumulus complex pick-up by the oviduct *in vitro* independent of ciliary beat frequency. *Reprod. Toxicol.* **12,** 57–68.

Knoll, M., Shaoulian, R., Magers, T., and Talbot, P. (1995). Ciliary beat frequency of hamster oviducts is decreased *in vitro* by exposure to solutions of mainstream and sidestream cigarette smoke. *Biol. Reprod.* **53,** 29–37.

Kogan, T. P., Revelle, B. M., Tapp, S., Scott, D., and Beck, P. J. (1995). A single amino acid residue can determine the ligand specificity of E-selectin. *J. Biol. Chem.* **270,** 14047–14055.

Koya, Y., Munehara, H., and Takano, K. (1997). Sperm storage and degradation in the ovary of a marine copulating sculpin, *Alcichthys alcicornis* (Teleosti: scorpaeniformes); role of intercellular junctions between inner ovarian epithelial cells. *J. Morphol.* **233,** 153–163.

Kunz, G., Beil, D., Deininger, H., Wildt, L., and Leyendecker, G. (1996). The dynamics of rapd sperm transport through the female genital tract: Evidence from vaginal sonography of uterine peristalsis and hysterosalpingoscintigraphy. *Hum. Reprod.* **11,** 627–632.

Kutteh, W. H., Prince, S. J., Hammond, K. R., Kutteh, C. C., and Mestecky, J. (1996). Variations in immunoglobulins and IgA subclasses of human uterine cervical secretions around the time of ovulation. *Clin. Exp. Immunol.* **104,** 538–542.

Lapointe, S., Sullivan, R., and Sirard, M.-A. (1998). Binding of a bovine oviductal fluid catalase to mammalian spermatozoa. *Biol. Reprod.* **58,** 747–753.

Lefebvre, R., and Suarez, S. S. (1996). Effect of capacitation on bull sperm binding to homologous oviductal epithelium. *Biol. Reprod.* **54,** 575–582.

Lefebvre, R., Chenoweth, P. J., Drost, M., LeClear, C. T., MacCubbin, M , Dutton, J. T., and Suarez, S. S. (1995a). Characterization of the oviductal sperm reservoir in cattle. *Biol. Reprod.* **53,** 1066–1074.

Lefebvre, R., DeMott, R. P., Suarez, S. S., and Samper, J. C. (1995b). Specific inhibition of equine sperm binding to oviductal epithelium. Equine Reproduction VI. *Biol. Reprod.* (Monogr. 1), 689–696.

Lefebvre, R., Lo, M. C., and Suarez, S. S. (1997). Bovine sperm binding to oviductal epithelium involves fucose recognition. *Biol. Reprod.* **56,** 1198–1204.

Mahi-Brown, C. A., and Yanagimachi, R. (1983). Parameters influencing ovum pickup by oviductal fimbria in the golden hamster. *Gamete Res.* **8,** 1–10.

Mahmoud, A. I., and Parrish, J. J. (1996). Oviduct fluid and heparin induce similar surface changes in bovine sperm during capacitation. *Mol. Reprod. Dev.* **43,** 554–560.

Manjunath, P., Chandonnet, L., Leblond, E., and Desnoyers, L. (1993). Major proteins of bovine seminal vesicles bind to spermatozoa. *Biol. Reprod.* **49,** 27–37.

Manning, J. T., and Chamberlain, A. T. (1994). Sib competition and sperm competitiveness: An answer to 'Why so many sperms?' and the recombination/sperm number correlation. *Proc. R. Soc. Lond. B* **256,** 177–182.

Matthur, S., Rosenlund, C., Carlton, M., Caldwell, J., Barber, M., Rust, P. F., and Williamson, H. O. (1988). Studies on sperm survival and motility in the presence of cytotoxic sperm antibodies. *Am. J. Reprod. Immunol. Microbiol.* **17,** 41–47.

Mattner, P. E. (1968). The distribution of spermatozoa and leukocytes in the female genital tract in goats and cattle. *J. Reprod. Fertil.* **17,** 253–261.

McComb, P., Langley, L., Villalon, M., and Verdugo, P. (1986). The oviductal cilia and Kartagener's syndrome. *Fertil. Steril.* **46,** 412–416.

Menge, A. C., and Edwards, R. P. (1993). Mucosal immunity of the reproductive tract and infertility. *In* "Immunology of Reproduction" (R. K. Naz, ed.), pp. 19–36. CRC Press, Boca Raton.

Mortimer, D., and Templeton, A. A. (1982). Sperm transport in the human female reproductive tract in relation to semen analysis characteristics and time of ovulation. *J. Reprod. Fertil.* **64,** 401–408.

Moyer, D. L., Rimdusit, S., and Mishell, D. R., Jr. (1970). Sperm distribution and degradation in the human female reproductive tract. *Obstet. Gynecol.* **35,** 831–840.

Mullins, K. J., and Saacke, R. G. (1989). Study of the functional anatomy of bovine cervical mucosa with special reference to mucus secretion and sperm transport. *Anat. Rec.* **226,** 106–117.

Murphy-Walker, S., and Haley, S. R. (1996). Functional sperm storage duration in female *Hemidactylus frenatus* (family gekkonidae). *Herpetologica* **52,** 365–373.

Murray, S. C., and Smith, T. T. (1997). Sperm interaction with Fallopian tube apical plasma membrane enhances sperm motility and delays capacitation. *Fertil. Steril.* **68,** 352–357.

Norwood, J. T., and Anderson, R. G. (1980). Evidence that adhesive sites on the tips of oviduct cilia membranes are required for ovum pickup *in situ*. *Biol. Reprod.* **23,** 788–791.

Overstreet, J. W., and Cooper, G. W. (1975). Reduced sperm motility in the isthmus of the rabbit oviduct. *Nature (London)* **258,** 718–719.

Overstreet, J. W., and Cooper, G. W. (1978). Sperm transport in the reproductive tract of the female rabbit: I. The rapid transit phase of transport. *Biol. Reprod.* **19,** 101–114.

Overstreet, J. W., Cooper, G. W., and Katz, D. F. (1978). Sperm transport in the reproductive tract of the female rabbit: II. The sustained phase of transport. *Biol. Reprod.* **19,** 115–132.

Overstreet, J. W., Katz, D. F., and Johnson, L. L. (1980). Motility of rabbit spermatozoa in the secretions of the oviduct. *Biol. Reprod.* **22,** 1083–1088.

Pacey, A. A., Freeman, S. L., and England, G. C. W. (2000). Contact of dog spermatozoa with homologous uterine tube epithelium prolongs flagellar activity in relation to the stage of the estrous cycle. *Theriogenology* **54,** 109–118.

Pacey, A. A., Hill, C. J., Scudamore, I. W., Warren, M. A., Barratt, C. L. R., and Cooke, I. D. (1995). The interaction *in vitro* of human spermatozoa with epithelial cells from the human uterine (Fallopian) tube. *Hum. Reprod.* **10,** 360–366.

Pandya, I. J., and Cohen, J. (1985). The leukocytic reaction of the human uterine cervix to spermatozoa. *Fertil. Steril.* **43,** 417–421.

Parkhurst, M. R., and Saltzman, W. M. (1994). Leukocytes migrate through three-dimensional gels of midcycle cervical mucus. *Cell. Immunol.* **156,** 77–94.

Perkins, M. J., and Palmer, B. D. (1996). Histology and functional morphology of the oviduct of an oviparous snake, *Diadophis punctatus*. *J. Morphol.* **227,** 67–79.

Phillips, D. M., and Mahler, S. (1977a). Phagocytosis of spermatozoa by the rabbit vagina. *Anat. Rec.* **189,** 61–72.

Phillips, D. M., and Mahler, S. (1977b). Leukocyte emigration and migration in the vagina following mating in the rabbit. *Anat. Rec.* **189,** 45–60.

Polge, C., Salamon, S., and Wilmut, I. (1970). Fertilizing capacity of frozen boar semen following surgical insemination. *Vet. Rec.* **87,** 424–428.

Pollard, J. W., Plante, C., King, W. A., Hansen, P. J., Betteridge, K. J., and Suarez, S. S. (1991). Fertilizing capacity of bovine sperm may be maintained by binding to oviductal epithelial cells. *Biol. Reprod.* **44,** 102–107.

Ralt, D., Goldenberg, M., Fetterolf, P. Thompson, D., Dor, J., Mashiachi, S., Garbers, D. L., and Eisenbach, M. (1991). Sperm attraction to a follicular fluid factor(s) correlates with human egg fertilizability. *Proc. Natl. Acad. Sci. U.S.A.* **88,** 2840–2844.

Raychoudhury, S. S., and Millette, C. F. (1997). Multiple fucosyltransferases and their carbohydrate ligands are involved in spermatogenic cell–Sertoli cell adhesion *in vitro* in rats. *Biol. Reprod.* **56,** 1268–1273.

Revah, I., Suarez, S. S., Flesch, F. M., Colenbrander, B., and Gadella, B. M. (2000). Changes in capacity of bull sperm to bind fucose. *Biol. Reprod.* **62,** 1010–1015.

Revelle, B. M., Scott, D., and Beck, P. J. (1996). Single amino acid residues in the E- and P-selectin epidermal growth factor domains can determine carbohydrate binding specificity. *J. Biol. Chem.* **271,** 16160–16170.

Roberts, S. J. (1986). "Veterinary Obstetrics and Genital Diseases," 3rd Ed. Stephen Roberts, Woodstock, VT.

Roldan, E. R. S., Gomendio, M., and Vitullo, A. D. (1992). The evolution of eutherian spermatozoa and underlying selective forces: Female selection and sperm competition. *Biol. Rev.* **67,** 551–593.

Rubenstein, B. B., Strauss, H., Lazarus, M. L., and Hankin, H. (1951). Sperm survival in women. *Fertil. Steril.* **2,** 15–19.

Shalgi, R., Smith, T. T., and Yanagimachi, R. (1992). A quantitative comparison of the passage of capacitated and uncapacitated hamster spermatozoa through the uterotubal junction. *Biol. Reprod.* **46,** 419–424.

Sinowatz, F., Töpfer-Petersen, E., and Calvete, J. J. (1997). Glycobiology of fertilization. *In* "Glycosciences" (H.-J. Gabius and S. Gabius, eds.), pp. 595–610. Chapman & Hall, Weinheim.

Smith, T. T., and Nothnick, W. B. (1997). Role of direct contact between spermatozoa and oviductal epithelial cells in maintaining rabbit sperm viability. *Biol. Reprod.* **56,** 83–89.

Smith, T. T., and Yanagimachi, R. (1990). The viability of hamster spermatozoa stored in the isthmus of the oviduct: The importance of sperm–epithelium contact for survival. *Biol. Reprod.* **42,** 450–457.

Smith, T. T., and Yanagimachi, R. (1991). Attachment and release of spermatozoa from the caudal isthmus of the hamster oviduct. *J. Reprod. Fertil.* **91**, 567–573.

Smith, T. T., Koyanagi, F., and Yanagimachi, R. (1987). Quantitative comparison of the passage of homologous and heterologous spermatozoa through the uterotubal junction of the golden hamster. *Gamete Res.* **19,** 227–234.

Smith, T. T., Koyanagi, F., and Yanagimachi, R. (1988). Distribution and number of spermatozoa in the oviduct of the golden hamster after natural mating and artificial insemination. *Biol. Reprod.* **37,** 225–234.

Sobrero, A. J., and McLeod, J. (1962). The immediate postcoital test. *Fertil. Steril.* **13,** 184–189.

Solomon, E. P., Schmidt, R. R., and Adragna, P. J. (1990). "Human Anatomy and Physiology," 2nd Ed. Saunders College Publishing, Philadelphia.

Srinivas, S. R., Shivanandappa, T., Hegde, S. N., and Sarkar, H. B. D. (1995). Sperm storage in the oviduct of the tropical rock lizard, *Psammophilus dorsalis. J. Morphol.* **224,** 293–301.

Suarez, S. S. (1987). Sperm transport and motility in the mouse oviduct: Observations *in situ. Biol. Reprod.* **36,** 203–210.

Suarez, S. S., and Oliphant, G. (1982). The interaction of rabbit spermatozoa and serum complement proteins. *Biol. Reprod.* **27,** 473–483.

Suarez, S. S., Brockman, K., and Lefebvre, R. (1997). Distribution of mucus and sperm in bovine oviducts after artificial insemination. *Biol. Reprod.* **56,** 447–453.

Suarez, S. S., Drost, M., Redfern, K., and Gottlieb, W. (1990). Sperm motility in the oviduct. *In* "Fertilization in Mammals" (B. D. Bavister, J. Cummins, and E. R. S. Roldan, eds.), pp. 111–124. Serono Symposia, Norwell.

Suarez, S. S., Redfern, K., Raynor, P., Martin, F., and Phillips, D. M. (1991). Attachment of boar sperm to mucosal explants of oviduct *in vitro:* Possible role in formation of a sperm reservoir. *Biol. Reprod.* **44,** 998–1004.

Suarez, S. S., Revah, I., Lo, M., and Kölle, S. (1998). Bull sperm binding to oviductal epithelium is mediated by a Ca^{2+}-dependent lectin on sperm which recognizes Lewis-A trisaccharide. *Biol. Reprod.* **59,** 39–44.

Taggart, D. A. (1994). A comparison of sperm and embryo transport in the female reproductive tract of marsupial and eutherian mammals. *Reprod. Fertil. Dev.* **6,** 451–472.

Talbot, P., DiCarlantonio, G., Knoll, M., and Gomez, C. (1998). Identification of cigarette smoke components that alter functioning of hamster (*Mesocricetus auratus*) oviducts *in vitro. Biol. Reprod.* **58,** 1047–1053.

Taylor, N. J. (1982). Investigation of sperm-induced cervical leukocytosis by a double mating study in rabbits. *J. Reprod. Fertil.* **66,** 157–160.

Thomas, P. G. A., Ball, B. A., and Brinsko, S. P. (1994). Interaction of equine spermatozoa with oviduct epithelial cell explants is affected by estrous cycle and anatomic origin of explant. *Biol. Reprod.* **51,** 222–228.

Thompson, L. A., Barratt, C. L. R., Bolton, A. E., and Cooke, I. D. (1992). The leukocytic reaction of the human uterine cervix. *Am. J. Reprod. Immunol.* **28,** 85–89.

Tyler, K. R. (1977). Histological changes in the cervix of the rabbit after coitus. *J. Reprod. Fert.* **49,** 341–345.

Ulcova-Gallova, Z. (1997). Ten-year experience with antispermatozool activity in ovulatory cervical mucus and local hydrocortisone treatment. *Am. J. Reprod. Immunol.* **38,** 231–234.

Varki, A. (1992). Selectins and othe mammalian sialic acid-binding lectins. *Curr. Opin. Cell Biol.* **4,** 257–266.

Weiss, W. I. (1994). Recognition of cell surface carbohydrates by C-type animal lectins. *In* "Cellular Adhesion" (B. W. Metcalf, B. J. Dalton, and G. Poste, eds.), Plenum Press, New York.

Williams, M., Thompson, L. A., Li, T. C., Mackenna, A., Barratt, C. L., and Cooke, I. D. (1993a). Uterine flushing: A method to recover sperm and leukocytes. *Hum. Reprod.* **8,** 925–928.

Williams, M., Hill, C. J., Scudamore, I., Dunphy, B., Cooke, I. D., and Barratt, C. L. R. (1993b). Sperm numbers and distribution within the human Fallopian tube around ovulation. *Hum. Reprod.* **8,** 2019–2026.

Wrobel, K.-H., Kujat, R., and Fehle, G. (1993). The bovine tubouterine junction: General organization and surface morphology. *Cell Tissue Res.* **271,** 227–239.

Yanagimachi, R. (1994). Mammalian fertilization. *In* "The Physiology of Reproduction" (E. Knobil and J. D. Neill, eds.), pp. 189–317. Raven Press, New York.

Yanagimachi, R., and Chang, M. C. (1963). Sperm ascent through the oviduct of the hamster and rabbit in relation to the time of ovulation. *J. Reprod. Fertil.* **6,** 413–420.

Yeung, W. S. B., Ng, V. K. H., Lau, E. Y. L., and Ho, P. C. (1994). Human oviductal cells and their conditioned medium maintain the motility and hyperactivation of human spermatozoa *in vitro. Hum. Reprod.* **9,** 656–660.

Zamboni, L. (1972). Fertilization in the mouse. *In* "Biology of Mammalian Fertilization and Implantation" (K. S. Moghissi and E. S. E. Hafez, eds.), pp. 213–262. Charles C. Thomas, Springfield.

Zhang, Q., Young, T. F., and Ross, R. F. (1994). Glycolipid receptors for attachment of *Mycoplasma hyopneumoniae* to porcine respiratory ciliated cells. *Infect. Immun.* **62,** 4367–4373.

2

Sperm Motility Activation and Chemoattraction

Timothy A. Quill*,† and David L. Garbers*,†,‡

*Cecil H. and Ida Green Center for Reproductive Biology Sciences and †Department of Pharmacology, University of Texas Southwestern Medical Center, Dallas; and ‡Howard Hughes Medical Institute, Dallas, Texas

I. Introduction

Under natural circumstances in sexually reproducing species, sperm motility is critical to fertilization and thus the continuation of a species. Several events important for successful fertilization in many species rely on adequate sperm motility: penetration of the extracellular matrix surrounding eggs, directed motility in response to factors released from the egg or closely associated structures, and, finally, migration through the female reproductive tract or within an environment such as water or seawater. In contrast, the absence of sperm motility, as found in immotile cilia syndrome, results in male sterility (Afzelius, 1985). In addition, human clinical observations clearly show that even in cases of reduced sperm motility, or asthenozoospermia, fertility potential is lowered (Chemes *et al.,* 1998)

Two basic modes of animal sperm locomotion have been observed in nature. Nematodes such as *Caenorhabditis elegans* have amoeboid spermatozoa that crawl to the site of fertilization by means of the polymerization and depolymerization of the major sperm protein (Roberts and Stewart, 1995). Alternatively, most other animals have flagellated spermatozoa; the interaction between the axonemal microtubules and the attached dynein arms results in flagellar beating (Eddy and O'Brien, 1994). The mechanical force driving the flagellar beat is provided by ATPase activity of the heavy chain component of the dynein arm complexes, which appears regulated by the light chain components of the dynein arms (Smith and Lefebvre, 1997). Some species, including many insects, produce spermatozoa with multiple flagella that propel the cell, whereas numerous animals, including all species of echinoderms and vertebrates, produce spermatozoa that are propelled by a single flagellum (Miller, 1985).

Sperm motility is regulated at several points between the completion of spermiogenesis in the testis and fertilization of the egg. Testicular spermatozoa, though morphologically mature, are immotile and therefore require a motility-activating mechanism. In the case of marine invertebrates, sperm motility is activated at spawning and, subsequently, can be modulated by egg-associated factors. For species in which fertilization is internal, sperm motility regulation is more complex because of posttesticular interactions with both the male and the female reproductive tracts. The focus of this chapter is on the identity and function of the molecular components that regulate sperm motility in response to environmental cues encountered during the productive existence of a spermatozoon.

II. ASSAYS

Before undertaking an analysis of the molecular regulation of sperm motility, we present a brief discussion of the methods used to assess motility, along with their limitations. Two fundamental methods have been used to measure sperm motility: spectroscopy and microscopy. Spectroscopic methods provide a measure of the percent motility and progressiveness of motility in a sperm population (Deana *et al.,* 1986). However, these methods do not differentiate subpopulation or individual cell motility characteristics. Consequently, microscopic methods have become the predominant approach used for sperm motility evaluation. Prior to the introduction of image-recording technology about 20 years ago, microscopic analysis of sperm motility was largely subjective. Early reports of increased or decreased sperm motility may have reflected a change in cell velocity, the percentage of motile cells, or the flagellar beating pattern. In clinical circumstances, efforts have been made to standardize these observations using semiquantitative five-point scales to assess percent motility and progressive motility (defined as a distance of 1 sperm head length/second) in an ejaculate by random sampling of a total of at least 200 spermatozoa in several view fields (Yeung *et al.,* 1997). Unfortunately, when examining the effects of a given treatment on sperm motility, the subjective nature of direct microscopic observation can cause variability in motility scoring between labs, leading to contradictory conclusions (Krause *et al.,* 1993). A more sophisticated measurement of sperm motility is obtained using computer-assisted sperm motility analysis (CASA) technology. This method records sperm motion and calculates cellular velocities and sperm head displacements resulting from flagellar beating. For a detailed discussion of CASA, one or more of several reviews on the topic can be read (Krause and Viethen, 1999; Krause, 1995; Mortimer *et al.,* 1995). Suffice it to note that as for manual microscopy, the motility measurements obtained remain dependent on many contributing factors, including sample preparation, the physical environment in which the measurement is made (e.g., temperature, buffer composition, depth of chamber), and, in addition, the recording parameters set by the operator (e.g., velocity threshold, sperm head size estimate).

III. MOTILITY ACQUISITION

The testicular spermatozoa of marine invertebrate species are immotile within the testis, and in most species quickly initiate vigorous movement when released from the testis (see Section IV). This is not true for mammalian spermatozoa. Only a small number of mammalian spermatozoa display any motion (characteristically a nonprogressive, low-frequency, and low-amplitude flagellar beat) when obtained from the testis and transferred into a medium that supports the motility of more mature spermatozoa isolated from the cauda epididymis (Yanagimachi, 1994). As mammalian spermatozoa migrate through the epididymal duct, they acquire the capacity for motility coincident with passage from the distal corpus to the proximal caudal region. The epididymis secretes into the lumen many factors that may affect sperm cell physiology (Jones, 1999). But the nature of such factors and how they alter sperm motility potential are largely unknown. The c-*ros* knockout mouse provides a model for the interaction of the epididymal environment with spermatozoa that affects motility (Yeung *et al.,* 1999). The c-*ros* protein is an orphan tyrosine kinase receptor that is expressed in the initial segment of the adult epididymal epithelium. The absence of c-*ros* disrupts the development of the volume regulation mechanism of spermatozoa, resulting in kinked flagella. The null males are sterile as a result of the compromised motility preventing cell migration into the oviduct. It seems likely that other interactions between epididymal factors and spermatozoa that affect motility remain to be discovered.

The biochemical signals that contribute to the acquisition of the capacity for mammalian sperm motility are thought to control the phosphorylation state, and thus activity, of the motor proteins of the axoneme. In most cases, these biochemical changes are correlated with motility changes, and a direct cause-and-effect relationship has not been demonstrated. Many studies suggest that sperm cyclic

AMP (cAMP) elevations participate in this process. For example, caput epididymal spermatozoa treated with various cAMP analogs or phosphodiesterase inhibitors initiate nonprogressive flagellar beating (Garbers and Kopf, 1980). Similarly, detergent-extracted testicular sperm cells become motile with the addition of cAMP and ATP; the level of motility approximates that observed with caudal epididymal spermatozoa (Yanagimachi, 1994). The latter experiment suggests that the axoneme of testicular spermatozoa is capable of motility, but that an undefined regulatory component, either membrane associated or cytosolic, which may mediate protein kinase A (PKA) activation, has not yet been appropriately processed or localized at the time of testicular release.

cAMP synthesis in spermatozoa is regulated by a novel adenylyl cyclase that appears to be conserved across an evolutionarily diverse group of animals. Compared to the widely expressed somatic cell adenylyl cyclase isoforms, the enzyme present in spermatozoa is unique based on (1) a preference for Mn^{2+} over Mg^{2+} as a cofactor, (2) an insensitivity to forskolin activation, and (3) an apparent lack of G-protein regulation (Garbers and Kopf, 1980; Forte *et al.,* 1983). Levin and colleagues have reported the purification and cDNA cloning of a rat sperm cell cytosolic adenylyl cyclase with a specific activity of approximately 20 μmol/min/mg (Buck *et al.,* 1999; Chen *et al.,* 2000). Expression of a recombinant polypeptide corresponding to the purified M_r 48,000 protein produces an active adenylyl cyclase with the properties noted above. Interestingly, the complete cDNA predicts a M_r 187,000 protein with a topology substantially different from that of the known somatic cell adenylyl cyclases. The somatic cell adenylyl cyclases are predicted as integral membrane proteins with twelve transmembrane segments and two homologous cytoplasmic catalytic domains, one located between the sixth and seventh and the other after the last transmembrane segment (Sunahara *et al.,* 1996). The testicular adenylyl cyclase is predicted to be a soluble protein with two tandem cyclase catalytic domains near the amino terminus. The amino acid sequences of these catalytic domains appear more similar to those found in bacteria and yeast than to those of the integral membrane adenylyl cyclases. Based on structural studies of the mammalian cyclase catalytic domains, the two domains found in the soluble adenylyl cyclase likely dimerize, with the active site formed at the resulting interface (Tesmer *et al.,* 1997). Sequence alignment of the catalytic domains of the integral membrane and soluble adenylyl cyclases demonstrates that the testicular cyclase lacks several conserved residues believed important for both G-protein and forskolin interactions, likely explaining its insensitivity to these factors (Figure 2.1).

The function of the remaining carboxy-terminal portion of the testicular adenylyl cyclase's coding sequence is not known. Catalysis does not require this region of the protein because both the purified protein (M_r 48,000) and an expressed truncation mutant (residues M1–V469) are enzymatically active (Buck *et al.,* 1999). A leucine zipper motif is found in the carboxy-terminal region, suggesting a possible protein–protein interaction. Interestingly, heterologous expression of the complete cDNA results in significant particulate-associated enzyme activity,

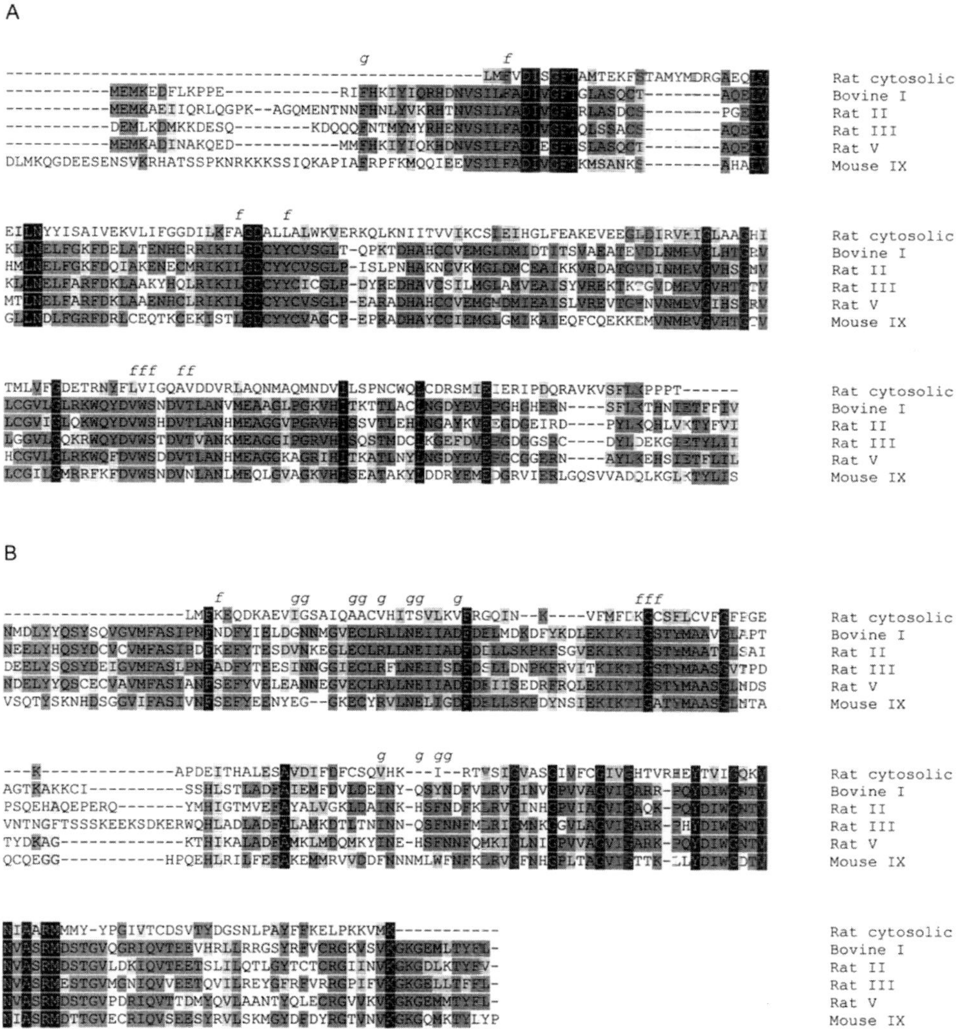

FIGURE 2.1 Alignment of the catalytic domains (A) C1a and (B) C2a of representative somatic membrane adenylyl cyclases and the testicular cytosolic adenylyl cyclase. Residues involved in forskolin (*f*) and Gsα (*g*) contact based on the crystal structure of the type V C1 and type II C2 heterodimers are indicated. Identical residues in all sequences are shaded black, residues identical in three or more sequences are shaded dark gray, and conserved residues are light gray. GenBank accession numbers: bovine ACI, P19754; rat ACII, P26769; rat ACIII, P21932; rat ACV, Q04400; mouse ACIX, NP033754; rat soluble AC, AAD04035.

whereas the truncated mutant is predominantly in the soluble fraction. In comparison, a M_r 190,000 protein found in sea urchin spermatozoa is also membrane associated (Bookbinder *et al.*, 1990) and may be the sperm adenylyl cyclase, based on the ability of monoclonal antibodies that recognize only this protein to inhibit sea urchin and equine sperm adenylyl cyclase activity. Thus, the intact sperm adenylyl cyclase may partition to the membrane compartment through its carboxy-terminal region. Other than this leucine zipper and a P-loop of uncertain physiological significance, no other known structural motif or domain is present in this region of the soluble adenylyl cyclase.

A variety of factors have been implicated in the regulation of the sperm cell adenylyl cyclase and, consequently, sperm motility. One such factor is calcium. Both sea urchin and equine sperm adenylyl cyclases bind calmodulin and are inhibited by calmodulin antagonists (Bookbinder *et al.,* 1990; Gross *et al.,* 1987). Furthermore, after removal of endogenous calmodulin, the abalone sperm enzyme is activated by exogenous calmodulin (Kopf and Vacquier, 1984). Bicarbonate ion enhances the calcium-dependent activation of mammalian sperm adenylyl cyclase (Garbers *et al.,* 1982). This effect is independent of pH effects, and appears to involve a direct interaction of bicarbonate with the enzyme (Chen *et al.,* 2000). The similarity of the novel cytosolic adenylyl cyclase to some bacterial forms of this enzyme further suggests that this interaction is of potential importance, because the bacterial adenylyl cyclases are stimulated by pyruvate and other α-keto acids (Peters *et al.,* 1991). Thus, the activation of the sperm adenylyl cyclase by bicarbonate (millimolar concentrations) may indicate a related regulatory mechanism by a structurally similar metabolite in spermatozoa. The sea urchin sperm adenylyl cyclase may also be regulated by both pH and membrane potential (Cook and Babcock, 1993a; Beltran *et al.,* 1996). Continuing expression studies using the novel soluble adenylyl cyclase cDNA should clarify our understanding of the regulation of this enzyme. Similarly, the expressed protein could be screened for specific and potent agonists/antagonists useful for enhancing or reducing sperm fertility potential. Additionally, protein–protein interaction assays (e.g., immunoprecipitations, yeast two-hybrid) may provide other insights into the sperm adenylyl cyclase signaling pathway.

In addition to synthesis, cellular cAMP levels are dependent on phosphodiesterase (PDE) activity. A threefold reduction in PDE activity from spermatozoa obtained from different regions of the epididymis correlates with both the increase in cAMP and the acquisition of the capacity for motility (Jaiswal and Majumder, 1996). Analysis of the PDE isoforms of human spermatozoa identified a PDE1 subtype (calcium/calmodulin dependent) and PDE4α subtype enzyme (Wasco and Orr, 1984; Fisch *et al.,* 1998). Sperm motility was selectively enhanced, though modestly (around 5–10%), in the presence of a selective PDE4 inhibitor (rolipram), but not with a PDE1-selective inhibitor [8-methoxy-isobutylmethylxanthine (IBMX)], suggesting that PDE4α contributes to the regulation of sperm motility. In addition to these PDE isoforms, the incomplete inhibition of PDE activity obtained with the selective inhibitors suggests that other PDE isoforms are also involved in regulating sperm cyclic nucleotide levels and consequently cellular functions.

As in all cells, cAMP effects in spermatozoa are mediated at least in part by

PKA. The RIIα subunit of PKA localizes to both the principal piece and the midpiece of the mammalian sperm flagellum, and therefore likely modulates flagellar activity (Vijayaraghavan *et al.,* 1997b). But disruption of the RIIα gene in mice produces no apparent sperm motility defects, and the null animals are fertile (Burton *et al.,* 1999). Cellular RIα does increase severalfold in the RIIα gene-null animals, but PKA activity is primarily present in the cytoplasmic droplet rather than associated with the detergent-insoluble components of the flagella. One interpretation of these results is that only a small portion of the total PKA activity normally associated with flagellar structures is required for motility. The localization of the RI subunits of PKA in spermatozoa is less clear because it is reported by different groups to be predominantly present in the head or the tail (Moos *et al.,* 1998; Vijayaraghavan *et al.,* 1997b). These distinct localizations probably reflect the use of different RI-specific antibodies (monoclonal vs. peptide) in combination with different fixation/permeabilization protocols, and suggest that RI is found throughout spermatozoa. No mouse sperm motility/fertility defects were reported for the RIIβ knockout mice (Cummings *et al.,* 1996). In the future, conditional mouse knockout models of the PKA catalytic subunit genes using the Cre–Lox recombination system under the control of a late-stage spermatogenic cell-specific promoter may provide additional information on the role of PKA in sperm cell physiology.

At least some of the PKA (both type I and type II) present in spermatozoa is associated with cytoskeletal structures through A kinase anchor proteins (AKAPs). AKAPs interact with PKA through an amphipathic helix (Lester and Scott, 1997). This interaction is hypothesized to increase PKA signaling specificity by localizing the enzyme near relevant substrates (e.g., L-type calcium channels in somatic cells) (Gao *et al.,* 1997). At least three unique AKAPs have been identified in mature mammalian spermatozoa. Using a standard RII subunit overlay method for detecting AKAPs, a M_r 110,000 protein that binds RIIα and RIIβ was identified in bovine, mouse, human, and monkey sperm cells (Vijayaraghavan *et al.,* 1999). This protein localizes to the acrosomal cap and flagellum. In mouse and rat spermatozoa, proteins of approximately M_r 80,000 (FSC1/AKAP 82 and TAKAP-80) were also found to bind to the regulatory subunit of PKA (Visconti *et al.,* 1997; Mei *et al.,* 1997). FSC1/AKAP 82 is the major component of the mouse fibrous sheath, suggesting a potential role in motility (Johnson *et al.,* 1997). Analysis of this protein using the yeast two-hybrid approach has identified two separate binding sites for PKA regulatory subunits; both sites demonstrate high apparent affinity for RI, and the second site also weakly binds RII (Miki and Eddy, 1998). A similar dual specificity for PKA regulatory subunits is predicted for AKAP 110 based on sequence similarity with FSC1/AKAP 82 (Vijayaraghavan *et al.,* 1999). In addition to sperm AKAP association with the fibrous sheath, it seems likely that flagellar AKAPs may interact with other structures, because the PKA activity in sea urchin flagella, which lack both a fibrous sheath and outer dense fibers, is also resistant to detergent extraction (Yokota and Mabuchi, 1990).

Another report indicates that mammalian (bovine and primate) sperm motility is rapidly arrested with cell-permeable peptides that competitively block the interaction of AKAP 110 with PKA RII subunits (Vijayaraghavan *et al.,* 1997a). Inhibitors that block PKA catalytic activity (stearyl-PKI peptide and H89) had little if any effect on sperm motility. These results may suggest that the localization of the PKA R subunit has a physiological significance independent of catalytic activity, or perhaps alternative protein interactions of the R subunit. However, the data from the RIIα knockout model suggest that association with flagellar structures is not critical to motility, at least in the mouse (Burton *et al.,* 1999). Possibly the competitive peptide disrupted additional protein interactions with sperm cell AKAPs (Carr *et al.,* 2001). Alternatively, other unknown protein interactions may have been disrupted.

Protein phosphatases also regulate sperm motility acquisition during epididymal transit. Incubation of immotile caput epididymal spermatozoa with the phosphatase inhibitors calyculin A (low nanomolar range) or okadaic acid (low micromolar range) increases the percent motile spermatozoa approximately fivefold above controls (Smith *et al.,* 1999). In addition, comparing caput to cauda epididymal spermatozoa revealed a correlation between a two- to threefold decrease in cellular phosphatase activity as measured by phosphorylase dephosphorylation and increased motility capacity. Based on inhibitor sensitivities and immunoblotting, this protein phosphatase is PP1γ. The activity of PP1γ appears regulated by glycogen synthase kinase 3 (GSK3), and I2, a protein inhibitor, both of which are also detected in spermatozoa (Smith *et al.,* 1999). In somatic cells, GSK3 activity phosphorylates I2, which consequently dissociates from PP1γ, resulting in activation of phosphatase activity (Puntoni and Villa-Moruzzi, 1995). As spermatozoa transit the epididymis, GSK3 activity drops, leading to decreased PP1γ activity, likely due to association with the dephosphorylated form of I2. In many somatic cells, GSK3 is controlled by receptor tyrosine kinase-mediated stimulation of PI3 kinase and the subsequent activation of PKB/Akt, which inactivates GSK3 by phosphorylation (Shaw *et al.,* 1998). The possible existence of a similar regulatory mechanism or a novel one in spermatozoa has not been determined.

IV. MOTILITY ACTIVATION

In general, sperm motility is activated at ejaculation as a consequence of changes in intracellular ion concentrations. The external signals that initiate these events, and thus flagellar beating, differ among species according to the environment in which fertilization occurs. In sea urchins and other marine invertebrates, spawning into seawater leads to elevations in pH_i from about 7.2 to 7.6, activating dynein ATPase, which is inactive below pH 7.3 (Christen *et al.,* 1983). This cellular alkalinization results from a reduction in the surrounding $[CO_2]$, and the activation of a voltage-dependent Na^+/H^+ exchange due to membrane hyperpolarization induced by a decrease in $[K^+]_o$ (Darszon *et al.,* 1999). The motility of salmonid fish spermatozoa is also activated by a K^+-mediated hyperpolarization, and can be blocked with K^+ channel inhibitors such as tetraethylammonium ion (Tanimoto *et al.,* 1994). Subsequent increases in $[Ca^{2+}]_i$ and cAMP lead to the rapid phosphorylation of tyrosine(s) on the flagellar motility initiation phosphoprotein (MIPP), which is hypothesized to be required for motility initiation (Morisawa, 1994). Teleost fish sperm motility is initiated by hypoosmolality or hyperosmolality for freshwater or marine fertilizing species, respectively (Darszon *et al.,* 1999). The changes in cell volume are thought to alter intracellular ion concentrations, leading to activation. Elevations in $[Ca^{2+}]_i$ and/or pH_i appear to be involved in this process, because artificial methods of increasing these factors also activate motility. Mammalian sperm motility is suppressed in some species because of a low pH environment in the epididymis and cell membrane permeability to lactic acid (Carr *et al.,* 1985). On release, pH_i increases along with both $[Ca^{2+}]_i$ and cAMP to contribute to motility activation. Alternatively, in species such as rat and hamster, sperm motility is suppressed by the viscosity of components present in the secretions of the male accessory gland (e.g., semenogelin) (Eksittikul and Chulavatnatol, 1986). Only after dispersal of these factors is full motility displayed.

Although the initial environmental signals that activate sperm motility appear diverse, the conserved structure of the axoneme across species implies that the signal transduction pathways converge to produce the same effect. In sea urchins, several flagellar proteins (p29/32, p45, p130, p500) are phosphorylated within 1 minute of sperm cell motility activation (Bracho *et al.,* 1998). Based on solubility properties, these proteins appear to be dynein components. Similarly, Tash and colleagues have identified a M_r 120,000 protein in mouse and human spermatozoa that is rapidly phosphorylated on motility initiation (Tash and Bracho, 1998). The identity of these proteins and the consequences of their phosphorylation remain to be determined. However, the phosphorylation of light and intermediate dynein chains in protozoa is known to regulate microtubule sliding and beat frequency (Satir *et al.,* 1995; Habermacher and Sale, 1997).

V. MOTILITY MODULATION IN THE FEMALE REPRODUCTIVE TRACT

In mammals, spermatozoa undergo a functional maturation known as capacitation as they migrate through the female reproductive tract prior to fertilization (see Chapter 3). The spermatozoa maintain progressive motility as they move through the cervix and/or uterus, and the uterotubulal junction into the oviducts. Once reaching the isthmic oviduct, motility is greatly reduced as spermatozoa attach to the epithelium (see Chapter 1). Because active motility is restored by washing the spermatozoa free of oviductal fluid, it appears that a motility inhibitor is present in the isthmic environment (Overstreet and Katz, 1990).

Near the time of ovulation, a form of sperm motility, *hyperactivation,* appears. Hyperactivation is characterized by vigorous flagellation and a flagellar beat with high curvature and a long wavelength (Mortimer, 1997). Based on *in situ* observations of hamster and mouse spermatozoa, the oviduct is the primary site of hyperactivation (approximately 90% of the cells at the site of fertilization) (Smith and Yanagimachi, 1990; Suarez and Osman, 1987). This form of motility is hypothesized to be important for one or more of the following processes: (1) release from the oviductal epithelium, (2) enhanced progressive motility through the oviduct, and (3) penetration of the cellular and/or acellular matrices around the egg (Mortimer, 1997).

What is the molecular basis of sperm hyperactivation? Extracellular calcium, Ca_e^{2+}, is required to initiate hyperactivated motility in the spermatozoa of all species displaying this form of motion (Yanagimachi, 1994). The factors that regulate the effects of Ca_e^{2+}, and thus hyperactivation, are largely unknown. *In vitro* capacitated human spermatozoa hyperactivate in response to progesterone, and the apparent localization of novel membrane progesterone receptors over the sperm head suggests that the local activation of these receptors and the consequent increase in Ca_i^{2+} can be relayed throughout the cell, having distant effects on flagellar function (Blackmore and Lattanzio, 1991). In fact, head- and midpiece-associated Ca_i^{2+} oscillations that approximated the flagellar bend initiation frequency have been observed in hamster spermatozoa hyperactivated *in vitro* (Suarez *et al.*, 1993). Alternative flagellar membrane-associated mechanisms may also regulate Ca_e^{2+} entry in human spermatozoa, because hyperactivation can occur prior to ovulation, when progesterone levels would be expected to be very low. Possibly changes associated with capacitation—for example, membrane cholesterol depletion—may "destabilize" the sperm membrane, causing increased Ca^{2+} permeability and therefore the initiation of hyperactivation (Chapter 3). Calmodulin is one calcium-binding protein in spermatozoa that appears to regulate hyperactivity. Various structurally distinct calmodulin inhibitors block hyperactivated motility (Ahmad *et al.*, 1995). These effects are probably mediated at least in part by a distinct sperm isoform of PP2B (Tash *et al.*, 1988). This phosphatase may associate with the flagellar axoneme, where it would presumably dephosphorylate a subset of axoneme proteins, leading to the increased flagellar curvature observed during hyperactivation. This proposed calcium-dependent covalent regulatory system is consistent with the observed maintanence of hyperactivated motility following its initiation and the removal of Ca_e^{2+} (Mortimer, 1997). Sperm cell hyperactivation is also associated with Ca_e^{2+}-dependent increases in cAMP. The elevation of cAMP appears either independent of, or perhaps upstream of, calmodulin activation because calmodulin inhibitors have no effect on this event (White and Aitken, 1989). In addition to flagellar PKA, another potential cAMP target in sperm flagella is a cyclic nucleotide-gated (CNG) ion channel (Wiesner *et al.*, 1998). The principal piece localization of the CNG channel suggests that it could mediate the early Ca^{2+} entry associated with hyperactivated motility in this region of the cell. Alternatively, or in addition to CNG channel activity, voltage-dependent calcium channels associated with the flagellar membrane have been identified and may be regulated by PKA (Westenbroek and Babcock, 1999). PKA-dependent tyrosine phosphorylations of several proteins (M_r 40,000 to 120,000) have been demonstrated in mouse, human, and bovine spermatozoa during *in vitro* capacitation (Visconti *et al.,* 1995). One of these phosphorylated proteins is the fibrous sheath-associated AKAP in human spermatozoa, but, interestingly, not the equivalent protein in mouse spermatozoa. The majority of these proteins appear to be localized to the fibrous sheath, suggesting a role in motility, specifically in hyperactivation due to the temporal association of the event with capacitation (Si and Okuno, 1999). However, whereas the sperm protein tyrosine phosphorylations in the mouse are dependent on the presence of sufficient bovine serum albumin (BSA) in the medium to allow capacitation, mouse sperm hyperactivation has been reported to develop without BSA in the medium, although the timing is delayed (Neill and Olds-Clarke, 1987). Thus, the relationship of these PKA-dependent tyrosine phosphorylations to hyperactivated motility may not represent one of cause and effect.

VI. EGG-ASSOCIATED MOTILITY STIMULATION

A. EFFECTORS

The eggs and their closely associated cellular and acellular matrices from a variety of vertebrate and invertebrate species release factors that modulate sperm motility. In mammals, several studies have demonstrated that exposure to cumulus cells or cumulus cell-conditioned medium alters both sperm velocity and flagellar amplitude (Bradley and Garbers, 1983; Tesarik *et al.,* 1990; Fetterolf *et al.,* 1994). The active cumulus cell component responsible for these effects has not been identified. Herring sperm motility is regulated by two distinct egg-associated factors. One factor, sperm motility initiation factor (SMIF), is a M_r 105,000 glycoprotein that is tightly associated with the micropyle of the egg chorion (Yanagimachi *et al.,* 1992). The second egg-associated factor, herring sperm activating peptide (HSAP), consists of a group of small polypeptides of approximately M_r 8000 that are homologous to Kazal-type protease inhibitors (Oda *et al.,* 1998). These proteins are synthesized in the follicular cells and secreted into the developing egg chorion during oogenesis. The mechanism of these factors' action on sperm motility has not been elucidated. Many invertebrate species also produce egg-associated factors that modify sperm motility. An extreme example is the horseshoe crab, whose spermatozoa are immotile until detecting the sperm motility initiation peptide released from the egg (Clapper and Epel, 1985). The most completely developed model of this phenomenon is that of the sea urchin egg peptides.

The first sea urchin egg peptides, speract (GFDLNGGGVG) and resact (CVTGAPGCVGGGRL-NH_2), from *Strongylocentrotus purpuratus* and *Arbacia punctulata,* respectively, were isolated almost 20 years ago (Suzuki *et al.,* 1981; Hansbrough and Garbers, 1981; Bradley *et al.,* 1984). These peptides stimulate sperm motility and respiration at subnanomolar concentrations in a species-specific manner (Garbers *et al.,* 1982; Shimomura *et al.,* 1985). Analysis of synthetic egg peptide analogs identified the carboxy-terminal amino acids as the critical structure responsible for bioactivity. In the case of resact, this activity is potentiated by the amino-terminal half of the peptide. Numerous additional egg peptides from many sea urchin species have now been characterized by Suzuki and colleagues (Suzuki and Yoshino, 1992). All of these peptides can be sorted into six groups based on structural similarity. These groups exactly correlate with the taxonomic order/suborder designations of the various species of sea urchin from which the peptides were isolated. Not surprisingly, peptides classified in the same structural group can stimulate sperm motility and respiration, albeit with lower efficacies, across species only within the same order of sea urchins.

B. RECEPTORS

The observation that the egg peptides' amino termini could be modified without losing specific binding or motility/respiratory stimulation led to the identification of potential receptors using radioiodinated peptide analogs and cross-linking reagents (Dangott and Garbers, 1984; Shimomura *et al.,* 1986). Radiolabeled speract cross-links to an *S. purpuratus* protein of M_r 77,000 [sodium dodecyl sulfate–polyacrylamide gel electrophoresis (SDS–PAGE), reducing conditions]. No cross-linking of this peptide is detected in the presence of excess unlabeled speract, nor does it cross-link to *A. punctulata* sperm proteins, indicating that the interaction is specific. Molecular cloning of a cDNA encoding this M_r 77,000 protein from a testis cDNA library predicts a type I transmembrane protein with four tandem extracellular scavenger receptor cysteine-rich (SRCR) domains and a 12-amino acid intracellular region (Dangott *et al.,* 1989; Resnick *et al.,* 1994). SRCR domains have been identified in several proteins, many of which are expressed by mammalian lymphocytes. Among the SRCR-containing proteins, a function for this domain has been demonstrated solely in the T cell protein CD6 (Bowen *et al.,* 1996). In this protein, the membrane-proximal SRCR domain binds to the amino-terminal immunoglobulin domain of activated lymphocyte cell adhesion molecule (ALCAM), modulating T cell receptor signaling. Further studies of more proteins with SRCR domains will be needed to determine if binding to immunoglobulin domains is a general property of this structure. Whether the SRCR domains of the M_r 77,000 sperm protein actually bind speract and thereby initiate a signaling cascade or are simply closely apposed to the true peptide receptor is not known.

In *A. punctulata,* a similar approach using a radiolabeled resact analog identified a membrane guanylyl cyclase as the receptor (Shimomura *et al.,* 1986). This observation is of particular interest because an increase in cellular cGMP is one of the earliest detected events in response to the egg peptides (Cook and Babcock, 1993b). Using peptide sequence from the purified guanylyl cyclase, a clone was isolated from an *A. punctulata* testis cDNA library (Singh *et al.,* 1988). The predicted topology indicates an extracellular domain (approximately 500 amino acids), and an intracellular region with a protein kinase homology domain near the transmembane segment and a more distal catalytic domain (Chinkers *et al.,* 1989).

A similar guanylyl cyclase sequence was subsequently obtained from an *S. purpuratus* testis cDNA library (Thorpe and Garbers, 1989). Comparison of these two sequences shows a highly conserved intracellular region with a more divergent extracellular domain. Additionally, each of the sea urchin sperm membrane guanylyl cyclases localizes to the entire length of the flagellum using a carboxy-terminal peptide antibody (Quill and Garbers, 1998). The same distribution is seen for a sperm-bound fluorescent analog of speract (Cardullo *et al.,* 1994). The early elevation of cellular cGMP, the correlation of egg peptide specificity with the sequence diversity of the guanylyl cyclase extracellular domain, and the localization data are all consistent with the function of guanylyl cyclase as a species-specific egg peptide receptor. Alternatively, the guanylyl cyclase may function in a receptor complex with the M_r 77,000 protein because both membrane proteins are present in each sea urchin species and the physiological response of the spermatozoa to the egg peptides is nearly identical across species (Quill and Garbers, 1998).

The cloning of cDNAs encoding the sea urchin sperm guanylyl cyclases led directly to cloning of mammalian homologs. The family of mammalian membrane guanylyl cyclases now consists of seven members, designated GC-A through GC-G; these cyclases function in a variety of physiological processes, including blood pressure regulation and visual perception (Foster *et al.,* 1999). GC-A, -B, and -C are receptors for various peptide hormones that stimulate guanylyl cyclase activity, comparable to the postulated function of the sea urchin sperm guanylyl cyclases. Ligands for the remaining mammalian membrane guanylyl cyclases have yet to be identified. The conserved topology and enzymatic characteristics of all of the membrane guanylyl cyclases suggest that the regulatory mechanisms are also shared. Each of these proteins possesses a potential amphipathic helix located between the intracellular protein kinase homology and catalytic domains (Quill and Garbers, 1998). This region appears to be primarily responsible for a ligand-independent oligomerization of two to four identical guanylyl cyclases. Based on studies of deletion and dominant-negative mutations of this region, the formation of at least a dimer is necessary for catalytic activity (Chinkers and Wilson, 1992; Thompson and Garbers, 1995). This is supported by the reported structures of enzymatically active homologous adenylyl cyclase catalytic domains, which indicate that residues from each monomer contribute to the catalytic pocket (Tesmer *et al.,* 1997). On ligand binding in the presence of ATP, an apparent kinase homology domain-mediated enzyme inhibition is relieved, resulting in guanylyl cyclase activation (Foster *et al.,* 1999). The activation of guanylyl cyclase is transient, with ligand binding inducing a rapid (seconds) dephosphorylation of the enzyme (e.g., from 15 to 2 mol PO_4/mol enzyme for *S. purpuratus* sperm guanylyl cyclase) that coincides with decreased cGMP synthesis (Quill and Garbers, 1998). Several phosphorylated amino acids in the kinase homology domain of GC-A have been identified (Potter and Hunter, 1998). Mutation of these residues to alanine, singly or as a group, reduces or eliminates, respectively, ligand-stimulated GC-A activity. In addition, the sensitization of GC-A in crude membrane preparations to ligand stimulation correlates with the guanylyl cyclase phosphorylation state, and glutamate substitution for all of the kinase homology domain phosphorylated residues of GC-A produces a receptor that is resistant to ligand-dependent decreases in cyclase activity (Foster and Garbers, 1998; Potter and Hunter, 1999). Thus, the dephosphorylation observed in sea urchin sperm guanylyl cyclase likely reflects a desensitization to ligand stimulation. The dephosphorylation mechanism in sea urchin spermatozoa appears sensitive to intracellular alkalinization, because both egg peptides and artificially raising pH_i can cause dephosphorylation (Quill and Garbers, 1998). In contrast, preventing alkalinization slows this process. The relative rate of guanylyl cyclase dephosphorylation caused by elevated pH in the presence or absence of speract suggests that a ligand-dependent guanylyl cyclase conformational change may increase access to a constitutive phosphatase. Alternatively, a regulated phosphatase may be activated as a consequence of egg peptide binding.

Several unanswered questions concerning guanylyl cyclase signal transduction remain. For example, the physiologically relevant kinases and phosphatases that regulate the phosphorylation state of these receptors have not been identified in any organism. Analysis of these proteins may provide insights into potential links with other signaling pathways. ATP modulates ligand-dependent activation of guanylyl cyclases, presumably by binding to the kinase homology domain, but how this interaction produces its effects is unknown. In addition, there are examples of signaling proteins that possess only an intracellular domain similar to the kinase homology domain of membrane guanylyl cyclase (Foster *et al.,* 1999). Thus, the guanylyl cyclases may signal through other mechanisms in addition to cGMP synthesis. Finally, although the guanylyl cyclases from sea urchin spermatozoa and *C. elegans* are structurally well conserved with the mammalian guanylyl cyclases, heterologous expression of the nonmammalian enzymes in either mammalian or insect cell lines has consistently failed to yield activity, despite detectable protein synthesis. These observations suggest the possibility that guanylyl cyclase activity may be regulated by other associated proteins. Precedent for such regulatory proteins exists in the retina, where photoreceptor guanylyl cyclase activity appears modulated in a calcium-dependent manner by guanylyl cyclase-associated protein (GCAP-2) (Gorczyca *et al.,* 1995).

C. SIGNAL TRANSDUCTION

A current model of the signal transduction pathway activated by the egg peptides is shown in Figure 2.2. In this model, peptide binding stimulates a transient increase in cGMP production and an associated K^+-dependent membrane hyperpolarization (Cook and Babcock, 1993b). The molecular mechanism that opens the K^+ channel, allowing K^+ efflux, remains unclear, although patch-clamp experiments suggest that a diffusible second messenger is involved (Babcock *et al.,* 1992). One candidate is cGMP, because elevations in extracellular K^+ inhibit all of the known egg peptide responses except for increases of cGMP (Harumi *et al.,* 1992). Additionally, a correlation between the level of cellular cGMP and the open

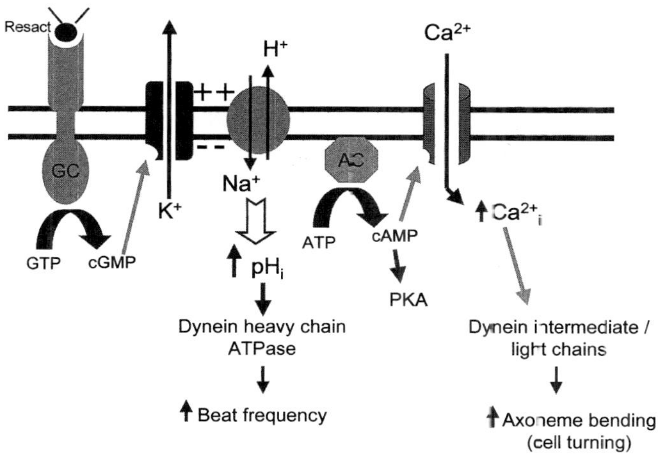

FIGURE 2.2 The egg peptide-activated signaling cascade in sea urchin spermatozoa. Gray arrows indicate hypothetical interactions. Membrane hyperpolarization is indicated by the membrane-associated plus and minus symbols.

state of the K^+ channel is found in the presence of IBMX (Cook and Babcock, 1993b). Thus, this sperm K^+ channel may be similar to a putative cGMP-regulated K^+ channel cloned from rabbit genomic DNA (Yao *et al.,* 1995). Expression of the cDNA corresponding to the rabbit genomic clone produces a K^+-selective ion channel that is preferentially activated by cGMP relative to cAMP, and that is inhibited by tetraethylammonium ion. These characteristics are shared with the sea urchin sperm K^+ channel activated by the egg peptides. As a consequence of the K^+-dependent hyperpolarization, a Na^+/H^+ exchanger is activated, causing the alkalinization of the spermatozoon (Lee and Garbers, 1986). This elevation of pH_i is thought to stimulate both motility and respiration, because artificial methods of raising cellular pH (e.g., NH_4^+, monensin) have the same effects (Repaske and Garbers, 1983; Hansbrough and Garbers, 1981). One potential target for the increased pH_i is the dynein arms, as indicated previously. The change in pH_i may also result in the activation of the sperm adenylyl cyclase and the opening of a calcium channel (Beltran *et al.,* 1996). The consequent increases in cAMP and Ca_i^{2+} are likely to regulate sperm motility further through the actions of various effectors of these second messengers (e.g., isoforms of PKA, PP2B, PDE, and calmodulin) on the axoneme proteins. Additional evidence suggests that cAMP may also regulate sea urchin sperm ion channels. Specifically, Cook and Babcock (1993a) observed a close correlation between cellular cAMP levels and the open state of a Ca^{2+}-permeable channel, suggesting a regulatory interaction. Another ion channel with homology to the cyclic nucleotide-gated ion channel family has been cloned from a sea urchin testicular cDNA library (Gauss *et al.,* 1998). Expression of this clone in HEK 293 cells produces a voltage-dependent, K^+-selective channel with an open-state probability that is enhanced by cAMP but not by cGMP. This channel, designated SPIH, localizes to the sperm flagellum and may correspond to a secondary K^+ channel activated following the elevations of Ca_i^{2+} and cAMP induced by speract binding (Cook and Babcock, 1993a).

VII. CHEMOATTRACTION

Sperm chemoattraction is defined as the movement of spermatozoa in response to a chemical signal toward the source of that signal. This form of motility is reported in many evolutionarily diverse species (e.g., marine invertebrates, aquatic vertebrates, and mammals) (Miller, 1985; Eisenbach, 1999). This suggests that there is a selectable advantage for chemoattraction during gamete interaction. One possible advantage is that the effective size of the spermatozoon's target would be increased, leading to an enhanced fertilization rate despite the limited number of spermatozoa near the egg at the time of fertilization. As detailed below, the chemoattractant source usually appears to be the egg, or in some cases a cell(s) or structure in close proximity to the egg. Because of the effect of water currents during the spawning of externally fertilizing species, and the adovarian flow of oviductal fluid of internally fertilizing species, it seems likely that the effective range of sperm chemoattractants is limited to perhaps a few hundred micrometers around the source (Battalia and Yanagimachi, 1979).

A. ASSAYS

The demonstration of a chemoattractive effect relies on the accumulation of spermatozoa to a localized source of the potential attractant. Assays designed to test for chemoattraction face two basic problems. First, the random movement of live spermatozoa produces a high background of cells in a defined area that makes it difficult to detect a direction-oriented movement caused by a potential chemoattractant. Second, a number of alternative factors, such as enhanced or reduced velocity, hyperactivation, medium viscosity, or medium flow, can result in a localized accumulation of spermatozoa. These factors have contributed to earlier incorrect claims of mammalian sperm chemoattraction (Jaiswal *et al.,* 1999). Thus, it is clear that distinguishing true directional motility from experimental artifacts can be difficult.

A variety of assays have been used to measure a chemoattractant effect on spermatozoa. Each of them can be classified as either "tracking" (continuous) or "accumulation" (end-point) assays. All tracking assays continuously monitor sperm motility, using a microscope, during exposure to the potential chemoattractant. Examples include adding the chemoattractant as a small bolus or as an impregnated agar plug at the end of a capillary tube into a suspension of spermatozoa on a slide (Ward *et al.,* 1985; Miller, 1985). Alternatively, a slide with two or more wells can be used to detect chemoattraction by placing the spermatozoa in one well and the chemoattractant or control solutions in the other well(s) (Cohen-Dayag *et al.,*

1994). The movement of the cells in response to the attractant relative to controls is analyzed either manually from a video recording, or by computer motion analysis. This type of assay measures both sperm accumulation near the potential chemoattractant and the actual turning of the cells in response to the signal. Consequently, tracking assays can distinguish a chemoattractive response from other effects on sperm motility (e.g., a cell velocity or progressiveness effect). Accumulation assays measure the number of spermatozoa near or in a well, chamber, or capillary tube containing a potential chemoattractant relative to a control containing medium alone after a given period of time. In those assays using either a well or chamber format, the compartments are connected through a small channel/tube or across a cell-permeable membrane, whereas the capillary tube format places the end of a capillary tube into a sperm cell suspension (Ralt *et al.,* 1991; Villanueva-Diaz *et al.,* 1992; Cohen-Dayag *et al.,* 1994). For these assays, spermatozoa are placed in one compartment, and the attractant is placed in either an adjacent or the same compartment to establish an increasing or decreasing gradient of chemoattractant. The effect of the potential chemoattractant on sperm motility in the absence of a gradient can also be examined by adding it to both compartments in the assay. Given appropriate controls, accumulation assays can distinguish between effects on cell velocity and chemoattraction. However, alone, sperm accumulation in the presence of a gradient does not unequivocally demonstrate attraction, because multiple factors can cause the observed accumulation, as noted above. Therefore, microscopic analysis of sperm motility in the presence of the potential chemoattractant must also be examined to support chemoattraction claims based on accumulation assays.

B. CHEMOATTRACTANTS

Sperm chemoattractants have been isolated and identified in only a few species. In the sea urchin *A. punctulata* the egg peptide resact (see Section VI,A) acts as a potent species-specific and calcium-dependent sperm chemoattractant (Ward *et al.,* 1985). The other sea urchin egg peptides have not been shown to attract spermatozoa. However, additional experiments with speract and *S. purpuratus* spermatozoa in the presence of IBMX do demonstrate a modified flagellar and cellular motion identical to that of the chemoattractant response to resact in *A. punctulata* (Cook *et al.,* 1994). Thus, many or all of the sea urchin egg peptides may be chemoattractants. In the coral *Montipora digitata* an aliphatic alcohol attracts spermatozoa in a species-specific manner (Babcock, 1995). This attractant is synthesized by the egg. A sperm chemoattractant, named startrak, was purified from starfish egg extracts (Miller and Vogt, 1996). A synthetic amino-terminal 32-amino acid fragment of startrak attracts conspecific spermatozoa.

In vertebrate species, no confirmed sperm chemoattractant has been isolated and identified. A recent study in *Xenopus laevis* suggests that spermatozoa of this species are attracted to a M_r 3000–10,000, heat-stable peptide present in egg-conditioned medium (al-Anzi and Chandler, 1998). The source of this factor could be the egg proper or the jelly secreted onto the egg during its passage through the oviduct. Notably, unlike sperm chemoattraction described in many invertebrate species, *Xenopus* sperm chemoattraction appears to be independent of external calcium.

The existence of chemoattraction by mammalian spermatozoa is still enthusiastically debated. Comparisons of the increased number of spermatozoa at the oviductal site of fertilization (ampulla) in humans relative to the same region in the contralateral oviduct suggest at least the potential of chemoattraction (Williams *et al.,* 1993). In this case, the egg, accompanying follicular fluid, or the ipsilateral oviductal epithelial cells at the ampulla/isthmus boundary could be a source of a potential sperm cell chemoattractant. Other experiments with human spermatozoa demonstrate an accumulation of cells in diluted follicular fluid, but not in diluted serum (Ralt *et al.,* 1991; Cohen-Dayag *et al.,* 1994). Interestingly, the potency of the follicular fluid samples varies substantially and displays a positive correlation between accumulation potency and subsequent *in vitro* fertilization (IVF) success. The accumulation of spermatozoa in diluted follicular fluid appears to be at least partially due to a chemoattractant response, because the cell population preferentially moved up an ascending gradient, remained at the site of highest attractant concentration in a descending gradient, and moved in a random manner in the absence of a gradient (with or without follicular fluid) (Cohen-Dayag *et al.,* 1994). Additional studies by Eisenbach and colleagues suggest that (1) only 2–12% of a sperm population is responsive to follicular fluid, (2) the capacity of individual spermatozoa to detect the chemoattractant is transient, (3) there is a continuous turnover of responsive spermatozoa in a given cell population, and (4) only capacitated spermatozoa (based on sensitivity to acrosome reaction stimulation by phorbol esters) can respond to the chemoattractant (Cohen-Dayag *et al.,* 1994, 1995). These findings have resulted in the hypothesis that sperm chemoattraction in humans may serve as an efficient means of providing fertilization-competent spermatozoa (i.e., capacitated) to the ovulated egg over an extended period of time.

Potential chemoattractants reported for human spermatozoa include *N*-formyl-Met-Leu-Phe (fMLP) and the follicular fluid components progesterone and atrial natriuretic peptide (ANP) (Gnessi *et al.,* 1985; Villanueva-Diaz *et al.,* 1992; Zamir *et al.,* 1993). The apparent effects of fMLP and progesterone have been established as not being the result of chemoattraction (Jaiswal *et al.,* 1999). Similarly, the lack of a correlation in the measured ANP concentration in various follicular fluid samples and the apparent chemoattractant potency of the same samples suggests that atrial natriuretic peptide also is not a sperm chemoattractant (Anderson *et al.,* 1995). The most recent human follicular fluid chemoattractant factor candidate proposed by Eisenbach and colleagues is a small, heat-stable peptide (Cohen-Dayag *et al.,* 1994). This factor has not been isolated or identified.

C. SIGNAL TRANSDUCTION

The molecular basis of sperm chemoattraction is partially understood only in the sea urchin. On binding to its receptor on *A. punctulata* spermatozoa, resact initiates the signal transduction pathway shown in Figure 2.2. The subsequent opening of a calcium-permeable channel leads to increased Ca_i^{2+} solely in the presence of external Ca^{2+} (Ward *et al.,* 1985). Although both motility and respiration increase in the absence of Ca_e^{2+}, the elevation of Ca_i^{2+} is required for the initiation of the chemoattractant response. Calcium causes asymmetric flagellar beating *in vitro* similar to that observed during chemoattraction (Brokaw and Nagayama, 1985). However, the targets through which calcium causes its effect on flagellar waveform have yet to be defined. One potential target would be homologs of the *Chlamydomonas* dynein light chain (M_r 18,000) protein that belongs to the calmodulin superfamily (King and Patel-King, 1995). Cook *et al.* (1994) have proposed a hypothesis that is consistent with what is known about sea urchin sperm chemoattraction. As spermatozoa move through an increasing resact concentration gradient, the cell membrane potential is hyperpolarized due to the activation of the K^+ channel. If the spermatozoon failed to encounter a continuously increasing gradient of resact (e.g., moved away from the egg), the K^+ channel-regulated hyperpolarization would be down-regulated by the increased pH_i, and under these circumstances a Ca^{2+}-permeable channel would be activated, causing the cell to turn. Ca_i^{2+} would then return to basal levels, perhaps through the activity of a Na^+/Ca^{2+} exchange, allowing the cell to restore a symmetric flagellar waveform and linear movement. Thus, in the presence of resact, sperm movement in any direction other than toward the source would lead to turning (Figure 2.3). Such a chemoattractive response would be similar to that observed during bacterial chemotaxis (Grebe and Stock, 1998).

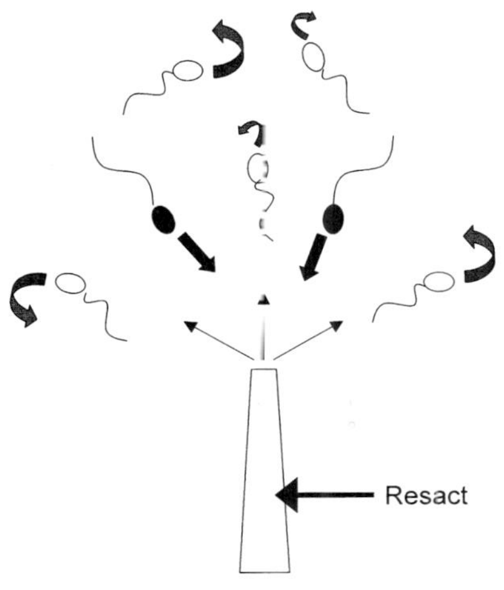

FIGURE 2.3 A proposed model for sperm chemoattraction. Spermatozoa with black heads, indicating low intracellular calcium, are shown moving up the resact concentration gradient. Spermatozoa with white heads, indicating elevated intracellular calcium, are not moving up the resact concentration gradient and consequently turn. Linear black arrows indicate straight movement. Curved gray arrows indicate turning.

Considerable speculation has been raised concerning a potential chemoattractant-activated signal transduction pathway in mammalian spermatozoa. This speculation is based on the finding in spermatozoa from a variety of mammalian species of several signaling components also found in olfactory neurons, where they are thought to form the odorant-responsive signaling module (Buck, 1996). The signal transduction pathway components identified in spermatogenic cells or mature spermatozoa by reverse transcriptase–polymerase chain reaction (RT–PCR) and/or immunoblots include G-protein-coupled receptors (GPCRs) of the large olfactory receptor subgroup, G_{olf}, adenylyl cyclase III (ACIII), GRK3, β-arrestin 2, an apparent cyclic nucleotide-gated ion channel, and the IP3 receptor (Wiesner *et al.*, 1998; Walensky *et al.*, 1995; Defer *et al.*, 1998; Parmentier *et al.*, 1992). Similar to olfactory tissue, for which there are estimated to be about 1000 different olfactory subtype GPCRs, there appear to be at least 40–50 of these receptors expressed by spermatozoa, as estimated by RT-PCR with dog, rat, mouse, and human spermatogenic cell cDNA (Vanderhaeghen *et al.*, 1997). Sequence comparison of the receptors found in testis to those present in the olfactory neurons does not indicate a distinct testicular group of these receptors. It is not known why there are so many testicular olfactory subtype GPCRs, nor whether an individual spermatozoon expresses one or more of these receptors. Using specific antibodies, a few of these receptors have been localized to the midpiece of dog and rat mature sperm cells (Vanderhaeghen *et al.*, 1993; Walensky *et al.*, 1995). In rat spermatozoa, both GRK3 and β-arrestin 2, components important for desensitization in the olfactory signal transduction pathway, also localize to the midpiece. In contrast, the cyclic nucleotide-gated channel in bovine spermatozoa is present on the flagellar principal piece (Wiesner *et al.*, 1998). In addition, there is no enzymatic or immunological evidence for the presence of ACIII or G_{olf}, a G_s-like protein, in mature spermatozoa, and a G_{olf} gene knockout mouse model has no apparent defects in either spermatogenesis or fertilization (Quill and Garbers, 1998; Belluscio *et al.*, 1998). Thus, the function of these signaling components in spermatogenic/sperm cells remains unknown. One possibility is that some of these receptors and effectors may participate in spermatogenesis, because this is a form of cellular regeneration that is also a property shared by the olfactory neurons (Morrison and Costanzo, 1995). Alternatively, only a portion of a sperm population may rely on this signal transduction pathway to fertilize an egg successfully. The answer to these questions will require identification and localization of the ligands for these receptors. One promising approach to this problem has been reported by Krautwurst *et al.* (1998). To identify olfactory GPCR ligands, these investigators heterologously expressed a fusion GPCR construct containing the central transmembrane II through VII regions of various olfactory GPCRs flanked by the rhodopsin 5′ untranslated region (UTR) sequence, including the first 20 translated codons to facilitate membrane localization, and the N and C termini of the olfactory M4 receptor. This construct coupled through the promiscuous $G_{15,16}$ protein to activate phospholipase C and the IP3 receptor, producing an increase in Ca_i^{2+}, which was measured as an increase in cellular fluorescence of fura-2-loaded cells. Using this generally applicable approach, they managed to identify only the third putative ligand for this subgroup of GPCRs. Perhaps this approach using extracts from the testis, oviduct, cumulus cells, or eggs as potential ligand sources would provide some insight to the function of the testicular olfactory subtype GPCRs found on spermatogenic/sperm cells.

VIII. CONCLUDING REMARKS

From the preceding discussion, it is clear that sperm motility is a complex phenomenon that is responsive to the external cellular environment. Factors regulating motility, such as changes in extracellular ion concentrations and secreted products from the male and/or female reproductive systems, activate sperm cell signaling involving changes in cyclic nucleotides, calcium, and protein phosphorylation/dephosphorylation in diverse species. In this conceptual model of sperm motility, the aspects describing both the plasma membrane components initially responsible for detecting and transducing the external signals and the final axoneme target components that alter the flagellar stroke are poorly defined. By analogy with the sea urchin sperm flagellar membrane, it seems likely that additional receptors will be discovered in the mammalian flagellar membrane. Similarly, proteomic studies may define, in addition to the dynein complex, axoneme proteins that modulate flagellar activity. Ultimately, a more comprehensive understanding of the signaling events that regulate sperm motility should lead to better methods to enhance or reduce motility and thus fertility.

REFERENCES

Afzelius, B. A. (1985). The immotile-cilia syndrome: A microtubule-associated defect. *CRC Crit. Rev. Biochem.* **19,** 63–87.

Ahmad, K., Bracho, G. E., Wolf, D. P., and Tash, J. S. (1995). Regulation of human sperm motility and hyperactivation components by calcium, calmodulin, and protein phosphatases. *Arch. Androl.* **35,** 187–208.

al-Anzi, B., and Chandler, D. E. (1998). A sperm chemoattractant is released from *Xenopus* egg jelly during spawning. *Dev. Biol.* **198,** 366–375.

Anderson, R. A. J., Feathergill, K. A., Rawlins, R. G., Mack, S. R., and Zaneveld, L. J. (1995). Atrial natriuretic peptide: A chemoattractant of human spermatozoa by a guanylate cyclase-dependent pathway. *Mol. Reprod. Dev.* **40,** 371–378.

Babcock, R. (1995). Synchronous multispecific spawning on coral reefs: Potential for hybridization and roles of gamete recognition. *Reprod. Fertil. Dev.* **7,** 943–950.

Babcock, D. F., Bosma, M. M., Battaglia, D. E., and Darszon, A. (1992). Early persistent activation of sperm K^+ channels by the egg peptide speract. *Proc. Natl. Acad. Sci. U.S.A.* **89,** 6001–6005.

Battalia, D. E., and Yanagimachi, R. (1979). Enhanced and co-ordinated movement of the hamster oviduct during the periovulatory period. *J. Reprod. Fertil.* **56,** 515–520.

Belluscio, L., Gold, G. H., Nemes, A., and Axel, R. (1998). Mice deficient in G(olf) are anosmic. *Neuron* **20,** 69–81.

Beltran, C., Zapata, O., and Darszon, A. (1996). Membrane potential regulates sea urchin sperm adenylylcyclase. *Biochemistry* **35,** 7591–7598.

Blackmore, P. F., and Lattanzio, F. A. (1991). Cell surface localization of a novel non-genomic progesterone receptor on the head of human sperm. *Biochem. Biophys. Res. Commun.* **181,** 331–336.

Bookbinder, L. H., Moy, G. W., and Vacquier, V. D. (1990). Identification of sea urchin sperm adenylate cyclase. *J. Cell. Biol.* **111,** 1859–1866.

Bowen, M. A., Bajorath, J., Siadak, A. W., Modrell, B., Malacko, A. R., Marquardt, H., Nadler, S. G., and Aruffo, A. (1996). The amino-terminal immunoglobulin-like domain of activated leukocyte cell adhesion molecule binds specifically to the membrane-proximal scavenger receptor cysteine-rich domain of CD6 with a 1:1 stoichiometry. *J. Biol. Chem.* **271,** 17390–17396.

Bracho, G. E., Fritch, J. J., and Tash, J. S. (1998). Identification of flagellar proteins that initiate the activation of sperm motility *in vivo. Biochem. Biophys. Res. Commun.* **242,** 231–237.

Bradley, M. P., and Garbers, D. L. (1983). The stimulation of bovine caudal epididymal sperm forward motility by bovine cumulus-egg complexes *in vitro. Biochem. Biophys. Res. Commun.* **115,** 777–787.

Bradley, M. P., Suzuki, N., and Garbers, D. L. (1984). The modulation of sperm metabolism and motility by factors associated with eggs. *Ann. N.Y. Acad. Sci.* **438,** 142–155.

Brokaw, C. J., and Nagayama, S. M. (1985). Modulation of the asymmetry of sea urchin sperm flagellar bending by calmodulin. *J. Cell. Biol.* **100,** 1875–1883.

Buck, J., Sinclair, M. L., Schapal, L., Cann, M. J., and Levin, L. R. (1999). Cytosolic adenylyl cyclase defines a unique signaling molecule in mammals. *Proc. Natl. Acad. Sci. U.S.A.* **96,** 79–84.

Buck, L. B. (1996). Information coding in the vertebrate olfactory system. *Annu. Rev. Neurosci.* **19,** 517–544.

Burton, K. A., Treash-Osio, B., Muller, C. H., Dunphy, E. L., and McKnight, G. S. (1999). Deletion of type II alpha regulatory subunit delocalizes protein kinase A in mouse sperm without affecting motility or fertilization. *J. Biol. Chem.* **274,** 24131–24136.

Cardullo, R. A., Herrick, S. B., Peterson, M. J., and Dangott, L. J. (1994). Speract receptors are localized on sea urchin sperm flagella using a fluorescent peptide analog. *Dev. Biol.* **162,** 600–607.

Carr, D. W., Usselman, M. C., and Acott, T. S. (1985). Effects of pH, lactate, and viscoelastic drag on sperm motility: A species comparison. *Biol. Reprod.* **33,** 588–595.

Carr, D. W., Fujita, A., Stente, C. L., Liberty, G. A., Olson, G. E., and Narumiya, S. (2001). Identification of sperm-specific proteins that interact with A-kinase anchoring proteins in a manner similar to the type II regulatory subunit of PKA. *J. Biol. Chem.* **276,** 17332–17338.

Chemes, H. E., Olmedo, S. B., Carrere, C., Oses, R., Carizza, C., Leisner, M., and Blaquier, J. (1998). Ultrastructural pathology of the sperm flagellum: Association between flagellar pathology and fertility prognosis in severely asthenozoospermic men. *Hum. Reprod.* **13,** 2521–2526.

Chen, Y., Cann, M. J., Litvin, T. N., Iourgenko, V., Sinclair, M. L., Levin, L. R., and Buck, J. (2000). Soluble adenylyl cyclase as an evolutionarily conserved bicarbonate sensor. *Science* **289,** 625–628.

Chinkers, M., and Wilson, E. M. (1992). Ligand-independent oligomerization of natriuretic peptide receptors. Identification of heteromeric receptors and a dominant negative mutant. *J. Biol. Chem.* **267,** 18589–18597.

Chinkers, M., Garbers, D. L., Chang, M. S., Lowe, D. G., Chin, H. M., Goeddel, D. V., and Schulz, S. (1989). A membrane form of guanylate cyclase is an atrial natriuretic peptide receptor. *Nature* **338,** 78–83.

Christen, R., Schackmann, R. W., and Shapiro, B. M. (1983). Metabolism of sea urchin sperm. Interrelationships between intracellular pH, ATPase activity, and mitochondrial respiration. *J. Biol. Chem.* **258,** 5392–5399.

Clapper, D. L., and Epel, D. (1985). The *Limulus* sperm motility-initiating peptide initiates acrosome reactions in sea water lacking potassium. *J. Exp. Zool.* **236,** 211–217.

Cohen-Dayag, A., Ralt, D., Tur-Kaspa, I., Manor, M., Makler, A., Dor, J., Mashiach, S., and Eisenbach, M. (1994). Sequential acquisition of chemotactic responsiveness by human spermatozoa. *Biol. Reprod.* **50,** 786–790.

Cohen-Dayag, A., Tur-Kaspa, I., Dor, J., Mashiach, S., and Eisenbach, M. (1995). Sperm capacitation in humans is transient and correlates with chemotactic responsiveness to follicular factors. *Proc. Natl. Acad. Sci. U.S.A.* **92,** 11039–11043.

Cook, S. P., and Babcock, D. F. (1993a). Activation of Ca^{2+} permeability by cAMP is coordinated through the pHi increase induced by speract. *J. Biol. Chem.* **268,** 22408–22413.

Cook, S. P., and Babcock, D. F. (1993b). Selective modulation by cGMP of the K^+ channel activated by speract. *J. Biol. Chem.* **268,** 22402–22407.

Cook, S. P., Brokaw, C. J., Muller, C. H., and Babcock D. F. (1994). Sperm chemotaxis: Egg peptides control cytosolic calcium to regulate flagellar responses. *Dev. Biol.* **165,** 10–19.

Cummings, D. E., Brandon, E. P., Planas, J. V., Motamed, K., Idzerda, R. L., and McKnight, G. S. (1996). Genetically lean mice result from targeted disruption of the RII beta subunit of protein kinase A [see comments]. *Nature* **382,** 622–626.

Dangott, L. J., and Garbers, D. L. (1984). Identification and partial characterization of the receptor for speract. *J. Biol. Chem.* **259,** 13712–13716.

Dangott, L. J., Jordan, J. E., Bellet, R. A., and Garbers, D. L. (1989). Cloning of the mRNA for the protein that crosslinks to the egg peptide speract. *Proc. Natl. Acad. Sci. U.S.A.* **86,** 2128–2132.

Darszon, A., Labarca, P., Nishigaki, T., and Espinosa, F. (1999). Ion channels in sperm physiology. *Physiol. Rev.* **79,** 481–510.

Deana, R., Foresta, C., Bonaga, G., and Rigoni, F. (1986). A simple nephelometric method for measuring the progressive motility and collecting motile spermatozoa. *Andrologia* **18,** 37–41.

Defer, N., Marinx, O., Poyard, M., Lienard, M. O., Jegou, B., and Hanoune, J. (1998). The olfactory adenylyl cyclase type 3 is expressed in male germ cells. *FEBS Lett.* **424,** 216–220.

Eddy, E. M., and O'Brien, D. A. (1994). The spermatozoon. *In* "The Physiology of Reproduction" (E. Knobil and J. D. Neill, eds.), Vol. 2, pp. 29–77. Raven Press, New York.

Eisenbach, M. (1999). Sperm chemotaxis. *Rev. Reprod.* **4,** 56–66.

Eksittikul, T., and Chulavatnatol, M. (1986). Increase in fluid viscosity during epididymal transit and the immobilization of rat epididymal spermatozoa. *Int. J. Androl.* **9,** 229–240.

Fetterolf, P. M., Jurisicova, A., Tyson, J. E., and Casper, R. F. (1994) Conditioned medium from human cumulus oophorus cells stimulates human sperm velocity. *Biol. Reprod.* **51,** 184–192.

Fisch, J. D., Behr, B., and Conti, M. (1998). Enhancement of motility and acrosome reaction in human spermatozoa: Differential activation by type-specific phosphodiesterase inhibitors [published erratum appears in *Hum. Reprod.* **14**(2), 575 (1999)]. *Hum. Reprod.* **13,** 1248–1254.

Forte, L. R., Bylund, D. B., and Zahler, W. L. (1983). Forskolin does not activate sperm adenylate cyclase. *Mol. Pharmacol.* **24,** 42–47.

Foster, D. C., and Garbers, D. L. (1998). Dual role for adenine nucleotides in the regulation of the atrial natriuretic peptide receptor, guanylyl cyclase-A. *J. Biol. Chem.* **273,** 16311–16318.

Foster, D. C., Wedel, B. J., Robinson, S. W., and Garbers, D. L. (1999). Mechanisms of regulation and functions of guanylyl cyclases. *Rev. Physiol. Biochem. Pharmacol.* **135,** 1–39.

Gao, T., Yatani, A., Dell'Acqua, M. L., Sako, H., Green, S. A., Dascal, N., Scott, J. D., and Hosey, M. M. (1997). cAMP-dependent regulation of cardiac L-type Ca^{2+} channels requires membrane targeting of PKA and phosphorylation of channel subunits. *Neuron* **19,** 185–196.

Garbers, D. L., and Kopf, G. S. (1980). The regulation of spermatozoa by calcium and cyclic nucleotides. *Adv. Cyclic Nucleotide. Res.* **13,** 251–306.

Garbers, D. L., Tubb, D. J., and Hyne, R. V. (1982). A requirement of bicarbonate for Ca^{2+}-induced elevations of cyclic AMP in guinea pig spermatozoa. *J. Biol. Chem.* **257,** 8980–8984

Gauss, R., Seifert, R., and Kaupp, U. B. (1998). Molecular identification of a hyperpolarization-activated channel in sea urchin sperm. *Nature* **393,** 583–587.

Gnessi, L., Ruff, M. R., Fraioli, F., and Pert, C. B. (1985). Demonstration of receptor-mediated chemotaxis by human spermatozoa. A novel quantitative bioassay. *Exp. Cell. Res.* **161,** 219–230.

Gorczyca, W. A., Polans, A. S., Surgucheva, I. G., Subbaraya, I., Baehr, W., and Palczewski, K. (1995). Guanylyl cyclase activating protein. A calcium-sensitive regulator of phototransduction. *J. Biol. Chem.* **270,** 22029–22036.

Grebe, T. W., and Stock, J. (1998). Bacterial chemotaxis: The five sensors of a bacterium. *Curr. Biol.* **8,** R154–R157

Gross, M. K., Toscano, D. G., and Toscano, W. A. J. (1987). Calmodulin-mediated adenylate cyclase from mammalian sperm. *J. Biol. Chem.* **262,** 8572–8676.

Habermacher, G., and Sale, W. S. (1997). Regulation of flagellar dynein by phosphorylation of a 138-kD inner arm dynein intermediate chain. *J. Cell. Biol.* **136,** 167–176.

Hansbrough, J. R., and Garbers, D. L. (1981). Sodium-dependent activation of sea urchin spermatozoa by speract and monensin. *J. Biol. Chem.* **256,** 2235–2241.

Harumi, T., Hoshino, K., and Suzuki, N. (1992). Effects of sperm-activating peptide I on *Hemicentrotus pulcherrimus* spermatozoa in high potassium seawater. *Dev. Growth Differ.* **34,** 163–172.

Jaiswal, B. S., and Majumder, G. C. (1996). Cyclic AMP phosphodiesterase: A regulator of forward motility initiation during epididymal sperm maturation. *Biochem. Cell Biol.* **74,** 669–674.

Jaiswal, B. S., Tur-Kaspa, I., Dor, J., Mashiach, S., and Eisenbach, M. (1999). Human sperm chemotaxis: Is progesterone a chemoattractant? *Biol. Reprod.* **60,** 1314–1319.

Johnson, L. R., Foster, J. A., Haig-Ladewig, L., VanScoy, H., Rubin, C. S., Moss, S. B., and Gerton, G. L. (1997). Assembly of AKAP82, a protein kinase A anchor protein, into the fibrous sheath of mouse sperm. *Dev. Biol.* **192,** 340–350.

Jones, R. C. (1999). To store or mature spermatozoa? The primary role of the epididymis. *Int. J. Androl.* **22,** 57–67.

King, S. M., and Patel-King, R. S. (1995). Identification of a Ca(2+)-binding light chain within *Chlamydomonas* outer arm dynein. *J. Cell Sci.* **108,** 3757–3764.

Kopf, G. S., and Vacquier, V. D. (1984). Characterization of a calmodulin-stimulated adenylate cyclase from abalone spermatozoa. *J. Biol. Chem.* **259,** 7590–7596.

Krause, W. (1995). The significance of computer-assisted semen analysis (CASA) for diagnosis in andrology and fertility prognosis. *Int. J. Androl.* **18,** 32–35.

Krause, W., and Viethen, G. (1999). Quality assessment of computer-assisted semen analysis (CASA) in the andrology laboratory. *Andrologia* **31,** 125–129.

Krause, W., Schonharl, G., and Brake, A. (1993). The variability of measuring sperm concentration and motility as determined by computer assisted image analysis and visual estimation. *Andrologia* **25,** 181–187.

Krautwurst, D., Yau, K. W., and Reed, R. R. (1998). Identification of ligands for olfactory receptors by functional expression of a receptor library. *Cell* **95,** 917–926.

Lee, H. C., and Garbers, D. L. (1986). Modulation of the voltage-sensitive Na^+/H^+ exchange in sea urchin spermatozoa through membrane potential changes induced by the egg peptide speract. *J. Biol. Chem.* **261,** 16026–16032.

Lester, L. B., and Scott, J. D. (1997). Anchoring and scaffold proteins for kinases and phosphatases. *Recent. Prog. Horm. Res.* **52,** 409–429.

Mei, X., Singh, I. S., Erlichman, J., and Orr, G. A. (1997). Cloning and characterization of a testis-specific, developmentally regulated A-kinase-anchoring protein (TAKAP-80) present on the fibrous sheath of rat sperm. *Eur. J. Biochem.* **246,** 425–432.

Miki, K., and Eddy, E. M. (1998). Identification of tethering domains for protein kinase A type I alpha regulatory subunits on sperm fibrous sheath protein FSC1. *J. Biol. Chem.* **273,** 34384–34390.

Miller, R. L. (1985). Sperm chemo-orientation in the metazoa. *Biol. Fert.* **2,** 276–340.

Miller, R. L., and Vogt, R. (1996). An N-terminal partial sequence of the 13 kDa *Pycnopodia helianthoides* sperm chemoattractant 'startrak' possesses sperm-attracting activity. *J. Exp. Biol.* **199,** 311–318.

Moos, J., Peknicova, J., Geussova, G., Philimonenko, V., and Hozak, P. (1998). Association of protein kinase A type I with detergent-resistant structures of mammalian sperm cells. *Mol. Reprod. Dev.* **50,** 79–85.

Morisawa, M. (1994). Cell signaling mechanisms for sperm motility. *Zoolog. Sci.* **11,** 647–662.

Morrison, E. E., and Costanzo, R. M. (1995). Regeneration of olfactory sensory neurons and reconnection in the aging hamster central nervous system. *Neurosci. Lett.* **198,** 213–217.

Mortimer, D., Aitken, R. J., Mortimer, S. T., and Pacey, A. A. (1995). Workshop report: Clinical CASA—the quest for consensus. *Reprod. Fertil. Dev.* **7,** 951–959.

Mortimer, S. T. (1997). A critical review of the physiological importance and analysis of sperm movement in mammals. *Hum. Reprod. Update* **3,** 403–439.

Neill, J. M., and Olds-Clarke, P. (1987). A computer-assisted assay for mouse sperm hyperactivation demonstrates that bicarbonate but not bovine serum albumin is required. *Gamete Res.* **18,** 121–140.

Oda, S., Igarashi, Y., Manaka, K., Koibuchi, N., Sakai-Sawada, M., Sakai, K., Morisawa, M., Ohtake, H., and Shimizu, N. (1998). Sperm-activating proteins obtained from the herring eggs are homologous to trypsin inhibitors and synthesized in follicle cells. *Dev. Biol.* **204,** 55–63.

Overstreet, J. W., and Katz, D. F. (1990). Interaction between the female reproductive tract and spermatozoa. *In* "Controls of Sperm Motility: Biological and Clinical Aspects" (C. Gagnon, ed.), pp. 63–75. CRC Press, Boca Raton.

Parmentier, M., Libert, F., Schurmans, S., Schiffmann, S., Lefort, A., Eggerickx, D., Ledent, C., Mollereau, C., Gerard, C., and Perret, J. (1992). Expression of members of the putative olfactory receptor gene family in mammalian germ cells. *Nature* **355,** 453–455.

Peters, E. P., Wilderspin, A. F., Wood, S. P., Zvelebil, M. J., Sezer, O., and Danchin, A. (1991). A pyruvate-stimulated adenylate cyclase has a sequence related to the fes/fps oncogenes and to eukaryotic cyclases. *Mol. Microbiol.* **5,** 1175–1181.

Potter, L. R., and Hunter, T. (1998). Phosphorylation of the kinase homology domain is essential for activation of the A-type natriuretic peptide receptor. *Mol. Cell. Biol.* **18,** 2164–2172.

Potter, L. R., and Hunter, T. (1999). A constitutively "phosphorylated" guanylyl cyclase-linked atrial natriuretic peptide receptor mutant is resistant to desensitization. *Mol. Biol. Cell* **10,** 1811–1820.

Puntoni, F., and Villa-Moruzzi, E. (1995). Phosphorylation of the inhibitor-2 of protein phosphatase-1 by cdc2-cyclin B and GSK3. *Biochem. Biophys. Res. Commun.* **207,** 732–739.

Quill, T. A., and Garbers, D. L. (1998). Fertilization: Common molecular signaling pathways across the species. *In* "Hormones and Signaling" (B. W. O'Malley, ed.), pp. 167–207. Academic Press, San Diego.

Ralt, D., Goldenberg, M., Fetterolf, P., Thompson, D., Dor, J., Mashiach, S., Garbers, D. L., and Eisenbach, M. (1991). Sperm attraction to a follicular factor(s) correlates with human egg fertilizability. *Proc. Natl. Acad. Sci. U.S.A.* **88,** 2840–2844.

Repaske, D. R., and Garbers, D. L. (1983). A hydrogen ion flux mediates stimulation of respiratory activity by speract in sea urchin spermatozoa. *J. Biol. Chem.* **258,** 6025–6029.

Resnick, D., Pearson, A., and Krieger, M. (1994). The SRCR superfamily: A family reminiscent of the Ig superfamily. *Trends Biochem. Sci.* **19,** 5–8.

Roberts, T. M., and Stewart, M. (1995). Nematode sperm locomotion. *Curr. Opin. Cell Biol* **7,** 13–17.

Satir, P., Barkalow, K., and Hamasaki, T. (1995). Ciliary beat frequency is controlled by a dynein light chain phosphorylation. *Biophys. J.* **68,** 222S

Shaw, M., Cohen, P., and Alessi, D. R. (1998). The activation of protein kinase B by H_2O_2 or heat shock is mediated by phosphoinositide 3-kinase and not by mitogen-activated protein kinase-activated protein kinase-2. *Biochem. J.* **336** , 241–246.

Shimomura, H., Dangott, L. J., and Garbers, D. L. (1986). Covalent coupling of a resact analogue to guanylate cyclase. *J. Biol. Chem.* **261,** 15778–15782.

Si, Y., and Okuno, M. (1999). Role of tyrosine phosphorylation of flagellar proteins in hamster sperm hyperactivation. *Biol. Reprod.* **61,** 240–246.

Singh, S., Lowe, D. G., Thorpe, D. S., Rodriguez, H., Kuang, W. J., Dangott, L. J., Chinkers, M., Goeddel, D. V., and Garbers, D. L. (1988). Membrane guanylate cyclase is a cell-surface receptor with homology to protein kinases. *Nature* **334,** 708–712.

Smith, E. F., and Lefebvre, P. A. (1997). The role of central apparatus components in flagellar motility and microtubule assembly. *Cell Motil. Cytoskeleton* **38,** 1–8.

Smith, G. D., Wolf, D. P., Trautman, K. C., and Vijayaraghavan, S. (1999). Motility potential of macaque epididymal sperm: The role of protein phosphatase and glycogen synthase kinase-3 activities. *J. Androl.* **20,** 47–53.

Smith, T. T., and Yanagimachi, R. (1990). The viability of hamster spermatozoa stored in the isthmus of the oviduct: The importance of sperm-epithelium contact for sperm survival. *Biol. Reprod.* **42,** 450–457.

Suarez, S. S., and Osman, R. A. (1987). Initiation of hyperactivated flagellar bending in mouse sperm within the female reproductive tract. *Biol. Reprod.* **36,** 1191–1198.

Suarez, S. S., Varosi, S. M., and Dai, X. (1993). Intracellular calcium increases with hyperactivation in intact, moving hamster sperm and oscillates with the flagellar beat cycle. *Proc. Natl. Acad. Sci. U.S.A.* **90,** 4660–4664.

Sunahara, R. K., Dessauer, C. W., and Gilman, A. G. (1996). Complexity and diversity of mammalian adenylyl cyclases. *Annu. Rev. Pharmacol. Toxicol.* **36,** 461–480.

Suzuki, N., and Yoshino, Y. (1992). The relationship between amino acid sequences of sperm-activating peptides from the jelly coat of sea urchin eggs. *Comp. Biochem. Physiol.* **102,** 679–690.

Suzuki, N., Nomura, K., Ohtake, H., and Isaka, S. (1981). Purification and the primary structure of sperm-activating peptides from the jelly coat of sea urchin eggs. *Biochem. Biophys. Res. Commun.* **99,** 1238–1244.

Tanimoto, S., Kudo, Y., Nakazawa, T., and Morisawa, M. (1994). Implication that potassium flux and increase in intracellular calcium are necessary for the initiation of sperm motility in salmonid fishes. *Mol. Reprod. Dev.* **39,** 409–414.

Tash, J. S., and Bracho, G. E. (1998). Identification of phosphoproteins coupled to initiation of motility in live epididymal mouse sperm. *Biochem. Biophys. Res. Commun.* **251,** 557–563.

Tash, J. S., Krinks, M., Patel, J., Means, R. L., Klee, C. B., and Means, A. R. (1988). Identification, characterization, and functional correlation of calmodulin-dependent protein phosphatase in sperm. *J. Cell Biol.* **106,** 1625–1633.

Tesarik, J., Mendoza, O. C., and Testart, J. (1990). Effect of the human cumulus oophorus on movement characteristics of human capacitated spermatozoa. *J. Reprod. Fertil.* **88,** 665–675.

Tesmer, J. J., Sunahara, R. K., Gilman, A. G., and Sprang, S. R. (1997). Crystal structure of the catalytic domains of adenylyl cyclase in a complex with $\alpha \cdot \gamma$. *Science* **278,** 1907–1916.

Thompson, D. K., and Garbers, D. L. (1995). Dominant negative mutations of the guanylyl cyclase-A receptor. Extracellular domain deletion and catalytic domain point mutations. *J. Biol. Chem.* **270,** 425–430.

Thorpe, D. S., and Garbers, D. L. (1989). The membrane form of guanylate cyclase. Homology with a subunit of the cytoplasmic form of the enzyme. *J. Biol. Chem.* **264,** 6545–6549.

Vanderhaeghen, P., Schurmans, S., Vassart, G., and Parmentier, M. (1993). Olfactory receptors are displayed on dog mature sperm cells. *J. Cell Biol.* **123,** 1441–1452.

Vanderhaeghen, P., Schurmans, S., Vassart, G., and Parmentier, M. (1997). Specific repertoire of olfactory receptor genes in the male germ cells of several mammalian species. *Genomics* **39,** 239–246.

Vijayaraghavan, S., Goueli, S. A., Davey, M. P., and Carr, D. W. (1997a). Protein kinase A-anchoring inhibitor peptides arrest mammalian sperm motility. *J. Biol. Chem.* **272,** 4747–4752.

Vijayaraghavan, S., Liberty, G. A., Mohan, J., Winfrey, V. P., Olson, G. E., and Carr, D. W. (1999). Isolation and molecular characterization of AKAP110, a novel, sperm-specific protein kinase A-anchoring protein. *Mol. Endocrinol.* **13,** 705–717.

Vijayaraghavan, S., Olson, G. E., NagDas, S., Winfrey, V. P., and Carr, D. W. (1997b). Subcellular localization of the regulatory subunits of cyclic adenosine 3′,5′-monophosphate-dependent protein kinase in bovine spermatozoa. *Biol. Reprod.* **57,** 1517–1523.

Villanueva-Diaz, C., Arias-Martinez, J., Bustos-Lopez, H., and Vadillo-Ortega, F. (1992). Novel model for study of human sperm chemotaxis. *Fertil. Steril.* **58,** 392–395.

Visconti, P. E., Bailey, J. L., Moore, G. D., Pan, D., Olds-Clarke, P., and Kopf, G. S. (1995). Capacitation of mouse spermatozoa. I. Correlation between the capacitation state and protein tyrosine phosphorylation. *Development* **121,** 1129–1137.

Visconti, P. E., Johnson, L. R., Oyaski, M., Fornes, M., Moss, S. B., Gerton, G. L., and Kopf, G. S. (1997). Regulation, localization, and anchoring of protein kinase A subunits during mouse sperm capacitation. *Dev. Biol.* **192,** 351–363.

Walensky, L. D., Roskams, A. J., Lefkowitz, R. J., Snyder, S. H., and Ronnett, G. V. (1995). Odorant receptors and desensitization proteins colocalize in mammalian sperm. *Mol. Med.* **1,** 130–141.

Ward, G. E., Brokaw, C. J., Garbers, D. L., and Vacquier, V. D. (1985). Chemotaxis of *Arbacia punctulata* spermatozoa to resact, a peptide from the egg jelly layer. *J. Cell Biol.* **101,** 2324–2329.

Wasco, W. M., and Orr, G. A. (1984). Function of calmodulin in mammalian sperm: Presence of a calmodulin-dependent cyclic nucleotide phosphodiesterase associated with demembranated rat caudal epididymal sperm. *Biochem. Biophys. Res. Commun.* **118,** 636–642.

Westenbroek, R. E., and Babcock, D. F. (1999). Discrete regional distributions suggest diverse functional roles of calcium channel alpha1 subunits in sperm. *Dev. Biol.* **207,** 457–469.

White, D. R., and Aitken, R. J. (1989). Relationship between calcium, cyclic AMP, ATP, and intracellular pH and the capacity of hamster spermatozoa to express hyperactivated motility. *Gamete Res.* **22,** 163–177.

Wiesner, B., Weiner, J., Middendorff, R., Hagen, V., Kaupp, U. B., and Weyand, I. (1998). Cyclic nucleotide-gated channels on the flagellum control Ca^{2+} entry into sperm. *J. Cell Biol.* **142,** 473–484.

Williams, M., Hill, C. J., Scudamore, I., Dunphy, B., Cooke, I. D., and Barratt, C. L. (1993). Sperm numbers and distribution within the human fallopian tube around ovulation. *Hum. Reprod.* **8,** 2019–2026.

Yanagimachi, R. (1994). Mammalian fertilization. *In* "The Physiology of Reproduction" (E. Knobil and J. D. Niell, eds.), pp. 189–317. Raven Press, New York.

Yanagimachi, R., Cherr, G. N., Pillai, M. C., and Baldwin, J. D. (1992) Factors controlling sperm entry into the micropyle of salmonid and herring eggs. *Dev. Growth Differ.* **34,** 447–461.

Yao, X., Segal, A. S., Welling, P., Zhang, X., McNicholas, C. M., Engel, D., Boulpaep, E. L., and Desir, G. V. (1995). Primary structure and functional expression of a cGMP-gated potassium channel. *Proc. Natl. Acad. Sci. U.S.A.* **92,** 11711–11715.

Yeung, C. H., Cooper, T. G., and Nieschlag, E. (1997). A technique for standardization and quality control of subjective sperm motility assessments in semen analysis. *Fertil. Steril.* **67,** 1156–1158.

Yeung, C. H., Sonnenberg-Riethmacher, E., and Cooper, T. G. (1999). Infertile spermatozoa of c-ros tyrosine kinase receptor knockout mice show flagellar angulation and maturational defects in cell volume regulatory mechanisms. *Biol. Reprod.* **61,** 1062–1069.

Yokota, E., and Mabuchi, I. (1990). The cAMP-dependent protein kinase in sea urchin sperm tails: Association of the enzyme with the flagellar axonemes. *J. Biochem. (Tokyo)* **108,** 1–3.

Zamir, N., Riven-Kreitman, R., Manor, M., Makler, A., Blumberg, S., Ralt, D., and Eisenbach, M. (1993). Atrial natriuretic peptide attracts human spermatozoa *in vitro*. *Biochem. Biophys. Res. Commun.* **197,** 116–122.

3

CAPACITATION

BIJAY S. JAISWAL AND MICHAEL EISENBACH

Department of Biological Chemistry, The Weizmann Institute of Science, Rehovot, Israel

I. INTRODUCTION

Mammalian spermatozoa undergo a process of maturation and become motile while being transported in the epididymis (Yanagimachi, 1994). However, this maturation is not sufficient for fertilization. Chang (1951) and Austin (1951; 1952) found that rabbit and rat spermatozoa cannot penetrate the eggs immediately after coitus, but rather that they require a period of about 2 hours in the female genital tract to acquire the ability to penetrate the egg and to fertilize it. They concluded that the spermatozoa must undergo an additional "maturation" process, termed "capacitation," to acquire fertilizing potential. (Note that, unlike sperm maturation, which occurs in the male reproductive tract, the process of capacitation occurs in the female genital tract. Like sperm maturation, sperm capacitation is unique to mammals.) Later it was shown that sperm capacitation is a prerequisite for the acrosome reaction (a release of proteolytic enzymes enabling sperm penetration through the egg coat; see Chapters 11 and 13). Therefore, the broad definition of capacitation, i.e., acquisition of fertilizing potential, became more focused—acquisition of a potential to undergo the acrosome reaction (Kopf and Gerton, 1991; Yanagimachi, 1989). [A repeated claim of a single research group that capacitation is not a prerequisite for acrosome-reaction induction in human spermatozoa (Anderson *et al.,* 1992; Bielfeld *et al.,* 1994a; De Jonge *et al.,* 1989) has been shown to be mistaken (Jaiswal *et al.,* 1998).] Since the discovery of Chang and Austin, many processes have been identified as being involved in sperm capacitation (Cohen-Dayag and Eisenbach, 1994; de Lamirande *et al.,* 1997b; Visconti *et al.,* 1998; Yanagimachi, 1994), but the roles of these processes and the overall molecular mechanism of sperm capacitation are still obscure. In this chapter we review these processes and propose a model for the molecular mechanism. We also review what is known of the physiology of sperm capacitation and indicate the central role that sperm capacitation has in fertilization. The reader is referred to Baldi *et al.* (1996), Harrison (1996), de Lamirande *et al.* (1997b), Nolan and Hammerstedt (1997), Cross (1998), and Visconti *et al.* (1998) for reviews on sperm capacitation with other emphases.

II. INITIATION, PROPAGATION, AND TERMINATION OF CAPACITATION

A. SPERM CAPACITATION *IN VITRO*

1. Initiation of Capacitation

a. The Trigger for Capacitation

Generally speaking, capacitation can be initiated by incubating spermatozoa in any medium that retains them in a competent state, provided that it is free of seminal fluid (defined as a capacitating medium). The removal of seminal fluid is essential, because it contains decapacitating factors that inhibit capacitation. This is possibly one of the reasons for the observation that ejaculated spermatozoa of many species are more resistant to *in vitro* capacitation than are epididymal spermatozoa (Yanagimachi, 1994). In some species, the binding of some of these decapacitating factors to the surface of ejaculated spermatozoa appears to be so firm that they may not be readily removed from the sperm surface by repeated washings with ordinary physiological solutions. *In vivo,* the female genital tract appears to have efficient mechanisms to remove or alter the decapacitating factors (Yanagimachi, 1994).

The inhibitory effect of seminal fluid on capacitation became evident when it was shown that, on exposure to seminal fluid, capacitated spermatozoa lose both their ability to undergo induced acrosome reaction and their fertilizing potential, i.e., they become "decapacitated" [(Bedford, 1970; Chang, 1957; Cross, 1996b) and references cited there]. This decapacitation process is reversible with respect to the sperm population: a decapacitated sperm population becomes recapacitated as a consequence of seminal fluid removal and appropriate incubation time. However, there is evidence which suggests that, at least *in vitro,* capacitation is a unidirectional process with respect to individual spermatozoa: a spermatozoon that has become decapacitated apparently cannot become capacitated again (see Section II,A,2,c). If so, recapacitation of a sperm population apparently involves capacitation of cells that have never been capacitated.

Over the years, a number of seminal fluid constituents have been proposed as "decapacitating factors." These involve factors of both epididymal and seminal origin. The former factors include acrosomal-stabilizing proteins ranging from 125 to 259 kDa (Oliphant *et al.,* 1985; Wendy *et al.,* 1986) and a 40-kDa anionic polypeptide (Fraser *et al.,* 1990). The latter factors include 5- and 10-kDa Caltrin (Coronel and Lardy, 1992; Florman and Babcock, 1991), 6.4-kDa protein with proteinase-inhibiting activity (Boettger-Tong *et al.,* 1993), 15-, 16-, and 23-kDa glycoproteins (Parry *et al.,* 1992; Watanabe *et al.,* 1991), spermine (Rubinstein and Breitbart, 1991), and Zn^{2+} (Andrews *et al.,* 1994; Aonuma *et al.,* 1981; Riffo *et al.,* 1992). However, additional studies suggested that cholesterol is the major and perhaps sole decapacitating factor [(Cross, 1996b; Cross and Mahasreshti, 1997) and references cited there]. The issue of whether one or the other is involved has yet to be resolved.

b. Requirements for the Initiation of Capacitation

The composition of the *in vitro* capacitating medium varies from species to species, but it commonly contains appropriate ions (primarily Ca^{2+} and HCO_3^-), energy substrates,[1] and albumin (Yanagimachi, 1994). The *in vitro* capacitating medium of hamster spermatozoa should also contain taurine or hypotaurine for maintaining sperm viability (Lui *et al.,* 1979; Mrsny *et al.,* 1979), and that of bovine spermatozoa should contain heparin (Miller and Ax, 1990; Parrish *et al.,* 1988). Except for heparin in the case of bovine spermatozoa, the initiation of sperm capacitation *in vitro* does not require any external factor. As a matter of fact, many reports [reviewed by Yanagimachi (1994)] have indicated that there is no component whose presence in the medium is absolutely necessary for initiation of capacitation, i.e., a component whose absence totally prevents capacitation. This has been demonstrated for K^+, Ca^{2+}, HCO_3^-, albumin, and other components (Yanagimachi, 1994). On the other hand, a number of other studies have found (Bhattacharyya, 1992; Visconti *et al.,* 1995a) or assumed (Anderson *et al.,* 1992; Baldi *et al.,* 1991; Bielfeld *et al.,* 1994a; De Jonge *et al.,* 1989; Shams-Borhan and Harrison, 1981) that spermatozoa do not become capacitated unless albumin is present. This was also demonstrated in mouse spermatozoa with respect to both Ca^{2+} (Visconti *et al.,* 1995a) and HCO_3^- (Neill and Olds-Clarke, 1987; Shi and Roldan, 1995; Visconti *et al.,* 1995a). The reason for the apparent conflict may be, at least in part, the fact that different studies measured capacitation by different techniques that measure different stages of capacitation (Section II,B). When this fact became apparent, it could be demonstrated that partial capacitation of human spermatozoa

[1] The commonly used energy substrate, glucose, appears to be inhibitory for heparin-induced sperm capacitation in bulls (Parrish *et al.,* 1994; Uguz *et al.,* 1994), probably because of inhibiting protein tyrosine phosphorylation (Galantino-Homer *et al.,* 1997). This inhibition can be overcome by cAMP agonists (Galantino-Homer *et al.,* 1997). In other mammals, where heparin is not required for capacitation, glucose does not inhibit protein tyrosine phosphorylation (Visconti *et al.,* 1995a) and it supports capacitation (Fraser and Herod, 1990; Mahadevan *et al.,* 1997; Rogers and Perreault, 1990).

does not require albumin, but complete capacitation does (Jaiswal *et al.*, 1998). Similarly, it was demonstrated that capacitation of hamster spermatozoa involves a bovine serum albumin (BSA)-independent phase followed by a BSA-dependent phase (Stewart-Savage, 1993).

2. Timing and Propagation of Sperm Capacitation

a. Capacitation Time

The average time required to complete the capacitation process, judged by the acquisition of fertilizing potential, varies among species, from ~1 hour (*in vivo*) or ~2 hours (*in vitro*) in the mouse to 5–6 hours in the rabbit (*in vivo*) and bull (*in vitro*) (Austin, 1985). In humans, the time is ~1 hour *in vitro,* although a period of 30 minutes (depending on the sperm sample) may be sufficient for partial capacitation (Cohen-Dayag *et al.*, 1995; Jaiswal *et al.*, 1998; Overstreet *et al.*, 1980). Due to lack of agreement on how female-derived factors affect the process of capacitation (see Section II,C,2), it is not clear whether the time required for completion of the capacitation process is different *in vitro* and *in vivo.*

b. Level of Capacitated Cells in a Capacitated Sperm Population

The acquisition of fertilizing potential by a sperm population, both *in vitro* and *in vivo,* does not mean that the whole population is capacitated. On the contrary, with the exception of heparin-induced capacitation of bull spermatozoa mentioned above (Parrish *et al.*, 1988), this is usually not the case at any time point. Apparently, even though capacitation can be initiated as a result of seminal fluid removal, not all the spermatozoa are capacitated simultaneously. Such a scenario was proposed by Eisenbach and Ralt (1992) and Yanagimachi (1994), and later demonstrated experimentally by Stewart-Savage (1993) in hamster spermatozoa and by Cohen-Dayag *et al.* (1995) in human spermatozoa.

There appear to be three time-dependent phases with respect to the level of human capacitated spermatozoa *in vitro* (judged by the potential to undergo the acrosome reaction): a period during which the fraction of capacitated cells gradually rises from 0 to ~10% on the average, a prolonged period during which the level of capacitated spermatozoa is maintained more or less constant, and a period of gradual decrease in the fraction of capacitated spermatozoa (Cohen-Dayag *et al.*, 1995). The duration of each phase varies from semen to semen, even within semen samples of the same individual. The first phase usually lasts 30–120 minutes in human spermatozoa. During the second phase, the level of capacitated cells in a sperm population ranges in humans between 2 and 44%, usually around 10%, depending on the sperm sample (Cohen-Dayag *et al.*, 1995; Jaiswal *et al.*, 1999a). (Here, too, the variations are between individuals and even between different semen samples of the same individual.) The finding that the fraction of capacitated cells in a sperm population is relatively low is in line with findings that only a small proportion of a sperm population is able to fertilize the egg *in vivo* (Cohen and Adeghe, 1987), to undergo the zona pellucida-stimulated acrosome reaction [the zona pellucida (ZP) is the coat that surrounds the egg] (Cross *et al.*, 1988), to bind mannose [a suggested molecular marker of human sperm capacitation (Benoff, 1993; Cohen-Dayag and Eisenbach, 1994)] under capacitating conditions (Benoff *et al.,* 1993a,b), or to exhibit hyperactivated motility (see Section IV,C) (Burkman, 1984; Grunert *et al.,* 1990; Morales *et al.,* 1988; Robertson *et al.,* 1988).

c. Continuous Replacement of Capacitated Spermatozoa

The low level of capacitated spermatozoa, discussed in the preceding paragraph, is a reflection of continuous replacement of capacitated spermatozoa. It was found that, at least *in vitro,* the capacitated state of a human spermatozoon is transient, with a life span of 50–240 minutes, and that there is a continuous process of replacement of capacitated spermatozoa within a sperm population (Cohen-Dayag *et al.,* 1995).

It is not known what causes the continuous replacement of capacitated spermatozoa. This might be the consequence of asynchronous capacitation that is a reflection of physiological differences between individual cells in a sperm population, or a reflection of the different absolute ages of individual spermatozoa in an ejaculate (Bedford, 1970) [well known in a number of mammals (Amann *et al.,* 1965, 1976; Orgebin-Crist, 1965)]. If the latter possibility is correct, it implies that the kinetics of capacitation at a given set of conditions is prewired in each sperm cell and it depends on the age of the cell, not only on the time of removal of the decapacitating factors. Yet another possibility is that there are physiological cues that affect the kinetics of capacitation, and the sperm's ability to respond to these cues may depend on the sperm's age. Although this possibility is very logical *in vivo* (Section II,C,2), it is less likely *in vitro,* where all the cells appear to be simultaneously exposed to the same cues.

The nature of the postcapacitated state is still obscure. (The term *postcapacitated* is used here to indicate sperm cells that did not undergo the acrosome reaction but stopped being capacitated.) It is known that a postcapacitated spermatozoon has an intact acrosome, as evident from the observation that the level of acrosome-less spermatozoa does not increase during the continuous replacement of capacitated spermatozoa (Cohen-Dayag *et al.,* 1995). We also know that, once a cell becomes postcapacitated, it is a dead end, and the cell will not undergo the acrosome reaction when the appropriate stimulus appears:

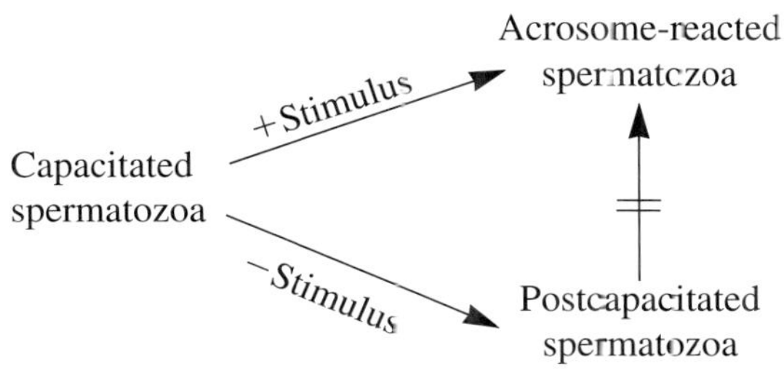

This is evident from the observation that a sperm subpopulation rich in postcapacitated spermatozoa does not acquire with time the ability to undergo the acrosome reaction on stimulation (Cohen-Dayag *et al.,* 1995).

The consequence of this continuous replacement of capacitated spermatozoa—which may have an essential physiological role in fertilization (Section IV,B)—is a heterogeneous sperm population. Such a population includes cells that have not yet started the process of capacitation, cells at various stages of capacitation, fully capacitated cells [the only ones that can undergo complete acrosome reaction (Jaiswal *et al.,* 1999a)], postcapacitated cells, and acrosome-reacted cells (Jaiswal *et al.,* 1998). If this heterogeneity holds also in other mammals, it is clearly reflected in the heterogeneity of a sperm population with respect to albumin requirement for the capacitation of hamster spermatozoa (Stewart-Savage, 1993), changes in intracellular pH during capacitation of bull spermatozoa (Vredenburgh-Wilberg and Parrish, 1995), acrosome reaction-associated changes in intracellular Ca^{2+} of bovine spermatozoa in response to solubilized zona pellucida (Florman, 1994), and HCO_3^--induced changes in lectin binding in boar and ram spermatozoa (Ashworth *et al.,* 1995). The heterogeneity also has implications for the measurement of the level of capacitated spermatozoa (Section II,B).

B. MEASUREMENT OF THE LEVEL OF CAPACITATED CELLS IN A SPERM POPULATION *IN VITRO*

The facts that capacitation is not an "all-or-none" phenomenon and that, following the initiation of capacitation, a sperm population almost always contains spermatozoa at various stages of capacitation complicate the experimental determination of the level of capacitated spermatozoa. This is because different probes may identify different stages of capacitation. Therefore, any quantitative comparison of the level of capacitated spermatozoa should take into account the measuring technique (Jaiswal *et al.,* 1999a).

1. Empirical Methods

There are many capacitation-associated changes, reviewed in detail by Cohen-Dayag and Eisenbach (1994), that potentially could be developed into direct methods for the assessment of capacitated spermatozoa. These involve redistribution and appearance or disappearance of surface antigens, receptors, membrane proteins, and phospholipids, as well as changes in other cell characteristics (Section III). However, none of these changes has, thus far, been developed into a reliable assay for capacitation. One of the reasons for this situation is that it is not yet known which of these changes will occur at the very end of the capacitation process. The only method that is used for measurement of capacitated spermatozoa, without employing the functional definition of capacitation, is the chlortetracycline (CTC) fluorescence assay (Table 3.1). CTC is a fluorescent antibiotic whose fluorescence changes when it chelates membrane-associated divalent cations (mainly Ca^{2+}) (Hallett *et al.,* 1972). The method is based on formation of a com-

TABLE 3.1 Methods for Identifying Capacitated Spermatozoa

Method	Species	Advantage	Disadvantage	References
Empirical methods				
Chlortetracycline (CTC) assay	Human, monkey, bull, goat, mouse	Distinguishes not only between capacitated and noncapacitated cells, but also between acrosome-reacted and acrosome-intact cells	Molecular mechanism not fully understood. Potential interference from compounds that modulate Ca^{2+}–phospholipid complexes, affect CTC absorption spectrum, or quench the fluorescence intensity	Amin *et al.* (1996); DasGupta *et al.* (1993); Fraser (1995b); Kaul *et al.* (1997); Kholkute *et al.* (1990); Lee *et al.* (1987); Neill and Olds-Clarke (1987); Otoole *et al.* (1996); Perry *et al.* (1995); Saling and Storey (1979); Visconti *et al.* (1998); Ward and Storey (1984)
Functional methods				
Ability to undergo *in vitro* fertilization	Bull, mouse	Applies the most physiological definition of capacitation	Difficult to determine whether a fertilization failure results from capacitation failure or from some other process involved in fertilization; time consuming; dependent on sperm/egg ratio; nonapplicable to humans because of egg unavailability	Aonuma *et al.* (1982); Fraser (1985); Go and Wolf (1985); McNutt and Killian (1991); Neill and Olds-Clarke (1987); Parrish *et al.* (1988); Visconti *et al.* (1998)
Ability to undergo an induced acrosome reaction	Human, bull	Easy to perform	—	Bielfeld *et al.* (1994a); Cohen-Dayag *et al.* (1995); de Lamirande *et al.* (1997b); Jaiswal *et al.* (1998); McNutt and Killian (1991), Parrish *et al.* (1989)
Ability to bind and penetrate the zona pellucida	Hamster, mouse	—	Time consuming; nonapplicable to humans because of egg unavailability	Andrews and Bavister (1989); Boatman *et al.* (1988); McMaster *et al.* (1978); Overstreet *et al.* (1980); Shur and Hall (1982); Stewart-Savage (1993); Yanagimachi (1994); Yanagimachi *et al.* (1976)

plex between CTC and membrane-bound Ca^{2+}, giving rise to enhanced fluorescence that changes with the modification of the plasma membrane during capacitation and acrosome reaction (Caswell and Hutchison, 1971; Saling and Storey, 1979). Distinction between noncapacitated spermatozoa with an intact acrosome, capacitated spermatozoa, and acrosome-reacted spermatozoa is made according to the different fluorescence patterns in these cell types (Lee *et al.,* 1987). The molecular mechanism that underlies these changes in fluorescence patterns is not fully understood. It is believed, however, that changes in the distribution of Ca^{2+}–CTC complexes bound to phospholipids in the plasma membrane are responsible for the different patterns observed (Visconti *et al.,* 1998). A disadvantage of this method is that the patterns vary with the species (Lee *et al.,* 1987) and that a distinction between them is not always easy.

2. Functional Methods

Because there are many degrees of partial capacitation along the route to full capacitation, the most reliable approach currently to identify such cells is by examining those sperm functions that can be performed only by fully capacitated cells. Such functions are *in vitro* fertilization, including the basic steps involved in fertilization, e.g., sperm adhesion to the zona pellucida that surrounds the egg, the acrosome reaction, egg penetration, and fusion with the egg (Table 3.1). As indicated in Table 3.1, although *in vitro* fertilization measures the actual physiological definition of the capacitated state, it has a number of serious disadvantages that make it less suitable for experimental purposes. Perhaps the most popular approach is the measurement of the sperm's ability to undergo an induced acrosome reaction. This approach, which exploits the operational definition of capacitation (Florman and Babcock, 1991; Yanagimachi, 1994), measures the increase in the number of acrosome-reacted spermatozoa in response to a proper acrosome-reaction inducer. Commonly used inducers are listed in Table 3.2. The number of acrosome-reacted spermatozoa prior to and subsequent to the induction is determined by probes sensitive to the acrosomal content (released during the acrosome reaction) or to the acrosomal membrane (exposed during the acrosome reaction). Table 3.3 presents a list of these probes. Assessment of the acrosomal status with these probes can be made by fluorescence/optical microscopy (Aitken and Brindle, 1993; Cross *et al.,* 1986; Holden *et al.,* 1990; Jaiswal *et al.,* 1999a; Kinger and Rajalakshmi, 1995; Köhn *et al.,* 1997; Moore *et al.,* 1987; Mortimer *et al.,* 1987) or by flow cytometry (Carver-Ward *et al.,* 1994; D'Cruz and Hass, 1992; Fenichel *et al.,* 1989; Jaiswal *et al.,* 1999a; Tao *et al.,* 1993a,b). The acrosomal status can also be determined directly (without probes) by electron microscopy (Köhn *et al.,* 1997; Nagae *et al.,* 1986; Stock and Fraser, 1987). As shown in Tables 3.2 and 3.3, not all the inducers can bring the acrosome reaction to completion (Jaiswal *et al.,* 1999a), and not all the probes can distinguish a fully acrosome-reacted cell from a partially reacted cell (Aitken and Brindle, 1993; Amin *et al.,* 1996; Jaiswal *et al.,* 1998, 1999a; Kinger and Rajalakshmi, 1995; Köhn *et al.,* 1997). Therefore, the acrosome-reaction approach appears to be highly reliable, provided that proper

TABLE 3.2 Acrosome-Reaction Inducers

Inducer	Species	Advantage	Disadvantage	References
Natural inducers[a]				
Solubilized ZP[b]	Human, hamster, mouse	The physiological inducer	Commercially unavailable	Bielfeld *et al.* (1994b); Bleil and Wassarman (1983); Cherr *et al.*, 1986; Cross *et al.* (1988); Florman and Storey (1982); Kopf and Gerton (1991); Miller *et al.* (1993); Moller *et al.* (1990); Saling and Storey (1979); Uto *et al.* (1988)
Progesterone	Human	—	Induces only partial AR	Aanesen *et al.* (1996); Baldi *et al.* (1991); Cross (1993); Jaiswal *et al.* (1999a); Meizel (1995); Osman *et al.* (1989); Parinaud *et al.* (1992); Tesarik *et al.* (1992)
Follicular fluid	Human	—	Induces only partial AR	Bielfeld *et al.* (1994a); Cross (1993); De Jonge *et al.* (1993); Jaiswal *et al.* (1999a); Meizel *et al.* (1990); Mortimer and Camenzind (1989); Parinaud *et al.* (1995); Suarez *et al.* (1986)
Platelet-activating factor	Human	—	—	Angle *et al.* (1993); Krausz *et al.* (1994)
Lysophosphatidyl choline	Human, bull, guinea pig	—	—	de Lamirande and Gagnon (1993b); de Lamirande *et al.* (1997b); Leclerc *et al.* (1997); McNutt and Killian (1991); Parrish *et al.* (1988); Yanagimachi and Suzuki (1985)

(*continues*)

TABLE 3.2 *(continued)*

Inducer	Species	Advantage	Disadvantage	References
Pharmacological inducers[c]				
Ca^{2+} ionophore (A23187, ionomycin)	Human, hamster, mouse	Induces complete AR; the preferred inducer according to the consensus	—	Aitken and Brindle (1993); Aitken *et al.* (1984); Amin *et al.* (1996); Baldi *et al.* (1991); Bleil and Wassarman (1983); Byrd and Wolf (1986); Carver-Ward *et al.* (1994); Cross (1993); Cross *et al.* (1986); Cummins *et al.* (1991); Emiliozzi and Fenichel (1997); Fenichel *et al.* (1989); Holden *et al.* (1990); Jaiswal *et al.* (1999a); Köhn *et al.* (1997); Meizel and Turner (1996); Miller *et al.* (1993); Moller *et al.* (1990); Mortimer and Fraser (1996); Rotem *et al.* (1992); Tesarik *et al.* (1993a)
PKC stimulator (PMA)	Human	Can be used in the absence of extracellular Ca^{2+}	Induces only partial AR	Bielfeld *et al.* (1994a,b); De Jonge *et al.* (1991b, 1993); Jaiswal *et al.* (1999a); Parinaud *et al.* (1995); Rotem *et al.* (1992)
PKA and PKG stimulators (e.g., dbcAMP and dbcGMP)	Human	Can be used in the absence of extracellular Ca^{2+}		Bielfeld *et al.* (1994a); De Jonge *et al.* (1991a); Parinaud *et al.* (1995)

[a] By "natural" we mean that the inducer is known to be present in the body.

[b] Abbreviations: AR, acrosome reaction; dbcAMP, dibutyl cAMP; dbcGMP, dibutyl cGMP; PKA, protein kinase A; PKC, protein kinase C; PKG, protein kinase G; PMA, phobol 12-myristate 13-acetate [termed also 12-*O*-tetradecanoyl-phorbol-13-acetate (TPA)]; ZP, zona pellucida.

[c] Pentoxyfylline, sometimes considered as an AR inducer (DasGupta *et al.*, 1994b), is not included in the table. This is because it does not induce the AR but it rather sensitizes the response to Ca^{2+} ionophores (Carver-Ward *et al.*, 1994; Mortimer and Fraser, 1996).

TABLE 3.3 Probes Used to Identify Acrosome-Reacted Spermatozoa

Probe[a]	Target	Comments	References
FITC–PSA[b]	An α-methyl mannoside residue localized in the acrosomal contents	Can detect both partially and completely acrosome-reacted cells, i.e., useful even for early stages of AR; requires sperm permeabilization	Aitken and Brindle (1993); Cohen-Dayag *et al.* (1995); Cross *et al.* (1986); Jaiswal *et al,* (1999a); Kinger and Rajalakshmi (1995); Köhn *et al.* (1997); Mendoza *et al.* (1992); Tesarik *et al.* (1993a)
FITC–ConA	A D-mannose/D-glucose residue in the inner acrosomal membrane	Identifies only completely acrosome-reacted cells	Fierro *et al.* (1996); Holden *et al.* (1990); Köhn *et al.* (1997)
FITC-labeled[c] monoclonal antibodies against the acrosomal contents	Antigens within the acrosomal contents (e.g., CRB9, C11H, HS-19, HS-21, HS-63)	Can detect both partially and completely acrosome-reacted cells, i.e., useful even for early stages of AR; requires sperm permeabilization	Aitken and Brindle (1993); Dorjee *et al.* (1997); Kallajoki *et al.* (1986); Moore *et al.* (1987); Wolf *et al.* (1985)
The above-mentioned probes coupled to paramagnetic beads instead of FITC	As above	In addition to the above-mentioned comments, the probe-coupled paramagnetic beads can sort out acrosome-reacted cells	Köhn *et al.* (1996); Ohashi *et al.* (1992); Okabe *et al.* (1992)
Chlortetracycline	Outer acrosomal membrane	Can identify acrosome-reacted cells simultaneously with capacitated cells; molecular mechanism not fully understood; potential interference from compounds that modulate Ca^{2+}–phospholipid complexes, affect CTC absorption spectrum, or quench the fluorescence intensity; does not allow one to distinguish live spermatozoa from dead ones	Amin *et al.* (1996); Kholkute *et al.* (1990); Lee *et al.* (1987); Saling and Storey (1979); Visconti *et al.* (1998); Yanagimachi (1994)

(*continues*)

TABLE 3.3 *(continued)*

Probe[a]	Target	Comments	References
Double or triple stain	Acrosomal contents	Technically difficult, often capricious	De Jonge *et al.* (1989); Köhn *et al.* (1997); Mortimer and Fraser (1996); Talbot and Chacon (1981)
FITC–PNA	A β-D-galactose residue localized in the outer acrosomal membrane	Can detect even earlier stages of AR than those detected by FITC–PSA; detects both partially and completely acrosome-reacted cells; requires sperm permeabilization	Aitken and Brindle (1993); Kallajoki *et al.* (1986); Kinger and Rajalakshmi (1995); Krausz *et al.* (1994); Luconi *et al.* (1995); Mortimer *et al.* (1987)
FITC-labeled[c] monoclonal antibodies against the inner acrosomal membrane	Antigens in the inner acrosomal membrane (e.g., CD46, GB24, HSA-10)	Can detect only completely acrosome-reacted cells; suitable for objective determination by flow cytometry	Carver-Ward *et al.* (1994); D'Cruz and Hass (1992); Dorjee *et al.* (1997); Fenichel *et al.* (1989); Jaiswal *et al.* (1999a); Parinaud *et al.* (1993)
Coomassie blue	Proteins associated with the acrosomal matrix		Aarons *et al.* (1993); Miller *et al.* (1993); Moller *et al.* (1990)
FITC-labeled mannosylated BSA	D-Mannose-specific lectins on the sperm head		Benoff *et al.* (1993d)
Quinacrine	Acrosomal contents	Fluorescent due to the acid pH within the intact acrosome; its fluorescence is lost at the onset of the AR as a result of equilibration between the acrosomal compartment and the suspending medium, with a consequent alkalization of the acrosomal compartment	Amin *et al.* (1996)

[a] All the probes were used with human spermatozoa. Chlortetracycline was also used with monkey, bull, goat, and mouse spermatozoa. Coomassie blue was used with mouse and hamster spermatozoa in addition to human spermatozoa.

[b] Abbreviations: AR, acrosome reaction; ConA, concanavalin A; CTC, chlortetracycline; FITC, fluorescein isothiocyanate; PNA, peanut agglutinin; PSA, *Pisum sativum* agglutinin.

[c] The labeling was either by direct coupling or by using secondary antibodies labeled with FITC.

TABLE 3.4 Changes That Occur in Sperm During Capacitation

Sort of change[a]	Molecule involved	Species studied	Observation during capacitation	Capacitation confirmed or not[b]	References
Surface changes					
Glycoprotein expression and distribution	HS1A.1 antigen	Human	Redistribution and increased expression	–	Villarroya and Scholler (1987)
	20-kDa sialo-glycoprotein	Human	Redistribution and release from sperm surface	–	Focarelli *et al.* (1990, 1995, 1998)
	Fibronectin	Human	Appearance on the sperm surface	+	Fusi and Bronson (1992)
	P86/5 antigen	Boar	Redistribution	–	Topfer-Petersen *et al.* (1990)
	Membrane glycoprotein antigen (MGP)	Rabbit	Restriction of mobility	–	O'Rand (1977)
	2B1 antigen	Rat	Redistribution	–	Jones *et al.* (1990); Shalgi *et al.* (1990)
	Galactosyltransferase	Mouse	Becoming unmasked by shedding epididymally derived glycosides from the sperm surface	–	Shur and Hall (1982); Youakim *et al.* (1994)
	Sperm maturation antigen (SMA 4)	Mouse	Released from sperm surface	+	Vernon *et al.* (1982)

(*continues*)

TABLE 3.4 (*continued*)

Sort of change[a]	Molecule involved	Species studied	Observation during capacitation	Capacitation confirmed or not[b]	References
Carbohydrate distribution	Mannose-like residues	Human, ram, guinea pig, rabbit, rat, mouse	Modified ability to bind ConA[c]	–	Courtens and Fournier-Delpech (1979); Cross and Overstreet (1987); Gordon *et al.* (1975); Koehler and Sato (1978); Lewin *et al.* (1979); Schwarz and Koehler (1979)
		Hamster	Modified ability to bind ConA	+	Kinsey and Koehler (1978)
	Galactose-like residues	Human	Modified ability to bind MPA	–	Cross and Overstreet (1987)
		Guinea pig	Modified ability to bind SBA	+	Schwarz and Koehler (1979; Talbot and Franklin (1978)
	N-Acetylglucosamine residues	Bovine, guinea pig	Loss of ability to bind WGA	+	Medeiros and Parrish (1996); Schwarz and Koehler (1979)
Lectin distribution and availability	Mannose receptor	Human	Increased expression and externalization	+	Benoff (1997) and references cited therein
Distribution and level of other proteins	Antigens for antisperm antibodies	Human	Appearance	+	Margalioth *et al.* (1992)
		Rabbit	Loss of sperm agglutination by antibodies against seminal plasma	+	Oliphant and Brackett, (1973)
	PT-1 antigen	Guinea pig	Redistribution	–	Myles and Primakoff (1984)
	Intramembranous particles in the plasma membrane, observed in FFEM	Human	Redistribution	–	Tesarik (1984)

		Hamster, guinea pig	Redistribution and disappearance	+	Friend (1980); Kinsey and Koehler (1978); Koehler and Gaddum-Rosse (1975); Suzuki and Yanagimachi (1989); Yanagimachi (1988)
	Acceptor of seminal proteinase inhibitor	Mouse	Loss of ability to bind the proteinase inhibitor	−	Boettger *et al,* (1989)
Membrane phospholipids—level and distribution	Cardiolipin, PC, PE, PI	Boar	Changes in level (mostly temporal) and redistribution	−	Harrison and Gadella (1995); Snider and Clegg (1975)
	Anionic phospholipids (e.g., cardiolipin, PA)	Guinea pig	Synthesis and redistribution	+	Bearer and Friend (1982)
Level of sterols in the membrane	Cholesterol	Human, bull, mouse	Efflux from the plasma membrane, presumably resulting in membrane destabilization	+	Cross (1998); de Lamirande *et al.* (1997b) and references cited therein
		Rabbit, hamster, rat	Decrease in level	−	Davis (1982); Davis *et al.* (1979); Suzuki and Yanagimachi (1989)
Lateral diffusion of membrane lipids	—	Hamster, mouse	Local changes	−	Smith *et al,* (1998); Wolf *et al.* (1986)
Surface charge	—	Human, ram, rabbit	Decrease in total surface charge and in net negative surface charge	−	Courtens and Fournier-Delpech (1979); Rosado *et al.* (1973); Vaidya *et al.* (1971)
Membrane potential	—	Bull, mouse	Hyperpolarization[d]	+	Zeng *et al.* (1995) and references cited therein

(*continues*)

TABLE 3.4 *(continued)*

Sort of change[a]	Molecule involved	Species studied	Observation during capacitation	Capacitation confirmed or not[b]	References
Chemical modifications					
Phosphorylation	Proteins	Human, bull, mouse	Increased tyrosine phosphorylation	+	Aitken *et al.* (1996); Carrera *et al.* (1996); de Lamirande *et al.* (1997b); Duncan and Fraser (1993); Emiliozzi and Fenichel (1997); Galantino-Homer *et al.* (1997); Leclerc *et al.* (1996); Leyton and Saling (1989a); Luconi *et al.* (1998a,b, 1995); Naz *et al.* (1991); Tesarik *et al.* (1993b); Visconti and Kopf (1998); Visconti *et al.* (1995b)
Methylation	Membrane lipids	Hamster	Methylation of PE, PC, and an additional unidentified lipid	+	Llanos and Meizel (1983)
Changes in enzymatic activities					
Adenylyl cyclase activity	Adenylyl cyclase	Boar, mouse	Increased activity	−	Berger and Clegg (1983); Stein and Fraser (1984)
		Hamster	Increased activity	+	Morton and Albagli (1973); White and Aitken (1989)
Ca^{2+}-ATPase activity	Ca^{2+}-ATPase	Human, bull, guinea pig, mouse	Decreased Ca^{2+} extrusion	+	Adeoya-Osiguwa and Fraser (1996); DasGupta *et al.* (1994a); Fraser *et al.* (1995); Roldan and Fleming, (1989)
$Na^{+}K^{+}$-ATPase activity	$Na^{+}K^{+}$-ATPase	Guinea pig, hamster	Increased activity	−	Garcia *et al.* (1991); Mrsny *et al.* (1984)
Na^{+}–Ca^{2+} exchange	Na^{+}–Ca^{2+} exchanger	Bull	Increased activity	−	Rufo *et al.* (1984)

NADPH oxidase activity	NADPH oxidase	Human	Possibly increased activity	+	de Lamirande and Gagnon (1993a)
Phosphodiesterase activity	Phosphodiesterase	Mouse	Decreased activity	−	Monks and Fraser (1987)
Protein kinase activity	PKA	Mouse	Increased activity	+	Visconti *et al.* (1997)
	PKC	Human, hamster, mouse	Perhaps increased activity[e]	+	Furuya *et al.* (1992, 1993b); Leclerc *et al.* (1996); Visconti and Tezon (1989)
	ERK-1 and ERK-2	Human	Increased activity	+	Luconi *et al.* (1998a)
Protein phosphatase activity	—	Human	Decreased activity[e]	+	Furuya *et al.* (1993a); Leclerc *et al.* (1996)
Changes in intracellular composition					
Ions	Ca^{2+}	Human, goat, guinea pig[f]	Increased level	+	Baldi *et al.* (1991); Coronel and Lardy (1987); Kaul *et al.* (1997); Singh *et al.* (1978)
		Bull, boar, pig, hamster, mouse	Increased level	−	Adeoya-Osiguwa and Fraser (1993); Harrison and Miller (1991); Ruknudin and Silver (1990); Suarez and Dai (1995); Suarez *et al.* (1993); Triana *et al.* (1980); White and Aitken (1989); Zhou *et al.* (1990)
	H^+/OH^-	Human, bull, mice	Increased pH	+	Cross and Razyfaulkner (1997); Parrish *et al.* (1994); Uguz *et al.* (1994); Vredenburgh-Wilberg and Parrish (1995); Zeng *et al.* (1996)

(continues)

TABLE 3.4 *(continued)*

Sort of change[a]	Molecule involved	Species studied	Observation during capacitation	Capacitation confirmed or not[b]	References
	K^+	Bull, mouse	Decreased level[g]	+	Chou *et al.* (1989); Zeng *et al.* (1995)
		Guinea pig	Decreased level	−	Hyne *et al.* (1985)
	Na^+	Guinea pig	Increased level	−	Hyne *et al.* (1985)
	Zn^{2+}	Hamster	Decreased level	+	Andrews *et al.* (1994)
Proteins	Calmodulin (CaM)	Bull	Release to the extracellular medium	+	Leclerc *et al.* (1992)
Nucleotides	ATP	Hamster	Decreased level	+	Rogers and Morton (1973); White and Aitken (1989)
	cAMP	Bull, hamster	Increased level[h]	+	Morton and Albagli (1973); Parrish *et al.* (1994); White and Aitken (1989)
		Mouse	Decreased level	−	Stein and Fraser (1984)
Reactive oxygen species	O_2^- and H_2O_2	Human, hamster	Increased level	+	Aitken (1995); Bize *et al.* (1991); de Lamirande and Gagnon (1993a,b); Griveau *et al.* (1994); Leclerc *et al.* (1997); for review see (de Lamirande *et al.* (1997a)

Other changes					
CaM binding	CaM-binding proteins (28, 30, and 49 kDa)	Bull	Decreased ability to bind CaM	+	Leclerc *et al.* (1990)
Motility	—	Human, bull, rabbit, hamster, mouse	Acquisition of hyperactivated motility	+	Boatman and Robbins (1991); Burkman (1984, 1990); de Lamirande and Gagnon (1993a,b); Kervancioglu *et al.* (1994); McNutt *et al.* (1994); Mortimer and Mortimer (1990); Neill and Olds-Clarke (1987); Robertson *et al.* (1988); Suarez *et al.* (1991); Yanagimachi (1970); Young and Bodt (1994); see Suarez (1996); Yanagimachi (1994) for reviews

[a] Claimed changes during capacitation, for which the experimental basis was not provided, were not included in the table.

[b] A plus sign means that the presence of capacitated spermatozoa was verified by an independent technique. A minus sign means that spermatozoa were exposed to capacitating conditions and were therefore presumed to be capacitated; however, this presumption was not verified experimentally.

[c] Abbreviations: CaM, calmodulin; ConA, concanavalin A; ERK, extracellular signal-regulated kinase; FFEM, freeze-fracture electron microscopy; MPA, *Maclura pomifera* agglutinin; PA, phosphatidic acid; PC, phosphatidylcholine; PE, phosphatidylethanolamine; PI, phosphatidylinositol; PKA, protein kinase A; PKC, protein kinase C; SBA, soybean agglutinin; WGA, wheat germ agglutinin.

[d] The hyperpolarization is due, at least in part, to enhanced K^+ permeability (Zeng *et al.*, 1995).

[e] The evidence is indirect.

[f] In one study employing rabbit spermatozoa (Mahanes *et al.*, 1986), no change in Ca^{2+}_{in} during capacitation was found.

[g] In one study (Zeng *et al.*, 1995), the decrease in the K+ level was not statistically significant.

[h] The level appears to be unaltered in human spermatozoa (Perreault and Rogers, 1982).

acrosome-reaction inducers and probes are used. Other sperm functions and properties that change during capacitation are listed in Table 3.4 and will be discussed in the section on the molecular mechanism of capacitation (Section III). Such changes may be exploited in the future for the identification of capacitated spermatozoa (Cohen-Dayag and Eisenbach, 1994). In principle, behavioral changes of spermatozoa [e.g., acquisition of chemotactic responsiveness (Section IV,B) and hyperactivation—a motility pattern thought to be associated with sperm capacitation (de Lamirande and Gagnon, 1993a,b; Kervancioglu *et al.,* 1994; McNutt and Killian, 1991; Neill and Olds-Clarke, 1987; Suarez, 1996; Suarez *et al.,* 1991; Yanagimachi, 1970, 1994)—Section IV,C] may also serve as potential assays for capacitation; these possibilities, however, have not yet been exploited. Of all the currently available assays, measurement of the sperm's ability to undergo an induced acrosome reaction appears to be the most convenient and popular assay for determining the level of capacitated cells.

C. SPERM CAPACITATION *IN VIVO*

1. Initiation Site of Capacitation

As indicated in Section II,A, capacitation appears to be a spontaneously propagating process that is initiated by removal of decapacitating factors. The sites where decapacitating factors are likely removed from the sperm depend, in turn, on the sites at which the spermatozoa are deposited. Accordingly, in species whose spermatozoa are deposited in the vagina at coitus (e.g., rabbit, cattle, sheep, and humans), the initiation site of capacitation appears to be the cervix or the cervical mucus (Figure 3.1) (Gould *et al.,* 1985; Lambert *et al.,* 1985; Zinaman *et al.,* 1989). In species whose spermatozoa are deposited in the uterus (e.g., mouse, rat, hamster, and pig), the lower segment of the oviductal isthmus appears to be the main initiation site of capacitation (Yanagimachi, 1994).

Although the initiation of capacitation may be in the cervix or the oviductal isthmus, the process of capacitation is apparently completed in the isthmus (Hunter, 1996, 1997), convincingly demonstrated to serve as a sperm reservoir or storage site (Barratt and Cooke, 1991; Harper, 1994; Hunter, 1993; Overstreet and Drobnis, 1991; Suarez, 1998) (see Section IV,A). [In humans, the location is not yet certain (Barratt and Cooke, 1991; Bedford, 1994; Williams *et al.,* 1993; Zinaman *et al.,* 1989).] This is because spermatozoa released from the storage site at ovulation appear to be capacitated [(Cummins, 1982; Cummins and Yanagimachi, 1982; Smith and Yanagimachi, 1991); see Eisenbach and Ralt (1992) and Suarez (1998) for reviews], as reflected by their fertilizing potential or hyperactivated motility.

2. Effect of Female-Originated Factors on Capacitation

The rate of capacitation *in vivo* appears to be modulated by female-originated factors. Thus, spermatozoa passing from uterus to oviduct capacitate much sooner than any remaining in the uterus (Bedford, 1969; Hunter and Hall, 1974) [see Bedford (1983) for review], and a number of female-originated factors have been

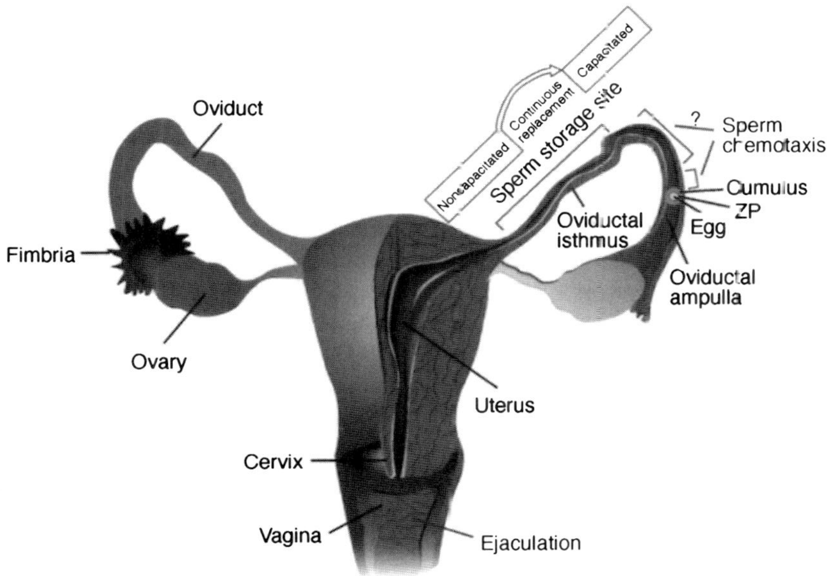

FIGURE 3.1 Diagram of the female genital tract (human). Spermatozoa move through the cervix and the uterus, find the opening of the oviduct, and enter into it. If they arrive prior to ovulation, the spermatozoa are retained with reduced motility in the sperm reservoir within the oviductal isthmus. During this time there is continuous replacement of capacitated spermatozoa. The steady-state level of capacitated spermatozoa is low (usually around 10% in the case of human spermatozoa *in vitro*). It appears that, when ovulation occurs, a few capacitated spermatozoa are detached from the oviductal epithelium at the storage site; they respond to attractant(s) secreted from the egg or the surrounding cumulus oophorus and thereby they are recruited to the egg. The exact location of sperm chemotaxis is not known. Only capacitated spermatozoa can penetrate through the cumulus oophorus, and only these spermatozoa—according to most studies—adhere to the zona pellucida (ZP) of the egg. A spermatozoon bound to the zona pellucida undergoes the acrosome reaction, penetrates the egg, and fertilizes it. The dimensions of the egg and the cumulus oophorus surrounding it are exaggerated in order to make them visible. Adapted with permission from Eisenbach (1999).

found in recent years to possess the ability to modulate the initiation and propagation of sperm capacitation. Human fallopian tube epithelium was found to increase, among other things, the level of presumably capacitated spermatozoa, reflected in the level of hyperactivated spermatozoa (Kervancioglu *et al.,* 1994). Membrane contact between spermatozoa and oviductal epithelial cells *in vitro* was demonstrated in humans and horses to enhance sperm motility, prolong sperm viability, and delay capacitation (Dobrinski *et al.,* 1997; Murray and Smith, 1997). This observation of delayed capacitation appears to be in contrast with the observations mentioned just above. The reason for the apparent contrast is not clear. In addition, a number of fluids and soluble factors therein were found to facilitate sperm capacitation: bovine (Chian *et al.,* 1995; Mahmoud and Parrish, 1996; Parrish *et al.,* 1988), porcine (Kim *et al.,* 1997), and canine (Kawakami *et al.,* 1998) oviductal fluid; ultrafiltrates of human fetal cord serum, follicular fluid, and seminal plasma (possibly by promoting capacitation-associated tyrosine phosphorylation) (de Lamirande *et al.,* 1998); a purified estrus-associated glycoprotein (85–95 kDa), secreted at estrus by the oviductal epithelium and present in cannula-derived bovine oviductal fluid (King *et al.,* 1994); bovine estrual isthmic conditioned medium (Anderson and Killian, 1994); lipid transfer protein I, known to be present and be active in the female reproductive tract (Ravnik *et al.,* 1995); progesterone—one of the substances that spermatozoa encounter during their migration in the female genital tract (Barboni *et al.,* 1995; DasGupta *et al.,* 1994b; de Lamirande *et al.,* 1998; Foresta *et al.,* 1992; Kay *et al.,* 1994; Revelli *et al.,* 1994; Roldan *et al.,* 1994); in the case of bull spermatozoa—heparin and high-density lipoprotein known to be present in the female reproductive tract (Parrish *et al.,* 1988; Thérien *et al.,* 1997); and a 54-kDa endometrial secretory sialic acid-binding glycoprotein in humans (Banerjee and Chowdhury, 1995, 1997). These observations substantiate the possibility that, *in vivo,* unlike *in vitro,* the propagation of the capacitation process may not be completely spontaneous. The female-derived soluble factors and the sperm–membrane interaction may provide a mechanism through which the timing of capacitation *in vivo* is modulated (Dobrinski *et al.,* 1997; Smith, 1998).

An interesting finding was made by Roldan *et al.* (1994), who demonstrated that, when mouse spermatozoa are exposed first to progesterone and then to the zona pellucida—as is presumably the case in the female genital tract—the number of cells that undergo the acrosome reaction is higher than that seen when progesterone and the zona pellucida are presented together or in the inverse order. This suggested that progesterone exerts a priming effect on spermatozoa. It thus seems reasonable that substances present in the female genital tract, primarily in the oviduct, may be involved in regulating capacitation. The finding that, at least in the hamster, the completion of sperm capacitation appears to be synchronous with the time of ovulation (Smith and Yanagimachi, 1989) indirectly supports this concept. It is also possible that different factors affect spermatozoa sequentially as they migrate in the female genital tract. Thus, it was shown that capacitation of rabbit spermatozoa is accomplished most efficiently when they are sequentially exposed to the uterus and the oviduct. Under such conditions the spermatozoa capacitate within 5–6 hours, compared to 10–11 hours required for capacitation of spermatozoa in the uterus or in the oviduct (Bedford, 1970).

3. Species Specificity of Capacitation

The fact that female-originated factors are not essential for capacitation (though they may be required for regulation or enhancement of capacitation) indicates that capacitation is not a species-specific process. This was demonstrated for mouse, hamster, rabbit, and other mammals by finding that functional sperm capacitation (i.e., fertilizing potential) can be achieved in a heterologous female tract [Saling and Bedford (1981) and references cited therein].

III. MOLECULAR MECHANISM OF CAPACITATION

In spite of extensive studies of capacitation for several decades, the molecular mechanism of this phenomenon is not fully understood. Many processes and cellular changes have been found to be involved in this phenomenon, and a number of factors and processes that modulate and perhaps regulate capacitation have been identified. These involve changes in the sperm surface, changes in the level of intracellular constituents, changes in enzymatic activities, chemical modifications, and other factors. A detailed list of these changes is given in Table 3.4. Below we review a few key processes that occur during capacitation and that may be involved in its regulation, as well as some processes found to modulate capacitation. We then attempt to propose a model for the molecular mechanism of capacitation and to show how these key processes may function in concert.

A. CHANGES IN THE SPERM SURFACE DURING CAPACITATION

Sperm surface changes during capacitation appear to be essential for a number of processes prior to and during the sperm–egg contact, such as the formation of a sperm reservoir and the release from the reservoir (Suarez, 1998), chemotaxis (Eisenbach, 1999; Eisenbach and Tur-Kaspa, 1999), penetrating the cumulus oophorus (Yanagimachi, 1994), adhesion to the egg's zona pellucida, and the acrosome reaction (see Section IV,D). Accordingly, major changes in the sperm surface during capacitation should be ones that expose or express proper receptors and ones that prepare the plasma membrane to undergo fusion with the underlying outer acrosomal membrane during the acrosome reaction. These involve changes in the plasma membrane composition and architecture, primarily changes in the distribution and the level of phospholipids and cholesterol within the membrane.

1. Changes in Phospholipid Composition and Distribution

The most abundant lipids present in the mammalian sperm anterior head membrane are, in decreasing order, phoshatidylcholine, phosphatidylethanolamine, sphingomyelin, cardiolipin, and traces of phosphatidylserine and phosphatidylinositol (Nolan and Hammerstedt, 1997; Poulos and White, 1973; Snider and Clegg, 1975). Like in other mammalian plasma membranes, phosphatidylethanolamine and phosphatidylserine are essentially confined to the inner leaflets of the membrane, and phoshatidylcholine and sphingomyelin are mostly confined to the outer leaflets [Harrison and Gadella (1995) and Nolan and Hammerstedt (1997) and references cited therein]. This asymmetric distribution of phospholipids in the membrane is believed to be carried out by an ATP-dependent aminophospholipid translocase, thus far identified in ram (Müller *et al.,* 1994) and bull (Nolan *et al.,* 1995) spermatozoa, and by other transporters, termed *flippases* (Harrison and Gadella, 1995;

Menon, 1995). During capacitation, most of these phospholipids undergo changes (often temporal) in their level and in their distribution between the outer and inner leaflets of the plasma membrane (Table 3.4). For example, under capacitating conditions, there appears to be a transient decrease in the total level of phoshatidylcholine in the plasma membrane (Snider and Clegg, 1975) as well as an apparently partial translocation of phoshatidylcholine from the outer to the inner leaflet (Harrison and Gadella, 1995). It is not known how these changes in lipid architecture occur during capacitation. However, because the ATP-dependent aminophospholipid translocase is inhibited by elevated levels of intracellular Ca^{2+} (Nolan and Hammerstedt, 1997), one possibility is that the observed increase in intracellular Ca^{2+} level during capacitation inhibits this enzyme (and perhaps other flippases). Consequently the precapacitation asymmetric lipid distribution may be changed into a new equilibrium distribution of lipids, a distribution that makes the inner leaflet of the plasma membrane more fusogenic (Nolan and Hammerstedt, 1997).

2. Changes in Cholesterol Level

a. Cholesterol Efflux

One of the processes that have been recognized as fundamental for the occurrence of capacitation is a decrease in the cholesterol/phospholipid molar ratio due to cholesterol efflux from the plasma membrane [Cross (1998) and Langlais and Roberts (1985) for reviews]. Thus, a number of laboratories have demonstrated that facilitation of cholesterol efflux from the membrane (e.g., by suspending the spermatozoa in a medium containing proteins or β-cyclodextrins that bind cholesterol) induces capacitation [see, e.g., Choi and Toyoda (1998), Davis (1982), Davis *et al.* (1979), Langlais *et al.* (1988), and Visconti *et al.* (1999); see Cross (1998), de Lamirande *et al.* (1997b), and Parks and Ehrenwald (1990) for reviews]. Conversely, prevention of cholesterol efflux by incubating spermatozoa in a cholesterol-containing medium inhibits or delays capacitation (Cross, 1996a,b; Zarintash and Cross, 1996). Consistently with these observations, the cholesterol/phospholipid molar ratio is significantly higher in spermatozoa of infertile men than in spermatozoa of fertile men (Benoff *et al.,* 1993c; Kulka *et al.,* 1984; Sebastian *et al.,* 1987; Sugkraroek *et al.,* 1991). This ratio is also higher in spermatozoa that exhibit delayed capacitation than in "normal" spermatozoa (Benoff *et al.,* 1993d). Oviductal (Ehrenwald *et al.,* 1990) and follicular (Langlais *et al.,* 1988; Ravnik *et al.,* 1992) fluids as well as serum (Langlais *et al.,* 1988) have been shown to be effective in cholesterol removal from the membrane. The main cholesterol-adsorbing constituents in these fluids are apparently lipoproteins and albumin (Davis *et al.,* 1979; Langlais *et al.,* 1988), demonstrated to have the ability to remove cholesterol from the plasma membrane of human (Benoff *et al.,* 1993c; de Lamirande *et al.,* 1997b; Ravnik *et al.,* 1993), rabbit (Davis, 1982), rat (Davis *et al.,* 1979, 1980), and mouse (Go and Wolf, 1985) spermatozoa (as well as most other plasma membranes) and consequently to induce capacitation (de Lamirande *et al.,* 1997b; Langlais *et al.,* 1988; Parks and Ehrenwald, 1990).

In addition, β-cyclodextrins (cyclic heptasaccharides consisting of β(1–4)-glucopyranose units) were demonstrated to promote capacitation (Choi and Toyoda, 1998; Visconti *et al.,* 1999) and capacitation-associated tyrosine phosphorylation (Visconti *et al.,* 1999).

b. Necessity of Sterol Acceptors for Cholesterol Efflux and Sperm Capacitation

As discussed in Section II,A,1,b, albumin is commonly included in the medium used for capacitation *in vitro,* whereas lipoproteins are not used. Apparently, some cholesterol efflux can occur *in vitro* by washing, even in the absence of albumin (de Lamirande *et al.,* 1997b; Tanphaichitr *et al.,* 1996), and spermatozoa can be capacitated to some extent even if only a small proportion of cholesterol is released from the membrane (de Lamirande *et al.,* 1997b). It has been shown that capacitation in the absence of sterol acceptors such as albumin is not completely equivalent to capacitation in the presence of sterol acceptors (Jaiswal *et al*, 1998). Spermatozoa that became capacitated in the absence of sterol acceptors could not be induced to undergo a complete acrosome reaction; they could, however, be induced to undergo partial acrosome reaction. On the other hand, spermatozoa that became capacitated in the presence of albumin could be induced to undergo a complete acrosome reaction. Apparently, the limited cholesterol efflux that occurs *in vitro* even in the absence of albumin is sufficient for making a spermatozoon partially capacitated. For becoming fully capacitated, i.e., for acquiring the potential to undergo a complete acrosome reaction, sterol acceptors such as albumin must be present in order to expedite a greater depletion of membrane cholesterol (Jaiswal *et al.,* 1998).

c. Consequences of Cholesterol Efflux

Cholesterol loss from the sperm plasma membrane appears to be the cause of a number of processes, including increased membrane fluidity (Shinitzky, 1984), destabilization of the plasma membrane and increased fusion ability (Cross, 1998; Harrison, 1996; Nolan and Hammerstedt, 1997), increased protein tyrosine phosphorylation (Visconti *et al.,* 1999), exposure of the mannose receptor on the sperm surface (Benoff *et al.,* 1993d), and a rise in intracellular pH (pH_{in}) (Cross and Razyfaulkner, 1997). Thus, it was demonstrated that prevention of cholesterol efflux (by sperm incubation in a cholesterol-enriched medium) suppresses the rise in pH_{in} (Cross and Razyfaulkner, 1997) and eliminates the increase in protein tyrosine phosphorylation (Visconti *et al.,* 1999), both of which normally occur during capacitation, and the sperm's ability to respond to acrosome-reaction inducers (Cross and Razyfaulkner, 1997). The mechanism by which cholesterol affects pH_{in} as well as the function of the rise in pH_{in} are still obscure. Since artificial modulation of pH_{in} does not substitute for cholesterol efflux (Cross and Razyfaulkner, 1997), the change in pH_{in} alone may not have an essential function, suggesting that the other effects of cholesterol during capacitation are more important—for example, externalization or expression of receptors that can adhere to the zona pellucida and thereby induce the acrosome reaction [e.g., the mannose receptor (Benoff *et al.,* 1993d)], and the cholesterol-dependent changes in the membrane fluidity and architecture that render the plasma membrane more fusogenic (Cross, 1998; Harrison, 1996; Nolan and Hammerstedt, 1997). The presence of cholesterol in a membrane usually makes it more rigid, i.e., the molecular packing in the membrane is increased and the molecular motion throughout the bilayer is restricted (Shinitzky, 1984). Moreover, under normal conditions the absence of phoshatidylcholine in the inner leaflet (see Section III,A,1) also maintains rigidity and resistance to fusion. Consequently, the membrane may be less fusogenic (Nolan and Hammerstedt, 1997). The efflux of cholesterol from sperm plasma membrane results in increased phospholipid movement within and between the membrane leaflets, with a consequent loss of the asymmetric lipid distribution. Phoshatidylcholine is now loosely packed in the inner leaflet, the hydration layer of the membrane is disrupted, and the inner core of the membrane becomes exposed for fusion (Nolan and Hammerstedt, 1997). It is not yet known whether the proposed exposure of the mannose receptor and perhaps other receptors on the sperm surface is one of the consequences of these changes in membrane architecture. Alternatively, cholesterol efflus may have specific effects that have not yet been studied. Benoff (1997) and Benoff *et al.* (1993d) have proposed a model in which the increased membrane fluidity due to the cholesterol efflux causes the mannose receptor to translocate from beneath the plasma membrane, where it resides prior to capacitation, to the membrane.

B. CHANGES IN Ca^{2+} LEVEL DURING CAPACITATION

It is well established that, during capacitation, the intracellular level of Ca^{2+} (Ca^{2+}_{in}) is increased (at a rate slower than that observed during the acrosome reaction) (Table 3.4). Based on experiments with $^{45}Ca^{2+}$ (Adeoya-Osiguwa and Fraser, 1993; Fraser and McDermott, 1992; Kaul *et al.,* 1997; Singh *et al.,* 1978) and on the observation that extracellular Ca^{2+} is required for capacitation *in vitro* (Fraser, 1987; Kaul *et al.,* 1997; Yanagimachi, 1982),[2] the increase in C^{2+}_{in} appears to be the consequence of net Ca^{2+} influx (Adeoya-Osiguwa and Fraser, 1993; Fraser and McDermott, 1992; Kaul *et al.,* 1997; Singh *et al.,* 1978). A number of mechanisms may account for this influx:

1. Ca^{2+}–ATPase. A Ca^{2+}–ATPase activity that extrudes Ca^{2+} from the cell and can be stimulated by decapacitating factors was found in the mammalian sperm head (Adeoya-Osiguwa and Fraser, 1996). When the decapacitating factors

[2] Guinea pig spermatozoa were regarded as an exception because their capacitation seemed to be independent of extracellular Ca^{2+} (Yanagimachi and Usui, 1974). Later studies, however, found that guinea pig spermatozoa do need extracellular Ca^{2+} for capacitation, though the threshold concentration appears to be orders of magnitude lower than in other species (Roldan and Fleming, 1989; Singh *et al.,* 1978).

are removed at the initiation of capacitation, the level of Ca^{2+}_{in} is consequently increased (Adeoya-Osiguwa and Fraser, 1996; Fraser, 1995a). Inhibitors of somatic cell Ca^{2+}–ATPases were found to increase the Ca^{2+}_{in} level and to increase in the number of capacitated spermatozoa (DasGupta *et al.*, 1994a; Fraser *et al.*, 1995; Perry *et al.*, 1997; Roldan and Fleming, 1989), as do calmodulin inhibitors (Adeoya-Osiguwa and Fraser, 1993; DasGupta *et al.*, 1994a) [somatic cell Ca^{2+}–ATPases are calmodulin sensitive (Carafoli, 1987)]. These observations suggested that Ca^{2+}–ATPase is involved in sperm capacitation and may be one of the main mechanisms that modulate the Ca^{2+}_{in} level during capacitation (DasGupta *et al.*, 1994a; Fraser *et al.*, 1995; Roldan and Fleming, 1989).

2. Na^+–Ca^{2+} exchanger. A Na^+ (outward) and Ca^{2-} (inward) exchanger has been demonstrated in mammalian spermatozoa (Ashraf *et al.*, 1982; Bradley and Forrester, 1980; Rufo *et al.*, 1984). However, its role in controlling Ca^{2+}_{in} during capacitation is not clear. It was proposed that, at least in bull spermatozoa, the activity of this exchanger might be regulated by caltrin, a low-molecular-weight protein that is associated with ejaculated spermatozoa. This protein is thought to inhibit the exchanger and it appears to be released during capacitation, thus stimulating the exchanger (Rufo *et al.*, 1982, 1984).

3. Ca^{2+} channel. The presence of Ca^{2+} channels through which Ca^{2+} influx occurs has been demonstrated in mammalian spermatozoa (Beltran *et al.*, 1994; Florman, 1994; Florman *et al.*, 1992). However, these channels appear to function mainly during the acrosome reaction, which involves fast and large Ca^{2+} fluxes, rather than during capacitation, which involves relatively smaller and slower fluxes.

4. Ca^{2+}–P_i transporter. Experimental evidence suggests that the plasma membrane of mammalian spermatozoa contains a Ca^{2+}–P_i symporter (i.e., a carrier that transports both Ca^{2+} and P_i inwardly) (Breitbart *et al.*, 1990; Zarca *et al.*, 1988). Because it was found that Ca^{2+} influx during capacitation appears to require external phosphate (Babcock *et al.*, 1975; Kaul *et al.*, 1997; Singh *et al.*, 1978) and that goat spermatozoa do not undergo capacitation when phosphate is omitted from the medium (Kaul *et al.*, 1997), it was suggested that this symporter may be involved in Ca^{2+} influx during capacitation (Kaul *et al.*, 1997).

The direct outcome of the increase in Ca^{2+}_{in} level during capacitation is not known. It is known, however, that this increase occurs at an early stage of the capacitation process (Fraser, 1995b). It is assumed that the elevated level of Ca^{2+}_{in} triggers an increase in the intracellular cAMP concentration and in protein tyrosine phosphorylation (see Section III,C). Interestingly, it was found that the increase in Ca^{2+}_{in} during hyperactivation (probably reflecting capacitated spermatozoa) is much greater in the flagellum than in the head. The increase in Ca^{2+}_{in} during the acrosome reaction (see Chapter 11, this volume) is greater in the head than in the flagellum (Suarez and Dai, 1995), thus indicating two putative but distinct modes of activation via Ca^{2+}_{in} elevation.

C. MODULATION OF PROTEIN TYROSINE PHOSPHORYLATION DURING CAPACITATION

There have been a number of reports on protein phosphorylation during capacitation, most of which have dealt with phosphorylation on tyrosine residues. In contrast, direct studies of serine/threonine phosphorylation in relation to capacitation have not been published. However, this lack of information (resulting primarily from technical difficulties) should not be taken as evidence against the involvement of serine/threonine phosphorylation in capacitation.[3]

1. Proteins That Undergo Tyrosine Phosphorylation during Capacitation

The first evidence for protein phosphorylation on tyrosine residues during sperm capacitation was provided a decade ago by Leyton and Saling (1989a). They found three phosphoproteins of molecular sizes of 52, 75, and 95 kDa in mouse spermatozoa, of which the two smaller ones were phosphorylated only in capacitated spermatozoa. Subsequently, studying human, bull, and mouse spermatozoa, workers in several laboratories observed a time-dependent increase in tyrosine phosphorylation of a number of proteins, ranging from 40 to 200 kDa, during capacitation (Aitken *et al.,* 1996; Burks *et al.,* 1995; Emiliozzi *et al.,* 1996; Emiliozzi and Fenichel, 1997; Galantino-Homer *et al.,* 1997; Leclerc *et al.,* 1996, 1997; Luconi *et al.,* 1995; Naz *et al.,* 1991; Visconti *et al.,* 1995a,b). There is no general rule with respect to the location of the tyrosine-phosphorylated proteins: a 95-kDa protein [initially thought to be a sperm receptor for ZP3—the egg's zona pellucida glycoprotein responsible for inducing the acrosome reaction (Burks *et al.,* 1995; Leyton *et al.,* 1992; Leyton and Saling, 1989b)] was localized in the mouse and human sperm head (Burks *et al.,* 1995; Leyton and Saling, 1989a), 81- and 105-kDa proteins of human spermatozoa were localized in the flagellum (Leclerc *et al.,* 1997), and other human sperm proteins, primarily 94- and 46-kDa proteins, were localized in both the sperm head and flagellum, with a progressive change in their location from the flagellum to the acrosmal region during capacitation (Naz *et al.,* 1991). Inhibition of capacitation [in mouse spermatozoa by omitting albumin, HCO_3^-, or Ca^{2+} from the medium (Visconti *et al.,* 1995b), and in bull spermatozoa by the presence of glucose (Galantino-Homer *et al.,* 1997)] inhibited the tyrosine phosphorylation. Conversely, protein kinase A inhibitors that inhibit tyrosine phosphorylation inhibited capacitation of mouse and human spermatozoa (Leclerc *et al.,* 1996; Visconti *et al.,* 1995b). All these observations suggest that protein tyrosine phosphorylation has an essential role in the molecular mechanism of capacitation. Consistent with this notion, a peptide hormone of thymic origin, thymosin Tα1 [detected in follicular fluid and in the seminal plasma of fertile men, but not of infertile men (Naz *et al.,* 1992)], was shown to enhance human sperm capacitation by enhancing tyrosine phosphorylation of several proteins of nonca-

[3] Following the completion of this chapter, Naz (1999) found in human spermatozoa that there were at least eight proteins that undergo serine and threonine phosphorylation, which increases as capacitation proceeds. One of these proteins also undergoes tyrosine phosphorylation.

pacitated spermatozoa; such an enhancement was not observed in capacitated spermatozoa (Naz, 1995).

2. Modulation of Tyrosine Phosphorylation by Reactive Oxygen Species

High concentrations of reactive oxygen species (ROS), mainly hydrogen peroxide, are known to have detrimental effects on spermatozoa (Aitken and Clarkson, 1987; Alvarez *et al.,* 1987; de Lamirande and Gagnon, 1995b). However, two lines of evidence suggest that low concentrations of ROS, primarily superoxide anion (O_2^-) and hydrogen peroxide (H_2O_2), are involved in regulation of capacitation:

1. Stimulation of ROS generation (e.g., by a superoxide-generating system consisting of xanthine, xanthine oxidase, and catalase, or by incubation of spermatozoa aerobically in the absence of catalase) accelerates capacitation and capacitation-associated processes such as hyperactivation and tyrosine phosphorylation (Aitken *et al.,* 1995, 1998; Bize *et al.,* 1991; de Lamirande and Gagnon, 1993b; de Lamirande *et al.,* 1997b, 1998; Griveau *et al.,* 1994; Leclerc *et al.,* 1997).
2. Suppression of the ROS level either by scavenging them (e.g., by antioxidants such as superoxide dismutase or catalase) or by inhibiting their production (e.g., by inhibiting NADPH oxidase with the flavoprotein inhibitor diphenylene iodonium) inhibits capacitation and the capacitation-associated processes (Aitken, 1995; Aitken *et al.,* 1996, 1998; Bize *et al.,* 1991; de Lamirande and Gagnon, 1993a,b; Griveau *et al.,* 1994; Leclerc *et al.,* 1997).

In a consistent manner, progesterone and various biological fluids such as follicular fluid and seminal plasma were found to stimulate O_2^- generation, tyrosine phosphorylation, and capacitation (de Lamirande *et al.,* 1998). The precise mechanism by which ROS stimulate capacitation and the location of the system(s) that generates them in sperm are not known. It was proposed that ROS up-regulate protein tyrosine phosphorylation, perhaps by activation of adenylyl cyclase and thereby elevation of cAMP (Aitken *et al.,* 1998; de Lamirande *et al.,* 1997a,b; Leclerc *et al.,* 1997) (see the following discussion).

3. Modulation of Tyrosine Phosphorylation by cAMP

Mouse spermatozoa are apparently unable to become capacitated and to undergo tyrosine protein phosphorylation in the absence of albumin, HCO_3^-, or Ca^{2+} (Visconti *et al.,* 1995a) (Section II,A,1,b). However, in the presence of permeant cAMP analogs, these functions are restored (Visconti *et al.,* 1995a,b). It therefore seems that cAMP is a modulator of protein tyrosine phosphorylation and, consequently, sperm capacitation. Thus, factors that increase the intracellular cAMP level (e.g., the adenylyl cyclase activator forskolin, or phosphodiesterase inhibitors such as caffeine, pentoxyfylline, and isobutylmethylxanthine) were shown to stimulate capacitation and capacitation-associated protein tyrosine phosphorylation in both human and mouse spermatozoa (Aitken *et al.,* 1998; Fraser, 1979; Leclerc *et al.,* 1997; Visconti *et al.,* 1995a,b). It was also found that superoxide dismutase, known to inhibit tyrosine phosphorylation and capacitation (see Section III,C,2), is unable to do so in the presence of phosphodiesterase inhibitors that elevate the cAMP level in the cell (de Lamirande *et al.,* 1997b). Thus it is reasonable to conclude that ROS (Aitken *et al.,* 1998; de Lamirande and Gagnon, 1995a; Leclerc *et al.,* 1996, 1997) as well as Ca^{2+} (Baldi *et al.,* 1991; de Lamirande *et al.,* 1997b) stimulate capacitation by activating adenylyl cyclase, thereby increasing the intracellular cAMP level and stimulating tyrosine phosphorylation. However, the possibility that ROS directly affect tyrosine phosphorylation (by interacting with tyrosine kinase or protein phosphatases) cannot be ruled out.

4. Modulation of Tyrosine Phosphorylation by Ca^{2+}

There appears to be conflicting evidence with respect to the effect of Ca^{2+} on tyrosine phosphorylation. Thus, although Visconti *et al.* (1995a,b) found that tyrosine phosphorylation of epididymal mouse spermatozoa is dependent on Ca^{2+} (likely via adenylyl cyclase) and that it increases with the Ca^{2+} concentration, Luconi *et al.* (1996) found a negative effect of Ca^{2+} on tyrosine phosphorylation in human spermatozoa. They found that depletion of Ca^{2+} from the capacitating medium does not inhibit protein tyrosine phosphorylation, whereas an increased Ca^{2+}_{in} level (achieved by the Ca^{2+} ionophore A23187) inhibits protein tyrosine phosphorylation (Luconi *et al.,* 1996). The reasons for the difference are obscure. Luconi *et al.* (1996) suggested that the difference may be due to different proteins that undergo tyrosine phosphorylation in these cases, to species differences, or to differences between epididymal and ejaculated spermatozoa.

5. Modulation of Tyrosine Phosphorylation by Protein Kinases

a. Protein Kinase A

It was found that inhibitors of the cAMP-dependent protein kinase A (PKA) inhibit protein tyrosine phosphorylation and block capacitation of human (Leclerc *et al.,* 1996), bovine (Uguz *et al.,* 1994), and mouse (Visconti *et al.,* 1995b) spermatozoa incubated under capacitating conditions *in vitro.* This raised the possibility that, during capacitation, cAMP activates PKA with a consequent increase in tyrosine phosphorylation (Leclerc *et al.,* 1996; Uguz *et al.,* 1994; Visconti *et al.,* 1995b). The effect of PKA on tyrosine phosphorylation is almost certainly not direct. Rather, it seems likely that PKA phosphorylates serine/threonine residues of a protein(s), which then phosphorylates the tyrosine kinase (Leclerc *et al.,* 1996). Consistent with a role for phosphorylation, Visconti *et al.* (1997) found a time-dependent increase in PKA activity in cauda epididymal mouse spermatozoa *in vitro,* provided that the spermatozoa were suspended in a medium that supports capacitation. They also localized PKA to the anterior sperm head.

b. Protein Kinase C

Protein kinase C (PKC) was found to be present in the equatorial segment of the sperm head and in the flagellar principal piece (Breitbart *et al.,* 1992; Lax *et al.,* 1997; Naor and Breitbart, 1997; Rotem *et al.,* 1990). Just as Ca^{2+} influx was shown to occur during both capacitation and (faster and to a greater extent) during the acrosome reaction (Section III,B), it was suggested that PKC was involved in both of these processes. The evidence for its involvement in capacitation is indirect:

1. The PKC activator, 12-*O*-tetradecanoyl-phorbol-13-acetate (TPA; termed also PMA; see Table 3.2), was found to increase the capacitation level in mouse, hamster, and human spermatozoa (Furuya *et al.,* 1992, 1993b; Leclerc *et al.,* 1996; Visconti and Tezon, 1989), as well as tyrosine phosphorylation of human spermatozoa (Furuya *et al.,* 1993b; Leclerc *et al.,* 1996).

2. PKC inhibitors (staurosporine or H7) inhibit the incremental TPA-induced increase in capacitation (Furuya *et al.,* 1992, 1993b) [note, however, that staurosporine (Prade *et al.,* 1997; Toledo and Lydon, 1997) and H7 (Cervo *et al.,* 1997; Gao *et al.,* 1998) can also inhibit other kinases].

3. TPA increases the cAMP levels in hamster spermatozoa by 20- to 50-fold only in a medium that supports capacitation (Visconti *et al.,* 1990; Visconti and Tezon, 1989).

The evidence for the involvement of PKC in the acrosome reaction is also indirect and based mainly on the finding that TPA induces the acrosome reaction of capacitated human spermatozoa (Cohen-Dayag *et al.,* 1995; De Jonge *et al.,* 1991b; Jaiswal *et al.,* 1999a; Rotem *et al.,* 1992). The observation mentioned above—that the PKC inhibitors inhibit just the TPA-stimulated incremental increase in capacitation but do not reduce the prestimulus steady-state level of capacitated spermatozoa (Furuya *et al.,* 1993b)—sheds doubt, in our opinion, on the possibility that PKC plays a central role, if any, in capacitation.

c. Extracellular Signal-Regulated Kinases

Luconi *et al.* (1998a,b) found two proteins of 42 and 44 kDa that are tyrosine phosphorylated in a time-dependent manner during *in vitro* capacitation of human spermatozoa, and identified them as extracellular signal-regulated kinases (ERKs), ERK-2 and ERK-1, respectively. The increase in tyrosine phosphorylation of these proteins during capacitation was accompanied by their increased kinase activity. Furthermore, PD098059, an inhibitor of mitogen-activated protein kinase (MAPK or MEK) activity and thereby an inhibitor of phosphorylation and activation of ERKs, appeared to inhibit sperm capacitation (Luconi *et al.,* 1998a). These findings suggest that ERKs are involved in the regulation of capacitation.

6. Modulation of Tyrosine Phosphorylation by Protein Phosphatases

The observed increase in protein tyrosine phosphorylation during capacitation may, in principle, result from kinase stimulation or phosphatase inhibition. Indeed, indirect evidence for the involvement of phosphatases in capacitation of human spermatozoa was provided by Furuya *et al.* (1993a) and Leclerc *et al.* (1996), who demonstrated that phosphatase inhibitors (calyculin A and okadaic acid) stimulate capacitation and tyrosine phosphorylation.

To conclude, the findings reviewed herein suggest that tyrosine phosphorylation is involved in capacitation and that the phosphorylation level may be regulated by the activities of both kinases and phosphatases.

D. OTHER COMPOUNDS THAT MODULATE CAPACITATION

1. Progesterone

Progesterone, a constituent of the female reproductive tract, is known to stimulate both sperm hyperactivation (Oehninger *et al.,* 1994; Parinaud and Milhet, 1996) and capacitation (Barboni *et al.,* 1995; DasGupta *et al.,* 1994b; de Lamirande *et al.,* 1998; Foresta *et al.,* 1992; Kay *et al.,* 1994; Revelli *et al.,* 1994; Roldan *et al.,* 1994). Progesterone was found to stimulate ERK-2 (Luconi *et al.,* 1998b), protein tyrosine phosphorylation, and Ca^{2+} influx, and to increase the intracellular cAMP level (Emiliozzi *et al.,* 1996; Parinaud and Milhet, 1996). It is not known which of these processes is directly affected by progesterone. However, because elevation of Ca^{2+} level is known to activate adenylyl cyclase (Rojas *et al.,* 1992), it is reasonable to think that the primary effect of progesterone is to stimulate Ca^{2+} influx. Consequently, the cAMP level, kinase activity, and protein tyrosine phosphorylation may sequentially be increased.

2. HCO_3^-

HCO_3^- is an important constituent of any capacitating medium (see Section II,A,1,b). In the mouse, it even appears to be obligatory, because capacitation does not occur in its absence (Neill and Olds-Clarke, 1987; Shi and Roldan, 1995; Visconti *et al.,* 1995a). Harrison (1997) considers HCO_3^- as the primary agent of boar sperm capacitation *in vitro,* because it induces rapid changes in the processes known to be associated with capacitation—Ca^{2+} influx (Harrison *et al.,* 1993), changes in surface glycoproteins (Ashworth *et al.,* 1995), and changes in the plasma membrane lipid architecture (Harrison and Gadella, 1995). HCO_3^- is also thought to raise pH_{in} (Jaiswal and Majumder, 1998; Roos and Boron, 1981) and to stimulate directly the adenylyl cyclase (Garty and Salomon, 1987; Okamura *et al.,* 1985; Rojas *et al.,* 1992, 1993; Visconti *et al.,* 1990). Because elevated pH_{in} and elevated Ca^{2+}_{in} can each stimulate the adenylyl cyclase (Hyne and Garbers, 1979; Kopf and Vacquier, 1984; Rojas *et al.,* 1992, 1993), it appears that HCO_3^- may stimulate the cyclase both directly and indirectly. The resulting elevated level of cAMP is thought to stimulate PKA and thereby, directly or indirectly, tyrosine phosphorylation of sperm membrane proteins associated with sperm capacitation (Fraser, 1995a; Fraser *et al.,* 1995; Visconti *et al.,* 1995a).

E. A MODEL FOR THE MOLECULAR MECHANISM OF CAPACITATION

As indicated in Table 3.4, many processes occur during capacitation. For some, there is solid evidence for their involvement in capacitation; for others, the evidence is indirect or weak. It is impossible at this stage to propose a mechanism that involves all these processes. Instead, we put forward a model that includes those processes that appear to be central. It is based on data discussed in Sections III,A–III,D as well as on previously proposed models (Baldi *et al.,* 1996; de Lamirande *et al.,* 1997b; Visconti and Kopf, 1998). Our model, shown schematically in Figure 3.2, involves the following sequence of events:

1. Removal of decapacitating factors. Capacitation is initiated by the removal of a number of decapacitating factors, including cholesterol (Cross, 1996b; Cross and Mahasreshti, 1997) and a 40-kDa anionic polypeptide (Fraser *et al.,* 1990), from the sperm surface (Section II,A,1). This removal requires seminal-fluid-free medium and a sterol acceptor such as albumin (Section III,A,2,b).

2. Elevation of pH_{in}. The removal of cholesterol (step 1) and the presence of HCO_3^- each results in elevated pH_{in} (Cross and Razyfaulkner, 1997; Harrison *et al.,* 1993).

3. Elevation of Ca^{2+}_{in}. The removal of the 40-kDa anionic polypeptide (step 1) results in Ca^{2+}–ATPase inhibition and thereby reduced Ca^{2+} extrusion (Adeoya-Osiguwa and Fraser, 1996) and elevated Ca^{2+}_{in} (Section III,B, mechanism 1). Elevation of Ca^{2+}_{in} is also thought to be carried out by activation of Ca^{2+} influx through Ca^{2+} channels and carriers, caused by ROS such as O_2^- and H_2O_2 (Section III,C,2). The generation of these species appears to result from activation of a yet-unidentified plasma membrane oxidase, possibly caused by removal of a decapacitating factor (de Lamirande *et al.,* 1997b) or by progesterone (de Lamirande *et al.,* 1998). The presence of HCO_3^- also stimulates Ca^{2+} influx (Harrison *et al.,* 1993). External factors such as progesterone may also increase the Ca^{2+}_{in} level (Section III,D,1).

4. Adenylyl cyclase activation. Elevated pH_{in} (step 2), elevated Ca^{2+}_{in} (step 3), and HCO_3^-—each activates the adenylyl cyclase (Garty and Salomon, 1987; Hyne and Garbers, 1979; Kopf and Vacquier, 1984; Okamura *et al.,* 1985; Visconti *et al.,* 1990). However, because elevated pH_{in} is known to result in increased Ca^{2+}_{in} level (Santi *et al.,* 1998), the possibility that elevated pH_{in} activates the adenylyl cyclase indirectly, via a rise in Ca^{2+}_{in} level, cannot be excluded. ROS were also suggested to activate adenylyl cyclase (de Lamirande *et al.,* 1997b). The consequence of adenylyl cyclase activation is elevation of the intracellular level of cAMP.

5. Tyrosine phosphorylation. The elevated level of cAMP activates PKA (Section III,C,5,a). Because PKAs are known to phosphorylate seryl or threonyl hydroxyls, it is reasonable that this activation results in phosphorylation of serine/threonine residues of a protein(s) that phosphorylates a tyrosine kinase, or that is a tyrosine kinase itself. Consequently, the tyrosine kinase is activated and tyrosine phosphorylation of specific proteins occurs (Section III,C,1). Two of these tyrosine-phosphorylated proteins are ERK-1 and ERK-2 (Section III,C,5,c). The other substrates of the tyrosine kinase as well as the substrates of the ERKs are not yet known. The activation of the tyrosine kinase is probably accompanied by inhibition of specific protein phosphatases (Section III,C,6). It is also possible that ROS are involved in the regulation of tyrosine phosphorylation.

6. Lipid redistribution in the membrane and increase in membrane fluidity and

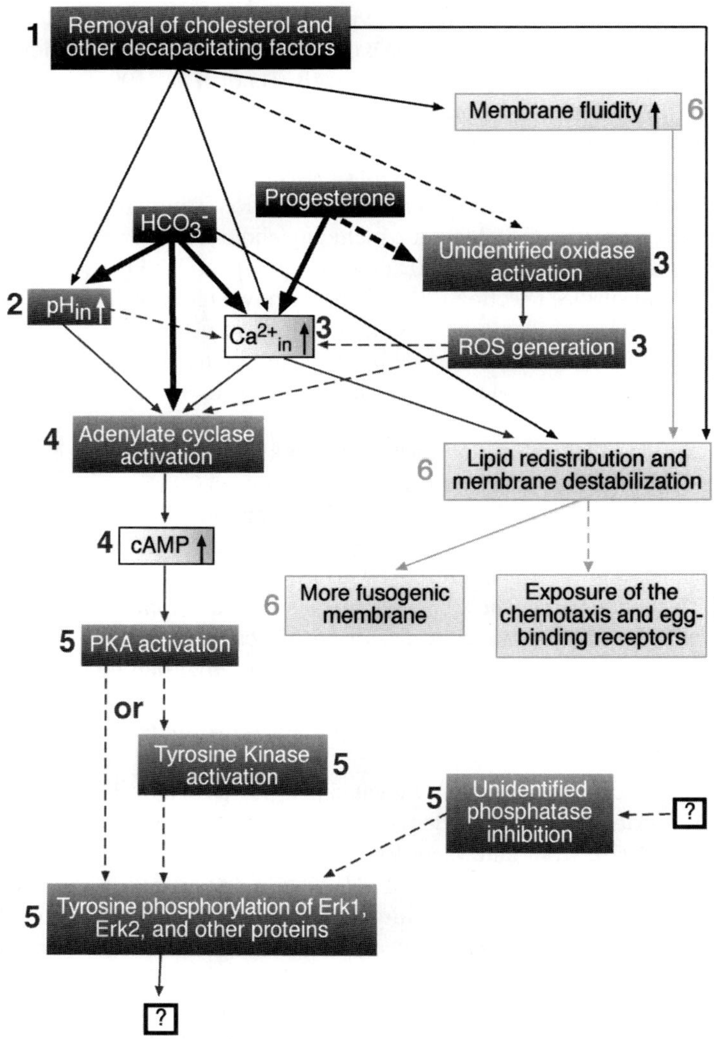

FIGURE 3.2 A schematic model for the sequence of the main molecular events that occur during capacitation. External processes and factors are shown in black. Dark and light gray represent sequences of events that presumably occur in parallel; it is not known, however, how these pathways are linked. Framed boxes represent key elements of signal transduction. Bold black arrows represent modulation of a process by external factors. Thin arrows represent the sequence of events. Solid and dashed arrows represent documented and hypothesized stimulatory pathways, respectively. The numbers correspond to the itemized sequence of events in Section III,E.

fusogenicity. Apparently, in parallel to the activation of adenylyl cyclase (step 4) and the consequent tyrosine phosphorylation (step 5), the elevation of Ca^{2+}_{in} (step 3) results in inhibition of the enzymes (flippases) that are responsible for the asymmetric lipid distribution in the plasma membrane leaflets (Section III,A,1). HCO_3^-

(Section III,D,2) and cholesterol efflux from the plasma membrane (step 1) also modulate the plasma membrane lipid architecture, resulting in loss of asymmetric lipid distribution (Sections III,A,2,c and III,D,2) and increased membrane fluidity. All these may make the inner leaflet of the plasma membrane more fusogenic and thereby enable membrane fusion during the acrosome reaction (Section III,A,1).

7. Exposure of receptors. The changes in the membrane fluidity and architecture (steps 6 and 7) may result in exposure of the chemotaxis receptors as well as of the zona pellucida-recognizing receptors, including the mannose receptor (Benoff, 1997; Benoff *et al.*, 1993d).

The model is certainly incomplete and should be considered as a first approximation. Not only does it not involve many of the processes known to occur during capacitation (Table 3.4), it does not connect the tyrosine phosphorylation (step 5) to the features that confer on a capacitated cell the potential to respond chemotactically, to adhere the zona pellucida, and to fuse the plasma membrane with the underlying outer acrosomal membrane during the acrosome reaction (steps 6–8). Further studies are clearly needed. The model does propose that a number of key steps have multiple regulatory mechanisms. Thus, there are a number of mechanisms that modulate the Ca^{2+}_{in} level (step 3), a number of mechanisms that regulate the activity of adenylyl cyclase (step 4), and a number of mechanisms that regulate tyrosine phosphorylation (step 5). In addition, a number of extracellular factors also appear to take part in this regulation—e.g., HCO_3^- (steps 2, 3, 4, and 6), progesterone (step 3), and others (Section II,C,2).

IV. PHYSIOLOGICAL MECHANISM AND ROLE OF CAPACITATION

Today, when mammalian eggs can be fertilized *in vitro* by injection of the sperm head (which is essentially the DNA content of the sperm), even of immature spermatozoa (Kimura and Yanagimachi, 1995; Uehara and Yanagimachi, 1977; Yanagimachi, 1997, 1998), it becomes evident that the role of capacitation *in vivo* is probably largely to ensure that ripe spermatozoa will meet the egg and that this encounter will result in sperm penetration and egg fertilization. Capacitation appears to have a global role in this task. Below we review the processes in the female genital tract for which the spermatozoa must be capacitated.

A. SPERMATOZOA WITHIN THE SPERM RESERVOIR

As reviewed in Section II,C,1, the oviductal isthmus seems to be the sperm storage site prior to ovulation (Figure 3.1). There appear to be mutual effects between the storage site and the process of capacitation. On the one hand, direct contacts between the spermatozoa and the oviductal epithelial cells apparently slow down capacitation (Dobrinski *et al.*, 1997; Murray and Smith, 1997; Smith, 1998). On the other hand, spermatozoa apparently must be capacitated to acquire the potential to detach from the oviductal epithelium. Thus, Smith and Yanagimachi (1991) found in hamsters that only noncapacitated spermatozoa can attach to the isthmic mucosa of the oviduct; Lefebvre and Suarez (1996) reported that when bull spermatozoa are capacitated *in vitro* by heparin, their binding to explants of oviductal epithelium is reduced; Dobrinski *et al.* (1997) compared the Ca^{2+}_{in} level in spermatozoa attached to and detached from the oviductal epithelium and found that the Ca^{2+}_{in} level is higher in the detached cells, in line with the notion that these cells are capacitated. Demott *et al.* (1995) demonstrated that only hyperactivated spermatozoa (i.e., apparently capacitated spermatozoa; see Section IV,C) are detached from the epithelium. Further studies have provided evidence that lectin–carbohydrate binding is involved in the attachment of noncapacitated spermatozoa to the oviductal epithelium (Suarez, 1998). The specificity of the lectin on the sperm surface and the identity of the carbohydrate on the oviductal epithelium appear to vary among species. It has been proposed that noncapacitated spermatozoa bind to carbohydrate moieties on glycoproteins or glycolipids on the surface of the oviductal epithelium via a lectin-like molecule, and that this molecule is lost or modified during capacitation, allowing the spermatozoa to be released (Suarez, 1998). To substantiate this model, it is necessary to demonstrate that the carbohydrate-binding ability of spermatozoa is lost on capacitation.

B. SPERM CHEMOTAXIS

Chemotaxis to follicular factors has been demonstrated in human and mouse spermatozoa [Eisenbach (1999) and Eisenbach and Tur-Kaspa (1999) for recent reviews]. In both species it was found that only a small fraction of a given sperm population is chemotactically responsive to a follicular factor(s) at any given time [2–12% in humans (Cohen-Dayag *et al.,* 1994) and ~10% in mice (Giojalas and Rovasio, 1998; Oliveira *et al.,* 1999)]. In human studies it was further found that the identity of the responsive spermatozoa continuously changes: chemotactic spermatozoa lose their activity while others acquire it (Cohen-Dayag *et al.,* 1994). This raised the possibility that, at least in humans, spermatozoa acquire their chemotactic responsiveness as part of the capacitation process, and lose this responsiveness when the capacitated state is terminated (Eisenbach and Ralt, 1992). The association of chemotactic responsiveness with the capacitated state was demonstrated experimentally, and it relied on the similar percentages of chemotactic and capacitated spermatozoa in a given sperm population, on the continuous replacement of capacitated and chemotactic spermatozoa with similar kinetics, and on the fact that deliberate depletion of capacitated spermatozoa results in total loss of chemotaxis and, vice versa, depletion of chemotactic spermatozoa results in depletion of capacitated spermatozoa (Cohen-Dayag *et al.,* 1995). The finding that, in mice as well, only ~10% of the spermatozoa are chemotactically responsive (Giojalas and Rovasio, 1998; Oliveira *et al.,* 1999) may indicate that the situation in the mouse is similar with respect to replacement of capacitated/chemotactically responsive spermatozoa.

The finding that, at least in humans, chemotactic responsiveness is restricted only to capacitated spermatozoa raised the possibility that, *in vivo,* the role of human sperm chemotaxis is to recruit a selective population of spermatozoa, i.e., capacitated spermatozoa, for fertilizing the egg, rather than to direct many spermatozoa to the egg [as is the case in spermatozoa of marine species (Eisenbach, 1999)]. A possible role for the continuous replacement of capacitated/chemotactic spermatozoa may be to ensure availability of capacitated spermatozoa for an extended period of time, despite the short life span of the capacitated state in any one spermatozoon (Cohen-Dayag *et al.,* 1995). It would be interesting to find out whether continuous replacement of capacitated spermatozoa also occurs in the rabbit, wherein ovulation is induced by copulation and therefore there is no obvious need for extended availability of capacitated spermatozoa.

Where and when sperm chemotaxis occurs *in vivo* are not known. One possibility is that chemotaxis is involved in directing the capacitated spermatozoa released from the storage site toward the egg (Figure 3.1). This may be important in view of the relatively small number of spermatozoa released from the storage site (Hunter, 1993). However, because the oviduct undergoes contractile movements that appear to move fluid in a direction opposite to that of follicular fluid transport (Battalia and Yanagimachi, 1979), it is unlikely that a chemical gradient can be established over a long range (Cohen-Dayag *et al.,* 1994). Thus, chemotaxis may function only in close proximity to the egg. It is possible that the spermatozoa respond chemotactically within the cumulus oophorus and at a short range in its vicinity (see below). Another possibility, based on the finding that, in mice, both oviductal and follicular fluids are chemotactically active, is that there are two sequential steps of chemotaxis, each to a different attractant (Oliveira *et al.,* 1999). Determination of the cellular origin of the sperm attractant(s) may distinguish between some of these open possibilities.

The mechanism that makes capacitated spermatozoa chemotactic is also not known. The changes in membrane architecture and properties that occur during capacitation and that result in exposure of the mannose receptor (Section III,A) may well result also in exposing the yet-unidentified chemotaxis receptors. Alternatively, receptors for chemotaxis may be present on both capacitated and noncapacitated spermatozoa, but only capacitated spermatozoa may have the potential to transduce the chemotactic signals from the receptors. For example, signal transduction in sperm chemotaxis is thought to involve changes in the level of Ca^{2+}_{in} (Eisenbach, 1999). It is possible that the higher Ca^{2+}_{in} level in capacitated spermatozoa (Section III,B) makes them chemotactically responsive.

C. SPERM HYPERACTIVATION

Hyperactivated motility of mammalian spermatozoa was first identified by Yanagimachi (1969, 1970) in hamsters, and then it was identified in a number of other mammalian species (Table 3.4). This motility pattern—characterized by increased velocity, decreased linearity,[4] increased amplitude of lateral head dis-

[4] Linearity is the ratio between (1) the straight line from the first position of the sperm head to its last position and (2) the actual trajectory made by the sperm head between these points (Davis and Siemers, 1995).

placement, and flagellar whiplash movement (Mortimer *et al.,* 1996, 1997; Suarez, 1996; Yanagimachi, 1994)—can be observed during capacitation *in vitro* when spermatozoa are flushed from the oviduct at a time close to fertilization, or when they are observed through the translucent walls of rodent oviducts (Suarez, 1996; Yanagimachi, 1994). However, the patterns of hyperactivated sperm movement vary to a large extent between species (Yanagimachi, 1994), resulting in lack of well-defined criteria for the identification of hyperactivated spermatozoa (Mortimer *et al.,* 1997). In addition, not all the spermatozoa in a given sample display hyperactivated motility at the same time of incubation, and the spermatozoa are constantly switching from one pattern of hyperactivated motility to another or from nonhyperactivated motility to hyperactivated motility (Burkman, 1990; de Lamirande and Gagnon, 1993a; Mortimer and Swan, 1995). Furthermore, in some media for mouse and hamster spermatozoa, completion of capacitation and hyperactivation is not synchronized [Suarez (1996) for review]. All these observations suggest that the correlation between capacitation and hyperactivation is not one-to-one: although a spermatozoon may have to be capacitated to become hyperactivated, being capacitated does not necessarily mean being hyperactivated. In line with this conclusion, it has been found that in chemotaxis assays capacitated human spermatozoa are not hyperactivated at all times; they acquire hyperactivation-like motility primarily when they approach progesterone (or progesterone-containing follicular fluid) (Jaiswal *et al.,* 1999b; Ralt *et al.,* 1994) [see also Mbizvo *et al.* (1990) and Uhler *et al.* (1992)]. Progesterone-depleted follicular fluid does attract spermatozoa, but hyperactivation-like motility is rarely observed in the absence of progesterone (Jaiswal *et al.,* 1999b). It thus seems that although capacitation confers on spermatozoa, among other things, the ability to become hyperactivated, and some cells do so spontaneously, additional female-derived cues such as progesterone enhance the transition to this motility pattern.

The role of sperm hyperactivation in fertilization is not known, but a few suggestions have been made. It was demonstrated that hyperactivated spermatozoa are more efficient than nonhyperactivated ones in penetrating viscous and viscoelastic substances (enabling them to penetrate mucoid oviductal secretions and the cumulus oophorus) (Suarez, 1996) and in penetrating the zona pellucida (Stauss *et al.,* 1995). Hyperactivation may also assist detachment of capacitated spermatozoa from the oviductal epithelium (DeMott and Suarez, 1992).

D. SPERM PENETRATION THROUGH THE CUMULUS OOPHORUS, ADHESION TO THE EGG'S ZONA PELLUCIDA, AND THE ACROSOME REACTION

Before reaching the egg, the spermatozoa must pass through the surrounding cumulus oophorus. Only capacitated spermatozoa can penetrate through this barrier; noncapacitated spermatozoa, or spermatozoa that have already undergone the acrosome reaction, can only attach to the cumulus surface but cannot proceed further (Austin, 1960; Cherr *et al.,* 1986; Cummins and Yanagimachi, 1986; Suarez *et al.,* 1984; Talbot, 1985) [Yanagimachi (1994) for review]. Because it was found that the first spermatozoon to enter the cumulus finds the egg very effectively (Bedford and Kim, 1993; Hunter, 1993), it is possible that the spermatozoa within the cumulus oophorus respond chemotactically to an attractant secreted from the egg. It is also possible that the cumulus cells are not homogeneous, and that those closest to the egg secrete the attractant. Such a situation, if correct, can form an attractant gradient within the cumulus even if the egg is not the cell that secretes the attractant.

When spermatozoa merge out of the cumulus oophorus toward the egg, they encounter the zona pellucida, a layer of glycoproteins surrounding the egg (Chapters 9 and 10, this volume). Before penetrating this layer, the spermatozoa bind firmly to its surface. ZP3, one of three glycoproteins in the zona pellucida, is the egg protein recognized by sperm (Bleil and Wassarman, 1980, 1983, 1986). It is well established that this binding to ZP3 initiates the acrosome reaction [Wassarman and Litscher (1995) and Yanagimachi (1994) for reviews]. Many sperm receptors for ZP3 have been proposed [Benoff (1997), Eisenbach (1995), and Tulsiani *et al.* (1997) for reviews]. Some of them [e.g., the human mannose receptor (Benoff *et al.,* 1993d) and the bull 70-kDa protein (Spungin *et al.,* 1995)] become exposed mainly or only during capacitation, suggesting that spermatozoa should be capacitated for this binding. In contrast, a few studies have provided evidence suggesting that, in the mouse, noncapacitated spermatozoa can also adhere to the zona pellucida (Benau and Storey, 1987; Heffner and Storey, 1982; Thaler and Cardullo, 1994; Ward and Storey, 1984). The difference might be a reflection of species differences or, more likely, differences between ejaculated and epididymal spermatozoa.

For penetrating an egg and fertilizing it *in vivo,* mammalian spermatozoa must undergo the acrosome reaction. This reaction involves fusion between the plasma membrane and the underlying outer acrosomal membrane, as a result of which the acrosomal contents, including a variety of hydrolytic enzymes that facilitate the penetration of the sperm through the zona pellucida, are released [see Chapters 11 and 13, this volume; see also Kopf and Gerton (1991), Wassarman (1987), and Yanagimachi (1994) for reviews]. In addition, the acrosome reaction confers on the spermatozoa the ability to fuse with the egg plasma membrane (Yanagimachi, 1994). Spermatozoa are induced to undergo the acrosome reaction *in vivo* when they adhere to the egg's zona pellucida[5]; *in vitro* they can also be stimulated by other inducers, both natural and pharmacological (Table 3.2). It is well established that spermatozoa cannot be induced (by any of these agents) to undergo the acrosome reaction, unless they are capacitated (Bedford, 1994; Jaiswal *et al.,* 1998; Yanagimachi, 1994). Even in a cell-free system, capacitation was demonstrated to

[5] A low level of acrosome-reacted spermatozoa is found also among ejaculated spermatozoa, defined as spontaneous acrosome-reacted cells (Stock and Fraser, 1987). These cells, however, cannot penetrate an egg and appear to be functionless in fertilization.

be a prerequisite for membrane fusion during the acrosome reaction (Spungin *et al.,* 1992). In apparent conflict, a number of studies proposed that, at least *in vitro,* capacitation is not a prerequisite for induced acrosome reaction (Anderson *et al.,* 1992; Bielfeld *et al.,* 1994a; De Jonge *et al.,* 1989). This notion was based on observations that human spermatozoa, preincubated in an albumin-free medium and considered noncapacitated, can be induced to undergo the acrosome reaction. However, the apparent conflict was resolved by demonstrating that, even in the absence of albumin, human spermatozoa can become partially capacitated and that the spermatozoa considered noncapacitated in those studies were, after all, partially capacitated (Jaiswal *et al.,* 1998).

E. HYPOTHETICAL SEQUENCE OF EVENTS *IN VIVO*

On the basis of the observations reviewed above, the following scenario may be proposed for what might be happening *in vivo* (Figure 3.3):

1. As soon as ejaculated spermatozoa, coated with epididymis- and seminal-plasma-originated decapacitating factors (Section II,A,1,a), reach the cervix (in species whose spermatozoa are deposited in the vagina at coitus) or the lower segment of the oviductal isthmus (in species whose spermatozoa are deposited in the uterus) (Figure 3.1), the decapacitating factors are removed and a unidirectional process of capacitation starts (Sections II,A,1,a and II,C,1).
2. As the not-yet capacitated spermatozoa move up the oviductal isthmus, they encounter the high mucus-containing narrow lumen, which impedes their forward progression, and they frequently come in contact with the oviductal epithelium, where they bind strongly to carbohydrate moieties of glycoproteins or glycolipids on the surface of the oviductal epithelium (Suarez, 1998; Suarez *et al.,* 1997). Consequently they become trapped and stored (Section IV,A).
3. The process of capacitation is completed in the oviductal isthmus (Section II,C,1). However, only a small fraction of the sperm population undergoes capacitation as soon as the decapacitating factors are removed. Moreover, only a small proportion is capacitated at any given time (Section II,A,2,b). This situation is the consequence of the short life span of the capacitated state (50–240 minutes), the continuous replacement of capacitated spermatozoa, and the fact that each spermatozoon can be capacitated only once (Section II,A,2,c). The continuous replacement may be caused by heterogeneous initiation of capacitation and perhaps also by a heterogeneous propagation rate. This heterogeneity might be prewired or might be a reflection of the different ages of individual spermatozoa within an ejaculate. In other words, the initiation of capacitation depends not only on the removal of decapacitating factors, but also on the age and/or on prewired, intrinsic properties of the individual spermatozoa. In addition, the propagation of capacitation appears to be affected by female-derived factors (Section II,C,2).
4. The continuous replacement of capacitated spermatozoa occurs primarily in the sperm reservoir in the oviductal isthmus (Figure 3.1). The contact with the

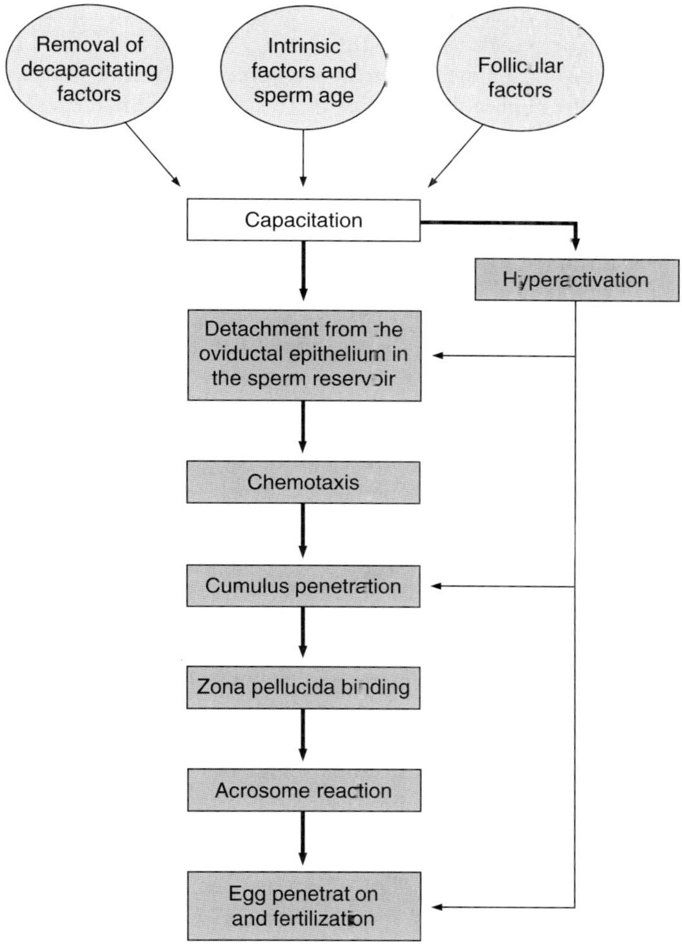

FIGURE 3.3 A simplified scheme of the events that affect, and the processes in the female genital tract that are regulated by, sperm capacitation. See text for details. Bold arrows represent the sequence of events. Thin arrows represent modulation of processes.

oviductal epithelium may slow down the rate of the replacement and prolong the sperm viability (Section II,C,2).

5. Spermatozoa that become capacitated are detached from the oviductal epithelium and are released from the isthmic reservoir (Section IV,A). Transient hyperactivation of motility, possibly in response to progesterone, may assist in this detachment. In the absence of a neighboring egg (e.g., prior to ovulation), these spermatozoa may be driven to the peritoneal cavity and may be attacked by phagocytes. In the presence of a neighboring egg (i.e., soon after ovulation), the spermatozoa in the reservoir or the detached capacitated spermatozoa are possibly sus-

ceptible to the priming effects of substances released from the follicle or from the egg and its surrounding cells (Section II,C,2).

6. The detached, capacitated spermatozoa approach the egg's zona pellucida through the cumulus oophorus by one or more steps of chemotaxis (Figure 3.1) (Section IV,B). In addition, during this journey the spermatozoa may encounter progesterone and consequently acquire hyperactivated motility temporarily. This motility pattern may assist them to penetrate the mucoid oviductal secretions and the cumulus oophorus (Section IV,C).

7. As the capacitated spermatozoa merge out of the cumulus oophorus they adhere to the surface of the zona pellucida, undergo the acrosome reaction, and penetrate this investment (Section IV,D). The latter process, as well, may benefit from hyperactivated motility of the spermatozoa (Section IV,C).

Thus, as schematically summarized in Figure 3.3, it appears that capacitation globally regulates the physiological steps that lead to fertilization: only capacitated spermatozoa can be detached from the oviductal epithelium at the storage site, they alone can respond to chemotactic factors secreted from the egg or the cumulus oophorus and thereby be recruited to the egg, only capacitated spermatozoa can penetrate through the cumulus oophorus, only they—according to most studies—can adhere to the zona pellucida, and only they can undergo the acrosome reaction, penetrate the zona pellucida, and fertilize the egg. Because the life spans of both the capacitated state and the ovulated egg are short (in humans, <4 and <24 hours, respectively), the chances to conceive would be disturbingly low in the absence of a promotion mechanism. Such a mechanism involves the replacement of capacitated spermatozoa, which ensures the availability of fertilizing spermatozoa for extended periods of time. Because the number of fertilizing spermatozoa is very small at any given moment, the fact that only capacitated spermatozoa can undergo the above-mentioned processes serves as a safeguard that ensures the arrival of these spermatozoa to the egg.

V. CONCLUSIONS

Mammalian spermatozoa should undergo two ripening stages for acquiring fertilization potential: maturation in the male reproductive tract and capacitation in the female genital tract or *in vitro.* Sperm motility and fertility are acquired during maturation, and the abilities to find the egg and penetrate it are acquired during capacitation. Capacitation is not involved in the actual transfer of the sperm genetic material to the egg. Its role is to prepare the conditions for such a transfer and to ensure that spermatozoa reaching the egg are those endowed with the capacity to penetrate and fertilize it. This is achieved by a number of sequential steps, each of them requiring capacitated spermatozoa. Thus, only capacitated spermatozoa can be detached from the oviductal epithelium and then be released from the sperm reservoir, and only they can respond to follicle-derived chemotactic stimuli, to penetrate the cumulus oophorus, to adhere to the zona pellucida, to undergo the acrosome reaction, and to penetrate the egg. This situation ensures that, if a noncapacitated spermatozoon succeeds in evading any of the first steps, it will probably be stopped in one of the subsequent steps. Because the life-span of the capacitated stage is short, the continuous replacement of capacitated spermatozoa compensates for this drawback and assures longer term of the availability of capacitated spermatozoa. Many molecular processes are involved in conferring on the spermatozoa the potential to accomplish all these steps. Most of the molecular processes are under the control of a number of regulators, both intrinsic and extrinsic, that make the capacitation process tightly regulated, reliable, and sensitive to the environment. Although the connections between some of these processes are being gradually revealed, the molecular mechanism as a whole is still obscure. Processes that lead to a more fusogenic membrane and to exposure of membrane receptors are becoming known, but it is not known how they are linked to many other molecular events that occur during capacitation. Sperm capacitation promises to remain an exciting field of research for years to come.

ACKNOWLEDGMENTS

We thank Drs. A. Cohen-Dayag and R. M. Johnstone for helpful comments.

REFERENCES

Aanesen, A., Fried, G., Andersson, E., and Gottlieb, C. (1996). Progesterone stimulates GABA uptake in human spermatozoa. *Biochem. Biophys. Res. Commun.* **226,** 88–93.

Aarons, D., Boettger-Tong, H., Biegler, B., George, B., and Poirier, G. R. (1993). The acrosomal status of human sperm evaluated by coomassie blue staining and electron microscopy. *Mol. Androl.* **5,** 31–37.

Adeoya-Osiguwa, S. A., and Fraser, L. R. (1993). A biphasic pattern of $^{45}Ca^{+2}$ uptake by mouse spermatozoa *in vitro* correlates with changing functional potential. *J. Reprod. Fertil.* **99,** 187–194.

Adeoya-Osiguwa, S. A., and Fraser, L. R. (1996). Evidence for Ca^{2+}-dependent ATPase activity, stimulated by decapacitation factor and calmodulin, in mouse sperm. *Mol. Reprod. Dev.* **44,** 111–120.

Aitken, R. J. (1995). Free radicals, lipid peroxidation and sperm function. *Reprod. Fertil. Dev.* **7,** 659–668.

Aitken, R., J., and Brindle, J. P. (1993). Analysis of the ability of three probes targeting the outer acrosomal membrane or acrosomal contents to detect the acrosome reaction in human spermatozoa. *Hum. Reprod.* **8,** 1663–1669.

Aitken, R. J., and Clarkson, J. S. (1987). Cellular basis of defective sperm function and its association with the genesis of reactive oxygen species by human spermatozoa. *J. Reprod. Fertil.* **81,** 459–469.

Aitken, R. J., Buckingham, D. W., Harkiss, D., Paterson, M., Fisher, H., and Irvine, D. S. (1996). The extragenomic action of progesterone on human spermatozoa is influenced by redox regulated changes in tyrosine phosphorylation during capacitation. *Mol. Cell. Endocrinol.* **117,** 83–93.

Aitken, R. J., Harkiss, D., Knox, W., Paterson, M., and Irvine, D. S (1998). A novel signal transduction cascade in capacitating human spermatozoa characterised by a redox-regulated, cAMP-mediated induction of tyrosine phosphorylation. *J. Cell Sci.* **111,** 645–656.

Aitken, R. J., Paterson, M., Fisher, H., Buckingham, D. W., and van Duin, M. (1995). Redox regulation of tyrosine phosphorylation in human spermatozoa and its role in the control of human sperm function. *J. Cell Sci.* **108,** 2017–2025.

Aitken, R. J., Ross, A., Hargreave, T., Richardson, D., and Best, F. (1984). Analysis of human sperm function following exposure to ionophore A23187. *J. Androl.* **5,** 321–329.

Alvarez, J. G., Touchstone, J. C., Blasco, L., and Storey, B. T. (1987). Spontaneous lipid peroxidation and production of hydrogen peroxide and superoxide in human spermatozoa. Superoxide dismutase as a major enzyme protectant against oxygen toxicity. *J. Androl.* **8,** 338–348.

Amann, R. P., Johnsen, L., Thompson, J., and Pickett, B. W. (1976). Daily spermatozoal production, epididymal spermatozoal reserves and transit time of spermatozoa through the epididymis of the rhesus monkey. *Biol. Reprod.* **15,** 586–592.

Amann, R. P., Koefoed-Johnsen, H. H., and Levi, H. (1965). Excretion pattern of labelled spermatozoa and the timing of spermatozoa formation and epididymal transit in rabbits injected with thymidine-3H. *J. Reprod. Fertil.* **10,** 169–183.

Amin, A.-H., Bailey, J. L., Storey, B. T., Blasco, L., and Heyner, S. (1996). A comparison of three methods for detecting the acrosome reaction in human spermatozoa. *Hum. Reprod.* **11,** 741–745.

Anderson, R. A., Feathergill, K. A., Dejonge, C. J., Mack, S. R., and Zaneveld, L. J. D. (1992). Facilitative effect of pulsed addition of dibutyryl cAMP on the acrosome reaction of noncapacitated human spermatozoa. *J. Androl.* **13,** 398–408.

Anderson, S. H., and Killian, G. J. (1994). Effect of macromolecules from oviductal conditioned medium on bovine sperm motion and capacitation. *Biol. Reprod.* **51,** 795–799.

Andrews, J. C., and Bavister, B. D. (1989). Capacitation of hamster spermatozoa with the divalent cation chelators D-penicillamine, L-histidine, and L-cysteine in a protein-free culture medium. *Gamete Res.* **23,** 159–170.

Andrews, J. C., Nolan, J. P., Hammerstedt, R. H., and Bavister, B. D. (1994). Role of zinc during hamster sperm capacitation. *Biol. Reprod.* **51,** 1238–1247.

Angle, M. J., Tom, R., Jarvi, K., and McClure, R. (1993). Effect of platelet-activating factor (PAF) on human spermatozoa-oocyte interactions. *J. Reprod. Fertil.* **98,** 541–548.

Aonuma, S., Okabe, M., Kawaguchi, M., and Kishi, Y. (1981). Zinc effects on mouse spermatozoa and *in vitro* fertilization. *J. Reprod. Fertil.* **63,** 463–466.

Aonuma, S., Okabe, M., Kishi, Y., Kawaguchi, M., and Yamada, H. (1982). Capacitation inducing activity of serum albumin in fertilization of mouse ova in vitro. *J. Pharmacobio-Dyn.* **5,** 980–987.

Ashraf, M., Peterson, R. N., and Russell, L. D. (1982). Evidence for $Ca^{2+}/Na+$ antiport in boar sperm plasma membrane vesicles. *Biol. Reprod.* **26,** 37A.

Ashworth, P. J. C., Harrison, R. A. P., Miller, N. G. A., Plummer, J. M., and Watson, P. F. (1995). Flow cytometric detection of bicarbonate-induced changes in lectin binding in boar and ram sperm populations. *Mol. Reprod. Dev.* **40,** 164–176.

Austin, C. R. (1951). Observations on the penetration of the sperm into mammalian egg. *Aust. J. Sci. Res. B.* **4,** 581–596.

Austin, C. R. (1952). The "capacitation" of the mammalian sperm. *Nature* **170,** 326.

Austin, C. R. (1960). Capacitation and the release of hyaluronidase from spermatozoa. *J. Reprod. Fertil.* **3,** 310–311.

Austin, C. R. (1985). Sperm maturation in the male and female genital tracts. *In* "Biology of Fertilization" (C. B. Metz, and A. Monroy, eds.), Vol. 2, pp. 121–155. Academic Press, New York.

Babcock, D. F., First, N. L., and Lardy, H. A. (1975). Transport mechanism for succinate and phosphate localized in the plasma membrane of bovine spermatozoa. *J. Biol. Chem.* **250,** 6488–6495.

Baldi, E., Casano, R., Falsetti, C., Krausz, C., Maggi, M., and Forti, A. (1991). Intracellular calcium accumulation and responsiveness to progesterone in capacitating human spermatozoa. *J. Androl.* **12,** 323–330.

Baldi, E., Luconi, M., Bonaccorsi, L., Krausz, C., and Forti, G. (1996). Human sperm activation during capacitation and acrosome reaction: Role of calcium, protein phosphorelation and lipid remodelling pathways. *Frontiers Biosci.* **1,** 189–205.

Banerjee, M., and Chowdhury, M. (1995). Induction of capacitation in human spermatozoa *in vitro* by an endometrial sialic acid-binding protein. *Hum. Reprod.* **10,** 3147–3153.

Banerjee, M., and Chowdhury, M. (1997). Localization of a 25 kDa human sperm surface protein: Its role in *in-vitro* human sperm capacitation. *Mol. Hum. Reprod.* **3,** 109–114.

Barboni, B., Mattioli, M., and Seren, E. (1995). Influence of progesterone on boar sperm capacitation. *J. Endocrinol.* **144,** 13–18.

Barratt, C. L. R., and Cooke, I. D. (1991). Sperm transport in the human female reproductive tract—a dynamic interaction. *Int. J. Androl.* **14,** 394–411.

Battalia, D. E., and Yanagimachi, R. (1979). Enhanced and co-ordinated movement of the hamster oviduct during the periovulatory period. *J. Reprod. Fert.* **56,** 515–520.

Bearer, E. L., and Friend, D. S. (1982). Modifications of anionic-lipid domains preceding membrane fusion in guinea pig sperm. *J. Cell Biol.* **92,** 604–615.

Bedford, J. M. (1969). Limitation of the uterus in the development of the fertility (capacitation) of spermatozoa. *J. Reprod. Fertil.* **8,** 19–26.

Bedford, J. M. (1970). Sperm capacitation and fertilization in mammals. *Biol. Reprod.* **2** (Suppl.), 128–158.

Bedford, J. M. (1983). Significance of the need for sperm capacitation before fertilization in eutherian mammals. *Biol. Reprod.* **28,** 108–120.

Bedford, J. M. (1994). The contraceptive potential of fertilization: A physiological perspective. *Hum. Reprod.* **9,** 842–858.

Bedford, J. M., and Kim, H. H. (1993). Cumulus oophorus as a sperm sequestering device, *in vivo. J. Exp. Zool.* **265,** 321–328.

Beltran, C., Darscon, A., Labarca, P., and Lievano, A. (1994). A high-conductive voltage-dependent multistate Ca^{2+} channels found in sea urchin and mouse spermatozoa. *FEBS Lett.* **338,** 23–26.

Benau, D. A., and Storey, B. T. (1987). Characterization of the mouse sperm plasma membrane zona-binding site sensitive to trypsin inhibitors. *Biol. Reprod.* **36,** 282–292.

Benoff, S. (1993). Preliminaris to fertilization. The role of cholesterol during capacitation of human spermatozoa. *Hum. Reprod.* **8,** 2001–2008.

Benoff, S. (1997). Carbohydrates and fertilization: An overview. *Mol. Hum. Reprod.* **3,** 599–637.

Benoff, S., Cooper, G. W., Hurley, I., Mandel, F. S., and Rosenfeld, D. L. (1993a). Antisperm antibody binding to human sperm inhibits capacitation-induced changes in the levels of plasma membrane sterols. *Am. J. Reprod. Immunol.* **30,** 113–130.

Benoff, S., Cooper, G. W., Hurley, I., Napolitano, B. Rosenfeld, D. L., Scholl, G. M., and Hershlag, A. (1993b). Human sperm fertilizing potential *in vitro* is correlated with differential expression of a head-specific mannose-ligand receptor. *Fertil. Steril.* **59,** 854–862.

Benoff, S., Hurley, I., Cooper, G. W., Mandel, F. S., Hershlag, A., Scholl, G. M., and Rosenfeld, D. L. (1993c). Fertilization potential *in vitro* is correlated with head-specific mannose-ligand receptor expression, acrosome status and membrane cholesterol content. *Hum. Reprod.* **8,** 2155–2166.

Benoff, S., Hurley, I., Cooper, G. W., Mandel, F. S., Rosenfeld, D. L., and Hershlag, A. (1993d). Head-specific mannose ligand receptor expression in human spermatozoa is dependent on capacitation-associated membrane cholesterol loss. *Hum. Reprod.* **8,** 2141–2154.

Berger, T., and Clegg, E. D. (1983). Adenylyl cyclase activity in porcine sperm in response to female reproductive tract secretions. *Gamete Res.* **7,** 169–177.

Bhattacharyya, A. (1992). Albumin is required for the guinea pig sperm capacitation but is not essential for acrosome reaction and fusion with eggs. *Arch. Androl.* **28,** 235–239.

Bielfeld, P., Anderson, R. A., Mack, S. R., De Jonge, C. J., and Zaneveld, L. J. D. (1994a). Are capacitation or calcium ion influx required for the human sperm acrosome reaction? *Fertil. Steril.* **62,** 1255–1261.

Bielfeld, P., Faridi, A., Zaneveld, L. J. D., and De Jonge, C. J. (1994b). The zona pellucida induced acrosome reaction of human spermatozoa is mediated by protein kinases. *Fertil. Steril.* **61,** 536–541.

Bize, I., Santander, G., Cabello, P., Driscoli, D., and Sharpe, C. (1991). Hydrogen peroxide is involved in hanster sperm capacitation *in vitro. Biol. Reprod.* **44,** 398–403.

Bleil, J. D., and Wassarman, P. M. (1980). Structure and function of zona-pellucida: Identification and characterization of the proteins of the mouse oocyte's zona-pellucida. *Dev. Biol.* **76,** 185–202.

Bleil, J. D., and Wassarman, P. M. (1983). Sperm–egg interaction in the mouse: Sequence of events and induction of the acrosome reaction by a zona pellucida glycoprotein. *Dev. Biol.* **95,** 317–324.

Bleil, J. D., and Wassarman, P. M. (1986). Autoradiographic visualization of the mouse egg's sperm receptor bound to sperm. *J. Cell Biol.* **102,** 1363–1371.

Boatman, D. E., and Robbins, R. S. (1991). Bicarbonate—carbon-dioxide regulation of sperm capacitation, hyperactivated motility, and acrosome reactions. *Biol. Reprod.* **44,** 806–813.

Boatman, D. E., Andrews, J. C., and Bavister, B. D. (1988). A quantitative assay for capacitation: Evaluation of multiple sperm penetration through the zona pellucida of salt-stored hamster eggs. *Gamete Res.* **19,** 19–29.

Boettger, H., Richardson, R., Free, D., Rushing, S., and Poirier, G. R. (1989). Effects of *in vitro* incubation on zona binding site found on murine spermatozoa. *J. Exp. Zool.* **249,** 90–98.

Boettger-Tong, H. L., Aarons, D. J., Biegler, B. E., George, B., and Poirier, G. R. (1993). Binding of a murine proteinase inhibitor to the acrosome region of the human sperm head. *Mol. Reprod. Devel.* **36,** 346–353.

Bradley, M. P., and Forrester, I. T. (1980). A sodium-calcium exchange mechanism in plasma membrane vesicles isolated from ram sperm flagella. *FEBS Lett.* **121,** 15–18.

Breitbart, H., Lax, J., Rotem, R., and Naor, Z. (1992). Role of protein kinase C in the acrosome reaction of mammalian spermatozoa. *Biochem. J.* **281,** 473–476.

Breitbart, H., Wehbie, R., and Lardy, H. (1990). Regulation of calcium transport in bovine spermatozoa. *Biochim. Biophys. Acta* **1027,** 72–78.

Burkman, L. J. (1984). Characterization of hyperactivated motility by human spermatozoa during capacitation: Comparison of fertile and oligozoospermic sperm populations. *Arch. Androl.* **13,** 153–165.

Burkman, L. J. (1990). Hyperactivated motility of human spermatozoa during *in vitro* capacitation and implications for fertility. *In* "Controls of Sperm Motility: Biological and Clinical Aspects" (C. Gagnon, ed.), pp. 303–329. CRC Press, Boca Raton.

Burks, D. J., Carballada, R., Moore, H. D. M., and Saling, P. M. (1995). Interaction of a tyrosine kinase from human sperm with the zona pellucida at fertilization. *Science* **269,** 83–86.

Byrd, W., and Wolf, D. P. (1986). Acrosomal status in fresh and capacitated human ejaculated sperm. *Biol. Reprod.* **34,** 859–869.

Carafoli, E. (1987). Calcium homeostasis. *Annu. Rev. Biochem.* **56,** 354–433.

Carrera, A., Moos, J., Ning, X. P., Gerton, G. L., Tesarik, J., Kopf, G. S., and Moss, S. B. (1996). Regulation of protein tyrosine phosphorylation in human sperm by a calcium/calmodulin-dependent mechanism: Identification of a kinase anchor proteins as major substrates for tyrosine phosphorylation. *Dev. Biol.* **180,** 284–296.

Carver-Ward, J. A., Jaroudi, K. A., Einspenner, M., Parhar, R. S., Al-Sediary, S. T., and Sheth, K. V. (1994). Pentoxifylline potentiates ionophore (A23187) mediated acrosome reaction in human sperm: Flow cytometric analysis using CD46 antibody. *Hum. Reprod.* **9,** 71–76.

Caswell, A. H., and Hutchison, J. D. (1971). Visualization of membrane bound cations by a fluorescent technique. *Biochem. Biophys. Res. Commun.* **42,** 43–49.

Cervo, L., Mukherjee, S., Bertaglia, A., and Samanin, R. (1997). Protein kinases A and C are involved in the mechanism underlying consolidation of cocaine place conditioning. *Brain Res.* **775,** 30–36.

Chang, M. C. (1951). Fertilizing capacity of spermatozoa deposited in Fallopian tubes. *Nature* **168,** 997–998.

Chang, M. C. (1957). A detrimental effect of rabbit seminal plasma on the fertilizing capacity of sperm. *Nature* **179,** 258–259.

Cherr, G. N., Lambert, H., Meizel, S., and Katz, D. F. (1986). *In vitro* studies of the golden hamster sperm acrosome reaction: Completion on the zona pellucida and induction by homologous soluble zonae pellucidae. *Dev. Biol.* **114,** 119–131.

Chian, R. C., Lapointe, S., and Sirard, M. A. (1995). Capacitation *in vitro* of bovine spermatozoa by oviduct epithelial cell monolayer conditioned medium. *Mol. Reprod. Dev.* **42,** 318–324.

Choi, Y.-H., and Toyoda, Y. (1998). Cyclodextrin removes cholesterol from mouse sperm and induces capacitation in a protein-free medium. *Biol. Reprod.* **59,** 1328–1333.

Chou, K., Chen, J., Yuan, S., and Haug, A. (1989). The membrane potential changes polarity during capacitation of murine epididymal sperm. *Biochem. Biophys. Res. Commun.* **165,** 58–64.

Cohen, J., and Adeghe, A. J.-H. (1987). The other spermatozoa: Fate and functions. *In* "New Horizons in Sperm Cell Research" (H. Mohri, ed.), pp. 125–134. Japan Sci. Soc. Press, Gordon and Breach Sci. Publ., Tokyo and New York.

Cohen-Dayag, A., and Eisenbach, M. (1994). Potential assays for sperm capacitation in mammals. *Am. J. Physiol.* **267** (*Cell Physiol.* **36**), C1167–C1176 [Errata: *Am. J. Physiol.* **268** (1995), issue 3, part 1].

Cohen-Dayag, A., Ralt, D., Tur-Kaspa, I., Manor, M., Makler, A., Dor, J., Mashiach, S., and Eisenbach, M. (1994). Sequential acquisition of chemotactic responsiveness by human spermatozoa. *Biol. Reprod.* **50,** 786–790.

Cohen-Dayag, A., Tur-Kaspa, I., Dor, J., Mashiach, S., and Eisenbach, M. (1995). Sperm capacitation in humans is transient and correlates with chemotactic responsiveness to follicular factors. *Proc. Natl. Acad. Sci. U.S.A.* **92,** 11039–11043.

Coronel, C. E., and Lardy, H. A. (1987). Characterization of Ca^{2+} uptake by guinea pig epididymal spermatozoa. *Biol. Reprod.* **37,** 1097–1107.

Coronel, C. E., and Lardy, H. A. (1992). Functional properties of caltrin proteins from seminal vesicle of the guinea pig. *Mol. Reprod. Dev.* **33,** 74–80.

Courtens, J. L., and Fournier-Delpech, S. (1979). Modifications in the plasma membranes of epididymal ram spermatozoa during maturation and incubation *in vitro*. *J. Ultrastruct. Res* **68,** 136–148.

Cross, N. L. (1993). Multiple effects of seminal plasma on the acrosome reaction of human sperm. *Mol. Reprod. Dev.* **35,** 316–323.

Cross, N. L. (1996a). Effect of cholesterol and other sterols on human sperm acrosomal responsiveness. *Mol. Reprod. Dev.* **45,** 212–217.

Cross, N. L. (1996b). Human seminal plasma prevents sperm from becoming acrosomally responsive to the agonist, progesterone: Cholesterol is the major inhibitor. *Biol. Reprod.* **54,** 138–145.

Cross, N. L. (1998). Role of cholesterol in sperm capacitation. *Biol. Reprod.* **59,** 7–11.

Cross, N. L., and Mahasreshti, P. (1997). Prostasome fraction of human seminal plasma prevents sperm from becoming acrosomally responsive to the agonist progesterone. *Arch. Androl.* **39,** 39–44.

Cross, N. L., and Overstreet, J. W. (1987). Glycoconjugates of the human sperm surface: Distribution and alterations that accompany capacitation *in vitro*. *Gamete Res.* **16,** 23–35.

Cross, N. L., and Razyfaulkner, P. (1997). Control of human sperm intracellular pH by cholesterol and its relationship to the response of the acrosome to progesterone. *Biol. Reprod.* **56,** 1169–1174.

Cross, N. L., Morales, P., Overstreet, J. W., and Hanson, F. W. (1986). Two simple methods for detecting acrosome-reacted human sperm. *Gamete Res.* **15,** 213–226.

Cross, N. L., Morales, P., Overstreet, J. W., and Hanson, P. W. (1988). Induction of acrosome reaction by the human zona-pellucida. *Biol. Reprod.* **38,** 235–244.

Cummins, J. M. (1982). Hyperactivated motility patterns of ram spermatozoa recovered from the oviducts of mated ewes. *Gamete Res.* **6,** 53–63

Cummins, J. M., and Yanagimachi, R. (1982). Sperm–egg ratios and the site of the acrosome reaction during *in vivo* fertilization in the hamster. *Gamete Res.* **5,** 239–256.

Cummins, J. M., and Yanagimachi, R. (1986). Development of ability to penetrate the cumulus oophorus by hamster spermatozoa capacitated *in vitro* in relation to the timing of the acrosome reaction. *Gamete Res.* **15,** 187–212.

Cummins, J. M., Pember, S. M., Jequier, A. M., Yovich, J. L., and Hartmann, P. E. (1991). A test of the human sperm acrosome reaction following ionophore challenge—relationship to fertility and other seminal parameters. *J. Androl.* **12,** 98–103.

DasGupta, S., Mills, C. L., and Fraser, L. R. (1993). Ca^{2+}-related changes in the capacitation state of human spermatozoa assessed by a chlorotetracycline fluorescence assay. *J. Reprod. Fertil.* **99,** 135–143.

DasGupta, S., Mills, C. L., and Fraser, L. R. (1994a). A possible role for Ca^{2+}-ATPase in human sperm capacitation. *J. Reprod. Fertil.* **102,** 107–116.

DasGupta, S., O'Toole, C., Mills, C. L., and Fraser, L. R. (1994b). Effect of pentoxifylline and progesterone on human sperm capacitation and acrosomal exocytosis. *Hum. Reprod.* **9,** 2103–2109.

Davis, B. K. (1982). Uterine fluid proteins bind sperm cholesterol during capacitation in the rabbit. *Experientia* **38,** 1063–1064.

Davis, B. K., Byrne, R., and Bedigian, K. (1980). Studies on the mechanism of capacitation: Albumin-mediated changes in plasma membrane lipids during *in vitro* incubation of rat sperm cells. *Proc. Natl. Acad. Sci. U.S.A.* **77,** 1546–1550.

Davis, B. K., Byrne, R., and Hungund, B. (1979). Studies on the mechanism of capacitation: II. Evidence for lipid transfer between plasma membrane of rat sperm and serum albumin during capacitation *in vitro. Biochim. Biophys. Acta* **558,** 257–266.

Davis, R. O., and Siemers, R. J. (1995). Derivation and reliability of kinematic measures of sperm motion. *Reprod. Fertil. Dev.* **7,** 857–869.

D'Cruz, O. J., and Hass, G. G. (1992). Flow cytometric quantitation of the expression of membrane cofactor protein as a marker for the human sperm acrosome reaction. *Fertil. Steril.* **58,** 633–636.

De Jonge, C. J., Barratt, C. L. R., Radwanska, E., and Cooke, I. D. (1993). The acrosome reaction-inducing effect of human follicular and oviductal fluid. *J. Androl.* **14,** 359–365.

De Jonge, C. J., Han, H. L., Lawrie, H., Mack, S. R., and Zaneveld, L. J. D. (1991a). Modulation of the human sperm acrosome reaction by effectors of the adenylyl cyclase/cyclic AMP second-messenger pathway. *J. Exp. Zool.* **258,** 113–125.

De Jonge, C. J., Han, H. L., Mack, S. R., and Zaneveld, L. J. D. (1991b). Effect of phorbol diesters, synthetic diacylglycerols, and a protein kinase-C inhibitor on the human sperm acrosome reaction. *J. Androl.* **12,** 62–70.

De Jonge, C. J., Mack, S. R., and Zaneveld, L. J. D. (1989). Synchronous assay for human sperm capacitation and the acrosome reaction. *J. Androl.* **10,** 232–239.

de Lamirande, E., and Gagnon, C. (1993a). Human sperm hyperactivation and capacitation as parts of an oxidative process. *Free Radic. Biol. Med.* **14,** 157–166.

de Lamirande, E., and Gagnon, C. (1993b). A positive role for the superoxide anion in triggering hyperactivation and capacitation of human spermatozoa. *Int. J. Androl.* **16,** 21–25.

de Lamirande, E., and Gagnon, C. (1995a). Capacitation-associated production of superoxide anion by human spermatozoa. *Free Radic. Biol. Med.* **18,** 487–495.

de Lamirande, E., and Gagnon, C. (1995b). Impact of reactive oxygen species on spermatozoa: A balancing act between benificial and detrimental effects. *Hum. Reprod.* **10,** 15–21.

de Lamirande, E., Harakat, A., and Gagnon, C. (1998). Human sperm capacitation induced by biological fluids and progesterone, but not by NADH or NADPH, is associated with the production of superoxide anion. *J. Androl.* **19,** 215–225.

de Lamirande, E., Jiang, H., Zini, A., Kodama, H., and Gagnon, C. (1997a). Reactive oxygen species and sperm physiology. *Rev. Reprod.* **2,** 48–54.

de Lamirande, E., Leclerc, P., and Gagnon, C. (1997b). Capacitation as a regulatory event that primes spermatozoa for the acrosome reaction and fertilization. *Mol. Hum. Reprod.* **3,** 175–194.

DeMott, R. P., and Suarez, S. S. (1992). Hyperactivated sperm progress in the mouse oviduct. *Biol. Reprod.* **46,** 779–785.

DeMott, R. P., Lefebvre, R., and Suarez, S. S. (1995). Carbohydrates mediate the adherence of hamster sperm to oviductal epithelium. *Biol. Reprod.* **52,** 1395–1403.

Dobrinski, I., Smith, T. T., Suarez, S. S., and Ball, B. A. (1997). Membrane contact with oviductal epithelium modulates the intracellular calcium concentration of equine spermatozoa *in vitro. Biol. Reprod.* **56,** 861–869.

Dorjee, S., Lai, C. L. B., Lee, A., and Lee, C. Y. G. (1997). Monoclonal antibodies as direct probes for human sperm acrosome reaction. *Am. J. Reprod. Immunol.* **37,** 283–290.

Duncan, A. E., and Fraser, L. R. (1993). Cyclic AMP-dependent phosphorylation of epididymal mouse sperm proteins during capacitation *in vitro:* Identification of an Mr 95000 phosphotyrosine-containing protein. *J. Reprod. Fertil.* **32,** 375–379.

Ehrenwald, E., Foote, R. H., and Parks, J. E. (1990). Bovine oviductal fluid components and their potential role in sperm cholesterol efflux. *Mol. Reprod. Dev.* **25,** 195–204.

Eisenbach, M. (1995). Sperm changes enabling fertilization in mammals. *Curr. Opin. Endocrinol. Diabetes* **2,** 468–475.

Eisenbach, M. (1999). Sperm chemotaxis. *Rev. Reprod.* **4,** 56–66.

Eisenbach, M., and Ralt, D. (1992). Precontact mammalian sperm–egg communication and role in fertilization. *Am. J. Physiol.* **262** (*Cell Physiol.* **31**), C1095–C1101.

Eisenbach, M., and Tur-Kaspa, I. (1999). Do human eggs attract spermatozoa? *BioEssays* **21,** 203–210.

Emiliozzi, C., and Fenichel, P. (1997). Protein tyrosine phosphorylation is associated with capacitation of human sperm *in vitro* but is not sufficient for its completion. *Biol. Reprod.* **56,** 674–679.

Emiliozzi, C., Cordonier, H., Guerin, J. F., Ciapa, B., Benchaib, M., and Fenichel, P. (1996). Effects of progesterone on human spermatozoa prepared for *in-vitro* fertilization. *Int. J. Androl.* **19,** 39–47.

Fenichel, P., Hsi, B. L., Farahifar, D., Donzeau, M., Barrier-Delpech, D., and Yeh, C. J. G. (1989). Evaluation of the human sperm acrosome reaction using a monoclonal antibody, GB24 and fluorescence-activated cell sorter. *J. Reprod. Fertil.* **87,** 699–706.

Fierro, R., Foliguet, B., Grignon, G., Daniel, M., Bene, M. C., Faure, G. C., and Barbarino-Monnier, P. (1996). Lectin-binding sites on human sperm during acrosome reaction: Modifications judjed by electron microscopy/flow cytometry. *Arch. Androl.* **36,** 187–196.

Florman, H. M. (1994). Sequential focal and global elevations of sperm intracellular Ca^{2+} are initiated by the zona pellucida during acrosomal exocytosis. *Dev. Biol.* **165,** 152–164.

Florman, H. M., and Babcock, D. F. (1991). Progress toward understanding the molecular basis of capacitation. *In* "Elements of Mammalian Fertilization" (P. M. Wassarman, ed.), Vol. 1, pp. 105–132. CRC Press, Boca Raton.

Florman, H. M., and Storey, B. T. (1982). Mouse gamete interactions: the zona pellucida is the site of the acrosome reaction leading to fertilization *in vitro. Dev. Biol.* **91,** 121–130.

Florman, H. M., Corron, M. E., Kim, T. D. H., and Babcock, D. F. (1992). Activation of voltage-dependent calcium channels of mammalian sperm is required for zona pellucida-induced acrosomal exocytosis. *Dev. Biol.* **152,** 304–314.

Focarelli, R., Giuffrida, A., Capparelli, S., Scibona, M., Fabris, F. M., Francavilla, F., Francavilla, S., Giovampaola, C. D., and Rosati, F. (1998). Specific localization in the equatorial region of gp20, a 20 kDa sialyglycoprotein of the capacitated human spermatozoon acquired during epididymal transit which is necessary to penetrate zona-free hamster eggs. *Mol. Hum. Reprod.* **4,** 119–125.

Focarelli, R., Giuffrida, A., and Rosati, F. (1995). Changes in the sialylglycoconjugate distribution on the human sperm surface during *in-vitro* capacitation: Partial purification of a 20 kDa sialylglycoprotein of capacitated spermatozoa. *Hum. Reprod.* **10,** 2755–2759.

Focarelli, R., Rosati, F., and Terrana, B. (1990). Sialyl glycoconjugates release during *in vitro* capacitation of human spermatozoa. *J. Androl.* **2,** 97–104.

Foresta, C., Rossato, M., Mioni, R., and Zorzi, M. (1992). Progesterone induces capacitation in human spermatozoa. *Andrologia* **24,** 33–35.

Fraser, L. R. (1979). Accelerated mouse sperm penetration *in vitro* in the presence of caffeine. *J. Reprod. Fertil.* **57,** 377–384.

Fraser, L. R. (1985). Albumin is required to support the acrosome reaction but not capacitation in mouse spermatozoa *in vitro. J. Reprod. Fertil.* **74,** 185–196.

Fraser, L. R. (1987). Minimum and maximum extracellular Ca^{2+} requirements during mouse sperm capacitation and fertilization *in vitro. J. Reprod. Fert.* **81,** 77–89.

Fraser, L. R. (1995a). Ionic control of sperm function. *Reprod. Fertil. Dev.* **7,** 905–925.

Fraser, L. R. (1995b). Mechanisms regulating capacitation and the acrosome reaction. *In* "Human Sperm Acrosome Reaction" (P. Fenichel and J. Parinaud, eds.), pp. 17–33. Inst. Natl. Sante Recherche Medicale, Paris.

Fraser, L. R., and Herod, J. E. (1990). Expression of capacitation-dependent changes in chlortetracycline fluorescence patterns in mouse spermatozoa requires a suitable glycolysable substrate. *J. Reprod. Fertil.* **88,** 611–621.

Fraser, L. R., and McDermott, C. A. (1992). Ca^{2+}-related changes in the mouse sperm capacitation state: A possible role for Ca^{2+}-ATPase. *J. Reprod. Fertil.* **96,** 363–377.

Fraser, L. R., Abeydeera, L. R., and Niwa, K. (1995). Ca^{2+}-regulating mechanisms that modulate bull sperm capacitation and acrosomal exocytosis as determined by chlortetracycline analysis. *Mol. Reprod. Dev.* **40,** 233–241.

Fraser, L. R., Harrison, R. A. P., and Herod, J. E. (1990). Characterization of a decapacitating factor associated with epididymal mouse spermatozoa. *J. Reprod. Fertil.* **89,** 135–148.

Friend, D. S. (1980). Freeze-fracture alterations in guinea pig sperm membranes preceding gamete fusion. *In* "Membrane-Membrane Interactions" (N. B. Gilula, ed.), pp. 153–165. Raven Press, New York.

Furuya, S., Endo, Y., Oba, M., Nozawa, S., and Suzuki, S. (1992). Effects of modulators of protein kinases and phosphatases on mouse sperm capacitation. *J. Assist. Reprod. Genet.* **9,** 391–399.

Furuya, S., Endo, Y., Osumi, K., Oba, M., Nozawa, S., and Suzuki, S. (1993a). Calyculin A, protein phosphatase inhibitor, enhances capacitation of human sperm. *Fertil. Steril.* **59,** 216–222.

Furuya, S., Endo, Y., Osumi, K., Oba, M., and Suzuki, S. (1993b). Effects of modulators of protein kinase-C on human sperm capacitation. *Fertil. Steril.* **59,** 1285–1290.

Fusi, F. M., and Bronson, R. A. (1992). Sperm surface fibronectin—expression following capacitation. *J. Androl.* **13,** 28–35.

Galantino-Homer, H. L., Visconti, P. E., and Kopf, G. S. (1997). Regulation of protein tyrosine phosphorylation during bovine sperm capacitation by a cyclic adenosine 3′,5′-monophosphate-dependent pathway. *Biol. Reprod.* **56,** 707–719.

Gao, J., Cohen, I. S., Mathias, R. T., and Baldo, G. J. (1998). The inhibitory effect of β-stimulation on the Na/K pump current in guinea pig ventricular myocytes is mediated by a cAMP-dependent PKA pathway. *Pflügers Arch.—Eur. J. Physiol.* **435,** 479–484.

Garcia, R., Martinez, R., Rabago, M., Hernandezperez, O., Reyes, A., and Rosado, A. (1991). Subcellular distribution of phospholipase-A_2 and ATPases during capacitation and acrosome reaction in guinea pig spermatozoa. *Arch. Androl.* **26,** 93–105.

Garty, N., and Salomon, Y. (1987). Stimulation of partially purified adenylyl cyclase from bull sperm by bicarbonate. *FEBS Lett.* **218,** 148–152.

Giojalas, L. C., and Rovasio, R. A. (1998). Mouse spermatozoa modify their dynamic parameters and chemotactic response to factors from the oocyte microenvironment. *Int. J. Androl.* **21,** 201–206.

Go, K. J., and Wolf, D. P. (1985). Albumin-mediated changes in sperm sterol content during capacitation. *Biol. Reprod.* **32,** 145–153.

Gordon, M., Dandekar, P. V., and Bartoszewicz, W. (1975). The surface coat of epididymal, ejaculated and capacitated sperm. *J. Ultrastruct. Res.* **50,** 199–207.

Gould, J. E., Overstreet, J. W., and Hanson, F. W. (1985). Interaction of human spermatozoa with the human zona pellucida and zona-free hamster oocyte following capacitation by exposure to human cervical mucus. *Gamete Res.* **12,** 47–54.

Griveau, J. F., Renard, P., and Le Lannou, D. (1994). An *in vitro* promoting role for hydrogen peroxide in human sperm capacitation. *Int. J. Androl.* **17,** 300–307.

Grunert, J.-H., De Geyter, C., and Nieschlag, E. (1990). Objective identification of hyperactivated human spermatozoa by computerized sperm motion analysis with the Hamilton-Thorn sperm motility analyser. *Hum. Reprod.* **5,** 593–599.

Hallett, M., Schneider, A. S., and Carbone, E. (1972). Tetracycline fluorescence as a calcium-probe for nerve membrane with some studies using erythrocyte ghosts. *J. Membr. Biol.* **10,** 31–44.

Harper, M. J. K. (1994). Gamete and zygote transport. *In* "The physiology of reproduction" (E. Knobil and J. D. Neill, eds.), Vol. 1, pp. 123–187, Raven Press, New York.

Harrison, R. A. P. (1996). Capacitation mechanisms, and the role of capacitation as seen in eutherian mammals. *Reprod. Fertil. Dev.* **8,** 581–594.

Harrison, R. A. P. (1997). Sperm plasma membrane characteristics and boar semen fertility. *J. Reprod. Fertil. Suppl.* 52, 195–211.

Harrison, R. A. P., and Gadella, B. M. (1995). Membrane changes during capacitation with special reference to lipid architecture. *In* "Human Sperm Acrosome Reaction" (P. Fenichel, and J. Parinaud, eds.), pp. 45–65. Inst. Natl. Sante Recherche Medicale, Paris.

Harrison, R. A. P., and Miller, N. G. A. (1991). Use of fluorescent intracellular probes to study capacitation. *In* "Comparative Spermatology 20 Years After" (B. Baccetti, ed.), Vol. 75, pp. 169–173. Raven Press, New York.

Harrison, R. A. P., Mairet, B., and Miller, N. G. A. (1993). Flow cytometric studies of bicarbonate-mediated Ca^{2+} influx in boar sperm populations. *Mol. Reprod. Dev.* **35,** 197–208.

Heffner, L. J., and Storey, B. T. (1982). Cold lability of mouse sperm binding to zona pellucida. *J. Exp. Zool.* **219,** 155–161.

Holden, C. A., Hyne, R. V., Sathananthan, A. H., and Trounson, A. O. (1990). Assessment of the human sperm acrosome reaction using concanavalin A lectin. *Mol. Reprod. Dev.* **25,** 247–257.

Hunter, R. H. F. (1993). Sperm:egg ratios and putative molecular signals to modulate gamete interactions in polytocous mammals. *Mol. Reprod. Dev.* **35,** 324–327.

Hunter, R. H. F. (1996). How, when and where do spermatozoa gain their fertilising ability *in vivo? Reprod. Domest. Anim.* **31,** 51–55.

Hunter, R. H. F. (1997). Sperm dynamics in the female genital tract: Interactions with Fallopian tube microenvironments. *In* "Microscopy of Reproduction and Development: A Dynamic Approach" (P. M. Motta, ed.), pp. 35–45. Antonio Delfino Editore Medicina Scienze, Rome.

Hunter, R. H. F., and Hall, J. F. (1974). Capacitation of boar spermatozoa: Synergism between uterine and tubal environments. *J. Exp. Zool.* **188,** 203–210.

Hyne, R. V., and Garbers, D. L. (1979). Regulation of guinea pig sperm adenylyl cyclase by calcium. *Biol. Reprod.* **21,** 1135–1142.

Hyne, R. V., Edwards, K. P., Lopata, A., and Smith, J. D. (1985). Changes in guinea-pig sperm intracellular sodium and potassium content during capacitation and treatment with monovalent ionophores. *Gamete Res.* **12,** 65–73.

Jaiswal, B. S., and Majumder, G. C. (1998). Biochemical parameters regulating forward motility initiation *in vitro* in goat immature epididymal spermatozoa. *Reprod. Fertil. Dev.* **10,** 299–307.

Jaiswal, B. S., Cohen-Dayag, A., Tur-Kaspa, I., and Eisenbach, M. (1998). Sperm capacitation is, after all, a prerequisite for both partial and complete acrosome reaction. *FEBS Lett.* **427,** 309–313.

Jaiswal, B. S., Eisenbach, M., and Tur-Kaspa, I. (1999a). Detection of partial and complete acrosome reaction in human sperm: Which inducers and probes to use? *Mol. Hum. Reprod.* **5,** 214–219.

Jaiswal, B. S., Tur-Kaspa, I., Dor, J., Mashiach, S., and Eisenbach, M. (1999b). Human sperm chemotaxis: Is progesterone a chemoattractant? *Biol. Reprod.* **60,** 1314–1319.

Jones, R., Shalgi, R., Hoyland, J., and Phillips, D. M. (1990). Topographical rearrangement of a plasma membrane antigen during capacitation of rat spermatozoa *in vitro. Dev. Biol.* **139,** 349–362.

Kallajoki, M., Virtanen, J., and Suominen, J. (1986) The fate of acrosomal staining during the acrosome reaction of human spermatozoa as revealed by a monoclonal antibody and PNA-lectins. *Int. J. Androl.* **9,** 184–194.

Kaul, G., Singh, S., Gandhi, K. K., and Anand, S. R. (1997). Calcium requirement and time course of capacitation of goat spermatozoa assessed by chlortetracycline assay. *Andrologia* **29,** 243–251.

Kawakami, E., Hori, T., and Tsutsui, T. (1998). Induction of dog sperm capacitation by oviductal fluid. *J. Vet. Med. Sci.* **60,** 197–202.

Kay, V. J., Coutts, J. R. T., and Robertson, L. (1994). Effects of pentoxifylline and progesterone on human sperm capacitation and acrosome reaction. *Hum. Reprod.* **9,** 2318–2323.

Kervancioglu, M. E., Djahanbakhch, O., and Aitken, R. J. (1994). Epithelial cell coculture and the induction of sperm capacitation. *Fertil. Steril.* **61,** 1103–1108.

Kholkute, S. D., Lian, Y., Roudebush, W. E., and Dukelow, W. R. (1990). Capacitation and the acrosome reaction in squirrel monkey (*Samiri sciureus*) spermatozoa evaluated by the chlortetracycline fluorescence assay. *Am. J. Primatol.* **20,** 115–125.

Kim, N. H., Day, B. N., Lim, J. G., Lee, H. T., and Chung, K. S. (1997). Effects of oviductal fluid and heparin on fertility and characteristics of porcine spermatozoa. *Zygote* **5,** 61–65.

Kimura, Y., and Yanagimachi, R. (1995). Mouse oocytes injected with testicular spermatozoa or round spermatids can develop into normal offspring. *Development* **121,** 2397–2405.

King, R. S., Anderson, S. H., and Killian, G. J. (1994). Effect of bovine oviductal estrus-associated protein on the ability of sperm to capacitate and fertilize oocytes. *J. Androl.* **15,** 468–478.

Kinger, S., and Rajalakshmi, M. (1995). Assessment of the vitality and acrosomal status of human spermatozoa using fluorescent probes. *Int. J. Androl.* **18,** 12–18.

Kinsey, W. H., and Koehler, J. K. (1978). Cell surface changes associated with *in vitro* capacitation of hamster sperm. *J. Ultrastruct. Res.* **64,** 1–13.

Koehler, J. K., and Gaddum-Rosse, P. (1975). Media induced alterations of the membrane associated particle of the guinea pig sperm tail. *J. Ultrastruct. Res.* **51,** 106–118.

Koehler, J. K., and Sato, K. (1978). Changes in lectin labeling pattern of mouse spermatozoa accompanying capacitation and the acrosome reaction. *J. Cell Biol.* **79,** 165a.

Köhn, F. M., Mack, S. R., Hashish, Y. A., Anderson, R. A., and Zaneveld, L. J. D. (1996). Paramagnetic beads coated with pisum sativum agglutinin bind to human spermatozoa undergoing the acrosome reaction. *Andrologia* **28,** 231–239.

Köhn, F. M., Mack, S. R., Schill, W. B., and Zaneveld, L. J. D. (1997). Detection of human sperm acrosome reaction: Comparison between methods using double staining, *Pisum sativum* agglutinin, concanavalin A and transmission electron microscopy. *Hum. Reprod.* **12,** 714–721.

Kopf, G. S., and Gerton, G. L. (1991). Mammalian sperm acrosome and the acrosome reaction. *In* "Elements of Mammalian Fertilization" (P. M. Wassarman, ed.), Vol. 1, pp. 153–203. CRC Press, Boca Raton.

Kopf, G. S., and Vacquier, V. D. (1984). Characterization of a calmodulin-stimulated adenylyl cyclase from abalone spermatozoa. *J. Biol. Chem.* **259,** 7590–7596.

Krausz, C., Gervasi, G., Forti, G., and Baldi, E. (1994). Effect of platelet-activating factor on motility and acrosome reaction of human spermatozoa. *Hum. Reprod.* **9,** 471–476.

Kulka, P., Nissen, H. P., and Kreysel, H. W. (1984). Triglycerides and phospholipids—relation to fertility. *Andrologia* **16,** 48–51.

Lambert, H., Overstreet, J. W., Morales, P., Hanson, F. W., and Yanagimachi, R. (1985). Sperm capacitation in the human female reproductive tract. *Fertil. Steril.* **43,** 325–327.

Langlais, J., Kan, F. W. K., Granger, L., Raymond, L., Bleau, G., and Roberts, K. D. (1988). Identification of sterol acceptors that stimulate cholesterol efflux from human spermatozoa during *in vitro* capacitation. *Gamate Res.* **20,** 185–201.

Langlais, U. K., and Roberts, K. D. (1985). A molecular membrane model of sperm capacitation and acrosome reaction of mammalian spermatozoa. *Gamete Res.* **12,** 183–224.

Lax, Y., Rubinstein, S., and Breitbart, H. (1997). Subcellular distribution of protein kinase c alpha and beta in bovine spermatozoa, and their regulation by calcium and phorbol esters. *Biol. Reprod.* **56,** 454–459.

Leclerc, P., de Lamirande, E., and Gagnon, C. (1996). Cyclic adenosine 3′,5′monophosphate-dependent regulation of protein tyrosine phosphorylation in relation to human sperm capacitation and motility. *Biol. Reprod.* **55,** 684–692.

Leclerc, P., de Lamirande, E., and Gagnon, C. (1997). Regulation of protein-tyrosine phosphorylation and human sperm capacitation by reactive oxygen derivatives. *Free Radic. Biol. Med.* **22,** 643–656.

Leclerc, P., Sirard, M.-A., Chafouleas, J. G., and Lambert, R. D. (1990). Decreased binding of calmodulin to bull sperm proteins during heparin-induced capacitation. *Biol. Reprod.* **42,** 483–489.

Leclerc, P., Sirard, M.-A., Chafouleas, J. G., and Lambert, R. D. (1992). Decrease in calmodulin concentrations during heparin-induced capacitation in bovine spermatozoa. *J. Reprod. Fertil.* **94,** 23–32.

Lee, M. A., Trucco, G. S., Bechtol, K. B., Wummer, N., Kopf, G. S., Blasco, L., and Storey, B. T. (1987). Capacitation and acrosome reactions in human spermatozoa monitored by a chlortetracycline fluorescence assay. *Fertil. Steril.* **48,** 649–658.

Lefebvre, R., and Suarez, S. S. (1996). Effect of capacitation on bull sperm binding to homologous oviductal epithelium. *Biol. Reprod.* **54,** 575–582.

Lewin, L. M., Weissenberg, R., Sobel, J. S., Marcus, Z., and Nebel, L. (1979). Differences in Concanavalin A-FITC binding to rat spermatozoa during epididymal maturation and capacitation. *Arch. Androl.* **2,** 279–281.

Leyton, L., and Saling, P. (1989a). 95 kD sperm proteins bind ZP3 and serve as tyrosine kinase substrates in response to zona binding. *Cell* **57,** 1123–1130.

Leyton, L., and Saling, P. (1989b). Evidence that aggregation of mouse sperm receptors by ZP3 triggers the acrosome reaction. *J. Cell Biol.* **108,** 2163–2168.

Leyton, L., LeGuen, P., Bunch, D., and Saling, P. M. (1992). Regulation of mouse gamete interaction by a sperm tyrosine kinase. *Proc. Natl. Acad. Sci. U.S.A.* **89,** 11692–11695.

Llanos, M. N., and Meizel, S. (1983). Phospholipid methylation increases during capacitation of golden hamster sperm *in vitro. Biol. Reprod.* **28,** 1043–1051.

Luconi, M., Barni, T., Vannelli, G. B., Krausz, C., Marra, F., Benedetti, P. A., Evangelista, V., Francavilla, S., Properzi, G., Forti, G., and Baldi, E. (1998a). Extracellular signal-regulated kinases modulate capacitation of human spermatozoa. *Biol. Reprod.* **58,** 1476–1489.

Luconi, M., Bonaccorsi, L., Krausz, C., Gervasi, G., Forti, G., and Baldi, E. (1995). Stimulation of protein tyrosine phosphorylation by platelet-activating factor and progesterone in human spermatozoa. *Mol. Cell. Endocrinol.* **108,** 35–42.

Luconi, M., Krausz, C., Barni, T., Vannelli, G. B., Forti G., and Baldi, E. (1998b). Progesterone stimulates p42 extracellular signal-regulated kinase (p42^{erk}) in human spermatozoa. *Mol. Hum. Reprod.* **4,** 251–258.

Luconi, M., Krausz, C., Forti, G., and Baldi, E. (1996). Extracellular calcium negatively modulates tyrosine phosphorylation and tyrosine kinase activity during capacitation of human spermatozoa. *Biol. Reprod.* **55,** 207–216.

Lui, C. W., Mrsny, R. J., and Meizel, S. (1979). Procedures for obtaining high percentages of viable *in vitro* capacitated hamster sperm. *Gamete Res.* **2,** 207–211.

Mahadevan, M. M., Miller, M. M., and Moutos, D. M. (1997). Absence of glucose decreases human fertilization and sperm movement characteristics *in vitro. Hum. Reprod.* **12,** 119–123.

Mahanes, M. S., Ochs, D. L., and Eng, L. A. (1986). Cell calcium of ejaculated rabbit spermatozoa before and following *in vitro* capacitation. *Biochem. Biophys. Res. Commun.* **134,** 664–670.

Mahmoud, A. I., and Parrish, J. J. (1996). Oviduct fluid and heparin induce similar surface changes in bovine sperm during capacitation. *Mol. Reprod. Dev.* **43,** 554–560.

Margalioth, E. J., Cooper, G. W., Taney, F. H., Scholl G. M., and Rosenfeld, D. L. (1992). Capacitated sperm cells react with different types of antisperm antibodies than fresh ejaculated sperm. *Fertil. Steril.* **57,** 393–398.

Mbizvo, M. T., Burkman, L. J., and Alexander, N. J. (1990). Human follicular fluid stimulates hyperactivated motility in human sperm. *Fertil. Steril.* **54,** 708–712.

McMaster, R., Yanagimachi, R., and Lopata, A. (1978). Penetration of human eggs by human spermatozoa *in vitro. Biol. Reprod.* **19,** 212–216.

McNutt, T. L., and Killian, G. J. (1991). Influence of bovine follicular and oviduct fluids on sperm capacitation *in vitro. J. Androl.* **12,** 244–252.

McNutt, T. L., Olds-Clarke, P., Way, A. L., Suarez, S. S., and Killian, G. J. (1994). Effect of follicular or oviductal fluids on movement characteristics of bovine sperm during capacitation *in vitro. J. Androl.* **15,** 328–336.

Medeiros, C. M. O., and Parrish, J. J. (1996). Changes in lectin binding to bovine sperm during heparin-induced capacitation. *Mol. Reprod. Dev.* **44,** 525–532.

Meizel, S. (1995). Initiation of human sperm acrosome reaction by progesterone. *In* "Human Sperm Acrosome Reaction" (P. Fénichel and J. Parinaud, eds.), Vol. 235, pp. 151–164. John Libbey Eurotext, Collioure.

Meizel, S., and Turner, K. O. (1996). Chloride efflux during the progesterone-initiated human sperm acrosome reaction is inhibited by lavendustin A, a tyrosine kinase inhibitor. *J. Androl.* **17,** 327–330.

Meizel, S., Pillai, M., Diaz-Perez, E., and Thomas, P. (1990). Initiation of the human sperm acrosome reaction by components of follicular fluid and cumulus secretions including steriods. *In* "Fertilization in Mammals" (B. D. Bavister, J. Cummins, and E. R. S. Roldan, eds.), pp. 205–222. Serono Symposia, Norwell, MA.

Mendoza, C., Carreras, A., Moos, J., and Tesarik, J. (1992). Distinction between true acrosome reaction and degenerative acrosome loss by a one-step staining method using *Pisum sativum* agglutinin. *J. Reprod. Fertil.* **95,** 755–763.

Menon, A. K. (1995). Flippases. *Trends Cell Biol.* **5,** 355–360.

Miller, D. J., and Ax, R. L. (1990). Carbohydrates and fertilization in animals. *Mol. Reprod. Dev.* **26,** 184–192.

Miller, D. J., Gong, X., and Shur, B. D. (1993). Sperm require β-*N*-acetylglucosaminidase to penetrate through the egg zona pellucida. *Development* **118,** 1279–1289.

Moller, C. C., Bleil, D. J., Klinoch, R. S., and Wassarman, P. M. (1990). Structural and functional relationships between mouse and hamster zona pellucida glycoproteins. *Dev. Biol.* **137,** 276–286.

Monks, N. J., and Fraser, L. R. (1987). Phosphodiesterase activity of mouse sperm incubated under conditions that modulate fertilizing potential *in vitro. Gamete Res.* **18,** 85–96.

Moore, H. D. M., Smith, C. A., Hartman, T. D., and Bye, A. P. (1987). Visualization and characterization of the acrosome reaction of human spermatozoa by immunolocalization with monoclonal antibody. *Gamate Res.* **17,** 245–259.

Morales, P., Overstreet, J. W., and Katz, D. F. (1988). Changes in human sperm motion during capacitation *in vitro. J. Reprod. Fertil.* **83,** 119–128.

Mortimer, D., and Camenzind, A. R. (1989). The role of follicular fluid in inducing the acrosome reaction of human spermatozoa incubated *in vitro. Hum. Reprod.* **4,** 169–174.

Mortimer, D., and Fraser, L. (1996). Consensus workshop on advanced diagnostic andrology techniques. *Hum. Reprod.* **11,** 1463–1479.

Mortimer, D., Curtis, E. F., and Miller, R. G. (1987). Specific labelling by peanut agglutinin of the outer acrosomal membrane of the human spermatozoon. *J. Reprod. Fertil.* **81,** 127–135.

Mortimer, S. T., and Mortimer, D. (1990). Kinematics of human spermatozoa incubated under capacitating conditions. *J. Androl.* **11,** 195–203.

Mortimer, S. T., and Swan, M. A. (1995). Kinematics of capacitating human spermatozoa analysed at 60 hz. *Hum. Reprod.* **10,** 873–879.

Mortimer, S. T., Schoevaert, D., Swan, M. A., and Mortimer, D. (1997). Quantitative observations of flagellar motility of capacitating human spermatozoa. *Hum. Reprod.* **12,** 1006–1012.

Mortimer, S. T., Swan, M. A., and Mortimer, D. (1996). Fractal analysis of capacitating human spermatozoa. *Hum. Reprod.* **11,** 1049–1054.

Morton, B., and Albagli, L. (1973). Modification of hamster sperm adenyl cyclase by capacitation *in vitro. Biochem. Biophys. Res. Commun.* **50,** 697–703.

Mrsny, R. J., Siiteri, J. E., and Meizel, S. (1984). Hamster sperm Na^+, K^+-adenosine triphosphatase: Increased activity during capacitation *in vitro* and its relationship to cyclic nucleotides. *Biol. Reprod.* **30,** 573–584.

Mrsny, R. J., Waxman, L., and Meizel, S. (1979). Taurine maintains and stimulates motility of hamster sperm during capacitation *in vitro. J. Exp. Zool.* **210,** 123–128.

Müller, K., Pomorsky, T., Muller, P., A., Z., and Herrmann, A. (1994). Protein-dependent translocation of aminophospholipids and asymmetric transbilayer distribution of phospholipids in the plasma membrane of ram sperm cells. *Biochemistry* **33,** 9968–9974.

Murray, S. C., and Smith, T. T. (1997). Sperm interaction with fallopian tube apical membrane enhances sperm motility and delays capacitation. *Fertil. Steril.* **68,** 351–357.

Myles, D. G., and Primakoff, P. (1984). Localized surface antigens of guinea pig sperm migrate to new regions prior to fertilization. *J. Cell Biol.* **99,** 1634–1641.

Nagae, P., Yanagimachi, R., Srivastava, P. N., and Yanagimachi, H. (1986). Acrosome reaction in human spermatozoa. *Fertil. Steril.* **45,** 701–707.

Naor, Z., and Breitbart, H. (1997). Protein kinase C and mammalian spermatozoa acrosome reaction. *Trends Endocrinol. Metabol.* **8,** 337–342.

Naz, R. K. (1995). Role of membrane tyrosine kinases in human sperm functions. *In* "Human Sperm Acrosome Reaction" (P. Fenichel and J. Parinaud, eds.), Vol. 236, pp. 245–255. Inst. Natl. Sante Recherche Medicale, Paris.

Naz, R. K. (1999). Involvement of protein serine and threonine phosphorylation in human sperm capacitation. *Biol. Reprod.* **60,** 1402–1409.

Naz, R. K., Ahmad, K., and Kumar, R. (1991). Role of membrane phosphotyrosine proteins in human spermatozoal function. *J. Cell. Sci.* **99,** 157–165.

Naz, R. K., Kaplan, P., and Goldstein, A. L. (1992). Thymosin alpha-1 enhances the fertilizing capacity of human sperm cell: Implication in diagnosis and treatment of male infertility. *Biol. Reprod.* **47,** 1064–1072.

Neill, J. M., and Olds-Clarke, P. (1987). A computer-assisted assay for mouse sperm hyperactivaion demonstrates that bicarbonate but not bovine serum albumin is required. *Gamete Res.* **18,** 121–140.

Nolan, J. P., and Hammerstedt, R. H. (1997). Regulation of membrane stability and the acrosome reaction in mammalian sperm. *FASEB J.* **11,** 670–682.

Nolan, J. P., Magargee, S. F., Posner, R. G., and Hammerstedt, R. H. (1995). Flow cytometric analysis of transmembrane phospholipid movement in bull sperm. *Biochemistry* **34,** 3907–3915.

Oehninger, S., Lanzendorf, S., Burkman, L. J., Sueldo, C., Mahony, M., Alexander, N. J., and Hodgen, G. D. (1994). A sequential analysis of the effect of progesterone on specific sperm functions crucial to fertilization *in vitro* in infertile patients. *Hum. Reprod.* **9,** 1322–1327.

Ohashi, K., Saji, F., Kato, M., Okabe, M., Mimura, T., and Tanizawa, O. (1992). Evaluation of acrosomal status using MH61-beads test and its clinical application. *Fertil. Steril.* **58,** 803–808.

Okabe, M., Matzno, S., Nagira, M., Ying, X., Kohama, Y., and Mimura, T. (1992). Collection of acrosome-reacted human sperm using monoclonal antibody-coated paramagnetic beads. *Mol. Reprod. Dev.* **32,** 389–393.

Okamura, N., Tajima, Y., Soejima, A., Masuda, H., and Sugita, Y. (1985). Sodium bicarbonate in seminal plasma stimulates the motility of mammalian spermatozoa through direct activation of adenylyl cyclase. *J. Biol. Chem.* **260,** 9699–9705.

Oliphant, G., and Brackett, B. G. (1973). Immunological assessment of surface changes of rabbit sperm undergoing capacitation. *Biol. Reprod.* **9,** 404–414.

Oliphant, G., Reynolds, A. B., and Thomas, T. S. (1985). Sperm surface components involved in the control of the acrosome reaction. *Am. J. Anat.* **174,** 269–283.

Oliveira, R. G., Tomasi, L., Rovasio, R. A., and Giojalas, L. C. (1999). Increased velocity and induction of chemotactic response in mouse spermatozoa by follicular and oviductal fluids. *J. Reprod. Fertil.* **115,** 23–27.

O'Rand, M. G. (1977). Restriction of a sperm surface antigen's mobility during capacitation. *Dev. Biol.* **55,** 260–270.

Orgebin-Crist, M. C. (1965). Passage of spermatozoa labelled with thymidine-3H through the ductus epididymidis of the rabbit. *J. Reprod. Fertil.* **10,** 241–251.

Osman, R. A., Andria, M. L., Jones, A. D., and Meizel, S. (1989). Steroid induced exocytosis: The human sperm acrosome reaction. *Biochem. Biophys. Res. Commun.* **160,** 828–833.

O'Toole, C. M. B., Roldan, E. R. S., and Fraser, L. R. (1996). Role for Ca^{2+} channels in the signal transduction pathway leading to acrosomal exocytosis in human spermatozoa. *Mol. Reprod. Dev.* **45,** 204–211.

Overstreet, J. W., and Drobnis, E. Z. (1991). Sperm transport in the female tract. *In* "Advances in Donor Insemination" (C. L. R. Barratt and I. D. Cooke eds.), pp. 33–49. Cambridge University Press, Cambridge.

Overstreet, J. W., Gould, J. E., Katz, D. F., and Hanson, F. W. (1980). *In vitro* capacitation of human spermatozoa after passage through a column of cervical mucus. *Fertil. Steril.* **34,** 604–606.

Parinaud, J., and Milhet, P. (1996). Progesterone induces Ca^{++}-dependent 3′,5′-cyclic adenosine monophosphate increase in human sperm. *J. Clin. Endocrinol. Metab.* **81,** 1357–1360.

Parinaud, J., Labal, B., and Vieitez, G. (1992). High progesterone concentrations induce acrosome reaction with a low cytotoxic effect. *Fertil. Steril.* **58,** 599–602.

Parinaud, J., Labal, B., Vieitez, G., Richoilley, G., and Grandjean, H. (1993). Comparison between fluoresence peanut agglutinin lectin and GB24 antibody techniques for the assessment of acrosomal status. *Hum. Reprod.* **8,** 1685–1688.

Parinaud, J., Vieitez, G., Moutaffian, H., Richoilley, G., and Labal, B. (1995). Variations in spontaneous and induced acrosome reaction: correlations with semen parameters and *in-vitro* fertilization results. *Hum. Reprod.* **10,** 2085–2089.

Parks, J. E., and Ehrenwald, E. (1990). Cholesterol efflux from mammalian sperm and its potential role in capacitation. *In* "Fertilization in Mammals" (B. D. Bavister, J. Cummins, and E. R. S. Rolden, eds.), pp. 155–167. Norwell, MA.

Parrish, J. J., Susko-Parrish, J. L., Handrow, R. R., Ax, R. L., and First, N. L. (1989). Effect of sulfated glycoconjugates on capacitation and the acrosome reaction of bovine and hamster spermatozoa. *Gamet. Res.* **24,** 403–413.

Parrish, J. J., Suako-Parrish, J. L., Uguz, C., and First, N. L. (1994). Differences in the role of cAMP during capacitation of bovine sperm by heparin or oviduct fluid. *Biol. Reprod.* **51,** 1099–1108.

Parrish, J. J., Suako-Parrish, J., Winer, M. A., and First, N. L. (1988). Capacitation of bovine sperm by heparin. *Biol. Reprod.* **38,** 1171–1180.

Parry, R. V., Barker, P. J., and Jones, R. (1992). Characterization of low *Mr* zona pellucida binding proteins from boar spermatozoa and seminal plasma. *Mol. Reprod. Dev.* **33,** 108–115.

Perreault, S. D., and Rogers, B. J. (1982). Relationship between fertilizing ability and c-AMP in human spermatozoa. *J. Androl.* **3,** 396–401.

Perry, R. L., Barratt, C. L. R., Warren, M. A., and Cooke, I. D. (1997). Response of human spermatozoa to an internal calcium ATPase inhibitor, 2,5-di(tertbutyl) hydroquinone. *J. Exp. Zool.* **279,** 284–290.

Perry, R. L., Naeeni, M., Barratt, C. L. R., Warren, M. A., and Cooke, I. D. (1995). A time course study of capacitation and the acrosome reaction in human spermatozoa using a revised chlortetracycline pattern classification. *Fertil. Steril.* **64,** 150–159.

Poulos, A., and White, I. G. (1973). The phospholipid composition of human spermatozoa and seminal plasma. *J. Reprod. Fertil.* **35,** 265–272.

Prade, L., Engh, R. A., Girod, A., Kinzel, V., Huber, R., and Bossemeyer, D. (1997). Staurosporine-induced conformational changes of cAMP-dependent protein kinase catalytic subunit explain inhibitory potential. *Structure* **5,** 1627–1637.

Ralt, D., Manor, M., Cohen-Dayag, A., Tur-Kaspa, I., Makler, A., Yuli, I., Dor, J., Blumberg, S., Mashiach, S., and Eisenbach, M. (1994). Chemotaxis and chemokinesis of human spermatozoa to follicular factors. *Biol. Reprod.* **50,** 774–785.

Ravnik, S. E., Albers, J. J., and Muller, C. H. (1993). A novel view of albumin-supported sperm capacitation: Role of lipid transfer protein-I. *Fertil. Steril.* **59,** 629–638.

Ravnik, S. E., Albers, J. J., and Muller, C. H. (1995). Stimulation of human sperm capacitation by purified lipid transfer protein. *J. Exp. Zool.* **272,** 78–83.

Ravnik, S. E., Zarutskie, P. W., and Muller, C. H. (1992). Purification and characterization of a human follicular fluid lipid transfer protein that stimulates human sperm capacitation. *Biol. Reprod.* **47,** 1126–1133.

Revelli, A., Modotti, M., Piffaretti-Yanez, A., Massobrio, M., and Balerna, M. (1994). Steroid receptors in human spermatozoa. *Hum. Reprod.* **9,** 760–766.

Riffo, M., Leiva, S., and Astudillo, J. (1992). Effect of zinc on human sperm motility and the acrosome reaction. *Int. J. Androl.* **15,** 229–237.

Robertson, L., Wolf, D. P., and Tash, J. S. (1988). Temporal changes in motility parameters related to acrosomal status: Identification and characterization of populations of hyperactivated human sperm. *Biol. Reprod.* **39,** 797–805.

Rogers, B. J., and Morton, B. (1973). ATP levels in hamster spermatozoa during capacitation *in vitro*. *Biol. Reprod.* **9,** 361–369.

Rogers, B. J., and Perreault, S. D. (1990). Importance of glycolysable substrates for *in vitro* capacitation of human spermatozoa. *Biol. Reprod.* **43,** 1064–1069.

Rojas, F. J., Bruzzone, M. E., and Moretti-Rojas, I. (1992). Regulation of cyclic adenosine monophosphate synthesis in human ejaculated spermatozoa. II. The role of calcium and bicarbonate ions on the activation of adenylyl cyclase. *Hum. Reprod.* **7,** 1131–1135.

Rojas, F. J., Patrizio, P., Do, J., Silber, S., Asch, R. H., and Moretti-Rojas, I. (1993). Evidence for a novel adenylyl cyclase in human epididymal sperm. *Endocrinology* **133,** 3030–3033.

Roldan, E. R. S., and Fleming, A. D. (1989). Is a Ca^{2+}-ATPase involved in Ca^{2+} regulation during capacitation and the acrosome reaction of guinea-pig spermatozoa? *J. Reprod. Fertil.* **85,** 297–308.

Roldan, E. R. S., Murase, T., and Shi, Q. (1994). Exocytosis in spermatozoa in response to progesterone and zona pellucida. *Science* **226,** 1578–1581.

Roos, A., and Boron, W. F. (1981). Intracellular pH. *Physiol. Rev.* **61,** 296–434.

Rosado, A., Velázquez, A., and Lara-Ricalde, R. (1973). Cell polarography. II. Effect of neuraminidase and follicular fluid upon the surface characteristics of human spermatozoa. *Fertil. Steril.* **24,** 349–354.

Rotem, R., Paz, G. F., Homonnai, Z. T., Kalina, M., Lax, J., Breitbart, H., and Naor, Z. (1992). Ca^{2+}-independent induction of acrosome reaction by protein kinase C in human sperm. *Endocrinology* **131,** 2235–2243.

Rotem, R., Paz, G. F., Homonnai, Z. T., Kalina, M., and Naor, Z. (1990). Protein kinase C is present in human sperm: Possible role in flagellar motility. *Proc. Natl. Acad. Sci. U.S.A.* **87,** 7305–7308.

Rubinstein, S., and Breitbart, H. (1991). Role of spermine in mammalian sperm capacitation and acrosome reaction. *Biochem. J.* **278,** 25–28.

Rufo, G. A., Schoff, P. K., and Lardy, H. A. (1984). Regulation of calcium content in bovine spermatozoa. *J. Biol. Chem.* **259,** 2547–2552.

Rufo, G. A., Singh, J. P., Babcock, D. F., and Lardy, H. A. (1982). Purification and characterization of a calcium transport inhibitor protein from bovine seminal plasma. *J. Biol. Chem.* **257,** 4627–4632.

Ruknudin, A., and Silver, I. A. (1990). Ca^{2+} uptake during capacitation of mouse spermatozoa and the effect of an anion transport inhibitor on Ca^{2+} uptake. *Mol. Reprod. Dev.* **26,** 63–68.

Saling, P. M., and Storey, B. T. (1979). Mouse gamete interactions during fertilization *in vitro*: Chlortetracycline as a fluorescent probe for the mouse sperm acrosome reaction. *J. Cell Biol.* **83,** 544–555.

Saling, P. M., and Bedford, J. M. (1981). Absence of species specificity for mammalian sperm capacitation *in vivo*. *J. Reprod. Fertil.* **63,** 119–123.

Santi, C. M., Santos, T., Hernandez-Cruz, A., and Darszon, A. (1998). Properties of a novel pH-dependent Ca^{2+} permeation pathway present in male germ cells with possible roles in spermatogenesis and mature sperm function. *J. Gen. Physiol.* **112,** 33–53.

Schwarz, M. A., and Koehler, J. K. (1979). Alterations in lectin binding to guinea pig spermatozoa accompanying *in vitro* capacitation and the acrosome reaction. *Biol. Reprod.* **21,** 1295–1307.

Sebastian, S. M., Selvarage, S., Aruldhas, M. M., and Govindarajulu, P. (1987). Pattern of natural lipids and phospholipids in semen of normospermic, oligospermic and azoospermic men. *J. Reprod. Fertil.* **79,** 373–378.

Shalgi, R., Matityahu, A., Gaunt, S. J., and Jones, R. (1990). Antigens on rat spermatozoa with a potential role in fertilization. *Mol. Reprod. Dev.* **25,** 286–296.

Shams-Borhan, G., and Harrison, R. A. P. (1981). Production, characterization, and use of ionophore-induced, calcium-dependent acrosome reaction in ram spermatozoa. *Gamete Res.* **4,** 407–432.

Shi, Q. X., and Roldan, E. R. S. (1995). Bicarbonate/CO_2 is not required for zona pellucida- or progesterone-induced acrosomal exocytosis of mouse spermatozoa but is essential for capacitation. *Biol. Reprod.* **52,** 540–546.

Shinitzky, M. (1984). Membrane fluidity and cellular functions. *In* "Physiology of Membrane Fluidity" (M. Shinitzky, ed.), Vol. 1, pp. 1–51. CRC Press, Boca Raton.

Shur, B. D., and Hall, N. G. (1982). Sperm surface galactosyltransferase activities during *in vitro* capacitation. *J. Cell Biol.* **95,** 567–573.

Singh, J. P., Babcock, D. F., and Lardy, H. A. (1978). Increased calcium-ion influx is a component of capacitation of spermatozoa. *Biochem. J.* **172,** 549–556.

Smith, T. T. (1998). The modulation of sperm function by the oviductal epithelium. *Biol Reprod.* **58,** 1102–1104.

Smith, T. T., and Yanagimachi, R. (1989). Capacitation status of hamster spermatozoa in the oviduct at various times after mating. *J. Reprod. Fertil.* **86,** 255–261.

Smith, T. T., and Yanagimachi, R. (1991). Attachment and release of spermatozoa from the caudal isthmus of the hamster oviduct. *J. Reprod. Fertil.* **91,** 567–573.

Smith, T. T., Mckinnon-Thompson, C. A., and Wolf, D. E. (1998). Changes in lipid diffusibility in the hamster sperm head plasma membrane during capacitation *in vivo* and *in vitro*. *Mol. Reprod. Dev.* **50,** 86–92.

Snider, D. R., and Clegg, E. D. (1975). Alteration of phospholipids in porcine spermatozoa during *in vivo* uterus and oviduct incubation. *J. Anim. Sci.* **40,** 269–274.

Spungin, B., Levinshal, T., Rubinstein, S., and Breitbart, H. (1992). A cell free system reveals that capacitation is a prerequisite for membrane fusion during the acrosome reaction. *FEBS Lett.* **311,** 155–160.

Spungin, B., Margalit, I., and Breitbart, H. (1995). A 70 kDa protein is transferred from the outer acrosomal to the plasma membrane during capacitation. *FEBS Lett.* **357,** 98–102.

Stauss, C. R., Votta, T. J., and Suarez, S. S. (1995). Sperm motility hyperactivation facilitates penetration of the hamster zona pellucida. *Biol. Reprod.* **53,** 1280–1285.

Stein, D. M., and Fraser, L. R. (1984). Cyclic nucleotide metabolism in mouse epididymal spermatozoa during capacitation *in vitro. Gamete Res.* **10,** 283–299.

Stewart-Savage, J. (1993). Effect of bovine serum albumin concentration and source on sperm capacitation in the golden hamster. *Biol. Reprod.* **49,** 74–81.

Stock, C. E., and Fraser, L. R. (1987). The acrosome reaction in human sperm from men of proven fertility. *Hum. Reprod.* **2,** 109–119.

Suarez, S. S. (1996). Hyperactivated motility in sperm. *J. Androl.* **17,** 331–335.

Suarez, S. S. (1998). The oviductal sperm reservoir in mammals: Mechanisms of formation. *Biol. Reprod.* **58,** 1105–1107.

Suarez, S. S., and Dai, X. B. (1995). Intracellular calcium reaches different levels of elevation in hyperactivated and acrosome-reacted hamster sperm. *Mol. Reprod. Dev.* **42,** 325–333.

Suarez, S. S., Brockman, K., and Lefebvre, R. (1997). Distribution of mucus and sperm in bovine oviducts after artificial insemination: The physical environment of the oviductal sperm reservoir. *Biol. Reprod.* **56,** 447–453.

Suarez, S. S., Katz, D. F., and Meizel, S. (1984). Changes in motility that accompany the acrosome reaction in hyperactivated hamster spermatozoa. *Gamete Res.* **10,** 253–265.

Suarez, S. S., Katz, D. F., Owen, D. H., Andrew, J. B., and Powell, R. L. (1991). Evidence for the function of hyperactivated motility in sperm. *Biol. Reprod.* **44,** 375–381.

Suarez, S. S., Varosi, S. M., and Dai, X. B. (1993). Intracellular calcium increases with hyperactivation in intact, moving hamster sperm and oscillates with the flagellar beat cycle. *Proc. Natl. Acad. Sci. U.S.A.* **90,** 4660–4664.

Suarez, S. S., Wolf, D. P., and Meizel, S. (1986). Induction of the acrosome reaction in human spermatozoa by a fraction of human follicular fluid. *Gamete Res.* **14,** 107–121.

Sugkraroek, P., Kates, M., Leader, A., and Tanphaichitr, N. (1991). Levels of cholesterol and phospholipids in freshly ejaculated sperm and percoll-gradient-pelleted sperm from fertile and eunexplained infertile men. *Fertil. Steril.* **55,** 820–827.

Suzuki, F., and Yanagimachi, R. (1989). Changes in the distribution of intramembranous particles and filipin-reactive membrane sterols during *in vitro* capacitation of golden hamster spermatozoa. *Gamete Res.* **23,** 335–347.

Talbot, P. (1985). Sperm penetration through oocyte investments in mammals. *Am. J. Anat.* **174,** 331–346.

Talbot, P., and Chacon, R. S. (1981). A triple-stain technique for evaluating normal acrosome reactions of human sperm. *J. Exp. Zool.* **215,** 201–208.

Talbot, P., and Franklin, L. E. (1978). Surface modification of guinea pig sperm during *in vitro* capacitation: An assessment using lectin-induced agglutination of living sperm. *J. Exp. Zool.* **203,** 1–14.

Tanphaichitr, N., Zheng, Y. S., Kates, M., Abdullah, N., and Chan, A. (1996). Cholesterol and phospholipid levels of washed and percoll gradient centrifuged mouse sperm: Presence of lipids possessing inhibitory effects on sperm motility. *Mol. Reprod. Dev.* **43,** 187–195.

Tao, J., Critser, E. S., and Critser, J. K. (1993a). Evaluation of mouse sperm acrosomal status and viability by flow cytometry. *Mol. Reprod. Dev.* **36,** 183–194.

Tao, J., Du, J., Crister, E. S., and Crister, J. K. (1993b). Assessment of the acrosomal status and viability of human spermatozoa simultaneously using flow cytometry. *Hum. Reprod.* **8,** 1879–1885.

Tesarik, J. (1984). Topographical relations of intramembrane particle distribution pattern in human sperm memebranes. *J. Ultrastruct. Res.* **89,** 42–55.

Tesarik, J., Mendoza, C., and Carreras, A. (1993a). Fast acrosome reaction measure—a highly sensitive method for evaluating stimulus-induced acrosome reaction. *Fertil. Steril.* **59,** 424–430.

Tesarik, J., Mendoza, C., Moos, J., and Carreras, A. (1992). Selective expression of a progesterone receptor on the human sperm surface. *Fertil. Steril.* **58,** 784–792.

Tesarik, J., Moos, J., and Mendoza, C. (1993b). Stimulation of protein tyrosine phosphorylation by a progesterone receptor on the cell surface of human sperm. *Endocrinology* **133,** 328–335.

Thaler, C. D., and Cardullo, R. A. (1994). Kinetics of mouse sperm binding to solubilized zonae pellucidae. *Mol. Biol. Cell* **5** (Suppl.), 224 (abstr.).

Thérien, I., Soubeyrand, S., and Manjunath, P. (1997). Major proteins of bovine seminal plasma modulate sperm capacitation by high-density lipoprotein. *Biol. Reprod.* **57,** 1080–1088.

Toledo, L. M., and Lydon, N. B. (1997). Structure of staurosporine bound to CDK2 and cAPK—new tools for structure-based design of protein kinase inhibitors. *Structure* **5,** 1551–1556.

Topfer-Petersen, E., Friess, A. E., Stoffel, M., and Schill, W.-B. (1990). Boar sperm membranes antigens. II. Reorganization of an integral membrane antigen during capacitation and acrosome reaction. *Histochemistry* **93,** 491–495.

Triana, L. R., Babcock, D. F., Lorton, S. P., First, N. L., and Lardy, H. A (1980). Release of acrosomal hyaluronidase follows increased membrane permeability to calcium in the presumptive capacitation sequence for spermatozoa of the bovine and other memmalian species. *Biol. Reprod.* **23,** 47–59.

Tulsiani, D. R. P., Yoshidakomiya, H., and Araki, Y. (1997). Mammalian fertilization: a carbohydrate-mediated event. *Biol. Reprod.* **57,** 487–494.

Uehara, T., and Yanagimachi, R. (1977). Behavior of nuclei of testicular, caput and cauda epididymal spermatozoa injected into hamster eggs. *Biol. Reprod.* **16,** 315–321.

Uguz, C., Vredenburgh, W. L., and Parrish, J. J. (1994). Heparine-induced capacitation but not intracellular alkalinization of bovine sperm is inhibited by Rp-adenosine-3′,5′-cyclic monophosphorothioate. *Biol. Reprod.* **51,** 1031–1039.

Uhler, M. L., Leung, A., Chan, S. Y. W., and Wang, C. (1992). Direct effects of progesterone and antiprogesterone on human sperm hyperactivated motility and acrosome reaction. *Fertil. Steril.* **58,** 1191–1198.

Uto, N., Yoshimatsu, N., Lopata, A., and Yanagimachi, R. (1988). Zona-induced acrosome reaction of hamster spermatozoa. *J. Exp. Zool.* **248,** 113–120.

Vaidya, R. A., Glass, R. H., Daudekar, P., and Johnson, K. (1971). Decrease in the electrophoretic mobility of rabbit spermatozoa following intrauterine incubation. *J. Reprod. Fertil.* **24,** 299–301.

Vernon, R. B., Muller, C. H., Herr, J. C., Feuchter, F. A., and Eddy, E. M. (1982). Epididymal secretion of a mouse surface component recognized by a monoclonal antibody. *Biol. Reprod.* **26,** 523–535.

Villarroya, S., and Scholler, R. (1987). Lateral diffusion of a human sperm-head antigen during incubation in a capacitation medium and induction of the acrosome reaction *in vitro*. *J. Reprod. Fert.* **80,** 545–562.

Visconti, P. E., and Kopf, G. S. (1998). Regulation of protein phosphorylation during sperm capacitation. *Biol. Reprod.* **59,** 1–6.

Visconti, P. E., and Tezon, J. G. (1989). Phorbol esters stimulate cyclic adenosine 3′,5′-monophosphate accumulation in hamster spermatozoa during *in vitro* capacitation. *Biol. Reprod.* **40,** 223–231.

Visconti, P. E., Bailey, J. L., Moore, G. D., Pan, D., Olds-Clarke, P., and Kopf, G. S. (1995a). Capacitation of mouse spermatozoa. I. Correlation between the capacitation state and protein tyrosine phosphorylation. *Development* **121,** 1129–1137.

Visconti, P. E., Galantino-Homer, H., Moore, G. D., Bailey, J. L., Ning, X. P., Fornes, M., and Kopf, G. S. (1998). The molecular basis of sperm capacitation. *J. Androl.* **19,** 242–248.

Visconti, P. E., Galantino-Homer, H., Ning, X. P., Moore, G. D., Valenzuela, J. P., Jorgez, C. J., Alvarez, J. G., and Kopf, G. S. (1999). Cholesterol efflux-mediated signal transduction in mammalian sperm. *J. Biol. Chem.* **274,** 3235–3242.

Visconti, P. E., Johnson, L. R., Oyaski, M., Fornes, M., Moss, S. B., Gerton, G. L., and Kopf, G. S. (1997). Regulation, localization, and anchoring of protein kinase a subunits during mouse sperm capacitation. *Dev. Biol.* **192,** 351–363.

Visconti, P. E., Moore, G. D., Bailey, J. L., Leclerc, P., Connors, S. A., Pan, D., Olds-Clarke, P., and

Kopf, G. S. (1995b). Capacitation of mouse spermatozoa. II. Protein tyrosine phosphorylation and capacitation are regulated by a cAMP-dependent pathway. *Development* **121,** 1139–1150.

Visconti, P. E., Muschietti, J. P., Flawia, M. M., and Tezon, J. G. (1990). Bicarbonate dependence of cAMP accumulation induced by phorbol esters in hamster spermatozoa. *Biochim. Biophys. Acta* **1054,** 231–236.

Vredenburgh-Wilberg, W. L., and Parrish, J. J. (1995). Intracellular pH of bovine sperm increases during capacitation. *Mol. Reprod. Dev.* **40,** 490–502.

Ward, C. R., and Storey, B. T. (1984). Determination of the time course of capacitation in mouse spermatozoa using a chlortetracycline fluorescence assay. *Dev. Biol.* **104,** 287–296.

Wassarman, P. M. (1987). The biology and chemistry of fertilization. *Science* **235,** 553–560.

Wassarman, P. M., and Litscher, E. S. (1995). Sperm-egg recognition mechanisms in mammals. *Curr. Top. Dev. Biol.* **30,** 1–19.

Watanabe, N., Hunt, T., Ikawa, Y., and Sagata, N. (1991). Independent activation of MPF and cytosolic factor (Mos) upon fertilization of *Xenopus* eggs. *Nature* **352,** 247–248.

Wendy, T. S., Wilson, L., Reynold, A. B., and Oliphant, G. (1986). Chemical and physical characterization of rabbit sperm acrosome stabilizing factor. *Biol. Reprod.* **35,** 691–703.

White, D. R., and Aitken, R. J. (1989). Relationship between calcium, cyclic AMP, ATP and intracellular pH and the capacity of hamster spermatozoa to express hyperactivated motility. *Gamete Res.* **22,** 163–177.

Williams, M., Hill, C. J., Scudamore, I., Dunphy, B., Cooke, I. D., and Barratt, C. L. R. (1993). Sperm numbers and distribution within the human Fallopian tube around ovulation. *Hum. Reprod.* **8,** 2019–2026.

Wolf, D. E., Hagopian, S. S., and Ishijima, S. (1986). Changes in sperm plasma membrane lipid diffusibility after hyperactivation during *in vitro* capacitation in the mouse. *J. Cell Biol.* **102,** 1372–1377.

Wolf, D. P., Boldt, J., Byrd, W., and Bechtol, K. B. (1985). Acrosomal status evaluation in human ejaculated sperm with monoclonal antibodies. *Biol. Reprod.* **32,** 1157–1162.

Yanagimachi, R. (1969). *In vitro* capacitation of hamster spermatozoa by follicular fluid. *J. Reprod. Fertil.* **18,** 275–286.

Yanagimachi, R. (1970). The movement of golden hamster spermatozoa before and after capacitation. *J. Reprod. Fertil.* **23,** 193–196.

Yanagimachi, R. (1982). Requirement of extracellular calcium ions for various stages of fertilization and fertilization-related phenomena in the hamster. *Gamete Res.* **5,** 323–344.

Yanagimachi, R. (1988). Mammalian fertilization. *In* "The Physiology of Reproduction" (E. Knobil and J. Neill, eds.), pp. 135–185. Raven Press, New York.

Yanagimachi, R. (1989). Sperm capacitation and gamete interaction. *J. Reprod. Fertil.* (Suppl.) **38,** 27–33.

Yanagimachi, R. (1994). Mammalian fertilization. *In* "The Physiology of Reproduction" (E. Knobil and J. Neill, eds.), pp. 189–317. Raven Press, New York.

Yanagimachi, R. (1997). The mammalian oocyte's response to intracytoplasmic sperm injection (ICSI). *In* "Microscopy of Reproduction and Development: A Dynamic Approach" (P. M. Motta, ed.), pp. 187–195. Antonio Delfino Editore Medicina Scienze, Rome.

Yanagimachi, R. (1998). Intracytoplasmic sperm injection experiments using the mouse as a model. *Hum. Reprod.* **13,** 87–98.

Yanagimachi, R., and Suzuki, F. (1985). A further study of the lysolecithin-mediated acrosome reaction of guinea pig spermatozoa. *Gamete Res.* **11,** 29–40.

Yanagimachi, R., and Usui, N. (1974). Calcium dependence of the acrosome—reaction and activation of guinea-pig spermatozoa. *Exp. Cell Res.* **89,** 161–174.

Yanagimachi, R., Yanagimachi, H., and Rogers, B. J. (1976). The use of zona-free animal ova as a test-system for the assessment of the fertilizing capacity of human spermatozoa. *Biol. Reprod.* **15,** 471–476.

Youakim, A., Hathaway, H. J., Miller, D. J., Gong, X. H., and Shur, B. D. (1994). Overexpressing sperm surface beta 1,4-galactosyltransferase in transgenic mice affects multiple aspects of sperm-egg interactions. *J. Cell Biol.* **126,** 1573–1583.

Young, R. J., and Bodt, B. A. (1994). Development of computer-directed methods for the identification of hyperactivated motion using motion patterns developed by rabbit sperm during incubation under capacitation conditions. *J. Androl.* **15,** 362–377.

Zarca, A., Rubinstein, S., and Breitbart, H. (1988). Transport mechanism for calcium and phosphate in ram spermatozoa. *Biochim. Biophys. Acta* **944,** 351–358.

Zarintash, R. J., and Cross, N. L. (1996). Unesterified cholesterol content of human sperm regulates the response of the acrosome to the agonist, progesterone. *Biol. Reprod.* **55,** 19–24.

Zeng, Y., Clark, E. N., and Florman, H. M. (1995). Sperm membrane potential: Hyperpolarization during capacitation regulates zona pellucida-dependent acrosomal secretion. *Dev. Biol.* **171,** 554–563.

Zeng, Y., Oberdorf, J. A., and Florman, H. M. (1996). pH regulation in mouse sperm: Identification of Na^+-, Cl^--, and HCO_3^--dependent and arylaminobenzoate-dependent regulatory mechanisms and characterization of their roles in sperm capacitation. *Dev. Biol.* **173,** 510–520.

Zhou, R., Shi, B., Chou, K. C. K., Oswalt, M. D., and Haug, A. (1990). Changes in intracellular calcium of porcine sperm during *in vitro* incubation with seminal plasma and a capacitating medium. *Biochem. Biophys. Res. Commun.* **172,** 47–53.

Zinaman, M., Drobnis, E. Z., Morales, P., Brazil, C., Kiel, M., Cross, N. L., Hanson, F. W., and Overstreet, J. W. (1989). The physiology of sperm recovered from human cervix: Acrosomal status and response to inducers of the acrosome reaction. *Biol. Reprod.* **41,** 791–797.

4

FUNCTION OF THE EGG'S EXTRACELLULAR MATRIX

RICHARD A. CARDULLO* AND
CATHERINE D. THALER†

*Department of Biology, The University of California, Riverside; and
†Department of Biology, University of Central Florida, Orlando

I. INTRODUCTION

Before membrane fusion between mammalian gametes can occur, the sperm must first negotiate two major extracellular layers that surround the egg. These layers are the cumulus layer, which is made up of cumulus cells, hyaluronic acid (HA), and a number of minor components, and the zona pellucida, which is a matrix made up of three to five glycoproteins (depending on the species) that are secreted by the developing oocyte. These two major layers may serve a number of distinct roles necessary for fertilization and early development, including the transport of oocytes from the ampulla to the oviduct and modulation of sperm motility, as well as forming a selective barrier for sperm, a recognition substrate for sperm adhesion and initiation of signal transduction pathways leading to acrosomal exocytosis, and a protective enclosure for the developing embryo prior to implantation in the uterus.

In this chapter we focus on these two egg-associated structures specifically in regard to their involvement in the process of fertilization. Because macromolecular structures ultimately determine function, we first briefly review what is known about the chemical composition and stoichiometry of relevant macromolecular structures before addressing their roles in fertilization.

II. THE CUMULUS OOPHORUS

A. STRUCTURE OF THE CUMULUS OOPHORUS

The cumulus oophorus is an extensive complex of cells and extracellular matrix (ECM) surrounding the mammalian oocyte. The cumulus cells produce the extensive ECM associated with the cumulus oophorus. The matrix is secreted by the cumulus cells at the time that meiosis is resumed during the final maturation steps prior to ovulation, in a process referred to as cumulus expansion (Ball *et al.,* 1985; Eppig, 1982, 1991a,b). The major ECM component of the cumulus complex is HA (Ball *et al.,* 1982; Talbot, 1984; Talbot and DiCarlontonio, 1984), a large polymer of alternating *N*-acetylglucosamine and glucuronic acid residues. The average molecular mass of the polymerized HA is approximately 2500 kDa (Piko, 1979). The HA of the cumulus ECM also penetrates into the large pores of the outer zona pellucida and in some cases may permeate pores traversing the entire zona pellucida into the perivitelline space (Talbot, 1984). The perivitelline space also contains HA (Dandekar and Talbot, 1992), but it is not known whether this is produced by the cumulus cells or the oocyte.

Several protein ECM components are present in the cumulus. Fibronectin, laminin, and tenascin-C have been detected in unique distributions by immunofluorescent staining (Familiari *et al.,* 1996). Laminin is uniformly distributed throughout the matrix whereas fibronectin is only present surrounding the corona radiata cells. Tenascin-C surrounds randomly scattered cells in the cumulus. Currently, the function of these components is not known. A number of additional protein components have been detected by transmission electron microscopy (TEM) (Talbot, 1984), electrophoretic analysis (Ball *et al.,* 1985; Virji *et al.,* 1990), or immunoblotting (Amiel *et al.,* 1993; Chen *et al.,* 1996). Several components appear to be incorporated during cumulus expansion (Ball *et al.,* 1985; Amiel *et al.,* 1993). For the most part, the function of these proteins is unknown, but some have been suggested to stabilize the cumulus ECM (Chen *et al.,* 1994, 1996; Camaioni *et al.,* 1996; Fulop *et al.,* 1997; Hirashima *et al.,* 1997). The serum factor inter-alpha-trypsin-inhibitor (IαI) protein has been shown to bind HA (Chen *et al.,* 1994). Disruption of this interaction by covalent modification of the inhibitor protein abolished the ability of the protein to stabilize the cumulus matrix (Chen *et al.,* 1994). The heavy chain of IαI can bind to HA and to the HA binding protein (HABP) (Hirashima *et al.,* 1997). The heavy chain of IαI is covalently cross-linked to HA in oocyte cumulus complexes (OCCs) matured *in vivo,* although intact IαI was noncovalently bound to the cumulus ECM on OCCs matured *in vitro* (Chen *et al.,* 1996). Incubation of OCCs with HA oligosaccharides *in vitro,* or treatment of animals by injection of HA oligosaccharides, resulted in decreased cumulus expansion and may suggest that serum IαI is the relevant matrix-stabilizing factor *in vivo* (Hess *et al.,* 1999). An additional stabilization mechanism has been proposed through ternary complex formation between HA, a dermatan sulfate proteoglycan, and the 46-kDa link protein, which has been shown to interact with HA and proteoglycans in other ECMs and has been isolated from mouse cumulus complexes (Camaioni *et al.,* 1996). The tumor necrosis factor (TNF)-stimulated gene-6 (TSG-6), a 35-kDa glycoprotein that binds HA and IαI, has also been implicated in matrix stabilization during cumulus expansion. The mRNA of TSG-6 is strongly up-regulated in OCCs stimulated to expand either *in vitro* or *in vivo* (Fulop *et al.,* 1997).

Although the ECM is completed by the time of ovulation, reports indicate that cumulus cells continue secreting progesterone and prostaglandins following ovulation (Schuetz and Dubin, 1981; Gurevich *et al.,* 1993; Viggiano *et al.,* 1995). In most eutherian mammals, the cumulus remains with the ovulated egg until fertilization is completed. However, marsupials and monotremes shed their cumulus layers during or shortly after ovulation (Rodger and Bedford, 1982), so that the only barrier to fertilization in these animals is the zona pellucida.

B. FUNCTIONS OF THE CUMULUS OOPHORUS

The cumulus oophorus plays many essential roles in maturation, ovulation, and fertilization of the mammalian oocyte. We review the functions of the cumulus as related to the overall success of fertilization and look specifically at three interrelated functions of the cumulus matrix: (1) production of soluble factors, (2) transport into the oviduct for fertilization, and (3) creation of a selective barrier or filter for sperm.

The cumulus cells provide a variety of soluble factors and hormones that affect the behavior of both oocyte and sperm at the time of fertilization. Cumulus factors act both before ovulation, to induce resumption of meiosis, and after ovulation, to stimulate sperm motility and possibly to potentiate acrosomal exocytosis. Soluble cumulus-derived factors also appear to influence the success of fertilization and early embryonic development. The presence of the mature cumulus complex is essential for ovum pickup and transport by the oviduct, and is therefore the egg component responsible for proper delivery of the oocyte to the site of fertilization. Finally, the cumulus affects not only sperm motility but also the ability of sperm to penetrate the cumulus matrix, as well as affecting the dynamics of acrosomal exocytosis. This complex matrix may also act as a selective barrier for sperm as they attempt to reach the egg plasma membrane.

C. EFFECTS OF SOLUBLE FACTORS PRODUCED BY THE CUMULUS CELLS

Mammalian oocytes are arrested at prophase I of meiosis until shortly before ovulation, when they resume meiotic activity and arrest at metaphase II until fertilization. This resumption of meiotic activity coincides with the cumulus expansion (Ball *et al.,* 1985; Eppig, 1982, 1991a,b). Interactions between cumulus cells and the oocyte appear to mediate these processes and both cell types may, in fact, produce soluble factors that affect the behavior of the other.

Numerous studies have investigated the effects of various hormones or growth factors, including extra- and intracellular growth factors (EGF, IGF-1, IGF-2), oxytocin, growth hormone (GH), and prostaglandin E_2 (PGE_2), on cumulus expansion and oocyte maturation. Both cumulus cells and oocytes have been documented to possess mRNAs for a number of these factors, but it is not clear which factors may be relevant *in vivo,* or if *in vivo* maturation depends solely on the cumulus expansion enabling factor (Eppig *et al.,* 1993).

Depending on the animal model, the growth factors can stimulate oocyte maturation, cumulus expansion, both, or have no effect at all. In the pig, both EGF (Singh *et al.,* 1995) and IGF-1 (Singh and Armstrong, 1997) are present, but only IGF-1 has been shown to induce cumulus expansion (Singh and Armstrong, 1999), whereas EGF appears to influence oocyte maturation (Abeydeera *et al.,* 1998). In the bovine system, EGF (Izadyar *et al.,* 1999a), GH (Izadyar *et al.,* 1997a,b), and oxytocin (Okuda *et al.,* 1999) have been shown to induce cumulus expansion *in vitro.* In mice, PGE_2 appears to be required for cumulus expansion, but the role of other factors was not investigated (Davis *et al.,* 1999). Both mouse and pig produce a high-molecular-weight protein factor, termed cumulus expansion enabling factor (CEEF), that is able to stimulate cumulus expansion (Eppig *et al.,* 1993; Prochazka *et al.,* 1998). Interestingly, it appears to be produced by cumulus cells, mural granulosa cells, and the oocyte during different stages of follicular growth in the pig (Prochazka *et al.,* 1998), but only by the oocyte in the mouse (Eppig *et al.,* 1993). In the mouse, the effects of CEEF can be mimicked by transforming growth factor β (TGF-β) (Tirone *et al.,* 1999).

Several studies have indicated that resumption of meiosis by the oocyte is induced by cumulus-derived factors. Cumulus-free mouse oocytes matured *in vitro* had a significantly lower fraction of oocytes reaching metaphase II than did cumulus-intact oocytes (Cross and Brinster, 1970; Cecconi *et al.,* 1996). Coculture with cumulus cells did not improve maturation rates of cumulus-free oocytes in one of the studies (Cecconi *et al.,* 1996). In contrast, cumulus cells cultured in the presence of follicle-stimulating hormone (FSH) produced a heat-stable, soluble factor that induced resumption of meiosis in cumulus-free oocytes that had not been exposed to FSH, cultured in the presence of the cumulus-conditioned medium (Byskov *et al.,* 1997). The factor was suggested to be similar to the meiosis-activating sterols isolated from testis or follicular fluid, but was not specifically identified in this study. In other work, culture of cumulus-free bovine oocytes with cumulus cells improved the maturation rate significantly. However, this improved rate was still significantly less than the maturation rate of cumulus-intact oocytes, suggesting that the effect is more complex than can be explained by a soluble factor alone (Zhang *et al.,* 1995). In bovine oocytes, meiosis can be induced by growth hormone, and this effect was shown to be due to the action of GH on the cumulus cells and not on the oocyte directly, again implicating a cumulus-derived factor as responsible for resumption of meiosis (Izadyar *et al.,* 1997a). Further studies suggested that GH-induced oocyte maturation follows a cAMP-dependent pathway, but EGF-stimulated oocyte maturation required tyrosine kinase activity (Izadyar *et al.,* 1997a).

Conflicting reports have appeared concerning the action of cAMP and whether it stimulates or inhibits meiosis. However, one study suggests that transient exposure of mouse oocyte cumulus complexes to forskolin or dibutyryl-cAMP enabled cumulus cells to stimulate meiosis in cumulus-free oocytes not previously exposed to cAMP (Guoliang *et al.,* 1994). Using isozyme-specific cAMP analogs, Downs and Hunzicker-Dunn (1995) showed that cumulus cells contained both RI and RII isozymes and that transient stimulation of the RII protein kinase A (PKA) increased the fraction of eggs resuming meiosis, whereas oocytes contained only the RI isotype, and stimulation of this enzyme blocked meiosis. These results may clarify the earlier conflicting data. Thus, the presence and proper functioning of the cumulus matrix in stimulating the final maturation of the oocyte are essential to obtain, either *in vivo* or *in vitro,* fertilization-competent oocytes.

The cumulus cells of the mature, ovulated oocyte produce additional factors that influence both sperm motility and fertilization rates. Additional studies have demonstrated a positive effect of the cumulus on early embryonic development (Gurevich *et al.,* 1993; Zhang *et al.,* 1995), but these observations are beyond the scope of the present discussion and will not be detailed further.

The cumulus matrix has also been implicated in changes in sperm motility. One study with human sperm has shown that the presence of HA in the culture medium increases both sperm velocity and the length of time sperm remained motile in culture (Sbracia *et al.,* 1997), but most studies have suggested that soluble factors produced by the cumulus cells are primarily responsible for effecting changes in sperm motility parameters. Early studies suggested a trend toward increased forward motility by sperm in the presence of oocyte cumulus complexes or cumulus cells, whereas cumulus-free oocytes had no effects on sperm motility (Bradley and Garbers, 1983). In coculture with mature cumulus cells, human sperm remained motile for greater lengths of time, and a higher percentage of motile sperm displayed greater linear progressive motility, characteristic of cumulus-related motility, than in the absence of cumulus cells (Mansour *et al.,* 1995). Sperm from rodents, including hamster, display abrupt hatchetlike head motions when penetrating the cumulus (Drobnis *et al.,* 1988; Katz *et al.,* 1989). In addition, the hyperactivated, high-amplitude, beat patterns characteristic of the sperm flagellum following capacitation are reduced to low amplitude beats once the sperm enters the cumulus. When sperm were incubated in cumulus cell-conditioned medium, two motility parameters were significantly increased, curvilinear velocity and lateral head displacement (Fetterolf *et al.,* 1994), indicating that the cumulus cells produce soluble factors that modulate sperm motility. Additional work has quantified the flagellar force generated by sperm exhibiting various types of motility. Cumulus-related motility induced by exposing capacitated sperm to a soluble cumulus matrix extract generated significantly greater force than either hyperactivated or noncapacitated motility in human sperm (Westphal *et al.,* 1993). Because the soluble extract in this study was a supernatant collected following hyaluronidase treatment of cumulus masses, it could contain both soluble and matrix-derived factors. These studies suggest that soluble factors produced by the cumulus cells, as well as components of the cumulus ECM, may modify sperm motility to enhance sperm penetration of the cumulus matrix.

Although acrosomal exocytosis is mediated by components of the zona pellucida, there is some indication that the cumulus matrix may exert a potentiating effect on acrosomal exocytosis. The fraction of human sperm undergoing acrosomal exocytosis increased in the presence of the cumulus matrix (Carrell *et al.,* 1993). Human sperm have also been noted to undergo acrosomal exocytosis at high rates when exposed to cumulus masses or follicular fluid (Tesarik, 1985), but it is not clear how this might relate to zona pellucida-induced acrosome reactions. A similar effect has been quantified in monkey sperm—preincubation of either sperm or cumulus-free oocytes with HA resulted in a significantly greater fraction of acrosome-reacted sperm bound to the zona pellucida, but the total number of bound sperm was unchanged (Vandevoort *et al.,* 1997). Sabeur *et al.* (1999) have shown that human sperm exposed to HA showed significantly higher acrosome reactions in response to zona glycoproteins or progesterone, compared to untreated sperm. This effect could be neutralized by preincubation of sperm with Fab fragments against the sperm glycosylphosphatidylinositol (GPI)-linked hyaluronidase. HA is present in the pores of the zona pellucida, and these studies suggest that it may influence sperm behavior at the surface of the zona pellucida.

An additional effect of the cumulus matrix is its influence on fertilization rates, variously measured by appearance of a male pronucleus or two cell embryos. Several studies have shown that the presence of the cumulus matrix increases fertilization rates. In some studies, this has been attributed to the quality of the cumulus matrix as measured by the degree of expansion. In these studies, groups of oocytes with larger cumulus masses achieved greater fertilization rates (Legendre and Stewart-Savage, 1993; Chen *et al.,* 1993). Although this correlation is intriguing, there is little information available addressing the mechanism by which the cumulus influences fertilization success. Because the effect was related to the amount of cumulus expansion, it is possible that the effect is purely mechanical: sperm simply cannot penetrate the unexpanded matrix efficiently (Kito and Bavister, 1999). However, as was discussed above, a variety of components are produced during cumulus expansion, and an alternative possibility is that some component produced at this stage of oocyte development enhances fertilization success.

There is some evidence to support the role of a soluble factor. The fertilization rate for mouse oocytes was significantly higher for oocyte populations that were cumulus intact during *in vitro* insemination, compared to cumulus-free oocytes (Cross and Brinster, 1970; Cecconi *et al.,* 1996; Zhang *et al.,* 1995). However, when cumulus-free oocytes were cocultured with cumulus cells prior to insemination, the fertilization rate improved significantly over that of cumulus-free oocytes, but never equaled that of cumulus-intact oocytes (Cecconi *et al.*, 1996). These data suggest that both soluble and nonsoluble factors may contribute to enhance fertilization. The authors suggest an additional hypothesis, that direct contact of cumulus cells and oocyte may be required to transfer factors influencing cytoplasmic maturation in the oocyte. Additional studies support the role of soluble factors. These studies demonstrated that cumulus cells produce prostaglandins during oocyte maturation, fertilization, and early embryo development (Gurevich *et al.*, 1993; Viggiano *et al.*, 1995). PGE_1, PGE_2, and $PGF_{2\alpha}$ were detected in mouse oocyte/cumulus complexes, and fertilization rates decreased in the presence of indomethacin, suggesting that one or more of these prostaglandins mediates this effect of the cumulus (Viggiano *et al.*, 1995). Addition of exogenous PGE_1 or PGE_2 overcame the indomethacin block, but $PGF_{2\alpha}$ had no effect (Viggiano *et al.*, 1995). Similarly, the effect of GH on increased developmental competence of bovine oocytes matured *in vitro* has been attributed to increased fertilization rates as a result of improved cytoplasmic maturation (Izadyar *et al.*, 1997b). The mechanism by which cumulus-derived factors influence cytoplasmic maturation and fertilization success remains a question for future studies.

D. THE CUMULUS DURING OVUM PICKUP

At the time of ovulation, the oocyte with surrounding cumulus matrix is extruded into the bursal cavity of the ovary or into the peritoneal cavity, depending on the species. It must be picked up and transported into the oviduct by the cilia of the fimbria. Studies have shown that proper transport is mediated by the cumulus. In hamsters (Mahi-Brown and Yanagimachi, 1983), rats (Blandau, 1967), and rabbits (Blandau, 1967; Norwood *et al.*, 1978; Norwood and Richardson, 1980), cumulus-free oocytes were not picked up or transported, or in some instances were transported only slowly (Blandau, 1967). The cilia of the fimbria have a negatively charged glycocalyx in rabbits (Norwood *et al.*, 1978) and humans (Kiss *et al.*, 1998). Transport can be blocked by treatment of the oviduct with polycations (Norwood *et al.*, 1978; Norwood and Richardson, 1980; Mahi-Brown and Yanagimachi, 1983), but these do not interfere with lectin binding to the oviductal glycocalyx (Kiss *et al.*, 1998) or with ciliary beating (Norwood *et al.*, 1978). Neuraminidase is also able to block transport in hamster oviducts (Mahi-Brown and Yanagimachi, 1983). In addition, only cumulus masses from preovulatory oocytes, with expanded cumulus matrix, and not earlier stages, were transported (Mahi-Brown and Yanagimachi, 1983), suggesting that the cumulus ECM is required for transport. Taken together, these data suggest that a specific interaction between the glycocalyx of the fimbrial cilia (Norwood and Richardson, 1980) and the ECM of the cumulus matrix is required for proper transport of the oocyte into the oviduct for fertilization. The molecular identity of the components mediating this transport are unknown, but development of a quantitative system to study ovum transport (Talbot *et al.*, 1999; Huang *et al.*, 1999) may provide the means to resolve some of these questions in future studies.

E. THE CUMULUS AS A SPERM BARRIER

Many studies have suggested that the cumulus matrix is a chemical and mechanical barrier to sperm progress toward the oocyte. Although this may seem contrary to the cumulus functions that appear to have a positive effect on sperm behavior, such as enhanced motility and acrosomal exocytosis, the barrier presented by the cumulus could have important consequences for the success of fertilization. It has been previously suggested that screening out abnormal sperm might be an important function of the cumulus matrix (Austin, 1960). When sperm morphology was compared between sperm that penetrated the cumulus and sperm remaining in the medium in human *in vitro* fertilization (IVF) trials, the population of sperm entering the cumulus had a significantly greater fraction of morphologically normal sperm compared to the sperm remaining outside the cumulus mass (Carrell *et al.,* 1993). Such observations may indicate that the cumulus does play an important role in excluding abnormal sperm from contact with the oocyte. If an abnormal sperm cell were to fertilize the oocyte, this would be likely to result in a nonviable embryo. A mechanism that excludes abnormal sperm would increase the chances of producing a viable offspring.

There are many ways in which a sperm might be considered abnormal. Although it remains to be determined just what types of abnormalities might be detected by the cumulus matrix as sperm attempt to gain access to the oocyte, there are some examples of exclusion by the cumulus of inappropriate sperm. One of the earliest definitions of capacitation was the ability of sperm to penetrate the cumulus layer (Austin, 1960). A number of studies have demonstrated that uncapacitated sperm are unable to penetrate the cumulus layer, whereas sperm that have been capacitated either *in vitro* or *in vivo* penetrate the matrix easily (Austin, 1960; Cummins and Yanagamachi, 1986; Suarez *et al.,*1984; Talbot, 1985; Cherr *et al.,* 1986; Corselli and Talbot, 1987). In addition, only sperm that are acrosome intact are able to penetrate the cumulus layer, because sperm that have undergone spontaneous acrosomal exocytosis stick to the cumulus layer and do not penetrate it (Cummins and Yanagamachi, 1986; Suarez *et al.,*1984; Talbot, 1985).

F. HOW DO SPERM GET THROUGH THE CUMULUS LAYER?

The high viscoelasticity of the HA surrounding the cumulus cells presents a formidable barrier for sperm penetration. Sperm are readily slowed and stopped by even moderate viscosities (Cardullo and Cone, 1986); however, these cells must somehow penetrate this matrix in order to proceed to the zona pellucida. Based on *in vitro* fertilization experiments, researchers proposed that many sperm were needed at the site of the cumulus layer to loosen the matrix and ultimately to allow entry of a single sperm (Yanagamachi, 1994). However, in contrast to the many sperm present during commonly used *in vitro* fertilization procedures, the situation *in vivo* is quite different, with the sperm:egg ratio being very small, even approaching ratios of 1:1 (Cummins and Yanagamachi, 1982; Shalgi and Phillips,

1988; Smith *et al.,*1987). In addition, fertilization can be achieved *in vitro* with small numbers of sperm (Corselli and Talbot, 1987). These observations suggest that a single sperm cell can penetrate the cumulus layer and to do so it must have a mechanism to overcome the effects of the viscoelastic extracellular matrix.

Sperm motility and flagellar output could be modulated during cumulus penetration, as discussed above (Drobnis *et al.,* 1988; Katz *et al.,* 1989; Cardullo and Cone, 1986; Westphal *et al.,* 1993). An additional mechanism by which sperm could penetrate the cumulus layer would be that sperm contain an enzyme that can break down the surrounding ECM. Numerous studies have shown conclusively that the sperm acrosome contains high concentrations of hyaluronidase (Hechter and Hadidian, 1947; Swyer, 1947; Borders and Raferty, 1968; Rogers and Morton, 1973; Brown, 1975; Talbot *et al.,* 1985; Zao *et al.,* 1985; Harrison, 1988a,b; Gadella *et al.,* 1991; Cherr *et al.,* 1996). However, observations by a number of researchers using hamster (Talbot, 1985; Cherr *et al.,* 1986) or mouse (Saling *et al.,* 1978) eggs have demonstrated that only acrosome-intact sperm penetrate the cumulus layer, and the hyaluronidase found within the acrosome would be unavailable. Another alternative, suggested by a number of investigators, is that sperm possess a plasma membrane-associated form of hyaluronidase that is distinct from the acrosomal form (Talbot, 1985; Talbot *et al.,* 1985; Zao *et al.,* 1985; Anand *et al.,* 1977; Joyce *et al.,*1985; Lewin *et al.,* 1981, 1982; Metz *et al.,* 1972). This cell surface hyaluronidase is one of several enzymes reported on the surface of sperm that may be required for cumulus penetration, including acrosin (Tesarik *et al.,* 1990), β-galactosidase (Farooqui and Srivastava, 1979), arylsulfatase (Nikolajczyk and O'Rand, 1992), and the serine protease guanidinobenzoatase (Pedoto *et al.,* 1997).

The GPI-linked guinea pig sperm surface antigen PH-20 has been shown to share significant sequence similarity with bee venom hyaluronidase (Gmachl and Kreil, 1993). A cloned PH-20 homolog isolated from a human testis cDNA library exhibits hyaluronidase activity (Gmachl *et al.,* 1993) and biochemical studies have now demonstrated the presence of GPI-linked hyaluronidases in several other species, including mice (Thaler and Cardullo, 1995), macaques (Cherr *et al.,* 1996), and humans (Sabeur *et al.,* 1997). PH-20 (Cowan *et al.,* 1987) and the mouse GPI–hyaluronidase (Thaler and Cardullo, 1995) are present on both acrosome-intact and acrosome-reacted sperm. Immunolocalization studies on mouse sperm showed that the enzyme is localized to the anterior portion of the mouse sperm head (Lin *et al.,* 1994). These data suggest that mammalian sperm possess on their surface a GPI-linked hyaluronidase that could allow acrosome-intact sperm to penetrate the cumulus layer by hydrolyzing the hyaluronic acid between the cumulus cells (see Figure 4.1). Studies have shown that inhibitors of hyaluronidase activity also inhibited cumulus penetration by acrosome-intact motile sperm in a dose-dependent manner (Meyers *et al.,* 1997; Li *et al.,* 1997).

In addition to the GPI-linked form there is also considerable evidence for a soluble hyaluronidase (Hechter and Hadidian, 1947; Swyer, 1947; Borders and Raferty, 1968; Rogers and Morton, 1973; Brown, 1975; Harrison, 1988a,b; Gadella *et*

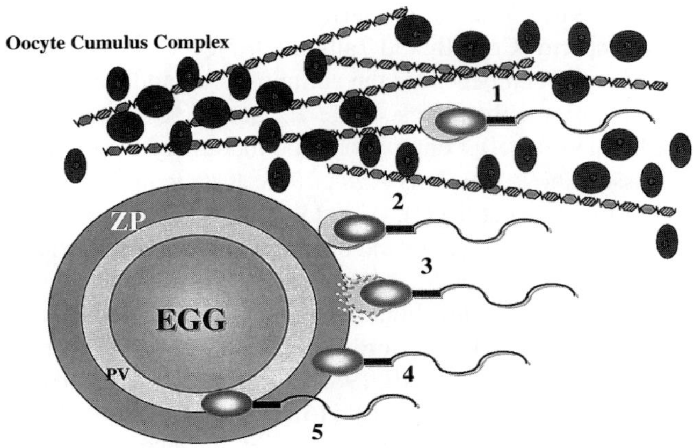

FIGURE 4.1 Diagram of the key steps of fertilization that involve the extracellular matrices surrounding the mammalian egg. (1) The sperm first encounters the cumulus layer, which must be penetrated by acrosome-intact sperm. The major extracellular matrix component between the cumulus cells is hyaluronic acid, a viscoelastic polymer that modifies sperm movements. (2) After negotioting the cumulus layer, acrosome-intact sperm then must adhere to the zona pellucida (ZP). (3) The zona pellucida also contains the secretegogue, which stimulates acrosomal exocytosis. (4) Following acrosomal exocytosis, the sperm engages in secondary binding events to the zona pellucida, allowing the sperm to penetrate the zona. (5) Following sperm–egg fusion and cortical granule exocytosis, the zona pellucida becomes modified, which prevents polyspermy. PV, Perivitelline space. See text for details.

al., 1991; Cherr *et al.,* 1996) that has biochemical and enzymatic properties distinct from the cell surface form. Given that the distribution of egg-associated HA extends from the cumulus layer into the zona pellucida and the perivitelline space (Ball *et al.,* 1982; Talbot, 1984; Dandekar and Talbot, 1992), a soluble isoform of hyaluronidase may aid sperm in penetration of the zona pellucida following adhesion to this matrix. Thus, at the same time that the viscoelasticity of the cumulus works to impede sperm progress (Cardullo and Cone, 1986), the action of a soluble factor in the cumulus may work to increase sperm motility and power output (Westphal *et al.,*1993) as the sperm GPI–hyaluronidase locally removes the viscoelastic matrix. Hydrolysis of the highly polymerized form of HA by the GPI-linked sperm surface hyaluronidase may provide optimal coupling of chemical and mechanical mechanisms for sperm penetration through the cumulus layer. The combination of these factors enables the cumulus matrix to act as a selective filter. Weakly motile sperm, sperm that are not able to respond to the soluble factor, or sperm lacking the GPI–hyaluronidase would be excluded from the matrix. Only sperm with all the proper motility, responsiveness, and surface components would be allowed to penetrate the matrix.

It is interesting to note, in the context of the cumulus serving as a barrier to abnormal sperm, that an infertility syndrome in mouse, the Rb(6.16) translocation, has been shown to involve a PH-20 homolog (Zheng and Martin-DeLeon, 1997). Homozygous mutant males show greatly reduced fertility and corresponding reductions in PH-20 mRNA levels. If this deficit results in lower protein and/or enzyme activity of the sperm GPI–hyaluronidase, the ability of these sperm to penetrate the cumulus could be compromised, contributing to the reduced fertility in homozygous mutant males.

Finally, although the cumulus may function in several ways as a barrier to abnormal sperm, there is some indication that the matrix may actually provide a reservoir for normal sperm. When the number of sperm present in ampullary fluid vs. cumulus matrix was measured at various times after mating, the ratio of sperm in fluid vs. matrix was 1:9. These data suggest that normal sperm rapidly enter the cumulus matrix, thereby facilitating early fertilization. This action of the cumulus matrix may be essential to producing viable offspring, because the incidence in abnormal/nonviable embryos increases substantially with age (i.e., time since ovulation) of the oocyte (Austin, 1970, 1982).

G. SUMMARY

The cumulus oophorus plays several roles in fertilization. Production of soluble factors by the cumulus cells appears to modulate the behavior of both oocyte, in resumption of meiosis, and sperm, in motility and acrosomal exocytosis. A more complex effect, that of enhancing fertilization success, has also been documented. The molecular mechanisms that generate this effect remain largely uncharacterized, although current data imply roles for both soluble and matrix/cumulus cell-associated factors. The targets of these factors—sperm, oocyte, or both—also remain unknown.

The cumulus oophorus may also function as a selective filter to block progress of abnormal sperm and enhance progress of normal sperm. Currently it is known that sperm that are not capacitated are acrosome-reacted, or if they lack GPI–hyaluronidase activity they are unable to penetrate the cumulus matrix. The molecular mechanisms underlying the exclusion of noncapacitated or acrosome-reacted sperm are currently unknown.

III. THE ZONA PELLUCIDA

The zona pellucida is the innermost extracellular matrix surrounding the mammalian egg. Relatively simple in its macromolecular composition, the zona pellucida plays a crucial role in fertilization by serving as a species-selective adhesive substrate for sperm as well as an agonist for regulated exocytosis of the sperm's acrosomal vesicle. In addition, modifications to the zona pellucida following fertilization prevent polyspermy and protect the early embryo in the oviduct prior to hatching and implantation in the uterus. Although much is known about the individual glycoproteins that make up the zona pellucida, little is known about the specific molecular interactions that regulate fertilization and early development. In this review we focus primarily on the roles of the zona pellucida in sperm adhesion, as an agonist for acrosomal exocytosis, and in changes following fertilization and egg activation.

A. STRUCTURE OF THE ZONA PELLUCIDA

The structure of the zona pellucida is covered in detail elsewhere in this volume. Here we briefly review relevant structural features of the zona pellucida in order to cogently present molecular details concerning the functional role of the zona pellucida in fertilization. Biochemical and molecular aspects of the zona pellucida have been studied extensively in the mouse, although there is a growing body of literature on studies in other mammals, including rabbits (Prasad *et al.*, 1996), pigs (Sacco *et al.*, 1989; Yurewicz *et al.*, 1987, 1993a,b,c), and humans (Barrat and Hornby, 1995). These studies have shown that the proteins that make up the zona pellucida are highly conserved among mammals, perhaps suggesting that the overall organization and function of the zona pellucida is conserved as well.

In the mouse, the zona pellucida is composed of three glycoproteins, designated mZP1, mZP2, and mZP3, with average molecular masses of 200, 120, and 83 kDa, respectively [for reviews see Wassarman (1988) and Wassarman and Litscher (1995)]. The primary sequence of these glycoproteins was deduced from full-length cDNAs, with the polypeptide chains of these three macromolecules accounting for only a fraction of the total molecular mass (Ringuette *et al.*, 1988). All three are heavily glycosylated proteins containing both N- and O-linked oligosaccharide chains. Gene sequences reveal a high degree of sequence similarity between members of the ZP3 gene family and to a lesser extent in the ZP1 and ZP2 gene families [for a review see McLeskey *et al.* (1998)].

Studies have shown that synthesis of these glycoproteins is temporally regulated during oogenesis, with the ZP genes transcribed exclusively by oocytes. In particular, mZP2 is expressed at low levels in resting oocytes but mZP1 and mZP3 are expressed exclusively by growing oocytes, with all three transcripts reaching maximal levels in midsized oocytes (Epifano *et al.*, 1995). The translation and secretion of these proteins occurs concomitantly, with the transcriptional activation of these genes leading to the assembly of the intact zona pellucida during growth of the developing oocyte (Epifano *et al.*, 1995). A number of reports have demonstrated that, in some mammals, the ZP genes may be expressed by granulosa cells as well (Lee and Dunbar, 1993; Grootenhuis *et al.*, 1996; Kolle *et al.*, 1996; Martinez *et al.*, 1996). If true, then posttranslational modifications to ZP polypeptides may lead to different structural components (e.g., glycosylation patterns) between oocyte-derived and granulosa-derived zona pellucida glycoproteins.

The zona pellucida is a compact, highly organized matrix that, in the mouse, is approximately 7 μm thick, with an outer diameter of approximately 110 μm (Figure 4.2). At the level of the electron microscope, the matrix has a lacy appearance

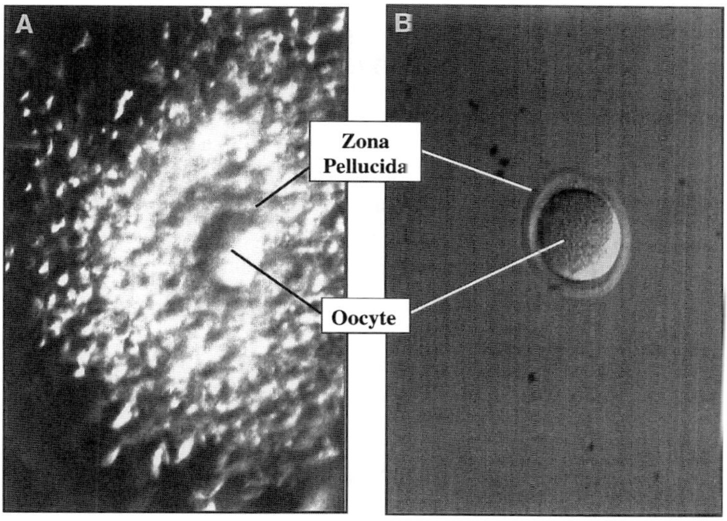

FIGURE 4.2 Micrographs of the cumulus layer and the zona pellucida surrounding a mouse oocyte. The cumulus matrix is a dense collection of cells surrounding the oocyte and the zona pellucida. (A) Low-magnification differential interference contrast (DIC) image (200×) of a zona-intact egg surrounded by the cumulus mass. (B) Following treatment with soluble hyaluronidase, the cumulus cells are dispersed, leaving only a zona pellucida surrounding the oocyte.

and presents a formidable barrier to sperm. Although little is known concerning the assembly of the different zona pellucida glycoproteins, current evidence suggests that the matrix is a noncovalently assembled structure made up of mZP2 and mZP3 dimers, which polymerize into filaments and are cross-linked by mZP1 homodimers to mZP2 (Greve and Wassarman, 1985).

In addition to studies in mammals, there has been some work concerning the composition of analogous structures in other vertebrate models. In *Xenopus laevis,* the vitelline envelope forms a structure that is similar to the zona pellucida and is composed of three glycoproteins (Doren *et al.,* 1999). These three glycoproteins are homologous (30–40% amino acid identity) to the three mouse zona pellucida glycoproteins. When full-length mRNAs from the mouse were injected into stage VI *Xenopus* oocytes, the mouse glycoproteins were expressed and secreted to the extracellular matrix of *Xenopus* eggs, suggesting that the mouse zona proteins have been sufficiently conserved in evolution to be integrated into the vitelline envelope (Doren *et al.,* 1999). In addition, some species of fish express ZP-like proteins, which are incorporated into their vitelline envelope (Lyons *et al.,* 1993; Murata *et al.,* 1995; Chang *et al.,* 1996). Interestingly, in the case of white flounder and medaka, these proteins are synthesized in the liver and then transported to the oocyte, where they are assembled in the vitelline envelope (Lyons *et al.,* 1993).

B. ADHESION ASSAYS USED TO ELUCIDATE THE FUNCTION OF THE ZONA PELLUCIDA DURING SPERM–ZONA INTERACTION

Assays for studying sperm adhesion to the zona pellucida fall into two categories: those that utilize the intact zona pellucida versus those that use solubilized proteins. The assay chosen depends on the particular question being asked. For instance, if information about adhesion at the cellular or biophysical level is needed, then using an intact zona pellucida would be warranted. If, however, molecular details about the interaction between a specific zona pellucida glycoprotein and its complementary receptor on the sperm surface are needed, then it is best to use solubilized and purified glycoproteins.

1. Intact Zona Pellucida Assays

The most utilized method for studying sperm–zona adhesion are light microscopic adhesion assays using living sperm and intact zona pellucidae. Sperm are incubated with eggs in the presence or absence of a competitor for some predetermined amount of time (e.g., 5–60 minutes), fixed, and attached sperm are subsequently counted under the light microscope. As a control for nonspecific binding, two-cell embryos containing zonae are included, because sperm do not bind to the zona pellucidae of embryos. Modifications of this assay include the stop–fix method in which eggs are centrifuged through fixative simultaneously to stop the assay and remove nonspecifically bound sperm (Saling *et al.,* 1978); another method analyzes the distribution of sperm bound per zona pellucida in populations of zona pellucidae (SamAth *et al.,* 1997).

The microscopic assays using intact zonae proved useful for identifying the key molecules involved in the sperm–zona interaction and were instrumental in identifying ZP3 as both the initial adhesion ligand and as the secretagogue for acrosomal exocytosis in mice (Bleil and Wassarman, 1980, 1980, 1986; Florman *et al.,* 1984; Florman and Wassarman, 1985). In addition, a number of putative receptors for ZP3 on the mouse sperm surface have been identified with this assay by using different antagonists against these sperm surface proteins. In the intact zona pellucida assay, sperm have a limited contact time and contact area for ligand–receptor interactions (Baltz and Cardullo, 1989). During this time period (~50–100 msec), sperm must successfully recognize and bind to at least one ZP3 molecule in order for adhesion to occur (Baltz *et al.,*1988). In competition assays in which sperm are coincubated with zona pellucidae and an appropriate antagonist, dissociation must occur within 50 msec. Because the lowest affinity for ZP3–ZP3 receptor interaction has been quantitatively measured to be around 50 n*M* (Thaler and Cardullo, 1996a), the molecular mass of a competing molecule would have to be of similar magnitude to successfully displace a sperm from the zona. Further exacerbating the problem is that the density of ZP3 receptors on the sperm surface is extremely high, making competition by simple molecular inhibitors unlikely (Thaler and Cardullo, 1996b). Specifically, concentrations of antagonists needed to even moderately inhibit binding most often exceed 1 m*M*. At these high concentrations, nonspecific effects are undoubtedly a significant problem.

2. Solubilized Zona Pellucida Assays

The use of solubilized zona pellucida glycoproteins has allowed investigators to study binding interactions in detail. The physiological role of the three glycoproteins has been investigated by a number of investigators (Bleil and Wassarman, 1980, 1983; Leyton and Saling, 1989; Miller *et al.,* 1992; Thaler and Cardullo, 1996a,b; Yurewicz *et al.,* 1998). Early studies by Bleil and Wassarman (1986), used ^{125}I-labeled ZP3 and cellular autoradiography to localize ZP3 receptors to the mouse sperm head. The use of solubilized components to characterize sperm–zona binding biochemically has been accomplished using conventional equilibrium, kinetic, and competition binding assays (Thaler and Cardullo, 1996a,b).

As with the intact zona pellucida assays, the use of solubilized components to study sperm–zona interactions is not without problems. In particular, because molecules are used in isolation, interactions reflecting the filamentous structure of the zona pellucida are lost. Using solubilized glycoproteins, investigators cannot approach the high density of these proteins in the intact matrix. This obviously has implications in understanding the basis of both adhesion and the signalling pathways responsible for acrosomal exocytosis.

C. TIME COURSE OF ADHESION TO THE ZONA PELLUCIDA

Several laboratories have described a number of separate steps in sperm-zona pellucida adhesion, which can be used to characterize both zona pellucidae and capacitated sperm *in vitro.* These steps include loose attachment, firm adhesion, induction of acrosomal exocytosis, and penetration of the zona pellucida (Hartmann *et al.,* 1972; Storey, 1991). Each step may reveal different molecular interactions between zona pellucida ligands and complementary sperm surface receptors. Understanding the time course of these cellular events is ultimately useful in identifying the molecules involved in primary and secondary adhesion events. In humans, a hemizona binding assay has been useful for studying the time course of binding as well as the dynamics of capacitation, hyperactivation, and the acrosome reaction (Burkman *et al.,* 1988).

Combining information at both the cellular and biochemical level has led to a molecular model for the role of the zona pellucida in sperm adhesion before and after acrosomal exocytosis. The sequence of events of this model include the following steps: (1) adhesion of acrosome-intact sperm to ZP3, (2) tight binding of acrosome-intact sperm to the zona pellucida followed by ZP3-induced activation of signal transduction pathways, leading to acrosomal exocytosis, (3) secondary adhesion of acrosome-reacted sperm mediated by ZP2, (4) penetration of the zona pellucida, and (5) molecular modifications to both ZP2 and ZP3 following sperm–egg fusion and egg activation.

D. EVIDENCE OF A SPECIFIC LIGAND–RECEPTOR INTERACTION

As alluded to in the previous section, a number of studies using different assays have implicated ZP3 as the primary adhesion molecule in the zona pellucidae in most mammals. In the mouse, mZP3 is an 83-kDa glycoprotein (Wassarman, 1988; Litscher *et al.,* 1995) having three or four N-linked oligosaccharide chains and an undetermined number of O-linked oligosaccharide chains (Wassarman and Litscher, 1995). Early experiments suggested that the bioactivity of mZP3 resides solely with one or more of the O-linked oligosaccharides on ZP3 (Bleil and Wassarman, 1988; Florman *et al.,* 1984; Florman and Wassarman, 1985), in the C-terminal half of the molecule (Rosiere and Wassarman, 1992). Subsequent experiments using mZP3 produced by transiently transfected embryonic carcinoma cells showed that mutated forms of mZP3 in which Ser-332 or Ser-334 was replaced with other amino acids were not capable of inducing acrosomal exocytosis (Chen *et al.,* 1998). Other changes in this region of the polypeptide had no effect. Among O-linked oligosaccharides from native ZP3, a small oligosaccharide, estimated at 3400–4600 Da using size exclusion chromatography, which retained sperm binding activity but did not induce the acrosome reaction, has been identified (Florman and Wassarman, 1985). This result suggested that the polypeptide chain may play an important role in the organization of oligosaccharides necessary for initiation of signal transduction pathways leading to acrosomal exocytosis.

Early characterization of critical biochemical and cell biological parameters was limited to crude microscopic methods such as binding of sperm to ZP3-conjugated beads (Vazquez *et al.,* 1989), cellular autoradiography (Bleil and Wassarman, 1986), and binding of colloidal gold-labeled ZP3 to acrosome-intact sperm (Mortillo and Wassarman, 1991). These methods showed that ZP3 binding sites localize to the sperm head and the cellular autoradiography estimated the number of ZP3 binding sites to be between 10,000 and 50,000 per sperm (Bleil and Wassarman, 1986). Although useful for confirming the localization of ZP3 receptors on the mouse sperm surface, these types of studies have done little to identify the complementary receptors for ZP3 or to establish the precise role of the zona pellucida in adhesion or the initiation of signal transduction pathways necessary for acrosomal exocytosis.

Thaler and Cardullo (1996a) performed a biochemical characterization of ZP binding on sperm using solubilized and ^{125}I-labeled ZP components in a standard biochemical assay. These studies demonstrated that the initial binding event between mZP3 and its complementary receptor on the mouse sperm surface is complex. Both equilibrium and kinetic analyses indicated that these high-affinity interactions cannot be explained by a single receptor–ligand interaction. These data suggest the presence of multiple ligands and/or receptors of varying affinities (Thaler and Cardullo, 1996a,b). Saturation binding revealed that there are approximately 30,000 mZP3 binding sites per sperm (Thaler and Cardullo, 1996a). Along with microscopic evidence that the initial binding event between acrosome-intact sperm and the zona pellucida occurs only over the approximately 10- to 15-μm² surface area overlying the acrosomal vesicle, the average mZP3 receptor density is moderately high at around 2000–3000 molecules/μm². Although more detailed experiments need to be performed, these results support the hypothesis that ZP3 is a multivalent ligand, consistent with earlier studies showing that monovalent ZP3 oligosaccharides bind to sperm but do not induce acrosomal exocytosis (Florman and Wassarman, 1985).

Experiments using solubilized ZP components may only partially address the interactions that occur on the intact zona pellucida. Theoretical arguments based on biophysical measurements of sperm adhesion have suggested that only a few sperm–zona bonds are needed to tether a sperm to an egg (Baltz *et al.,* 1988). The rate-limiting step for sperm adhesion to the zona pellucida is directly proportional to the surface density of ZP3 on the zona pellucida, the receptor density on the surface of sperm, the contact area between sperm and the zona pellucida, and the diffusion coefficient of the receptor on the sperm surface (Baltz and Cardullo, 1989). Estimates of average ZP3 receptor density on the zona approach 300 molecules/μm² and corresponding densities of the ZP3 receptor on the mouse sperm surface are about 2000 molecules/μm². These high densities, along with the assertion that few bonds are needed to tether a sperm, virtually ensure that sperm adhesion to the zona pellucida will occur if sperm contact the zona pellucida in the correct orientation.

Successful adhesion of mammalian sperm to the zona pellucida is a combination of achieving the correct molecular specificity between ligands in the zona pellucida and their complementary receptors on the sperm surface along with optimized structural and biophysical characteristics of these surfaces. Zona pellucidae that do not have the correct chemical modifications to their bioactive ligands would ultimately result in infertility. Similarly, if either the zona pellucidae or the sperm surfaces do not have optimized biophysical parameters to ensure adhesion (such as ligand density on the zona surface, receptor density or mobility on the sperm surface), then fertilization will not occur. In this context, it is possible that events upstream from the sperm interaction with the zona pellucida may play a critical role in the successful interaction of the zona pellucida with the sperm. In particular, capacitating factors in the oviduct or soluble and/or ECM factors within the cumulus layer may play a role in modifying chemical structures or biophysical parameters (such as increasing receptor diffusion coefficients), which will ensure successful fertilization.

E. EVIDENCE THAT THE BIOACTIVE COMPONENT ON ZP3 IS AN OLIGOSACCHARIDE

Although it is generally accepted that the bioactive component within ZP3 is related to its carbohydrate composition, there is considerable disagreement over the identity of the oligosaccharides that are recognized by mouse sperm. Only the identity of the terminal sugar residue has been investigated directly.

Competitive binding assays using intact zona pellucidae have implicated either a terminal α-linked galactose (Bleil and Wassarman, 1988; Litscher *et al.,* 1995) or a terminal *N*-acetylglucosamine (Shur and Hall, 1982a,b; Lopez *et al.,* 1985; Miller *et al.,* 1992; Youakim *et al.,* 1994) on ZP3. Shur and colleagues have argued that ZP3 must contain terminal *N*-acetylglucosamine (GlcNAc) residues because substrates for β-1,4-galactosyltransferase (GalTase) decrease the amount of sperm binding to zona-intact eggs (Shur and Hall, 1982a,b; Lopez *et al.,* 1985), consistent with their hypothesis that a sperm surface GalTase is the mouse sperm ZP3 receptor [for reviews see Shur (1993) and Dubois and Shur (1995)]. Miller *et al.* (1992) provided evidence that all ZP glycoproteins contain GlcNAc residues because all three can be used as substrates for bovine milk soluble GalTase. Further, the mouse sperm GalTase selectively galactosylates ZP3 (and not ZP1 or ZP2), consistent with subsequent work showing that the sperm surface GalTase is distinguishable from Golgi forms (Miller *et al.,* 1992). It has been suggested that GlcNAc residues cross-link GalTase on the mouse sperm surface, resulting in the activation of a heterotrimeric G_i protein (Gong *et al.,* 1995). These experiments suggest that a terminal GlcNAc on ZP3 is necessary for sperm zona adhesion and induction of acrosomal exocytosis.

Experiments using neoglycoproteins have shown that the presence of a protein backbone is necessary to induce acrosomal exocytosis (Loeser and Tulsiani, 1999). In these studies, bovine serum albumin (BSA) conjugated with different monosaccharides was used to test whether specific sugars were competent to induce acrosomal exocytosis. In this study, three neoglycoproteins (BSA-GlcNAc, BSA-Man, and BSA-Neu) induced acrosomal exocytosis whereas neoglycoproteins terminating in either Gal or Glu did not (Loesser and Tulsiani, 1999). Given that at least three different carbohydrates conjugated to a protein background are sufficient to induce acrosomal exocytosis, these studies suggest the possibility that sperm may recognize multiple sugars on ZP3.

To test for participation of Gal or GlcNAc in sperm–zona adhesion, a series of complex oligosaccharides, which varied in composition (terminal Gal or terminal GlcNAc) and in degree of branching (uni-, bi-, tri-, and tetraantennary structures), were used as competitors in a sperm–zona binding assay. Oligosaccharides that terminated in Gal (in either the α- or β-configuration), but not GlcNAc, significantly inhibited binding (Litscher *et al.,* 1995). Further, the effectiveness of inhibition increased as the degree of branching and the length of terminal Gal-containing oligosaccharides increased (Litscher *et al.,* 1995). Studies demonstrating that transgenic mice lacking all α-1,3-galactose residues are fertile indicate that a terminal α-galactose may not entirely account for sperm–zona adhesion (Thall *et al.,* 1995). Given the results of Litscher *et al.* showing that both α-Gal and β-Gal are effective inhibitors of binding, it is possible that the pertinent sugar is β-Gal. These studies have been further substantiated using different oligosaccharides in a sperm–zona binding assay (Johnston *et al.,* 1998). In this study, researchers concluded that terminal GlcNAc residues on ZP3 do not interact with complementary receptors on the sperm surface with high affinity and, in addition, provide evidence that a fucosyl residue is required for the high-affinity interaction (Johnston *et al.*, 1998).

Other work suggests that N-linked oligosaccharides are the bioactive components that determine adhesion or that a monosaccharide on O-linked oligosaccharides, other than galactose, may be the ligand for the complementary receptor on the sperm surface. In contrast to earlier reports (Bleil and Wassarman, 1988), Nagdas *et al.*, (1994) were unable to detect α-galactose residues on O-linked oligosaccharides of ZP3. These investigators suggested that α-Gal residues are on N-linked rather than O-linked oligosaccharide chains, although this assertion has not yet been demonstrated. When Nagdas *et al.* (1994) exhaustively treated ZP2 and ZP3 with endo-β-galactosidase, an enzyme that cleaves repeating units of acetyllactosamine (3Galβ1,4GlcNAcβ1), there was a significant reduction in the molecular masses of both ZP2 and ZP3 of 23 and 16 kDa, respectively. These poly(lactosaminoglycans) were found to be associated solely with N-linked, and not O-linked, oligosaccharides. In addition, these investigators treated de-N-glycosylated ZP3 with mild alkali in the presence of NaB^3H_4 and released a radiolabeled trisaccharide (GlcNAcβ → Galβ1,3GalNAcol). Significantly, this trisaccharide contains a GlcNAc residue that may serve as the ligand for the sperm surface GalTase, as suggested by Shur and colleagues.

A microscopic study using lectins and antibodies against mZP2 and mZP3 has revealed that the intact zona pellucida is heterogeneous, possessing both an inner and outer core, with ZP2 and ZP3 densities highest in the outer core. Further, as detected with different lectins, the carbohydrate composition in these two regions varies (Aviles *et al.*, 1997). Of particular relevance, the lectin $BSAIB_4$, which recognizes terminal α-Gal residues, was localized only to the inner half of the zona pellucida. This suggests that α-Gal residues would not be available for primary binding events but may be available for binding during sperm penetration of the zona pellucida. In contrast, terminal β-Gal residues do occur in the outer zona, as evidenced by peanut agglutinin (PNA) and RCA-I lectin binding (Aviles *et al.*, 1997).

Controversies centered around the different carbohydrate moieties on ZP3 that may act as ligands for a complementary sperm receptor, along with evidence that other zona pellucida molecules may be involved in primary adhesion and the identification of multiple putative receptors on the sperm surface, may simply reflect the complex nature of the initial adhesion event between sperm and egg. It is possible that different mammals may use different molecules or ligands to ensure successful adhesion and onset of acrosomal exocytosis. Alternatively, requiring multiple molecular interactions may serve as a selection filter that allows only competent sperm to undergo a complex binding event and penetrate the zona pellucida. In the context of a complex binding event that displays both high- and low-affinity states, it will be necessary to determine how these different states relate to adhesion and the initiation of signal transduction pathways leading to acrosomal exocytosis.

F. ZONA PELLUCIDA GLYCOPROTEINS OTHER THAN ZP3 MAY BE INVOLVED IN PRIMARY ADHESION

In some mammals, ZP1 has been suggested to play a role in primary binding between the zona pellucida and acrosome-intact sperm. In the current mouse model, mZP1 plays a purely structural role, whereas mZP2 and mZP3 serve as ligands for receptors on acrosome-reacted and acrosome-intact sperm, respectively (Wassarman, 1988). Evidence for molecules other than mZP3 serving a role in initial adhesion comes from work done in pigs, rabbits, and humans.

The porcine zona pellucida is composed of three glycoproteins identified as pZP1, pZP3α, and pZP3β, which are homologous to mZP2 (Yamasaki *et al.,* 1996), mZP1 (Yurewicz *et al.,* 1993a), and mZP3 (Harris *et al.,* 1994), respectively. Both pZP3α and pZP3β show identical electrophoretic mobilities on reducing sodium dodecyl sulfate–polyacrylamide gel elecIrophoresis (SDS-PAGE) and separation of these two molecules can be achieved only following chemical modification (Yurewicz *et al.,* 1987). pZP3α binds to sperm membranes via its N-linked oligosaccharides (Sacco *et al.,* 1989; Yurewicz *et al.,* 1993a,b; Yonezawa *et al.,* 1995; Nakano *et al.,* 1996). Initial studies suggested that binding of pZP3α was significantly enhanced by ZP3β. (Yurewicz *et al.,* 1993b). Further dissection of this system using highly purified ZP3α and ZP3β has demonstrated that both components must be present for high-affinity binding (i.e., nanomolar range) and that ZP3α or ZP3β alone does not interact with sperm because neither is able to compete for binding in a competitive assay with the ZP3α/ZP3β heterodimer ($IC_{50} > 1$ μm) (Yurewicz *et al.,* 1998). Additionally, antibodies against pZP3β inhibit binding of boar sperm to the zona pellucida (Bagavant *et al.,* 1993). In these studies, the antibody was raised against a 25-mer derived against pZP3β, which is putatively rich in O-linked oligosaccharides, suggesting that like mZP3, pZP3β recognition may involve O-linked carbohydrates, although this has yet to be demonstrated. Clearly, further studies must be performed to determine the role of pZP3α and pZP3β in sperm–zona adhesion and to see if pZP3β plays a direct role in adhesion or merely serves to coordinate the binding of pZP3α to its complementary receptor on the sperm surface.

Studies of rabbit and human ZP1 have shown that it may be involved in primary adhesion events (Prasad *et al.,* 1996). In the rabbit, primary adhesion is mediated by a 55-kDa zona pellucida glycoprotein (R55) that is 51% similar to mZP1 and shows little sequence similarity to mZP3 (Epifano *et al.,* 1995). However, a zona pellucida ligand for sperm adhesion in humans has greater than 75% sequence identity to pZP3α and R55 and only 50% sequence identity to mZP1 (Epifano and Dean, 1994), suggesting that members of the ZP1 family may have different functions in different mammals. Although further evidence is needed, these data suggest that in these species a homolog of mZP1, and not mZP3, may serve as the primary adhesion ligand for sperm interaction with the zona pellucida.

G. ROLE OF THE ZONA PELLUCIDA FOLLOWING ACROSOMAL EXOCYTOSIS—SECONDARY ADHESION

In contrast to experiments involving initial adhesion events, there are few data available on the role of the zona pellucida following acrosomal exocytosis. Once sperm have undergone acrosomal exocytosis, a new membrane surface is presented and sperm proceed to penetrate the zona matrix (Figure 4.1). To penetrate the matrix, it has been argued that acrosome-reacted sperm must bind to the zona pellucida using either receptors and/or ligands that are distinct from those involved in the primary adhesion. Monoclonal antibodies against mZP2 do not affect initial binding between acrosome-intact sperm and the zona pellucida, but inhibit the continued adhesion of acrosome-reacted sperm to the zona pellucida (Bleil *et al.,*1988). Later studies using colloidal gold-labeled mZP3 and mZP2 localized these molecules to either acrosome-intact or acrosome-reacted sperm, respectively, using transmission electron microscopy (Mortillo and Wassarman, 1991). Interestingly, low levels of mZP3 binding to the postacrosomal region of acrosome-reacted sperm were also detected.

Unfortunately, no biochemical analysis of mZP2 binding to acrosome-reacted sperm has yet been performed. This is somewhat surprising because (1) mZP2 generally presents a stronger signal than the other two glycoproteins, (2) it is easy to obtain enriched populations of acrosome-reacted sperm using calcium ionophores or other pharmacological agents such as thapsigargin, and (3) no physiological transformations, similar to the acrosome reaction, are thought to occur as a result of secondary binding. Once key experiments are performed, binding parameters such as molecular masses, the number of binding sites, and different affinity states can be ascertained.

In the pig, pZP1 has been suggested to be the secondary adhesion ligand for sperm. The pZP1 gene is homologous to mZP2, suggesting its role in secondary binding. In addition, fluorescently labeled, recombinant pZP1 (expressed in *Escherichia coli*) binds to the equatorial region on the head of sperm from five different mammals, including the boar (Tsubamoto *et al.,* 1996). The fluorescence pattern of this protein translocated from the equatorial segment to the posterior head over time, suggesting that pZP1 may assist sperm in penetration of the zona pellucida following the acrosome reaction. Further, using affinity blotting, this recombinant pZP1 bound to proacrosin and to an uncharacterized 40-kDa protein from sperm (Tsubamoto *et al.,* 1996).

A few mouse sperm surface proteins have been proposed to be involved in secondary adhesion events. One of these molecules is the GPI-linked hyaluronidase, PH-20, on guinea pig sperm. As discussed earlier, PH-20 may have a role in potentiating the acrosome reaction (Sabeur *et al.,* 1999), as well as in secondary adhesion. Some, but not all, antibodies directed against PH-20 block zona pellucida binding of acrosome-reacted, but not acrosome-intact, guinea pig sperm (Primakoff *et al.,* 1985; Myles *et al.,* 1987). Because PH-20 is localized to the posterior head of acrosome-intact sperm and to the anterior head of acrosome-reacted sperm, it has been suggested that these two populations play distinct physiological roles: penetration through the cumulus matrix for acrosome-intact sperm, and secondary adhesion for acrosome-reacted sperm (Primakoff, 1994),

Soybean trypsin inhibitor (SBTI) also binds to acrosome-reacted sperm and blocks secondary adhesion in a similar manner to anti-ZP2 (Bleil *et al.,* 1988). This infers that a protease may be involved in secondary adhesion events. Although the SBTI binding protein has not been identified, proacrosin, the precursor to the sperm's major serine protease, acrosin, has been implicated as a secondary adhesion molecule (Jones, 1990; Topfer-Petersen *et al.,* 1995; Topfer-Petersen, 1996). It has been shown that porcine proacrosin recognizes ZP2 (Tsubamoto *et al.,* 1996), supporting the hypothesis that both ZP2 and proacrosin are involved in secondary binding events. Biochemical and molecular analyses of proacrosin have identified potential binding domains in the proacrosin molecule (Topfer-Petersen *et al.,* 1990; Jansen *et al.,* 1995; Richardson and O'Rand, 1996). Interestingly, proacrosin may play two important roles in fertilization: first, in secondary adhesion events following acrosomal exocytosis, and second, in penetration of the zona matrix as proacrosin is converted to acrosin, a process that is triggered directly by the zona pellucida (Topfer-Petersen and Cechova, 1990). The importance of proacrosin in secondary binding has been challenged by experiments showing that proacrosin knockout mice are fertile (Baba *et al.,* 1994). However, a structurally similar, but distinct, molecule known as sp38 has been identified in boar (Mori *et al.,* 1995), perhaps indicating the presence of more than one class of secondary binding molecule in acrosome-reacted sperm.

H. MODIFICATIONS TO THE ZONA PELLUCIDA FOLLOWING FERTILIZATION

Following fertilization, molecular changes that occur in the zona pellucida prevent additional sperm from binding to, or penetrating, the matrix. This represents a slow block to polyspermy, analogous to the hardening of the vitelline envelope in marine invertebrates, and has been termed the zona reaction (Braden *et al.,* 1954; Gwatkin *et al.,* 1973; Gulyas, 1980). Direct and indirect evidence suggests that both mZP2 and mZP3 are modified following egg activation and exocytosis of the cortical granules by the egg. Specific biochemical modifications to mZP2 would render it ineffective as a ligand for secondary adhesion to acrosome-reacted sperm, whereas modifications to mZP3 would have a similar effect on acrosome-intact sperm. In addition to preventing polyspermy, the zona reaction may provide protection and support for the developing embryo as it passes through the oviduct prior to implantation. It is also possible that these specific molecular changes in the zona pellucida serve as recognition molecules necessary for disruption of the zona pellucida during hatching.

Modification in mZP2 was first detected as a change in electrophoretic mobility and isoelectric point on two-dimensional gels, denoted as a ZP2 to $ZP2_f$ transition (Bleil *et al.,* 1981). The modifications may be due to a protease released from cortical granules following egg activation, because mZP2 is readily converted to $mZP2_f$ in the presence of the calcium ionophore, A23187, which induces cortical granule exocytosis in the absence of fertilization (Bleil *et al.,* 1981; Kalab *et al.,* 1993). In addition, serine protease inhibitors block the conversion of mZP2 to $mZP2_f$ (Moller and Wassarman, 1989).

In addition to a change in mZP2 to $mZP2_f$ following egg activation, it has been shown that eggs incubated *in vitro* will undergo a precocious loss of cortical granules, releasing their contents into the perivitelline space (Kalab *et al.,* 1993). Although this represents only a minor fraction of the entire population of cortical granules within the egg, the spontaneous exocytosis of these vesicles is sufficient to convert mZP2 to $mZP2_f$ making it impossible for sperm to fertilize these eggs. Experiments have shown that this premature conversion of mZP2 to $mZP2_f$ *in vitro* can be prevented by adding a variety of serum components that are found in the oviduct, including fetuin, a known protease inhibitor (Kalab *et al.,* 1993). *In vivo,* it has been hypothesized that spontaneous fusion events do not lead to the conversion of mZP2 to $mZP2_f$ because components in oviductal fluid prevent this conversion. However, these oviductal components are ineffective following egg activation because the massive release of cortical granules is thought to overwhelm those inhibitors, leading to the conversion of mZP2 to $mZP2_f$.

In contrast to mZP2, mZP3 undergoes no detectable change in electrophoretic mobility following egg activation (Bleil *et al.,* 1981). However, acrosome-intact sperm are unable to bind to fertilized eggs or embryos. mZP3 may be modified by a cortical granule glycosidase that hydrolyzes terminal sugars needed for primary binding. Although those that believe that a terminal O-linked galactose is necessary for adhesion might argue that a galactosidase would be sufficient to render ZP3 inactive, Miller *et al.* (1993) have reported that a cortical granule-derived *N*-acetylglucosaminidase hydrolyzes the GlcNAc residue that is recognized by the sperm surface GalTase. Yet another possibility is that changes in the structure of ZP2 (due to conversion to $ZP2_f$) may lead to a change in conformation of adjacent ZP3 molecules. However, evidence for any structural modifications in mZP3, other than that reported by Miller *et al.* (1993), still awaits verification.

Additional studies provide evidence that both ZP2 and ZP3 are modified after fertilization (Aviles *et al.,* 1997). Immunoreactivity of ZP2 and ZP3 decreased after fertilization and the binding patterns of a number of lectins changed as well. In particular, lectin binding showed that terminal GlcNAc residues did not decrease following fertilization (Aviles *et al.,* 1997), in contrast to the observation by Miller *et al.* (1993) that GlcNAc on mZP3 is hydrolyzed by a *N*-acetylglucosaminidase following fertilization.

I. SUMMARY

The zona pellucida plays critical roles in sperm recognition and adhesion, initiation of acrosomal exocytosis, sperm penetration of the matrix, and, subsequently, protection of the fertilized egg and embryo. The structure of the zona pellucida, especially the carbohydrate composition of the individual glycoproteins and the precise arrangement of those molecules, is key in determining biological function. Biochemical assays have shown that the interaction between ZP3 and acrosome-intact sperm displays both high- and low-affinity components and suggest that there may be multiple ligands and/or receptors involved in the initial recognition events between sperm and the zona pellucida. Although the molecular determinants are currently being investigated, there is still considerable debate about the role of specific carbohydrates on ZP3 in forming the bioactive ligand of this molecule.

Subsequent to acrosomal exocytosis the zona pellucida plays a role in secondary adhesion during sperm penetration of the matrix. ZP2 is thought to mediate secondary adhesion and may interact with a number of acrosomal matrix proteins, including proacrosin.

Following sperm–egg fusion and egg activation, the zona pellucida undergoes molecular changes, collectively known as the zona reaction, which function to block polyspermy. The major biochemical modification is a proteolytic conversion of ZP2 to $ZP2_f$ following the release of egg cortical granule proteases. This conversion prevents binding to, and penetration of, the zona pellucida by acrosome-reacted sperm. Modifications of ZP3 to $ZP3_f$ have also been proposed, based on the inability of acrosome-intact sperm to bind to fertilized eggs or embryos, although evidence for any modifications is limited.

IV. FUTURE DIRECTIONS

Significant progress has been made in our understanding of mammalian egg extracellular matrices during fertilization. However, many questions remain about the role of these structures, especially in light of conflicting results from a number of different laboratories. In the future, novel approaches that involve genetic, molecular, and biophysical methodologies will be needed to identify specifically the role of molecules within the cumulus oophorus and the zona pellucida.

A. THE CUMULUS MATRIX

In the past decades, numerous components have been detected in the cumulus ECM, but only few have been studied in sufficient depth to understand their function. Most of the ECM-associated components of the cumulus have been identified only by their electrophoretic mobility, and it remains for future studies to determine whether these and other yet-unidentified proteins have structural roles or are components modulating sperm and oocyte behavior. Many intriguing studies have suggested the presence and activity of soluble components produced by the cumulus cells that affect sperm and oocyte functioning. If these factors can be characterized and isolated, future work can directly addresses the molecular effects of these factors. The molecular targets on the responding cells, either sperm or oocyte, could be identified once an isolated factor is in hand.

In addition to understanding individual components of the matrix and their effects, there is a need to understand the aspects of this matrix that are responsible for complex functioning, such as creating a selective barrier for sperm or enhancing fertilization success. Such functions are likely to involve multiple components of both sperm and cumulus matrix and may provide exciting clues to the mechanisms of intercellular interactions and communication.

B. THE ZONA PELLUCIDA

The molecular and biochemical characterizations of the individual glycoproteins that make up the zona pellucida have provided important insights into its role during fertilization. ZP gene families are being constructed, with the identification of ZP1, ZP2, and ZP3 genes in a number of different species, including the cow, human, mouse, and pig. Although the GP gene families share significant sequence similarities, questions regarding their regulation, secretion, and function still remain. At the molecular level, most ZP genes have the same number of exons and introns, although human ZP2 has an additional exon at its C-terminal end (Liang and Dean, 1993). In addition, all ZP3 genes are single-copy genes except for human ZP3 (Epifano *et al.,* 1995). These similarities and differences may provide important clues about the particular structural components that confer both matrix assembly and function of the zona pellucida in different mammals.

Further information about the role of these zona glycoproteins may be obtained by taking an evolutionary approach. Genes that encode the proteins making up the vitelline envelope from amphibians and fish share significant sequence similarities to the mammalian ZP genes. Additional gene sequences from other vertebrates, especially those that are internal fertilizers, may provide clues about the origins of species specificity conferred by the innermost extracellular matrix. Comparisons with vitelline envelope proteins in marine invertebrates, such as sea urchin and starfish, may also provide additional clues about gamete recognition and modification prior to sperm–egg fusion.

In addition, the physiological role of each of the zona pellucida glycoproteins needs to be examined. This is particularly true in light of evidence that ZP1 may play a role in primary binding and that ZP2 is involved in secondary binding events following acrosomal exocytosis. Further biochemical characterization of zona pellucida glycoproteins using recent advances in expression and genetic methodologies should be extremely helpful in elucidating their function. To date, a limiting step in performing detailed biochemical studies has been the small amount of ZP glycoproteins that can be obtained from mammalian zona pellucidae. A number of investigators have recently reported that recombinant ZP3 can be expressed in a variety of different cell types. Because it is widely believed that the bioactivity of these glycoproteins is primarily associated with the oligosaccharides, and not the polypeptide chain, choice of expression system and conditions is critical. Given the large amounts of protein that can be obtained from these expression systems, the secreted glycoproteins should allow researchers to determine the critical bioactive components in each of the glycoproteins.

Finally, further biophysical characterization of ZP glycoproteins before and after fertilization will provide important clues about the role of ZP1, ZP2, and ZP3 in matrix assembly, sperm recognition, modification following fertilization, and dissolution coincident with hatching and implantation. Methodologies such as X-ray crystallography, nuclear magnetic resonance, and mass spectrometry will be useful in identifying even subtle changes in these molecules, both within one animal during fertilization and development and between different animals.

REFERENCES

Abeydeera, L. R., Wang, W.-H., Cantley, T. C., Rieke, A., Prather, R. S., and Day, B. N. (1998). Presence of epidermal growth factor during in vitro maturation of pig oocytes and embryo culture can modulate blastocyst development after *in vitro* fertilization. *Mol. Reprod. Dev.* **51,** 395–401.

Amiel, M.-L., Moos, J., Tesarik, J., and Testart, J. (1993). Evidence of new antigens in the mouse cumulus oophorus during preovulatory cumulus expansion. *Mol. Reprod. Dev.* **34,** 81–86.

Anand, S. R., Kaur, S. P., and Chaudhry, P. S. (1977). Distribution of beta-*N*-acetylglucosaminidase, hyaluronoglucosaminidase and acrosin in buffalo and goat spermatozoa. *Hoppe-Seyler's Z. Physiol. Chem.* **358,** 685–688.

Austin, C. R. (1960). Capacitation and the release of hyaluronidase from spermatozoa. *J. Reprod. Fertil.* **3,** 310–311.

Austin, C. R. (1970). Ageing and reproduction: Post-ovulatory deterioration of the egg *J. Reprod. Fertil.* (Suppl.) **12,** 39–53.

Austin, C. R. (1982). Fertilization. *In:* "Reproduction in Mammals. I. Germ Cells and Fertilization" (C. R. Austin and R. V. Short, eds.), pp. 103–133. Cambridge University Press, Cambridge.

Aviles, M., Jaber, L., Castells, M. T., Ballesta, J., and Kan, F. W. K. (1997). Modifications of carbohydrate residues and ZP2 and ZP3 glycoproteins in the mouse zona pellucida after fertilization. *Biol. Reprod.* **57,** 1155–1163.

Baba, T., Azuma, S., Kashiwabara, S., and Toyoda, Y. (1994). Sperm from mice carrying a targeted mutation of the acrosin gene can penetrate the oocyte zone pellucida and effect fertilization. *J. Biol. Chem.* **269,** 31845–31849.

Bagavant, H., Yurewicz, E. C., Sacco, A. G., Talwar, G. P., and Gupta, S. K. (1993). Block in porcine gamete interaction by polyclonal antibodies to a pig ZP3 beta fragment having partial sequence homology to human ZP3. *J. Reprod. Immunol.* **25,** 277–283.

Ball, G. D., Bellin, M. E., Ax, R. L., and First, N. L. (1982). Glycosaminoglycans in bovine cumulus-oocyte complexes: Morphology and chemistry. *Mol. Cell. Endocrinol.* **28,** 113–122.

Ball, G. D., Wieben, E. D., and Byers, A. P. (1985). DNA, RNA, and proein synthesis by porcine oocyte-cumulus complexes during expansion. *Biol. Reprod.* **33,** 739–744.

Baltz, J. M., and Cardullo, R. A. (1989). On the number and rate of formation of sperm–zona bonds in the mouse. *Gamete Res.* **24,** 1–8.

Baltz, J. M., Katz, D. F., and Cone, R. A. (1988). Mechanics of sperm–egg interaction at the zona pellucida. *Biophys. J.* **54,** 643–654.

Barrat, C. L. R., and Hornby, D. P. (1995). *In* "The Human Acrosome Reaction" (P. Fenichel and J. Parinaud, eds.), pp. 105–122. John Libbey Eurotext, Montrouge, France.

Blandau, R. J. (1967). *In* "The Mammalian Oviduct: Comparative Biology and Methodology" (E. S. E. Hafez and R. J. Blandau, eds.), pp. 129–162. The University of Chicago Press, Chicago.

Bleil, J. D., and Wassarman, P. M. (1980). Mammalian sperm–egg interaction: Identification of a glycoprotein in mouse egg zonae pellucidae possessing receptor activity for sperm. *Cell* **20,** 873–882.

Bleil, J. D., and Wassarman, P. M. (1980). Synthesis of zona pellucida proteins by denuded and follicle-enclosed mouse oocytes during culture *in vitro*. *Dev. Biol.* **76,** 185–202.

Bleil, J. D., and Wassarman, P. M. (1986). Autoradiographic visualization of the mouse egg's sperm receptor bound to sperm. *J. Cell Biol.* **102,** 1363–1371.

Bleil, J. D., and Wassarman, P. M. (1988). Galactose at the nonreducing terminus of O-linked oligosaccharides of mouse egg zona pellucida glycoprotein ZP3 is essential for the glycoprotein's sperm receptor activity. *Proc. Natl. Acad. Sci. U.S.A.* **85,** 5778–6782.

Bleil, J. D., Greve, J. M., and Wassarman, P. M. (1988). Identification of a secondary sperm receptor in the mouse egg zona pellucida: Role in maintenance of binding of acrosome-reacted sperm to eggs. *Dev. Biol.* **128,** 376–385.

Bleil, J. D., Beall, C. F., and Wassarman, P. M. (1981). Mammalian sperm–egg interaction: Fertilization of mouse eggs triggers modification of the major zona pellucida glycoprotein, ZP2. *Dev. Biol.* **86,** 189–197.

Borders, C. L., and Raftery, M. A. (1968). Purification and partial characterization of testicular hyaluronidase. *J. Biol. Chem.* **243,** 3756–3762.

Braden, A. H. W., Austin, C. R., and David, H. A. (1954). The reaction of the pellucida to sperm penetration. *Aust. J. Biol. Sci.* **7,** 391–409.

Bradley, M. P., and Garbers, D. L. (1983). The stimulation of bovine caudal epididymal sperm forward motility by bovine cumulus-egg complexes *in vitro*. *Biochem. Biophys. Res. Commun.* **115,** 777–787.

Brown, C. R. (1975). Distribution of hyaluronidase in the ram spermatozoon. *J. Reprod. Fertil.* **45,** 537–539.

Burkman, L. J., Coddington, C. C., Franken, D. R., Krugen, T. F., Rosenwaks, Z., and Hogen, G. D. (1988). The hemizona assay (HZA): Development of a diagnostic test for the binding of human spermatozoa to the human hemizona pellucida to predict fertilization potential. *Fertil. and Steril.* **49,** 688–697.

Byskov, A. G., Yding Andersen, C., Hossaini, A., and Guoliang, X. (1997). Cumulus cells of oocyte–cumulus complexes secrete a meiosis-activating substance when stimulated with FSH. *Mol. Reprod. Dev.* **46,** 296–305.

Camaioni, A., Salustri, A., Yanagishita, M., and Hascall, V. C. (1996). Proteoglycans and proteins in the extracellular matrix of mouse cumulus cell-oocyte complexes. *Arch. Biochem. Biophys.* **325,** 190–1988.

Cardullo, R. A., and Cone, R. A. (1986). Mechanical immobilization of rat sperm does not change their oxygen consumption rate. *Biol. Reprod.* **34,** 820–830.

Carrell, D. T., Middleton, R. G., Peterson, C. M., Jones, K. P., and Urry, R. L. (1993). Role of the cumulus in the selection of morphologically normal sperm and induction of the acrosome reaction during human in vitro fertilization. *Arch. Androl* **31,** 133–1377.

Cecconi, S., D'Aurizio, R., and Colonna, R. (1996). Role of antral follicle development and cumulus cells on *in vitro* fertilization of mouse oocytes. *J. Reprod. Fertil.* **107,** 207–214.

Chamberlin, M. E., and Dean, J. (1990). Human homolog of the mouse sperm receptor. *Proc. Natl. Acad. Sci. U.S.A.* **87,** 6014–6018.

Chang, Y. S., Wang, S. C., Tsao, C. C., and Huang, F. L. (1996). Molecular cloning, structural analysis, and expression of carp ZP3 gene. *Mol. Reprod. Dev.* **44,** 295–304.

Chauhan, M. S., Singla, S. K., Palta, P., Manik, R. S., and Madan, M. L. (1999). Effect of epidermal growth factor on the cumulus expansion, meiotic maturation and development of buffalo oocytes *in vitro*. *Vet. Rec.* **144,** 266–267.

Chen, J., Litscher, E. S., and Wassarman, P. M. (1998). Inactivation of the mouse sperm receptor, mZP3, by site-directed mutagenesis of individual serine residues located at the combining site for sperm. *Proc. Natl. Acad. Sci. U.S.A.* **95,** 6193–6197.

Chen, L., Mao, S. J., McLean, L. R., Powers, R. W., and Larsen, W. J. (1994). Proteins of the inter-alpha-trypsin inhibitor family stabilize the cumulus extracellular matrix through their direct binding with hyaluronic acid. *J. Biol. Chem.* **269,** 28282–28287.

Chen, L., Russell, P. T., and Larsen, W. J. (1993). Functional significance of cumulus expansion in the mouse: Roles for the preovulatory synthesis of hyaluronic acid within the cumulus mass. *Mol. REprod. Dev.* **34,** 87–93.

Chen, L., Zhang, H., Powers, R. W., Russell, P. T., Larsen, W. J. (1996). Covalent linkage between proteins of the inter-alpha-inhibitor family and hyaluronic acid is mediated by a factor produced by granulosa cells. *J. Biol. Chem.* **271,** 19409–19014.

Cherr, G. N., Lambert, H., Meizel, S., and Katz, D. F. (1986). *In vitro* studies of the golden hamster sperm acrosome reaction: completion on the zona pellucida and induction by homologous soluble zonae pellucidae. *Dev. Biol.* **114,** 119–131.

Cherr, G. N., Meyers, S. A., Yudin, A. I., VandeVoort, C. A., Myles, D. G., Primakoff, P., and Overstreet, J. W. (1996). The PH-20 protein in cynomolgus macaque spermatozoa: Identification of two different forms exhibiting hyaluronidase activity. *Dev. Biol.* **175,** 142–153.

Corselli, J., and Talbot, P. (1987). *In vitro* pernetration of hamster oocyte–cumulus complexes using physiological numbers of sperm. *Dev. Biol.* **122,** 227–242.

Cowan, A. E., Myles, D. G., and Koppel, D. E. (1987). Lateral diffusion of the PH-20 protein on guinea pig sperm: Evidence that barriers to diffusion maintain plasma membrane domains in mammalian sperm. *J. Cell Biol.* **104,** 917–923.

Cross, P. C., and Brinster, R. L. (1970). *In vitro* development of mouse oocytes. *Biol. Reprod.* **3,** 298–307.

Cummins, J. M., and Yanagimachi, R. (1982). Sperm-egg ratios and the site of the acrosome reaction during *in vivo* fertilization in the hamster. *Gamete Res.* **5,** 239–256.

Cummins, J. M., and Yanagimachi, R. (1986). Development of ability to penetrate the cumulus oophorus by hamster spermatozoa capacitated in vitro in relation to the timing of the acrosome reaction. *Gamete Res.* **15,** 187–212.

Dandekar, P., and Talbot, P. (1992). Perivitelline space of mammalian oocytes: Extracellular matrix of unfertilized oocytes and formation of a cortical granule envelope following fertilization. *Mol. Reprod. Dev.* **31,** 135–143.

Davis, B. J., Lennard, D. E., Lee, C. A., Tiano, H. F., Morham, S. G., Wetsel, W. C., and Langenbach, R. (1999). Anovulation in cyclooxygenase-2-deficient mice is restored by prostaglandin E2 and interleukin-1β. *Endocrinology* **140,** 2685–2695.

Doren, S., Landsberger, N., Dwyer, N., Gold, L., Blanchette-Mackie, J., and Dean, J. (1999). Incorporation of mouse zona pellucida proteins into the envelope of *Xenopus laevis* oocytes. *Dev. Genes Evol.* **209,** 330–9.

Downs, S. M., and Hunzicker-Dunn, M. (1995). Differential regulation of oocyte maturation and cumulus expansion in the mouse oocyte-cumulus cell complex by site-selective analogs of cyclic adenosine monophosphate. *Dev. Biol.* **172,** 72–85.

Drobnis, E. Z., Yudin, A. I., Cherr, G. N., and Katz, D. F. (1988). Kinematics of hamster sperm during penetration of the cumulus matrix. *Gamete Res.* **21,** 367–383.

Dubois, D. H., and Shur, B. D. (1995). Cell surface β-1,4-galactosyltransferase. A signal transducing receptor? *Adv. Exp. Med. Biol.* **376,** 105–114.

Epifano, O., and Dean, J. (1994). Biology and structure of the zona pellucida. A target for immunocontraception. *Reprod. Fertil. Dev.* **6,** 319–330.

Epifano, O., Liang, L. F., Familiari, M., Moos, M. L., and Dean, J. (1995). Coordinate expression of the three zona pellucida genes during mouse oogenesis. *Development* **121,** 1947–1956.

Eppig, J. J. (1982). The relationship between cumulus cell-oocyte coupling, oocyte meiotic maturation, and cumulus expansion. *Dev. Biol.* **89,** 268–272.

Eppig, J. J. (1991a). *In* "Elements of Mammalian Fertilization" (P. M. Wassarman, ed.), Vol. 1, pp. 57–76. CRC Press, Boca Raton.

Eppig, J. J. (1991b). Intercommunication between mammalian oocytes and companion somatic cells. *BioEssays* **13,** 569–574.

Eppig, J. J., Peters, A. H., Tefler, E. E., and Wigglesworth, K. (1993). Production of cumulus expansion enabling factor by mouse oocytes grown *in vitro:* Preliminary characterization of the factor. *Mol. Reprod. Dev.* **34,** 450–456.

Familiari, G., Verlengia, C., Nottola, S. A., Renda, T., Micara, G., Aragona, C., Zardi, L., and Motta, P. M. (1996). Heterogeneous distribution of fibronectin, tenascin-C, and laminin immunoreactive material in the cumulus-corona cells surrounding mature human oocytes from IVF-EG protocols—evidence that they are composed of different subpopulations: An immunohistochemical study using scanning confocal laser and fluorescence microscopy. *Mol. Reprod. Dev.* **43,** 392–402.

Farooqui, A. A., and Srivastava, P. N. (1979). Isolation, characterization and the role of rabbit testicular arysulphatase A in fertilization. *Biochem. J.* **181,** 331–337.

Fetterolf, P. M., Jurisicova, A., Tyson, J. E., and Casper, R. F. (1994). Conditioned medium from human cumulus oophorus cells stimulates human sperm velocity. *Biol. Reprod.* **51,** 184–92.

Florman, H. M., and Wassarman, P. M. (1985). O-Linked oligosaccharides of mouse egg ZP3 account for its sperm receptor activity. *Cell* **41,** 313–324.

Florman, H. M., Bechtol, K. B., and Wassarman, P. M. (1984). Enzymatic dissection of the functions of the mouse egg's receptor for sperm. *Dev. Biol.* **106,** 243–255.

Fulop, C., Kamath, R. V., Li, Y., Otto, J. M., Salustri, A., Olsen, B. R., Glant, T. T., and Hascall, V. C. (1997). Coding sequence, exon–intron structure and chromosomal localization of murine TNF-stimulated gene 6 that is specifically expressed by expanding cumulus cell–oocyte complexes. *Gene* **202,** 95–102.

Gadella, B. M., Colenbrander, B., and Lopes-Cardozo, M. (1991). Arylsulfatases are present in seminal plasma of several domestic mammals. *Biol. Reprod.* **45,** 381–386.

Gmachl, M., and Kreil, G. (1993). Bee venom hyaluronidase is homologous to a membrane protein of mammalian sperm. *Proc. Natl. Acad. Sci. U.S.A.* **90,** 3569–3573.

Gmachl, M., Sagan, S., Ketter, S., and Kreil, G. (1993). The human sperm protein PH-20 has hyaluronidase activity. *FEBS Lett.* **336,** 545–548.

Gong, X. H., Dubois, D. H., Miller, D. J., and Shur, B. D. (1995). Activation of a G protein complex by aggregation of β-1,4-galactosyltransferase on the surface of sperm. *Science* **269,** 1718–1721.

Greve, J. M., and Wassarman, P. M. (1985). Mouse egg extracellular coat is a matrix of interconnected filaments possessing a structural repeat. *J. Mol. Biol.* **181,** 253–264.

Grootenhuis, A. J., Philipsen, H. L. A., de Breet-Grijsbach, J. T. M., and Van Duin, M. (1996). Immunocytochemical localization of ZP3 in primordial follicles of rabbit, marmoset, rhesus monkey and human ovaries using antibodies against human ZP3. *J. Reprod. Fertil.* (Suppl.) **50,** 43–54.

Gulyas, B. J. (1980). Cortical granules of mammalian eggs. *Int. Rev. Cytol.* **63,** 267–278.

Guoliang, X., Byskov, A. G., and Andersen, C. L. (1994). Cumulus cells secrete a meiosis-inducing substance by stimulation with forskolin and dibutyric cyclic adenosine monophospate. *Mol. Reprod. Dev.* **39,** 17–24.

Gurevich, M., Harel-Markowitz, E., Marcus, S., Shore, L. S., and Shemesh, M. (1993). Prostaglandin production by the oocyte cumulus complex around the time of fertilization and the effect of prostaglandin E on the development of the early bovine embryo. *Reprod. Fertil. Dev.* **5,** 281–283.

Gwatkin, R. B. L., Williams, D. T., Hartmann, J. F., and Kniazuk, M. (1973) The zona reaction of hamster and mouse eggs: production *in vitro* by a trypsin-like protease from cortical granules. *J. Reprod. Fertil.* **32,** 259–265.

Harris, J. D., Hibler, D. W., Fontenot, G. K., Hsu, K. T., Yurewicz, E. C., and Sacco, A. G. (1994). Cloning and characterization of zona pellucida genes and cDNAs from a variety of mammalian species: The ZPA, ZPB and ZPC gene families. *DNA Seq.* **4,** 361–393.

Harrison, R. (1988a). Hyaluronidase in ram semen. Quantitative determination, and isolation of multiple forms. *Biochem. J.* **252,** 865–874.

Harrison, R. (1988b). Preliminary characterization of the multiple forms of ram sperm hyaluronidase. *Biochem. J.* **252,** 875–882.

Hartmann, J. F., Gwatkin, R. B. L., and Hutchison C. F. (1972). Early contact interactions between mammalian gametes *in vitro:* Evidence that the vitellus influences adherence between sperm and zona pellucida. *Proc. Natl. Acad. Sci. U.S.A.* **69,** 2767–2769.

Hechter, O., and Hadidian, Z. (1947). Hyaluronidase activity of spermatozoa. *Endocrinology* **41,** 204–205.

Hess, K. A., Chen, L., and Larsen, W. J. (1999). Inter-alpha-inhibitor binding to hyaluronan in the cumulus extracellular matrix is required for optimal ovulation and development of mouse oocytes. *Biol. Reprod.* **61,** 436–443.

Hirashima, Y., Kobayashi, H., Gotoh, J., and Terao, T. (1997). Inter-alpha-trypsin inhibitor is concentrated in the pericellular environment of mouse granulosa cells through hyaluronan-binding. *Eur. J. Obstet. Gynecol. Repro. Biol.* **73,** 79–84.

Huang, S., Driessen, N., Knoll, M., and Talbot, P. (1999). *In vitro* analysis of oocyte cumulus complex pickup rate in the hamster *Mesocricetus auratus. Mol. Reprod. Dev.* **47,** 79–84.

Izadyar, F., Colenbrander, B., and Bevers, M. M. (1997a). Stimulatory effect of growth hormone on *in vitro* maturation of bovine oocytes is exerted through the cyclic adenosine 3′,5′-monophosphate signaling pathway. *Biol. Reprod.* **57,** 1484–1489.

Izadyar, F., Van Tol, H. T., Colenbrander, B., and Bevers, M. M. (1997b). Stimulatory effect of growth hormone on *in vitro* maturation of bovine oocytes is exerted through cumulus cells and not mediated by IGF-I. *Mol. Reprod. Dev.* **47,** 175–180.

Izadyar, F., Hage, W. J., Collenbrander, B., and Bevers, M. M. (1998). The promotory effect of growth hormone on the developmental competence of *in vitro* matured bovine oocytes is due to improved cytoplasmic maturation. *Mol. Reprod. Dev.* **49,** 444–453.

Jansen, S., Quigley, M., Reik, W., and Jones, R. (1995). Analysis of polysulfate-binding domains in procine proacrosin, a putative zona adhesion protein from mammalian spermatozoa. *Int. J. Dev. Biol.* **39,** 501–510.

Johnston, D. S., Wright, W. W., Shaper, J. H., Hokke, C. H., Van den Eijnden, D. H., and Joziasse, D. H. (1998). Murine sperm-zona binding, a fucosyl residue is required for a high affinity sperm-binding ligand. A second site on sperm binds a nonfucosylated, β-galacotsyl-capped oligosaccharide. *J. Biol. Chem.* **273,** 1888–1895.

Jones, R. (1990). Identification and functions of mammalian sperm–egg recognition molecules during fertilization. *J. Reprod. Fertil.* (Suppl.) **42,** 89–105.

Joyce, C., Jeyendran, J. S., and Zanevald, L. J. D. (1985). Release, extraction, and stability of hyaluronidase associated with human spermatozoa. Comparisons with the rabbit. *J. Androl.* **6,** 152–161.

Kalab, P., Schultz, R. M., and Kopf, G. S. (1993). Modifications of the mouse zona pellucida during oocyte maturation: Inhibitory effects of follicular fluid, fetuin, and α-2HS-glycoprotein. *Biol. Reprod.* **49,** 561–567.

Katz, D. F., Drobnis, E. Z., and Overstreet, J. W. (1989). Factors regulating mammalian sperm migration through the female reproductive tract and oocyte vestments. *Gamete Res.* **22,** 443–469.

Kiss, H., Walter, I., Lehner, R., Egarter, C., Breitnecker, G., and Bock, P. (1998). Lectin histochemistry of fallopian tube epithelial cells. Relation to ovum transport and ovum pickup. *J. Reprod. Med.* **43,** 535–540.

Kito, S., and Bavister, B. D. (1999). Kinetics of sperm penetration and fertilization *in vitro* in hamster follicular and oviductal ova. *J. Exp. Zool.* **274,** 373–383.

Kolle, S., Sinowatz, F., Boie, G., Totzauer, I., Amelsgruber, W., and Plendl, J. (1996). Localization of the mRNA encoding the zona protein ZP3 alpha in the porcine ovary, oocyte and embryo by non-radioactive *in situ* hybridization. *Histochem. J.* **28,** 441–447.

Lee, V. H., and Dunbar, B. S. (1993). Developmental expression of the rabbit 55-kDa zona pellucida protein and messenger RNA in ovarian follicles. *Dev. Biol.* **155,** 371–382.

Legendre, L. M., and Stewart-Savage, J. (1993). Effect of cumulus maturity on sperm penetration in the golden hamster. *Biol. Reprod.* **49,** 82–88.

Lewin, L. M., Nevo, Z., and Weissenberg, R. (1981). The role of sperm bound hyaluronidase in rat sperm–cumulus oophorus interaction. *Ann. N.Y. Acad. Sci.* **383,** 473–474.

Lewin, L. M., Nevo, Z., Gbsu, A., and Weissenberg, R. (1982). The role of sperm-bound hyaluronidase in the dispersal of the cumulus oophorus surrounding rat ova. *Int. J. Androl.* **5,** 37–44.

Leyton, L., and Saling, P. M. (1989). Evidence that aggregation of mouse sperm receptors by ZP3 triggers the acrosome reaction. *J. Cell Biol.* **108,** 2163–2168.

Li, M. W., Yudin, A. I., VandeVoort, C. A., Sabeur, K., Primakoff, P., and Overstreet, J. W. (1997). Inhibition of monkey sperm hyaluronidase activity and heterologous cumulus penetration by flavonoids. *Biol. Reprod.* **56,** 1383–1389.

Liang, L. F., and Dean, J. (1993). Conservation of mammalian secondary sperm receptor genes enables the promoter of the human gene to function in mouse oocytes. *Dev. Biol.* **156,** 399–408.

Lin, Y., Mahan, K., Lathrop, W. F., Myles, D. G., and Primakoff, P. (1994). A hyaluronidase activity of the sperm plasma membrane protein PH-20 enables sperm to penetrate the cumulus cell layer surrounding the egg. *J. Cell Biol.* **125,** 1157–1163.

Litscher, E. S., Juntunen, K., Seppo, A., Pentilla, L., Niemala, R., Renkonen, O., and Wassarman, P. M. (1995). Oligosaccharide constructs with defined structures that inhibit binding of mouse sperm to unfertilized eggs *in vitro*. *Biochemistry* **34,** 4662–4669.

Loeser, C. R., and Tulsiani, D. R. P. (1999). The role of carbohydrates in the induction of the acrosome reaction in mouse spermatozoa. *Biol. Reprod.* **60,** 94–101.

Lopez, L. C., Bayna, E. M., Litoff, D., Shaper, N. L., Shaper, J. H., and Shur, B. D. (1985). Receptor function of mouse sperm surface galactosyltransferase during fertilization *J. Cell Biol.* **101,** 1501–1510.

Lyons, C. E., Payette, K. L., Price, J. L., and Huang, R C. C. (1993). Expression and structural analysis of a teleost homolog of a mammalian zona pellucida gene. *J. Biol. Chem.* **268,** 21351–21358.

Mahi-Brown, C. A., and Yanagimachi, R. (1983). Parameters influencing ovum pickup by oviductal fimbria in the golden hamster. *Gamete Res.* **8,** 1–10.

Mansour, R. T., Aboulghar, M. A., Serour, G. I., Abbas, A. M., and Elattar, I (1995). The life span of sperm motility and pattern in cumulus coculture. *Fertil. Steril.* **63,** 660–662.

Martinez, M. L., Fortenot, G. K., and Harris, J. D. (1996). The expression and localization of zona pellucida glycoproteins and mRNA in cynomolgus monkeys (*Macaca fascicularis*). *J. Reprod. Fertil.* (Suppl.) **50,** 35–41.

McLeskey, S. B., Dowds, C., Carballada, R., White, R. R., and Saling, P. M. (1998). Molecules involved in mammalian sperm–egg interaction. *Int. Rev. Cytol.* **177,** 57–113.

Metz, C. B., Seiguer, A. C., and Castro, A. E. (1972). Inhibition of the cumulus dispersing and hyaluronidase activities of sperm by heterologous and isologous antisperm antibodies. *Proc. Soc. Exp. Biol. Med.* **140,** 776–781.

Meyers, S. A., Yudin, A. I., Cherr, G. N., VandeVoort, C. A., Myles, D. G., Primakoff, P., and Overstreet, J. W. (1997). Hyaluronidase activity of macaque sperm assessed by an *in vitro* cumulus penetration assay. *Mol. Reprod. Dev.* **46,** 392–400.

Miller, D. J., Macek, M. B., and Shur, B. D. (1992). Complementarity between sperm surface β-1,4-galactosyltransferase and egg-coat ZP3 mediates sperm–egg binding. *Nature* **357,** 589–593.

Miller, D. J., Gong, X., Decker, G., and Shur, B. D. (1993). Egg cortical granule *N*-acetylglucosaminidase is required for the mouse zona block to polyspermy. *J. Cell Biol.* **123,** 1431–1440.

Moller, C. C., and Wassarman, P. M. (1989). Characterization of a proteinase that cleaves zona pellucida glycoprotein ZP2 following activation of mouse eggs. *Dev. Biol.* **132,** 103–12.

Mori, E., Kashiwabara, S., Baba, T., Inagaki, Y., and Mori, T. (1995). Amino acid sequences of porcine Sp38 and proacrosin required for binding to the zona pellucida. *Dev. Biol.* **168,** 575–583.

Mortillo, S., and Wassarman, P. M. (1991). Differential binding of gold-labeled zona pellucida glycoproteins on mZP2 and mZP3 to mouse sperm membrane compartments. *Development* **113,** 141–149.

Murata, K., Sasaki, T., Yasumasu, S., Iuchi, I., Enami, J., Yamamasu, I., and Yamagami, K. (1995). Cloning of cDNAs for the precursor protein of a low-molecular-weight subunit of the inner layer of the egg envelope (chorion) of the fish *Oryzias latipes*. *Dev. Biol.* **167,** 9–17.

Myles, D. G., Hyatt, H., and Primakoff, P. (1987). Binding of both acrosome-intact and acrosome-reacted guinea pig sperm to the zona pellucida during *in vitro* fertilization. *Dev. Biol.* **121,** 559–567.

Nagae, T., Yanagimachi, R., Srivastava, P. N., and Yanagimachi, H. (1986). Acrosome reaction in human spermatozoa. *Fertil. Steril.* **45,** 701–707.

Nagdas, S. K., Araki, Y., Chayko, C. A., Orgebin-Crist, M. C., and Tulsiani, D. R. (1994). O-Linked trisaccharide and N-linked poly-*N*-acetyllactosaminyl glycans are present on mouse ZP2 and ZP3. *Biol. Reprod.* **51,** 262–272.

Nakano, M., Yonezawa, N., Hatanaka, Y., and Noguchi, S. (1996). Structure and function of the N-linked carbohydrate chains of pig zona pellucida glycoproteins. *J. Reprod. Fertil.* (Suppl.) **50,** 25–34.

Nikolajczyk, B. S., and O'Rand, M. G. (1992). Characterization of rabbit testis β-galactosidase and

arylsulfatase A: Purification and localization in spermatozoa during the acrosome reaction. *Biol. Reprod.* **46,** 366–478.
Norwood, J. T., and Anderson, R. G. W. (1980). Evidence that adhesive sites on the tips of oviduct cilia membranes are required for ovum pickup *in situ. Biol. Reprod.* **23,** 788–791.
Norwood, J. T., Hein, C. E., Halbert, S. A., and Anderson, R. G. W. (1978). Polycationic molecules inhibit cilia-mediated ovum transport in the rabbit oviduct. *Proc. Natl. Acad. Sci. U.S.A.* **75,** 4413–4416.
Okuda, K., Uenoyoma, Y., Fujita, U. Y., Iga, K., and Sakamoto, K. (1999). Functional oxytocin receptors in bovine granulosa cells. *Biol. Reprod.* **56,** 625–631.
Pedoto, S. Y., Aarons, D. J., Young, S. B., Poirer, G. R. (1997). Cumulus cell dispersal from murine oocytes by an epididymal guanidinobenzoatase. *J. Exp. Zool.* **279,** 185–188.
Piko, L. (1979). *In* "Fertilization" (C. B. Metz and A. Monroy, eds.), Vol. 2, pp. 325–403. Academic Press, New York.
Prasad, S. V., Wilkins, B., Skinner, S. M., and Dunbar, B. S. (1996). Evaluating zona pellucida structure and function using antibodies to rabbit 55 kDa ZP protein expressed in baculovirus expression system. *Mol. Reprod. Dev.* **43,** 519–529.
Primakoff, P. (1994). Sperm proteins being studied for use in a contraceptive vaccine. *Am. J. Reprod. Immunol.* **31,** 208–210.
Primakoff, P., Hyatt, H., and Myles, D. G. (1985). A role for the migrating sperm surface antigen PH-20 in guinea pig sperm binding to the egg zona pellucida. *J. Cell Biol.* **101,** 2239–2244.
Prochazka, R., Naguyova, E., Brem, G., and Schellander, K. (1998). Secretion of cumulus expansion-enabling factor (CEEF) in porcine follicles. *Mol. Reprod. Dev.* **49,** 141–149.
Richardson, R. T., and O'Rand, M. G. (1996). Site-directed mutagenesis of rabbit proacrosin. Identification of residues involved in zona pellucida binding. *J. Biol. Chem.* **271,** 24069–24074.
Ringuette, M. J., Chamberlin, M. E., Baur, A. W., Sobieski, D. A., and Dean, J. (1988). Molecular analysis of cDNA coding for ZP3, a sperm binding protein of the mouse zona pellucida. *Dev. Biol.* **127,** 287–295.
Rodger, J. C., and Bedford, J. M. (1982). Separation of sperm pairs and sperm–egg interaction in the opossum, *Didelphis virginiana. J. Reprod. Fertil.* **64,** 171–179.
Rogers, B. J., and Morton, B. E. (1973). The release of hyaluronidase from capacitating hamster spermatozoa. *J. Reprod. Fertil.* **35,** 477–487.
Rosiere, T. K., and Wassarman, P. M. (1992). Identification of a region of mouse zona pellucida glycoprotein mZP3 that possesses sperm receptor activity. *Dev. Biol.* **154,** 309–317.
Sabeur, K., Cherr, G. N., Yuddin, A. I., and Overstreet, J. W. (1999). Hyaluronic acid enhances induction of the acrosome reaction of human sperm through interaction with the PH-20 protein. *Zygote* **6,** 103–111.
Sabeur, K., Cherr, G. N., Yudin, A. I., Primakoff, P., Li, M. W., and Overstreet, J. W. (1997). The PH-20 protein in human spermatozoa. *J. Androl.* **18,** 151–158.
Sacco, A. G., Yurewicz, E. C., Subramanian, M. G., and Matzat, P. D. (1989). Porcine zona pellucida: Association of sperm receptor activity with the alpha-glycoprotein component of the Mr = 55,000 family. *Biol. Reprod.* **41,** 523–532.
Saling, P. M., Storey, B. T., and Wolf, D. P. (1978). Calcium-dependent binding of mouse epididymal spermatozoa to the zona pellucida. *Dev. Biol.* **65,** 515–525.
SamAth, V., Cardullo, R. A., and Thaler, C. D. (1997). Role of calcium in mous sperm capacitation and zona pellucida binding. *Mol. Biol. Cell* **8,** 110a.
Sbracia, M., Grasso, J., Sayme, N., Stronk, J., and Huszar, G. (1997). Hyaluronic acid substantially increases the retention of motility in cryopreserved/thawed human spermatozoa. *Hum. Reprod.* **12,** 1949–1954.
Schuetz, A. W., and Dubin, N. H. (1981). Progesterone and prostaglandin secretion by ovulated rat cumulus cell–oocyte complexes. *Endocrinology* **108,** 457–463.
Shalgi, R., and Phillips, D. M. (1988). Motility of rat spermatozoa at the site of fertilization. *Biol. REprod.* **39,** 1207–1213.
Shur, B. D. (1993). Glycosyltransferases as cell adhesion molecules. *Curr. Opin. Cell Biol.* **5,** 854–863.

Shur, B. D., and Hall, N. G. (1982a). Sperm surface galactosyltransferase activities during *in vitro* capacitation. *J. Cell Biol.* **95,** 567–573.

Shur, B. D., and Hall, N. G. (1982b). A role for mouse sperm surface galactosyltransferase in sperm binding to the egg zona pellucida. *J. Cell Biol.* **95,** 574–579.

Singh, B., and Armstrong, D. T. (1997). Insulin-like growth factor-1, a component of serum that enables porcine cumulus cells to expand in response to follicle-stimulating hormone *in vitro. Biol. Reprod.* **56,** 1370–1375.

Singh, B., Rutledge, J. M., and Armstrong, D. T. (1995). Epidermal growth factor and its receptor gene expression and peptide localization in porcine ovarian follicles. *Mol. Reprod. Dev.* **40,** 391–399.

Smith, T. T., Koyanagi, F., and Yanagamachi, R. (1987). Distribution and number of spermatozoa in the oviduct of the golden hamster after natural mating and artificial insemination. *Biol. Reprod.* **37,** 225–234.

Storey, B. T. (1991). Sperm capacitation and the acrosome reaction. *Ann. N.Y. Acad. Sci.* **637,** 459–473.

Suarez, S. S., Katz, D. F., and Meizel, S. (1984). Changes in motility that accompany the acrosome reaction in hyperactivated hamster spermatozoa. *Gamete Res.* **10,** 253–265.

Swyer, G. I. M. (1947). The release of hyaluronidase from spermatozoa. *Biochem. J.* **41,** 413–417.

Talbot, P. (1984). Hyaluronidase dissolves a component in the hamster zona pellucida. *J. Exp. Zool.* **229,** 309–316.

Talbot, P. (1985). Sperm penetration through oocyte investments in mammals. *Am. J. Anat.* **174,** 331–346.

Talbot, P., and DiCarlantonio, G. (1984). The oocyte-cumulus complex: Ultrastructure of the extracellular components in hamsters and mice. *Gamete Res.* **10,** 127–142.

Talbot, P., DiCarlantonio, G., Zao, P. Penkala, J., and Haimo, L. T. (1985). Motile cells lacking hyaluronidase can penetrate the haster oocyte cumulus complex. *Dev. Biol.* **108,** 387–398.

Talbot, P., Geiske, C., and Knoll, M. (1999). Oocyte pickup by the mammalian oviduct. *Mol. Biol. Cell* **10,** 5–8.

Tesarik, J. (1985). Comparison of acrosome reaction-inducing activities of human cumulus oophorus, follicular fluid and ionophore A23187 in human sperm populations of proven fertilizing ability *in vitro. J. Reprod. Fertil.* **74,** 383–388.

Tesarik, J., Drahorad, J., Testart, J., and Mendoza, C. (1990). Acrosin activation follows its surface exposure and precedes membrane fusion in human sperm acrosome reaction. *Development* **110,** 391–400.

Thaler, C. D., and Cardullo, R. A. (1995). Biochemical characterization of glycosylphosphatidylinositol-linked hyaluronidase on mouse sperm. *Biochemistry* **34,** 7788–7795.

Thaler, C. D., and Cardullo, R. A. (1996a). The initial molecular interaction between mouse sperm and the zona pellucida is a complex binding event. *J Biol. Chem.* **271,** 23289–23297.

Thaler, C. D., and Cardullo, R. A. (1996b). Defining oligosaccharide specificity for initial sperm-zona pellucida adhesion in the mouse. *Mol. Reprod. Dev.* **45,** 535–546.

Thall, A. D., Maly, P., and Lowe, J. B. (1995). Oocyte Gal alpha 1,3Gal epitopes implicated in sperm adhesion to the zona pelllucida glycoprotein ZP3 are not required for fertilization in the mouse. *J. Biol. Chem.* **270,** 21437–21440.

Tirone, E., D'Allesandris, C., Hascall, V. C., Siracusa, G., and Salustri, A. (1997). Hyaluronan synthesis by mouse cumulus cells is regulated by interactions between follicle-stimulating hormone (or epidermal growth factor) and a soluble oocyte factor (or transforming growth factor beta1). *J. Biol. Chem.* **272,** 4787–4794.

Topfer-Petersen, E. (1996). Molecular mechanisms of fertilization in the pig. *Reprod. Domest. Anim.* **31,** 93–100.

Topfer-Petersen, E., and Cechova, D. (1990). Zona pellucida induces conversion of proacrosin to acrosin. *Int. J. Androl.* **13,** 190–196.

Topfer-Petersen, E., Steinberger, M., von Eschenbach, C. E., and Zucker, A. (1990). Zona pellucida-binding of boar sperm acrosin is associated with the N-terminal peptide of the acrosin B-chain (heavy chain). *FEBS Lett.* **265,** 51–54.

Tsubamoto, H., Hasegawa, A., Inoue, M., Yamasaki, N., and Koyama, K. (1996). Binding of recombinant pig zona pellucida protein 1 (ZP1) to acrosome-reacted spermatozoa. *J. Reprod. Fertil.* (Suppl.) **50,** 63–67.

Vandevoort, C. A., Cherr, G. N., and Overstreet, J. W. (1997). Hyaluronic acid enhances the zona pellucida-induced acrosome reaction of macaque sperm. *J. Androl.* **18,** 1–5.

Vazquez, M. H., Phillips, D. M., Wassarman, P. M. (1980). Interaction of mouse sperm with purified sperm receptors covalently linked to silica beads. *J. Cell Sci.* **92,** 713–722.

Viggiano, J. M., Herrero, M. B., Cebral, E., Boquet, M. G., and de Gimeno, M. F. (1995). Prostaglandin synthesis by cumulus–oocyte complexes: Effects on *in vivo* fertilization in mice. *Prostaglandins Leukot. Essent. Fatty Acids* **53,** 261–265.

Virji, N., Phillips, D. M., and Dunbar, B. S. (1990). Identification of extracellular proteins in the rat cumulus oophorus. *Mol. Reprod. Dev.* **25,** 339–344.

Wassarman, P. M. (1988). Zona pellucida glycoproteins. *Annu. Rev. Biochem.* **57,** 415–442.

Wassarman, P. M., and Litscher, E. S. (1995). Sperm–egg recognition mechanisms in mammals. *Curr. Top. Dev. Biol.* **30,** 1–19.

Westphal, L. M., el Dansasouri, I., Shimizu, S., Tadir, Y., and Berns, M. W. (1993). Exposure of human spermatozoa to the cumulus oophorus results in increased relative force as measured by a 760 nm laser optical trap. *Hum. Reprod.* **8,** 1083–1086.

Yamasaki, T., Tsubamoto, H., Hagesawa, A., Inoue, M., and Koyama, K. (1996). Genomic organization of the gene for pig zona pellucida glycoprotein ZP1 and its expression in mammalian cells. *J. Reprod. Fertil.* **50,** 19–23.

Yanagamachi, R. (1994). Mammalian fertilization. *In* "The Physiology of Reproduction" (E. Knobil and J. D. Neill, eds.), Vol. 1, pp. 189–317. Raven Press, New York.

Yonezawa, N., Aoki, H., Hatanaka, Y., and Nakano, M. (1995). Involvement of N-linked carbohydrate chains of pig zona pellucida in sperm–egg binding. *Eur. J. Biochem.* **233,** 35–41.

Youakim, A., Hathaway, H. J., Miller, D. J., Gong, X. H., and Shur, B. D. (1994). Overexpressing sperm surface beta 1,4-galactosyltransferase in transgenic mice affects multiple aspects of sperm–egg interactions. *J. Cell Biol.* **126,** 1573–1583.

Yurewicz, E. C., Sacco, A. G., and Subramanian, M. G. (1987). Structural characterization of the Mr = 55,000 antigen (ZP3) of porcine oocyte zona pellucida. Purification and characterization of α- and β-glycoproteins following digestion of lactosaminoglycan with endo-beta-galactosidase. *J. Biol. Chem.* **262,** 564–571.

Yurewicz, E. C., Hibler, D., Fontenot, G. K., Sacco, A. G., and Harris, J. (1993a). Nucleotide sequence of cDNA encoding ZP3 alpha, a sperm-binding glycoprotein from zona pellucida of pig oocyte. *Biochim. Biophys. Acta* **1174,** 211–214.

Yurewicz, E. C., Pack, B. A., Armant, D. R., and Sacco, A. G. (1993b). Porcine zona pellucida ZP3 alpha glycoprotein mediates binding of the biotin-labeled M(r) 55,000 family (ZP3) to boar sperm membrane vesicles. *Mol. Reprod. Dev.* **36,** 382–389.

Yurewicz, E. C., Zhang, S., and Sacco, E. G. (1993c). Generation and characterization of site-directed antisera against an amino-terminal segment of a 55 kDa sperm adhesive glycoprotein from zona pellucida of pig oocytes. *J. Reprod. Fertil.* **98,** 147–152.

Yurewicz, E. C., Sacco, A. G., Gupta, S. K., Xu, N., and Gage, D. A. (1998). Hetero-oligomerization-dependent binding of pig oocyte zona pellucida glycoproteins ZPB and ZPC to boar sperm membrane vesicles. *J. Biol. Chem.* **273,** 7488–7494.

Zao, P. Z. R., Meizel, S., and Talbot, P. (1985). Release of hyaluronidase and β-*N*-acetylhexosaminidase during *in vitro* incubation of hamster sperm. *J. Exp. Zool.* **134,** 63–71.

Zhang, L., Jiang, S., Wozniak, P. J., Yang, X., and Godke, R. A. (1995). Cumulus cell function during bovine oocyte maturation, fertilization, and embryo development *in vitro*. *Mol. Reprod. Dev.* **40,** 338–44.

Zheng, Y., and Martin-DeLeon, P. A. (1997). The murine Spam1 gene: RNA expression pattern and lower steady-state levels associated with the RB(6.16) translocation. *Mol. Reprod. Dev.* **46,** 252–257.

5

Sperm Adhesion to the Extracellular Matrix of the Egg

Ming Bi, Michael J. Wassler,
and Daniel M. Hardy

Department of Cell Biology and Biochemistry, Texas Tech University Health Sciences Center, Lubbock

I. Introduction

Adhesion of the sperm cell to the egg's extracellular matrix is one of the most captivating events in biology. Easily viewed *in vitro* with a microscope, this cellular interaction represents the first direct physical contact between the male and female germ cells. Indeed, as early as 1851, Newport reported sperm penetration of an amphibian egg (Newport, 1851). Some 26 years later, Fol (1877) and Hertwig (1877) independently reported observing sperm–egg interactions using cells from sea urchins and other species. But these early successes belied the complex nature of sperm–egg recognition, which would not be fully appreciated until more than a century later.

Many investigators have sought to identify gamete recognition molecules in a variety of animal species ranging from marine invertebrates to man. Sea urchins and abalones have been the most extensively studied marine invertebrates, and the mouse has been the most popular mammal. These diverse studies have not yet yielded a complete view of the biochemistry subserving gamete interactions in any one species. However, they have established a number of important concepts that collectively provide a framework for ongoing research. In this review, we summarize the research results from which this framework of ideas was built. Our focus is on the generalizations that have emerged from comparative studies, and also the significant remaining gaps in our understanding. Most of the results presented here come from a few well-studied species, and where possible we synthesize these results into general concepts. We have not emphasized the functions of egg structures or the sperm acrosome, because these topics are covered in detail in Chapters 4, 6, and 7 in this volume. Not all published work could be represented in this short review; additional relevant information can also be found in various earlier reviews (Yanagimachi, 1977, 1984, 1994; Wassarman, 1995; Shur *et al.,* 1998).

II. BASIC BIOLOGY OF SPERM–EEM ADHESION

A. DEFINITION OF TERMS

Animal fertilization has attracted the interest of investigators in a wide range of disciplines, from animal sciences and marine biology to cell biology and biochemistry. One consequence of these researchers' different perspectives is that different names have been assigned both to functionally equivalent structures of gametes and to conceptually similar cellular processes in the many animal species that have been studied. Although the unique terms that have developed are important reminders of the extraordinary interspecies variety in gamete interactions, they can also serve as impediments to understanding these processes.

The terms for the major structures of the sperm cell are relatively consistent among species, but this is not true for the investments of the egg. Here we focus on adhesion of spermatozoa to the acellular investment closest to the plasma membrane of the egg.[1] This structure is called the vitelline layer in sea urchins, vitelline envelope in anuran amphibians, chorion in fishes, and zona pellucida (ZP) in most mammals. Similarities in the primary structures of ZP glycoproteins defined a new type of protein extracellular domain called the ZP domain (Prasad *et al.,* 2000). ZP domains are present in the major glycoprotein components of egg investments from all vertebrate species examined thus far. This observed conser-

[1] Even the meaning of "egg" is debatable. One view is that an oocyte becomes an egg when it completes meiosis II (MII). In most of the species covered in this review, including mammals, the sperm cell fertilizes an MII-arrested oocyte, and the second polar body is ejected only after sperm–egg fusion. However, sea urchins' oocytes do complete meiosis II prior to fertilization. Because some of the female gametes discussed here are truly eggs by this criterion, we will use "egg" throughout for simplicity. This usage is consistent with a looser definition of an egg as a female germ cell released from the ovary by ovulation.

vation of structure (gene products with ZP domains) established the common evolutionary origin of egg investments in diverse species. Together with known conservation of function (gamete adhesion), the developing evidence of shared biochemical properties supports the use of more general terms. Here, we use "egg extracellular matrix" (EEM) for the structure to which spermatozoa adhere in a species-specific manner (i.e., the functional equivalent of the mammalian ZP). In most animals used for fertilization research, the EEM is largely if not exclusively the product of the egg.

In addition to the various labels used for actual physical structures of the gametes, multiple terms have also developed for interactions between sperm and egg. In early studies of hamster spermatozoa, two types of contact between sperm cells and the EEM were described. These were an initial, loose interaction termed "attachment," and a subsequent, more secure interaction termed "binding" (Hartmann *et al.,* 1972; Hartmann and Hutchison, 1974; Yanagimachi, 1994). Since these studies were published, "binding" has become the most commonly used term for sperm–EEM interaction. However, this idiosyncratic usage of "binding" unnecessarily sets the field of fertilization research apart from the rest of biochemistry and cell biology. In this review, we use the term "adhesion" for the sustained interaction of sperm cells with the EEM that ultimately leads to fertilization. This term is preferable because of its standard usage in reference to interactions of somatic cells. "Binding" will instead refer to molecular events such as interactions of enzymes with their substrates, receptors with their ligands, and adhesion molecules with each other.

Finally, studies on the functional properties of mammalian EEM glycoproteins have led in the mouse to the identification of ZP3 as the "receptor" for spermatozoa (see below). However, ZP3 is certainly not a receptor in the same sense as the receptors for catecholamines, growth factors, and countless other ligands. Thus, in keeping with our desire to be consistent with the rest of biochemistry and cell biology, we will use "adhesion molecule" for mediators of cell–cell interactions, and reserve the terms "ligand" and "receptor," respectively, for the agonists and transducers of cell surface signals.

B. STAGES OF ADHESION AND INTERACTING STRUCTURES

Sperm–egg interactions are complicated by dramatic changes in the sperm head that occur during fertilization. Sperm populations present a continuum of physiological and morphological states to the egg. Sperm cells may be uncapacitated or capacitated, normomotile or hypermotile, acrosome intact or undergoing the acrosome reaction. A given cell may make contact with the EEM at any point in this continuum, and the likelihood that this cell is in a particular state depends on the timing of mating relative to ovulation. Morphologically, an interacting sperm cell may have an intact acrosome, or it may be in any of several stages of the acrosome reaction. In mammals, sperm structures that interact with the EEM include the

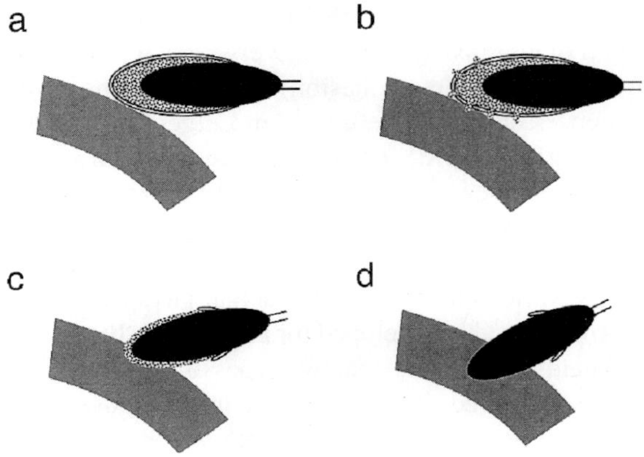

FIGURE 5.1 Morphology of the sperm apical head in interactions with the mammalian EEM. (a) Interaction of the plasma membrane overlying the head of a sperm cell that has an intact acrosome. (b) Interaction of the outer aspect of the acrosomal matrix that has emanated through discontinuities in the plasma and outer acrosomal membranes in the very early stages of acrosomal exocytosis. (c) Interaction of acrosomal matrix remnants later in the progression of acrosomal exocytosis. (d) Interaction of the inner acrosomal membrane after completion of acrosomal exocytosis.

plasma membrane, the outer zone of the acrosomal matrix (which first becomes exposed in initial stages of the acrosome reaction), the more interior regions of the acrosomal matrix, and the inner acrosomal membrane (Figure 5.1). When acrosome-intact spermatozoa adhere to the EEM, the obvious physical requirement that adhesion be sustained as the acrosome reaction progresses suggests strongly that multiple adhesion molecules function in this interaction. Indeed, this requirement for sustained adhesion during EEM penetration presents one of the most interesting biophysical problems of fertilization (see Chapter 13, this volume).

Adhesion molecules in each of the sperm structures described above likely interact with one or more component of the EEM. Wolf *et al.* (1976) first reported isolation and biochemical characterization of the EEM from a vertebrate species, the frog *Xenopus laevis.* Four years later, Dunbar *et al.* (1980) and Bleil and Wassarman (1980a) described the molecular composition of the EEM from two mammals (pig and mouse, respectively). Since then, the structures of the EEM glycoprotein components and the functions of these molecules in gamete interactions have been studied extensively. Efforts to ascribe discrete functions to individual EEM glycoproteins have been partly successful within certain species, but these findings do not seem to hold true even for most, let alone all, species (see Section II,E).

C. SPECIES DIVERSITY OF CELLULAR EVENTS

Research using sea urchin gametes established three fundamental concepts of sperm–egg adhesion in echinoderms: (1) egg substances induce acrosome reactions in spermatozoa; (2) the sperm cell's primary adhesive substance, bindin, is in the acrosome, so the acrosome reaction must occur before the gametes adhere; and (3) bindin interacts in a species-specific manner with its cognate ligand in the egg's vitelline layer. These well-established findings at once suggest possibilities for sperm–EEM adhesion in mammals, but when compared to results obtained in other species, also demonstrate existence of species differences (see below).

Many interactions of somatic cells are largely similar in all animal species (for example, signaling between neurons). In contrast, gamete interactions can differ markedly even between closely related species. In species that fertilize externally, such as sea urchins and amphibians, jelly layers applied to the surface of the egg are derived from somatic tissues in the reproductive tract. At least one jelly component appears to possess the acrosome reaction-inducing activity in these species, but as previously mentioned this activity resides in the EEM of mammalian eggs. Furthermore, the vitelline layer site of sea urchin eggs is the site of species-specific recognition, but this structure is not exactly analogous to the EEM in mammals, either morphologically or in the evolutionary origin of its polypeptide components. It is therefore clear that not all functions of the EEM proper are conserved among species. Such functional and molecular interspecies differences, along with the known dramatic species differences in sperm morphology, demonstrate that multiple species must be studied before valid generalizations about gamete interactions can be made. They also raise the possibility that no conclusions about fertilization processes will hold true for all species.

Early studies of mammalian gametes demonstrated that sperm–EEM interaction is more complex than the corresponding process in echinoderms. In contrast to echinoderm spermatozoa, which are fertile immediately on spawning, mammalian spermatozoa must undergo capacitation in the female to acquire fertility (see Chapter 3). In addition, the mammalian acrosome contains dozens if not hundreds of proteins (Myles *et al.,* 1981), whereas the sea urchin acrosome is composed almost entirely of bindin (Glabe and Vacquier, 1977). Furthermore, spermatozoa of several (and possibly most) mammals can adhere to the EEM both before and after they undergo acrosome reactions. Finally, no differing, progressive interactions between echinoderm gametes have been described that are comparable to the initial "attachment" and a subsequent "binding" that occurs in mammalian sperm adhesion. Although it is tempting to connect the two, the relationship between the strength of the interaction between mammalian gametes and acrosomal status of an interacting sperm cell has not been clarified.

In mammals, sperm adhesion to the EEM activates signaling cascades that control the sperm acrosome reaction. Solubilized EEM activates G-proteins in the mouse sperm plasma membrane (Ward and Kopf, 1993; see also Chapter 6). Pertussis toxin, which ADP-ribosylates the α subunits and inactivates G_i-like G-proteins, blocks both activation of sperm G-proteins and induction of the acrosome reaction by the EEM (Ward and Kopf, 1993). Sperm G-proteins are localized primarily to the apical sperm head, overlying the acrosome (Glassner *et al.,* 1991). Thus, signaling in the acrosome reaction proceeds through a path that involves

G-proteins of the G_i class (Ward *et al.*, 1992). In addition, tyrosine phosphorylation of a 95,000 M_r putative EEM receptor (p95) increases as a consequence of surface protein aggregation by the EEM (Leyton and Saling, 1989a), suggesting that tyrosine kinases are also involved in this process. Activation of phospholipase C and release of Ca^{2+} from intracellular stores by the inositol 1,4,5-triphosphate (IP_3) receptor may also be required (Walensky and Snyder, 1995). Finally, adenylyl cyclase is activated during the acrosome reaction, at least partly in response to increases in intracellular Ca^{2+} concentrations (Hyne and Garbers, 1979; Leclerc and Kopf, 1995). This unique enzyme, which was isolated and its cDNA characterized by Buck *et al.* (1999), has properties unlike those of all other mammalian cyclases, in that it is stimulated by bicarbonate but is unresponsive to forskolin or to regulation by G-proteins. These observations have not yet been integrated into a comprehensive understanding of cellular signaling during the mammalian acrosome reaction, largely because the receptor for the EEM that activates acrosome reaction signaling has not been identified unequivocally.

In sea urchins, one or more components of the egg jelly induce the acrosome reaction, but the signaling cascades have not been characterized as well as they have in mammals (Hardy and Garbers, 1993). As in mammals, adenylyl cyclase is present in sea urchin spermatozoa, but it is not clear whether the enzyme is the ortholog of the unique cyclase in mammalian spermatozoa (Hardy and Garbers, 1993). The major difference in cell signaling components between echinoderm and mammalian spermatozoa is guanylyl cyclase (Hardy *et al.*, 1994). Despite extensive efforts, this activity has never been clearly demonstrated in mammalian spermatozoa, yet in sea urchin sperm cells the amount of this enzyme, in the form of membrane guanylyl cyclase receptors for egg peptides (speract in *Strongylocentrotus purpuratus* and *Lytechinus pictus;* resact in *Arbacia punctulata*) (Hardy and Garbers, 1993), is higher than in any other cell type that has been tested.

D. THE ACROSOME REACTION CONTROVERSY

The relationship between the acrosome reaction and adhesion is of major importance to understanding gamete interactions. As described above, the morphological state of a mammalian sperm cell interacting with the EEM can vary markedly. The "acrosome reaction controversy" was born of debate over the timing of essential adhesion events relative to exocytosis of the acrosome. Huang *et al.* (1981) reported that, in guinea pig, only acrosome-reacted spermatozoa adhere to the EEM. These results were consistent with those obtained using sea urchin gametes, which had demonstrated that the adhesive substance bindin is located within the acrosome. However, data previously obtained using mouse gametes solidly supported the conflicting view that spermatozoa must adhere to the EEM before they undergo the acrosome reaction (Saling *et al.*, 1979; Saling and Storey, 1979). This view was further bolstered by evidence that guinea pig spermatozoa with intact acrosomes could indeed adhere productively to the EEM (Myles *et al.*, 1987), and that the EEM can induce the acrosome reaction (Bleil and Wassarman, 1983). Morales *et al.* (1989) then found that human spermatozoa are also capable of adhering to the EEM both before and after they undergo the acrosome reaction. Collectively, these and other studies (Yanagimachi and Phillips, 1984; Uto *et al.*, 1988; Cherr *et al.*, 1986; Shalgi *et al.*, 1989; Crozet and Dumont, 1984; Crozet *et al.*, 1987) suggest that spermatozoa from most species are fully capable of fertilizing the egg regardless of whether they make initial EEM contact with their acrosomes intact or in various stages of exocytosis. Nevertheless, because the mouse model dominates studies of mammalian fertilization, the notion has become entrenched that sperm plasma membrane–EEM adhesion represents an obligatory first physical contact between sperm and egg. This interaction is often called "primary" adhesion, which implies a greater importance to the overall process of fertilization than the "secondary" interaction of acrosome-reacting cells or cells that have completed exocytosis (Bleil *et al.*, 1988). However, notwithstanding the results of many studies on mouse gamete adhesion, it is entirely possible that "primary" adhesion is not a predominant interaction in many species. Partly as a result of this possibility, there is an ongoing evolution of thinking about acrosome function in fertilization (see Chapter 7).

E. SPECIES SPECIFICITY

Sperm adhesion to the EEM seems conceptually analogous to adhesion of somatic cells to the extracellular matrix. Both processes are mediated by interaction of cell surface proteins with complementary components of an acellular structure. However, sperm–EEM adhesion is fundamentally different from somatic adhesion processes in at least one way: it exhibits relative or absolute species specificity (Yanagimachi, 1981, 1994; Vacquier, 1998). Indeed, among the many steps in mammalian fertilization, adhesion of spermatozoa to the EEM exhibits the greatest degree of species specificity (Yanagimachi, 1981). Without the EEM, the direct exposure of the egg plasma membrane to spermatozoa permits heterologous fertilization between some species (Yanagimachi, 1972). The most striking example is the egg of the golden hamster, which is promiscuous when devoid of the EEM and can be penetrated by spermatozoa from many if not most mammals (Yanagimachi, 1977). Indeed, this property is the basis of a test that can be used clinically to assess the function of spermatozoa from the male partners of infertile couples (Longo and Yanagimachi, 1993; Overstreet *et al.*, 1980). Although EEM removal does not cause complete loss of species selectivity in all mammals (Yanagimachi, 1994), results from such loss-of-function studies support the view that the EEM serves as a major barrier for interspecific fertilization.

The species specificity of sperm–EEM interaction constitutes some of the most compelling evidence that spermatozoa possess adhesion molecules capable of specifically recognizing the EEM, because such specificity is unlikely to arise from a general adhesiveness on the part of the sperm, the EEM, or both. Indeed, the demonstration that heterologous spermatozoa do not adhere to the EEM of pig eggs was among the first evidence that mammalian sperm–EEM interaction was not merely a consequence of nonspecific adhesiveness (Peterson *et al.*, 1980). Rather, this and other similar observations support the prevailing view that adhesion between spermatozoa and the EEM is mediated by unique, complementary factors located on the exposed surfaces of these cells. In sea urchins, this view was confirmed by the discovery of bindin, which demonstrated that species specificity of sea urchin gamete adhesion may be conferred by the species-specific binding of a single sperm protein to one or more ligand in the EEM (Vacquier and Moy, 1977). Unfortunately, we have not achieved a comparable molecular understanding of species specificity in mammalian gamete adhesion.

How might Nature have produced such remarkable specificity in cell–cell interactions that recognition between gametes from closely related species does not occur? There is little doubt that such specificity must come from species diversity in the molecular components that mediate sperm–EEM interactions. Such molecular diversity could arise in at least four ways (Figure 5.2). Each species could have its own unique set of complementary adhesion molecules. In this scenario, completely different sets of gene products mediate adhesion in different species. However, the major glycoproteins of the EEM are encoded by essentially the same genes (ZP1, ZP2, and ZP3) in most if not all mammals (Dunbar *et al.,* 1994), and it seems unlikely that the active sperm molecules would be species-unique even though the genes for their complementary targets in the EEM are highly conserved. In addition, the sea urchin results described above, wherein a single gene product (bindin) in sperm cells mediates species-specific adhesion, as well as the species-specific interaction of abalone lysin with its cognate target VERL (see below), demonstrate that the required specificity may be achieved without species-unique gene products. Furthermore, some closely related animal species can cross-fertilize (albeit inefficiently), suggesting that there is at least some between-species similarity in the active sperm proteins. A second possibility is that functionally distinct adhesion molecules arose by evolutionary divergence of ancestral adhesion molecule genes. In this model, the structures of adhesion molecules on the sperm cell surface evolve in concert with changes in the EEM. Such changes can occur because the primary selective pressure is for a given species' complementary sperm and egg adhesion molecules to stay compatible with each other, and not necessarily to stay the same as the ancestral molecules. Hence the nature of the adhesion molecule pair would be relatively free to change so long as within-species compatibility was maintained. This process would lead ultimately to development of functionally unique sets of adhesion molecules in each species that are nevertheless recognizably similar in interspecies sequence comparisons. This model is strongly supported by the results from studies of sea urchin (mentioned above) and abalone fertilization (see below), but whether the concept applies to the more complex interactions of mammalian gametes is unclear. A third possibility is that the same gene products are present in gametes of multiple species, but that species-unique combinations of these molecules confer species specificity. Finally, a fourth possibility is that all of the above are true; gene products common to all mammals but also highly divergent between species may act along with species-unique gene products to mediate adhesion. To distinguish between these possible mechanisms of species specificity, interspecies comparisons of sperm–EEM adhesion molecules are required.

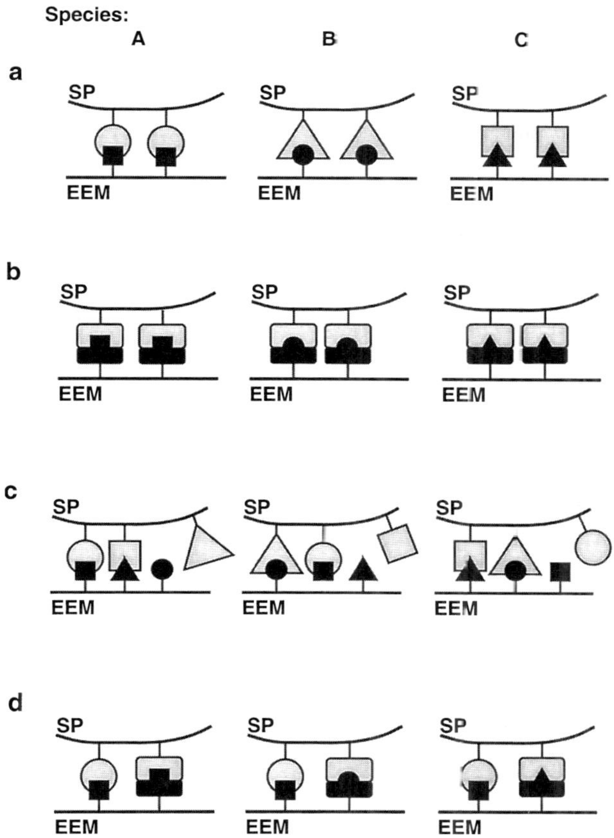

FIGURE 5.2 Hypothetical molecular mechanisms for species specificity. Interactions between spermatozoa (SP) and egg extracellular matrix (EEM) of three species (A, B, and C) are depicted. (a) Each species has a unique adhesion molecule pair. (b) Each species has the same adhesion molecule pair that has developed species-specific function through divergence from a common ancestral protein. (c) Each species uses a unique combination of proteins shared by the three species. (d) Combination of b and c.

Studies on the functions of individual EEM components have established that they can retain at least some of the activity of the intact structure, but that their activities vary among species. EEMs from a wide variety of mammals all comprise a limited number of glycoproteins that are closely related to mouse ZP1, ZP2, and ZP3 (Dunbar *et al.,* 1994). Even the EEMs of nonmammals such as fish, birds, and amphibians are composed of glycoproteins homologous to the mammalian ZP glycoproteins. In the mouse, the EEM component that mediates sperm adhesion is ZP3. This conclusion is based primarily on three observations. First, after disruption of the native structure of the EEM, only ZP3 retains the ability to block sperm adhesion (Bleil and Wassarman, 1980b). Such adhesion inhibition activity of mouse ZP3 has further been attributed to a specific subset of its O-linked oligosaccharides (Florman *et al.,* 1984; Florman and Wassarman, 1985). Second, purified ZP3 (but not ZP1 or ZP2) can induce the sperm acrosome reaction, an exocytotic event that is a physiological consequence of sperm–egg adhesion and is required for sperm penetration of the EEM (Bleil and Wassarman, 1983). Finally, ZP3 expressed by cultured cells transfected with a ZP3 cDNA exhibits both the adhesion-inhibition and acrosome-reaction-inducing activities of the glycoprotein isolated from disrupted ZP (Kinloch *et al.,* 1991; Beebe *et al.,* 1992). Similar results are obtained in the hamster, although this species has not been studied as extensively as the mouse (Moller *et al.,* 1990). Collectively these results represent major progress in our understanding of the rodent EEM and of sperm adhesion to it. However, spermatozoa *in vivo* encounter an intact EEM whose complex structure may be important in gamete interactions. Even though many different studies point to ZP3 as the adhesive component of the mouse EEM, it is quite possible that the other components support or modulate this activity in the context of the intact EEM. Indeed, quantitative studies (Thaler and Cardullo, 1996) have determined that adhesion of mouse spermatozoa to the EEM is a complex interaction that may not be explainable by the simple binding of ZP3 to a complementary adhesion molecule on the sperm cell.

Biochemical studies of nonrodent EEM glycoproteins have been difficult because of the heterogeneity of these molecules (Dunbar *et al.,* 1980, 1994). Nevertheless, sperm adhesion activity of the pig EEM has been attributed primarily to one of the M_r 55,000 glycoproteins, designated ZP3α (Sacco *et al.,* 1989). Surprisingly, cDNA cloning revealed that ZP3α is not the same gene product as the glycoprotein designated ZP3 that exhibits adhesion activity in mouse and hamster eggs (Yurewicz *et al.,* 1993). Rather, this activity appears to reside in the pig ortholog of the rabbit glycoprotein R55 (Schwoebel *et al.,* 1996) and the mouse glycoprotein ZP1 (Epifano *et al.,* 1996). Furthermore, glycans of the pig EEM that confer adhesion activity are N-linked (Yonezawa *et al.,* 1995), not O-linked as observed in rodents. Thus despite the limitations of these experiments (the necessary use of solubilized EEM or its constituent glycoproteins), it appears that the molecular basis of sperm–EEM adhesion may be significantly different in mice and pigs. This species variation in EEM glycoprotein function probably contributes to the species specificity of sperm–EEM adhesion.

The relationship of species specificity to evolution is perhaps the single most important question addressed by fertilization research. Reproductive isolation is a key criterion in most definitions of "species." But what are the primary determinants of reproductive isolation, and how do they arise? What is the relationship of reproductive isolation to the speciation process? Subpopulations of an ancestral species can diverge to become two new species, but this cannot occur if the populations interbreed and thereby homogenize genetic information among most or all members (Li, 1997). But when interbreeding is prevented by geographic isolation, speciation can and does occur. This is called allopatric speciation, and it requires that the two populations be separated long enough for them to acquire traits that prevent interbreeding if they are reunited. Such traits could be mating barriers (e.g., physical inability to copulate, or incompatible mating behaviors) or altered cellular or molecular interactions (e.g., failure of sperm–egg recognition). Thus in allopatric speciation, the species specificity of fertilization may be considered an expected *consequence* of the divergence that occurs in the absence of the homogenizing effect of interbreeding.

In contrast to allopatric speciation, sympatric speciation occurs in the absence of geographic isolation. Exactly how interbreeding is prevented in sympatric speciation is unclear, but results of several studies have shown that species-specific interactions of the gametes may not only contribute to, but actually drive this process (Vacquier, 1998). Marine invertebrates reproduce by broadcasting their gametes into the surrounding sea water, and opportunities for heterospecific fertilization are plentiful. For example, the habitats of seven species of California abalone overlap. Because anatomic or behavioral traits do not preclude chance contact between gametes of different species, mate choice is determined primarily by the species specificity of gamete interactions. The abalone sperm protein lysin mediates species-specific recognition and penetration of the EEM. The structure of lysin has diverged at a rate that is 2–50 times faster than the rate of rapidly evolving proteins that are not involved in gamete interactions (Vacquier, 1998). The EEM protein that binds lysin, VERL, is a mosaic protein with many nearly identical domains (Swanson and Vacquier, 1997). VERL proteins in the different species have diverged by a combination of rapid change between species, followed by concerted evolution within species to distribute changes among the repeated domains (Swanson and Vacquier, 1998; Vacquier, 1998). Vacquier has proposed a model for the evolution of species specificity wherein mutations in a single VERL domain occur first, then spread to the other domains by concerted evolution, thereby applying selective pressure that drives the adaptation of the lysin molecule. This "positive darwinian selection" is proving to be a hallmark not only of gamete recognition proteins, but also of other proteins that are required for proper gamete function (Wyckoff *et al.,* 2000). Though commonly called "positive" selection, it is perhaps more accurate to think of this as selecting against a lack of change. Simply put, abalone sperm cannot fertilize if their lysin proteins do not adapt and thereby retain ability to bind to VERL; consequently, the genetic material of nonadapted males is not passed on to the next generation. This phenomenon is not limited to marine invertebrates, as recent studies have found that the sequences of vertebrate reproductive genes, in both males and females, diverge at a higher rate than expected for neutral evolution (Wyckoff *et al.,* 2000; Swanson *et al.,* 2001).

III. IDENTIFICATION OF EEM ADHESION MOLECULES

A. ASSUMPTIONS

Adhesion of somatic cells to each other or to the extracellular matrix is mediated either by interactions between complementary extracellular protein domains, or by interaction of a protein domain with carbohydrate. Somatic cell adhesion molecules are generally plasma membrane proteins or extracellular matrix components containing one or more active protein domain (Kreis and Vale, 1993; Doolittle, 1995). The vast array of known adhesion molecules reflects the complexity of cell–cell interactions in multicellular organisms. Structural diversity of cell adhesion molecules in these organisms has arisen from the duplication of various extracellular domain types and the recombination of these duplicated domains into large, "mosaic" proteins (Doolittle, 1995). This process was especially pronounced during the metazoan radiation, and gave rise to many families of rapidly evolving adhesion molecules with a wide range of functional specificities. Given the precedents set by studies of somatic cell interactions, it seems likely that gamete adhesion molecules will be cell surface proteins with one or more identifiable adhesion domains. In addition, because of their unique functions, it also seems likely that gamete recognition molecules will be expressed solely in germ cells and not in somatic cells.

The relative simplicity of the sea urchin system, wherein bindin mediates sperm adhesion to the vitelline layer, suggests that a single mammalian sperm component might mediate EEM recognition and adhesion. Furthermore, one can consider the seemingly clear-cut evidence from the mouse that ZP3 serves a dual role as an adhesion molecule and a ligand for induction of the acrosome reaction as additional support for the "single-component" hypothesis. Perhaps for these reasons, much of the research for the past 15 years has been conducted under the apparent assumption that a single protein in the sperm plasma membrane mediates a "primary" adhesion event and, in response to the ZP3 agonist, transduces the signal that initiates the sperm acrosome reaction. However, one must also consider both the asynchrony of mating relative to ovulation in mammals, and the absolute necessity of fertilization for reproductive success. Taking account of these observations, it seems likely that gametes are equipped with multiple, redundant adhesion systems in order to maximize chances that fertilization will occur.

B. TYPES OF MOLECULES IDENTIFIED

The identity of the sperm protein(s) that mediate adhesion to the mammalian EEM is controversial. Conflicting reports in the literature reflect the inherent difficulties of the experiments and the diversity of methods used. Whether the conflicts also reflect fundamental interspecies differences in the underlying biochemistry of this process is still unclear. The amino acid sequences of several possible adhesion molecules are now known. These molecules can be thought of as falling into two categories: unique adhesion molecules and hijacked enzymes (see Section III,C). Those in the unique adhesion molecule category are often germ cell-specific proteins with no known catalytic activity. In contrast, the hijacked enzymes have known enzymatic activities, but presumably do double duty as adhesion molecules. Precedent for such adaptation of enzymes for nonenzymatic uses comes from the lens crystallines, which are glycolytic enzymes present at high concentrations (Piatigorsky and Wistow, 1989). It is important to note that three of the hijacked enzymes (hyaluronidase, hexokinase, and aryl sulfatase) were identified first as novel, putative adhesion molecules (PH-20, p95, and SLIP, respectively); their enzyme activities were discovered years after their first characterization as proteins. Of course, it is possible that some proteins currently considered to be unique, sperm-specific adhesion molecules will later be found also to be enzymes. Finally, nearly all of the known EEM adhesion molecules are sperm-specific proteins that are either products of genes uniquely expressed in germ cells, or they are variants produced by tissue-specific splicing. Both of these processes can readily give rise to molecules with sperm-specific activity.

The properties of several proposed EEM adhesion molecules from mammalian spermatozoa are summarized in the following section. Unfortunately, the methods typically used to identify candidate adhesion molecules can sometimes produce misleading results. Some of the proteins were identified by inhibition studies (biochemical "loss-of-function" experiments), which are only as selective as the inhibitors used. Others were identified by using denatured sperm proteins, denatured EEM glycoproteins, or both in assays for binding activity (e.g., blot overlays or cross-linking experiments). It is therefore uncertain whether these proteins can bind to the native EEM under physiological conditions. Most or all of these candidate proteins may be involved in EEM adhesion. However, it is important to note that no positive demonstration of adhesion activity (a "gain-of-function" experiment) has been reported for any putative adhesion molecule. Specifically, there has been no demonstration that addition of one of these molecules to the surface of a cell or other structure makes that cell/structure able to adhere to the native EEM. Thus multiple proteins may be necessary for adhesion, but none has yet been shown to be sufficient.

C. PROPERTIES OF ADHESION MOLECULE CANDIDATES

1. Hijacked Enzymes

a. Galactosyltransferase

Galactosyltransferase (GalTase) is perhaps the most well-characterized member of the glycosyltransferase family of enzymes (Shur, 1989). Glycosyltransferases are primarily located in the Golgi apparatus, where they function in the synthesis of complex glycoconjugates. These enzymes sequentially add monosaccharides to the nonreducing termini of growing polysaccharide chains. In this essential enzymatic capacity, they exhibit specificity both for their "donor" sugar nucleotide and their "recipient" oligosaccharide substrates. In the early 1970s glycosyltransferases were also found, albeit in small amounts, on the surfaces of a variety of cell types (Roseman, 1970; Roth *et al.*, 1971). Subsequent experiments verified the surface location of some glycosyltransferases. For example, intact cells glycosylated both high-molecular-weight glycoproteins and also glycosides bound to inert glass and plastic surfaces, consistent with a cell surface localization of these enzymes

(Turley and Roth, 1979). The adjunct surface localization of glycosyltransferases suggested they might mediate processes distinct from their known roles in biosynthesis of glycoconjugates. Consequently, the potential function of these enzymes, especially GalTase, in cell adhesion has been characterized extensively.

The GalTase gene encodes two proteins, both of which have a type II membrane protein configuration. Both isoforms possess a small amino-terminal cytoplasmic domain, a short transmembrane segment, and a large, carboxy-terminal catalytic domain. The two isoforms differ in length and are derived from distinct mRNAs generated by alternative transcription initiations. The short form is mainly located in the Golgi apparatus. The long form possesses an additional 13-amino acid peptide that targets GalTase to the cell surface. This plasma membrane form of GalTase has been found to be important in a variety of cellular functions, primarily adhesion events. The predominant GalTase transcript in spermatogenic cells encodes the long form of the enzyme (Lopez *et al.,* 1991). This transcript is synthesized by pachytene spermatocytes, and the nascent polypeptide is distributed during spermatogenesis, presumably by the actin cytoskeleton, to what will become the anterior aspect of the mature sperm cell's head (Scully *et al.,* 1987; Shur and Neely, 1988).

The function of GalTase as a gamete adhesion molecule has been studied primarily in the mouse. The possible importance of GalTase in mammalian fertilization was first suggested by studies on mice with mutant alleles in the *T/t* complex (Shur and Bennett, 1979). The relationship between *T/t* locus alleles and sperm fertility is covered in detail in Chapter 11 of this volume, so will not be addressed here. It is sufficient to note that spermatozoa from mice homozygous for complementing *t* alleles have increased GalTase activity that parallels increased levels of GalTase on the cell surface. This observation suggested that the enzyme might function in fertilization and thereby account, at least in part, for the unique fertility characteristics of spermatozoa from mice with different *T/t* alleles.

Results of many studies support the hypothesis that GalTase mediates sperm–EEM adhesion. The strongest evidence comes from three observations: (1) purified GalTase and antibody against GalTase inhibit, in a dose-dependent manner, the adhesion of spermatozoa to the EEM (Shur and Neely, 1988); (2) sperm GalTase can bind preferentially to ZP3 in solubilized EEM (Miller *et al.,* 1992); and (3) antibody against GalTase induces acrosome reactions that exhibit the same G-protein dependence as those induced by the EEM (Gong *et al.,* 1995). These findings are reinforced by other studies, including some showing that sperm–EEM adhesion can be blocked by a variety of inhibitors and by monoclonal antibodies raised against bacterially expressed GalTase (Shur *et al.,* 1998). In addition to G-protein activation, GalTase has also been implicated in the regulation of the acrosome reaction through actin depolymerization (Wassler and Shur, 2000). Collectively, the work on GalTase supports a model wherein the long form of the enzyme, as an integral plasma membrane protein of the sperm head, mediates gamete adhesion by binding to oligosaccharides in the EEM (Shur and Hall, 1982). In this model GalTase binds preferentially to the carbohydrate moiety of mouse ZP3, specifically to oligosaccharides that terminate with *N*-acetylglucosamine (GlcNAc), in an enzyme-substrate-binding manner (Miller *et al.*, 1992). The ligand-aggregated GalTase in turn activates a G-protein-dependent signaling cascade that initiates the acrosome reaction.

The GalTase model conflicts with a series of studies showing that terminal α-galactosyl residues of the EEM mediate sperm adhesion (Bleil and Wassarman, 1988). Such terminal galactosyl residues on O-linked oligosaccharides of ZP3 would block the binding of GalTase. However, further experiments showed that GlcNAc, but not galactose or glucose, is important for the acrosome reaction (Loeser and Tulsiani, 1999); only polymeric GlcNAc was able to induce the acrosome reaction, consistent with a clustering of the surface GalTase. Ultrastructural localization of GalTase-binding ligands in the EEM revealed that *N*-acetylglucosamine residues are evenly distributed in the EEM, whereas α-galactosyl residues are confined to the inner portions of the structure (Aviles *et al.*, 1997). Moreover, eggs from α-galactosyltransferase knockout mice are fertilized normally (Thall *et al.*, 1995). Collectively, these results support the view that EEM carbohydrate comprises at least part of the structure recognized by adhesion molecules in spermatozoa. Unfortunately, the controversy over the nature of the active saccharide has not been resolved. As a result, no clear-cut oligosaccharide binding criteria have been defined that could be used to exclude GalTase or any other candidate adhesion molecule identified in spermatozoa.

Despite the body of evidence supporting a function of GalTase in sperm–EEM adhesion, some puzzles remain. Sperm from transgenic mice that overexpress GalTase on the sperm surface bound more radiolabeled ZP3 than did wild-type cells (Youakim *et al.*, 1994). In addition, the transgenic cells underwent accelerated G-protein activation and precocious acrosome reactions, presumably because the surface GalTase was more readily aggregated when present in greatly increased amounts. However, the GalTase overexpressing cells displayed a reduced avidity for the whole EEM compared to wild-type cells (Youakim *et al.*, 1994), indicating that interaction with a single EEM glycoprotein does not exactly parallel the interaction with the whole EEM. The GalTase gene has also been knocked out by homologous recombination (Lu and Shur, 1997). Spermatozoa from the GalTase −/− mice could neither bind nor undergo acrosome reactions in response to solubilized, ovarian EEM. Nevertheless, the null mice were fertile, and more GalTase −/− spermatozoa adhered to the EEM of ovulated eggs *in vitro* than did wild-type spermatozoa. These results showed that GalTase cannot be the only protein capable of mediating adhesion of mouse spermatozoa to the intact EEM during fertilization *in vivo*. Indeed, although GalTase has been identified on the sperm surface in species other than the mouse, there is at present no evidence to suggest that GalTase binds to ZP3 in a species-specific manner. In fact, the putative GalTase binding site in the EEM is its enzymatic substrate which is unlikely to display enough species variation to confer the species specificity of sperm–EEM adhesion. One plausible synthesis of these data is that GalTase functions as a non-species-specific adhesion molecule that acts in concert with other sperm proteins to mediate adhesion to the EEM. In this view, GalTase represents one of potentially several redundant sperm–EEM adhesion molecules that collectively maximize chances that fertilization will succeed.

b. PH-20/Hyaluronidase

The PH-20 protein was originally defined with a monoclonal antibody that bound to the posterior head of guinea pig spermatozoa (Primakoff and Myles, 1983). This finding was particularly exciting because the binding of PH-20 antibody blocked guinea pig sperm–EEM adhesion (Primakoff *et al.,* 1985). The PH-20 protein was subsequently characterized extensively, and found to be a glycosyl phosphatidylinositol (GPI)-linked protein on the surface of acrosome-intact spermatozoa that migrated to the inner acrosomal membrane during the acrosome reaction (Phelps *et al.,* 1988). Immunizing either male or female guinea pigs with purified PH-20 protein produced temporary sterility (Primakoff *et al.,* 1988), further supporting a function for this protein in fertilization. This functional importance was confirmed by the discovery that PH-20 was homologous to hyaluronidase from bee venom (Gmachl and Kreil, 1993). Spermatozoa were known to possess high levels of hyaluronidase activity, and although the enzyme had been purified to homogeneity and extensively characterized (Harrison, 1988a,b; Harrison and Gaunt, 1988), it had not been sequenced. PH-20 protein was quickly demonstrated to have hyaluronidase activity (Hunnicutt *et al.,* 1996a), thereby confirming that it was indeed the long-studied sperm enzyme. Prior to this discovery, the primary function ascribed to hyaluronidase was to aid in the penetration of the cumulus oophorus. PH-20/hyaluronidase's probable function as an enzyme in of the cumulus matrix appears to be distinct from its function as an adhesion molecule at the surface of the EEM (Hunnicutt *et al.,* 1996b). The exact nature of these interactions will need to be determined before their relative importance to fertilization can be assessed.

c. Proacrosin

Proacrosin is the zymogen of acrosin, the major serine protease of the sperm acrosome. The primary function of acrosin was originally thought to be to digest the mammalian EEM and thereby facilitate penetration by motile spermatozoa (Urch *et al.,* 1985; Hedrick *et al.,* 1989). However, trypsin inhibitors were shown to interfere with sperm–EEM adhesion (Saling, 1981; Benau and Storey, 1987), so acrosin, as the dominant sperm protease, logically became a candidate adhesion molecule. Even though the original impetus for studying the possible adhesion activity of acrosin came from inhibition experiments that targeted the enzyme's active site, much of the ensuing work focused on the C-terminal part of proacrosin, which does not participate in catalysis (Williams and Jones, 1990, 1993; Jansen *et al.,* 1998; Barros *et al.,* 1996; Moreno *et al.,* 1998, 1999). This unique part of the zymogen is highly basic and proline rich (Baba *et al.,* 1989; Kashiwabara *et al.,* 1990). Of particular interest were observations that sulfated polysaccharides inhibited sperm–EEM adhesion (Huang and Yanagimachi, 1984) and also bound readily to proacrosin (Williams and Jones, 1990; Urch and Patel, 1991; Moreno *et al.*, 1998, 1999). These findings were important primarily because similar sulfated polysaccharides were known to block adhesion of sea urchin gametes by binding to bindin (DeAngelis and Glabe, 1987, 1988, 1990).

Although proacrosin's potential function as an adhesion molecule is still being investigated, the supportive evidence for this line of research is indirect, and several observations suggest that proacrosin is unlikely to be of primary importance in EEM adhesion. First, and perhaps foremost, is the very nature of proacrosin's binding to the EEM. Most published work has focused on *in vitro* binding studies using denatured EEM components, and in some instances, potentially nonnative fragments of proacrosin. It is therefore uncertain whether results of such studies are truly relevant to the fertilization process. EEM components are on average acidic, whereas proacrosin is very basic, especially in its C-terminal region. One would therefore expect proacrosin and the EEM to exhibit some minimal level of binding avidity solely on the basis of ionic interactions. Indeed, pig proacrosin readily binds directly to intact EEM from pig oocytes, but this binding is easily reversed by washing with a high-ionic-strength detergent solution (Hardy and Garbers, 1994). In addition, this direct binding of proacrosin to intact EEM is highly promiscuous, because pig proacrosin bound equally well to EEM from pig, bovine, mouse, and *X. laevis* oocytes (Hardy and Garbers, 1994). Thus it seems that ionic interactions between the basic regions of proacrosin and the acidic EEM could provide some general adhesiveness between the gametes, but such interactions cannot account for the known specificity of sperm–EEM adhesion.

Proacrosin's properties do suggest a function in the progressive penetration of the EEM that occurs after specific recognition and adhesion. In this process, adhesive forces must retain the sperm head in the penetration slit while at the same time permitting forward movement as the sperm tail propels the cell through the EEM. Proacrosin is a prominent component of the acrosomal matrix, and its activation to acrosin leads to proteolytic dissolution of the matrix (Noland *et al.*, 1989; Hardy *et al.*, 1991; Yamagata *et al.*, 1998). Ionic interactions between proacrosin and the EEM, coupled with acrosin activation and release from the acrosomal matrix, could provide a ratcheting mechanism for the progressive penetration of the EEM (see Chapters 8 and 13, this volume). This model of proacrosin function is supported in part by loss-of-function transgenic studies. Two research groups independently generated null alleles of the mouse proacrosin gene (Adham *et al.*, 1997; Baba *et al.*, 1994). Unexpectedly, the Acr −/− mice were fertile. *In vitro* fertilization demonstrated that Acr −/− spermatozoa could penetrate zona pellucida, but the process was significantly delayed compared to wild-type spermatozoa. Furthermore, the proteolytic dissolution of the acrosomal matrix in spermatozoa from Acr −/− mice was slow in comparison to wild-type cells (Yamagata *et al.*, 1998). Collectively, these findings suggest that proacrosin does not contribute significantly to the specific recognition of the EEM, but it may well function in EEM penetration by virtue of its proteolytic activity and relatively nonspecific affinity for the EEM.

d. p95/ZRK/Hexokinase

The putative adhesion molecule p95 was originally characterized as the major M_r 95,000 tyrosine phosphoprotein of mouse spermatozoa (Leyton and Saling, 1989b). Blot overlays, wherein both sperm proteins and the ZP had been denatured, suggested that it had ZP-binding activity (Leyton and Saling, 1989b). The phosphorylation of p95 correlated with capacitation and EEM-induced acrosome reactions (Leyton and Saling, 1989b), and aggregation of p95 appeared to induce acrosome reactions (Leyton and Saling, 1989a). These observations were exciting because they suggested the involvement of a receptor tyrosine kinase in the signal transduction pathway for the acrosome reaction. However, the predominant M_r 95,000 tyrosine phosphoprotein of mouse spermatozoa was later identified as a novel form of hexokinase (Kalab *et al.,* 1994). In addition, the putative human homolog of p95, ZRK (Burks *et al.,* 1995), has striking sequence similarities to the protooncogene c-mer. These observations have not yet been reconciled with the hypothesis that p95/ZRK is a ZP adhesion molecule.

e. Other Hijacked Enzymes

At least two other putative EEM adhesion molecules fall in the "hijacked enzymes" category. Sulfolipid immobilizing protein (SLIP1), named for its lipid-binding activity, also possesses affinity for the EEM (Tanphaichitr *et al.,* 1993). SLIP1 appears to comprise multiple polypeptides. The M_r 68,000 immunoreactive component of pig SLIP1 retains affinity for the EEM, although the binding is not species-specific (Tanphaichitr *et al.,* 1998). The M_r 68,000 polypeptide was reported to be identical to arylsulfatase-A, an enzyme previously characterized as an acrosomal constituent (Weerachatyanukul *et al.,* 2000). The relationship of this protein's enzymatic and EEM-binding activities has not been clarified. A sperm surface mannosidase potentially involved in sperm–EEM adhesion has also been described (Cornwall *et al.,* 1991).

2. Unique Adhesion Molecules

a. sp56

sp56 was identified by photoaffinity cross-linking to ZP3 (Bleil and Wassarman, 1990). The specificity of the interaction was demonstrated by binding competition with unlabeled ZP3. Evidence for a function of sp56 in EEM adhesion comes mainly from experiments that showed adhesion could be inhibited if spermatozoa were preincubated with purified sp56 or with sp56 antibodies (Cheng *et al.,* 1994). sp56 was originally reported to be a peripheral membrane protein (Bookbinder *et al.,* 1995), but a later study showed that it is largely, if not exclusively, located in the acrosomal matrix of both mouse and guinea pig spermatozoa (Foster *et al.,* 1997). Orthologs of sp56 originally were not detected in nonrodent species (e.g., human) (Bookbinder *et al.,* 1995; Snell and White, 1996), suggesting species differences potentially related to the species specificity of sperm–EEM adhesion. However, the species specificity of its ZP3 binding activity has not been thoroughly evaluated, and the protein's status as a prominent component of the mouse and guinea pig acrosomal matrix suggests strongly that it is present in the acrosomes of spermatozoa from most or all mammalian species. It is possible that sp56 functions in concert with proacrosin in the progressive penetration of spermatozoa through the EEM.

b. Zonadhesin

Zonadhesin was discovered based on its ability to bind directly and in a species-specific manner to native, particulate EEM (Hardy and Garbers, 1994, 1995). The rationale of this experimental approach was that the intact EEM is the structure a sperm cell recognizes *in vivo,* so true adhesion molecules should be able to bind directly to it. In most previous studies, the binding properties of EEM adhesion molecules were assessed using purified glycoproteins or denatured EEM instead of the intact structure. This could result in detection of interactions that do not occur in normal fertilization. In addition, the sperm ligand activity of EEM glycoproteins varies by species, so a single glycoprotein (for example, ZP3) does not function exclusively in this capacity for all mammals. Indeed, studies on the pig EEM have confirmed the importance of quaternary structure to its adhesion activity (Yurewicz *et al.,* 1998). Neither pig ZPB (ortholog of mouse ZP1) nor ZPC (ortholog of mouse ZP3) individually possessed binding activity, but these glycoproteins could form heteroduplexes *in vitro* that bound avidly to pig sperm membranes (Yurewicz *et al.,* 1998). Furthermore, results from human ZP3-rescued mZP3 −/− mice indicated that ZP3 alone does not determine the species-specific adhesion between spermatozoa and the EEM. As long as an intact EEM was assembled outside the egg, its original species specificity was retained even if it contained heterologous ZP3 (Rankin *et al.,* 1998). Collectively, these results show that full, species-specific sperm adhesion activity is probably conferred by the whole EEM structure, and not by a single glycoprotein component. Thus, in the search for adhesion molecules, proteins that bind directly to the native structure may be more likely to have a natural function in sperm–EEM adhesion than molecules identified by their binding to a single glycoprotein.

Zonadhesin was first detected in a membrane fraction isolated from pig spermatozoa (Hardy and Garbers, 1994). Although several proteins bound initially to the native EEM in a direct binding assay, only the two disulfide-bonded polypeptides (M_r 105,000 and 45,000) of zonadhesin remained bound after several rounds of detergent washing. Most importantly, pig zonadhesin bound with high apparent affinity to the pig EEM, but did not bind to the EEM of bovine, mouse, or *X. laevis* oocytes (Hardy and Garbers, 1994). Zonadhesin remains the only mammalian sperm protein that exhibits such species-specific EEM-binding activity.

A cDNA encoding pig zonadhesin has been cloned and fully sequenced. The deduced amino acid sequence revealed that pig zonadhesin comprised a novel N-terminal domain, a putative mucin domain, five domains homologous to the D domains of von Willebrand factor, a transmembrane segment, and a short C-terminal tail (Hardy and Garbers, 1995). The mRNA was expressed only in haploid germ cells of the testis, suggesting strongly that zonadhesin is a germ cell-specific protein. The predicted, nascent polypeptide had a calculated molecular mass of 267 kDa, and contained the sequences of the M_r 105,000 and 45,000 subunits, indicating that the protein isolated from spermatozoa had undergone processing to remove approximately 100 kDa of mass and generate the two subunits.

Orthologs of pig zonadhesin have been characterized in mouse (Gao and Garbers, 1998) and human (T. L. Cheung, M. J. Wassler, G. A. Cornwall, and D. M. Hardy, unpublished). Like pig, the mouse and human zonadhesin mRNAs were detected only in testis, consistent with the protein's putative sperm-specific function in gamete adhesion. The mouse and human cDNAs specify proteins with domain structures similar to that of pig zonadhesin. Analysis of the mouse zonadhesin sequence revealed that it too comprised mucin and von Willebrand D domains, and that the N-terminal regions of both the mouse and pig proteins comprised MAM (meprin, A5, μ) domains (Gao and Garbers, 1998). MAM, mucin, and von Willebrand D domains have each been shown to function as adhesion domains in other proteins (Varki, 1994; Ruggeri and Ware, 1992; Gao and Garbers, 1998), and their presence in zonadhesin suggests the potential for the protein to participate in multiple adhesive interactions. Importantly, mouse zonadhesin exhibited dramatic variation in its D domains, containing 20 additional truncated D domains between D3 and D4 (Gao and Garbers, 1998). The overall similarity yet substantial between-species variation of zonadhesin domain structure suggests that shuffled domains, as well as point mutations within domains, may determine the species specificity of zonadhesin's EEM binding activity. These and other observations support the hypothesis that zonadhesin mediates species-specific adhesion of spermatozoa to the EEM.

c. Other Unique Adhesion Molecules

In addition to the eight hijacked enzymes and nonenzyme adhesion molecules described above, several other proteins have been characterized as sperm–EEM adhesion molecules. These include the spermadhesins (Topfer-Petersen *et al.,* 1998), RSA/Sp17 (O'Rand *et al.,* 1988; O'Rand and Widgren, 1994), p26H (Boue *et al.,* 1994), and Sp38 (Mori *et al.,* 1993). Although substantial evidence supports the potential adhesive functions of these proteins, in some instances there is little comparative information on them, so it is unclear whether these proteins have the potential to be important general adhesion molecules in multiple species. Among these proteins, the spermadhesins have been heavily studied, and there is an extensive literature describing these results [reviewed in Topfer-Petersen *et al.* (1998)]. These proteins, studied primarily in the pig, are members of the CUB domain gene family. The various spermadhesins appear to be multifunctional, so their activities are not limited to sperm–EEM adhesion (Topfer-Petersen *et al.,* 1998). There is also a substantive body of literature on RSA/Sp17 (Yamasaki *et al.,* 1995, and references therein). Support for a function of RSA/Sp17 in EEM adhesion is primarily immunological (O'Rand *et al.,* 1988; O'Rand and Widgren, 1994), as is the evidence for p26H (Berube and Sullivan, 1994).

IV. FUTURE PROSPECTS AND DIRECTIONS

The dominant, recurrent themes of research on sperm–EEM adhesion are species diversity and species specificity. It will therefore be essential over the coming years to study the biochemical basis of this process in many species if we are to build a comprehensive understanding of it. Just as ecosystems dominated by a single species are easily recognized as unhealthy, so too is it clear that concentrating fertilization research on a single species can limit the process of discovery. The mouse has dominated this research area for some years. Ironically, the mouse now affords an opportunity for comparative studies. By combining loss-of-function and gain-of-function transgenic strategies, individual sperm–EEM adhesion molecules can be replaced with their homologs from other species. This has already been done for the major EEM glycoproteins, and it is only a matter of time before similar studies are done on sperm proteins. Perhaps the ultimate test of a sperm protein's function in EEM adhesion will be if spermatozoa from a knockout mouse do not recognize the mouse EEM, but the phenotype is rescued, with altered species specificity, by replacing the null allele with a transgene from another species.

As our understanding of the basic cell biology and biochemistry of sperm–EEM adhesion improves, there will be opportunities to develop practical applications. The interspecies differences in this fertilization event make it essential to study the species of interest if practical applications are a research goal. Humans will be increasingly important research subjects because of available genome sequence and the interest in human fertility. Agricultural species will also continue to be important models because of interest in increasing profits by improving fecundity and increasing genetic uniformity. Practical benefits of research on sperm–EEM adhesion include the possible development of molecular methods for diagnosing and treating some cases of human infertility, or increasing fecundity of livestock. In addition, an understanding of the molecular determinants of species-specific adhesion could enable development of antagonists that would provide a nonlethal means of controlling the population of pest species without affecting reproduction of other animals. Adhesion antagonists could also be the basis of new human contraceptives.

REFERENCES

Adham, I. M., Nayernia, K., and Engel, W. (1997). Spermatozoa lacking acrosin protein show delayed fertilization. *Mol. Reprod. Dev.* **46,** 370–376.

Aviles, M., Jaber, L., Castells, M. T., Ballesta, J., and Kan, F. W. (1997). Modifications of carbohydrate residues and ZP2 and ZP3 glycoproteins in the mouse zona pellucida after fertilization. *Biol. Reprod.* **57,** 1155–1163.

Baba, T., Azuma, S., Kashiwabara, S., and Toyoda, Y. (1994). Sperm from mice carrying a targeted mutation of the acrosin gene can penetrate the oocyte zona pellucida and effect fertilization. *J. Biol. Chem.* **269,** 31845–31849.

Baba, T., Watanabe, K., Kashiwabara, S., and Arai, Y. (1989). Primary structure of human proacrosin deduced from its cDNA sequence. *FEBS Lett.* **244,** 296–300.

Barros, C., Crosby, J. A., and Moreno, R. D. (1996). Early steps of sperm-egg interactions during mammalian fertilization. *Cell Biol. Int.* **20,** 33–39.

Beebe, S. J., Leyton, L., Burks, D., Ishikawa, M., Fuerst, T., Dean, J., and Saling, P. (1992). Recombinant mouse ZP3 inhibits sperm binding and induces the acrosome reaction. *Dev. Biol.* **151,** 48–54.

Benau, D. A., and Storey, B. T. (1987). Characterization of the mouse sperm plasma membrane zona-binding site sensitive to trypsin inhibitors. *Biol. Reprod.* **36,** 282–292.

Berube, B., and Sullivan, R. (1994). Inhibition of *in vivo* fertilization by active immunization of male hamsters against a 26-kDa sperm glycoprotein. *Biol. Reprod.* **51,** 1255–1263.

Bleil, J. D., and Wassarman, P. M. (1980a). Structure and function of the zona pellucida: Identification and characterization of the proteins of the mouse oocyte's zona pellucida. *Dev. Biol.* **76,** 185–202.

Bleil, J. D., and Wassarman, P. M. (1980b). Mammalian sperm-egg interaction: Identification of a glycoprotein in mouse egg zonae pellucidae possessing receptor activity for sperm. *Cell* **20,** 873–882.

Bleil, J. D., and Wassarman, P. M. (1983). Sperm-egg interactions in the mouse: Sequence of events and induction of the acrosome reaction by a zona pellucida glycoprotein. *Dev. Biol.* **95,** 317–324.

Bleil, J. D., and Wassarman, P. M. (1988). Galactose at the nonreducing terminus of O-linked oligosaccharides of mouse egg zona pellucida glycoprotein ZP3 is essential for the glycoprotein's sperm receptor activity [published erratum appears in *Proc. Natl. Acad. Sci. U.S.A.* **85**(24), 9600]. *Proc. Natl. Acad. Sci. U.S.A.* **85,** 6778–6782.

Bleil, J. D., and Wassarman, P. M. (1990). Identification of a ZP3-binding protein on acrosome-intact mouse sperm by photoaffinity crosslinking. *Proc. Natl. Acad. Sci. U.S.A.* **87,** 5563–5567.

Bleil, J. D., Greve, J. M., and Wassarman, P. M. (1988). Identification of a secondary sperm receptor in the mouse egg zona pellucida: Role in maintenance of binding of acrosome-reacted sperm to eggs. *Dev. Biol.* **128,** 376–385.

Bookbinder, L. H., Cheng, A., and Bleil, J. D. (1995). Tissue- and species-specific expression of sp56, a mouse sperm fertilization protein. *Science* **269,** 86–89.

Boue, F., Berube, B., De Lamirande, E., Gagnon, C., and Sullivan, R. (1994). Human sperm–zona pellucida interaction is inhibited by an antiserum against a hamster sperm protein. *Biol. Reprod.* **51,** 577–587.

Buck, J., Sinclair, M. L., Schapal, L., Cann, M. J., and Levin, L. R. (1999). Cytosolic adenylyl cyclase defines a unique signaling molecule in mammals. *Proc. Natl. Acad. Sci. U.S.A.* **96,** 79–84.

Burks, D. J., Carballada, R., Moore, H. D. M., and Saling, P. M. (1995). Interaction of a tyrosine kinase from human sperm with the zona pellucida at fertilization. *Science* **269,** 83–86.

Cheng, A., Le, T., Palacios, M., Bookbinder, L. H., Wassarman, P. M., Suzuki, F., and Bleil, J. D. (1994). Sperm–egg recognition in the mouse: Characterization of sp56, a sperm protein having specific affinity for ZP3. *J. Cell Biol.* **125,** 867–878.

Cherr, G. N., Lambert, H., Meizel, S., and Katz, D. F. (1986). *In vitro* studies of the golden hamster sperm acrosome reaction: Completion on the zona pellucida and induction by homologous soluble zonae pellucidae. *Dev. Biol.* **114,** 119–131.

Cornwall, G. A., Tulsiani, D. R. P., and Orgebin-Crist, M.-C. (1991). Inhibition of the mouse sperm surface alpha-D-mannosidase inhibits sperm–egg binding *in vitro*. *Biol. Reprod.* **44,** 913–921.

Crozet, N., and Dumont, M. (1984). The site of the acrosome reaction during *in vivo* penetration of the sheep oocyte. *Gamete Res.* **10,** 97–105.

Crozet, N., Theron, M. C., and Chemineau, P. (1987). Ultrastructure of *in vivo* fertilization in the goat. *Gamete Res.* **18,** 191–199.

DeAngelis, P. L., and Glabe, C. G. (1987). Polysaccharide structural features that are critical for the binding of sulfated fucans to bindin, the adhesive protein from sea urchin sperm. *J. Biol. Chem.* **262,** 13946–13952.

DeAngelis, P. L., and Glabe, C. G. (1988). Role of basic amino acids in the interaction of bindin with sulfated fucans. *Biochemistry* **27,** 8189–8194.

DeAngelis, P. L., and Glabe, C. G. (1990). Specific recognition of sulfate esters by bindin, a sperm adhesion protein from sea urchins. *Biochim. Biophys. Acta* **1037,** 100–105.

Doolittle, R. F. (1995). The multiplicity of domains in proteins. *Annu. Rev. Biochem.* **64,** 287–314.

Dunbar, B. S., Avery, S., Lee, V., Prasad, S., Schwahn, D., Schwoebel, E., Skinner, S., and Wilkins, B. (1994). The mammalian zona pellucida: Its biochemistry, immunochemistry, molecular biology, and developmental expression. *Reprod. Fertil. Dev.* **6,** 331–347.

Dunbar, B. S., Wardrip, N. J., and Hedrick, J. L. (1980). Isolation, physicochemical properties, and macromolecular composition of zona pellucida from porcine oocytes. *Biochemistry* **2,** 356–365.

Epifano, O., Liang, L. F., Familari, M., Moos, M. C., and Dean, J. (1996). Coordinate expression of the three zona pellucida genes during mouse oogenesis. *Development* **121,** 1947–1956.

Florman, H. M., and Wassarman, P. M. (1985). O-Linked oligosaccharides of mouse egg ZP3 account for its sperm receptor activity. *Cell* **41,** 313–324.

Florman, H. M., Bechtol, K. B., and Wassarman, P. M (1984). Enzymatic dissection of the functions of the mouse egg's receptor for sperm. *Dev. Biol.* **106,** 243–255.

Fol, H. (1877). Sur le commencement de l'hémogénie chez divers animaoux. *Arch. Zool. Exp. Gén.* **6,** 145–169.

Foster, J. A., Friday, B. B., Maulit, M. T., Blobel, C., Winfrey, V. P., Olson, G. E., Kim, K. S., and Gerton, G. L. (1997). AM67, a secretory component of the guinea pig sperm acrosomal matrix, is related to mouse sperm protein sp56 and the complement component 4-binding proteins. *J. Biol. Chem.* **272,** 12714–12722.

Gao, Z., and Garbers, D. L. (1998). Species diversity in the structure of zonadhesin, a sperm-specific membrane protein containing multiple cell adhesion molecule-like domains. *J. Biol. Chem.* **273,** 3415–3421.

Glabe, C. G., and Vacquier, V. D. (1977). Species specific agglutination of eggs by bindin isolated from sea urchin sperm. *Nature* **267,** 836–838.

Glassner, M., Jones, J., Kligman, I., Woolkalis, M. J., Gerton, G. L., and Kopf, G. S. (1991). Immunocytochemical and biochemical characterization of guanine nucleotide-binding regulatory proteins in mammalian spermatozoa. *Dev. Biol.* **146,** 438–450.

Gmachl, M., and Kreil, G. (1993). Bee venom hyaluronidase is homologous to a membrane protein of mammalian sperm. *Proc. Natl. Acad. Sci. U.S.A.* **90,** 3569–3573.

Gong, X., Dubois, D. H., Miller, D. J., and Shur, B. D. (1995). Activation of a G protein complex by aggregation of beta-1,4-galactosyltransferase on the surface of sperm. *Science* **269,** 1718–1721.

Hardy, D. M., and Garbers, D. L. (1993). Molecular basis of signaling in spermatozoa. *In* "Molecular Biology of the Male Reproductive System" (D. de Kretser, ed.), pp. 233–270. Academic Press, San Diego.

Hardy, D. M., and Garbers, D. L. (1994). Species-specific binding of sperm proteins to the extracellular matrix (zona pellucida) of the egg. *J. Biol. Chem.* **269,** 19000–19004.

Hardy, D. M., and Garbers, D. L. (1995). A sperm membrane protein that binds in a species-specific manner to the egg extracellular matrix is homologous to von Willebrand factor. *J. Biol. Chem.* **270,** 26025–26028.

Hardy, D. M., Harumi, T., and Garbers, D. L. (1994). Sperm receptors for egg peptides. *Sem. Dev. Biol.* **5,** 217–224.

Hardy, D. M., Oda, M. N., Friend, D. S., and Huang, T. T., Jr. (1991). A mechanism for differential release of acrosomal enzymes during the acrosome reaction. *Biochem. J.* **275,** 759–766.

Harrison, R. A. (1988a). Hyaluronidase in ram semen. Quantitative determination, and isolation of multiple forms. *Biochem. J.* **252,** 865–874.

Harrison, R. A. (1988b). Preliminary characterization of the multiple forms of ram sperm hyaluronidase. *Biochem. J.* **252,** 875–882.

Harrison, R. A., and Gaunt, S. J. (1988). Multiple forms of ram and bull sperm hyaluronidase revealed by using monoclonal antibodies. *J. Reprod. Fertil.* **82,** 777–785.

Hartmann, J. F., and Hutchison, C. F. (1974). Nature of the pre-penetration contact interactions between hamster gametes *in vitro. J. Reprod. Fertil.* **36,** 49–57.

Hartmann, J. F., Gwatkin, R. B., and Hutchison, C. F. (1972). Early contact interactions between mammalian gametes *in vitro:* Evidence that the vitellus influences adherence between sperm and zona pellucida. *Proc. Natl. Acad. Sci. U.S.A.* **69,** 2767–2769.

Hedrick, J. L., Urch, U. A., and Hardy, D. M. (1989). Structure-function properties of the sperm enzyme acrosin. *In* "Biocatalysis in Agricultural Biotechnology" (J. R. Whitaker and P. E. Sonnet, eds.), pp. 215–229. American Chemical Society, Washington, DC.

Hertwig, O. (1877). Beiträge zur Kenntniss der Bildung, Befruchtung, und Theilung des theirischen Eies. *Morphol. Jahr.* **1,** 347–452.

Huang, T. T., and Yanagimachi, R. (1984). Fucoidin inhibits attachment of guinea pig spermatozoa to the zona pellucida through binding to the inner acrosomal membrane and equatorial domains. *Exp. Cell Res.* **153,** 363–373.

Huang, T. T., Fleming, A. D., and Yanagimachi, R. (1981). Only acrosome-reacted spermatozoa can bind to and penetrate zona pellucida: A study using the guinea pig. *J. Exp. Zool.* **217,** 287–290.

Hunnicutt, G. R., Mahan, K., Lathrop, W. F., Ramarao, C. S., Myles, D. G., and Primakoff, P. (1996a). Structural relationship of sperm soluble hyaluronidase to the sperm membrane protein PH-20. *Biol. Reprod.* **54,** 1343–1349.

Hunnicutt, G. R., Primakoff, P., and Myles, D. G. (1996b). Sperm surface protein PH-20 is bifunctional: One activity is a hyaluronidase and a second, distinct activity is required in secondary sperm–zona binding. *Biol. Reprod.* **55,** 80–86.

Hyne, R. V., and Garbers, D. L. (1979). Calcium-dependent increase in adenosine 3′,5′-monophosphate and induction of the acrosome reaction in guinea pig spermatozoa. *Proc. Natl. Acad. Sci. U.S.A.* **76,** 5699–5703.

Jansen, S., Jones, R., Jenneckens, I., Marschall, B., Kriegesmann, B., Coadwell, J., and Brenig, B. (1998). Site-directed mutagenesis of boar proacrosin reveals residues involved in binding of zona pellucida glycoproteins. *Mol. Reprod. Dev.* **51,** 184–192.

Kalab, P., Visconti, P., Leclerc, P., and Kopf, G. S. (1994). p95, the major phosphotyrosine-containing protein in mouse spermatozoa, is a hexokinase with unique properties. *J. Biol. Chem.* **269,** 3810–3817.

Kashiwabara, S., Baba, T., Takada, M., Watanabe, K., Yano, Y., and Arai, Y. (1990). Primary structure of mouse proacrosin deduced from the cDNA sequence and its gene expression during spermatogenesis. *J. Biochem. (Tokyo)* **108,** 785–791.

Kinloch, R. A., Mortillo, S., Stewart, C. L., and Wassarman, P. M. (1991). Embryonal carcinoma cells transfected with ZP3 genes differentially glycosylate similar polypeptides and secrete active mouse sperm receptor. *J. Cell Biol.* **115,** 655–664.

Kreis, T., and Vale, R. (1993). "Guidebook to the Extracellular Matrix and Adhesion Proteins." Oxford University Press, Oxford.

Leclerc, P., and Kopf, G. S. (1995). Mouse sperm adenylyl cyclase: General properties and regulation by the zona pellucida. *Biol. Reprod.* **52,** 1227–1233.

Leyton, L., and Saling, P. (1989b). 95 kd sperm proteins bind ZP3 and serve as tyrosine kinase substrates in response to zona binding. *Cell* **57,** 1123–1130.

Leyton, L., and Saling, P. (1989a). Evidence that aggregation of mouse sperm receptors by ZP3 triggers the acrosome reaction. *J. Cell Biol.* **108,** 2163–2168.

Li, W.-H. (1997). "Molecular Evolution." Sinauer Associates, Sunderland, MA.

Loeser, C. R., and Tulsiani, D. R. (1999). The role of carbohydrates in the induction of the acrosome reaction in mouse spermatozoa. *Biol. Reprod.* **60,** 94–101.

Longo, F. J., and Yanagimachi, R. (1993). Detection of sperm-egg fusion. *Methods Enzymol.* **221,** 249–260.

Lopez, L. C., Youakim, A., Evans, S. C., and Shur, B. D. (1991). Evidence for a molecular distinction between Golgi and cell surface forms of beta 1,4-galactosyltransferase. *J. Biol. Chem.* **266,** 15984–15991.

Lu, Q., and Shur, B. D. (1997). Sperm from beta 1,4-galactosyltransferase-null mice are refractory to ZP3-induced acrosome reactions and penetrate the zona pellucida poorly. *Development* **124,** 4121–4131.

Miller, D. J., Macek, M. B., and Shur, B. D. (1992). Complementarity between sperm surface beta-1,4-galactosyltransferase and egg-coat ZP3 mediates sperm–egg binding. *Nature* **357,** 589–593.

Moller, C. C., Bleil, J. D., Kinloch, R. A., and Wassarman, P. M. (1990). Structural and functional relationships between mouse and hamster zona pellucida glycoproteins. *Dev. Biol.* **137,** 276–286.

Morales, P., Cross, N. L., Overstreet, J. W., and Hanson, F. W. (1989). Acrosome intact and acrosome-reacted human sperm can initiate binding to the zona pellucida. *Dev. Biol.* **133,** 385–392

Moreno, R. D., Hoshi, M., and Barros, C. (1999). Functional interactions between sulphated polysaccharides and proacrosin: Implications in sperm binding and digestion of zona pellucida. *Zygote* **7,** 105–111.

Moreno, R. D., Sepulveda, M. S., de Ioannes, A., and Barros, C. (1998). The polysulphate binding domain of human proacrosin/acrosin is involved in both the enzyme activation and spermatozoa–zona pellucida interaction. *Zygote* **6,** 75–83.

Mori, E., Baba, T., Iwamatsu, A., and Mori, T. (1993). Purification and characterization of a 38-kDa protein, sp38, with zona pellucida-binding property from porcine epididymal sperm [published erratum appears in Biochem. Biophys. Res. Commun. **198**(1), 408]. *Biochem. Biophys. Res. Comm.* **196,** 196–202.

Myles, D. G., Hyatt, H., and Primakoff, P. (1987). Binding of both acrosome-intact and acrosome-reacted guinea pig sperm to the zona pellucida during *in vitro* fertilization. *Dev. Biol.* **121,** 559–567.

Myles, D. G., Primakoff, P., and Bellve, A. R. (1981). Surface domains of the guinea pig sperm defined with monoclonal antibodies. *Cell* **23,** 433–439.

Newport, G. (1851). On the impregnation of the ovum in the amphibia. *Philos. Trans. R. Soc. Lond. (First Ser. Pt. 1)* **141,** 169–242.

Noland, T. D., Davis, L. S., and Olson, G. E. (1989). Regulation of proacrosin conversion in isolated guinea pig sperm acrosomal apical segments. *J. Biol. Chem.* **264,** 13586–13590.

O'Rand, M. G., and Widgren, E. E. (1994). Identification of sperm antigen targets for immunocontraception: B-cell epitope analysis of Sp17. *Reprod. Fertil. Dev.* **6,** 289–296.

O'Rand, M. G., Widgren, E. E., and Fisher, S. J. (1988). Characterization of the rabbit sperm membrane autoantigen, RSA, as a lectin-like zona binding protein. *Dev. Biol.* **129,** 231–240.

Overstreet, J. W., Yanagimachi, R., Katz, D. F., Hayashi, K., and Hanson, F. W. (1980). Penetration of human spermatozoa into the human zona pellucida and the zona-free hamster egg: A study of fertile donors and infertile patients. *Fertil. Steril.* **33,** 534–542.

Peterson, R. N., Russell, L., Bundman, D., and Freund, M. (1980). Sperm-egg interaction Evidence for boar sperm plasma membrane receptors for porcine zona pellucida. *Science* **207,** 73–74.

Phelps, B. M., Primakoff, P., Koppel, D. E., Low, M. G., and Myles, D. G (1988). Restricted lateral diffusion of PH-20, a PI-anchored sperm membrane protein. *Science* **240,** 1780–1782.

Piatigorsky, J., and Wistow, G. J. (1989). Enzyme/crystallins: Gene sharing as an evolutionary strategy. *Cell* **57,** 197–199.

Prasad, S. V., Skinner, S. M., Carino, C., Wang, N., Cartwright, J., and Dunbar, B. S. (2000). Structure and function of the proteins of the mammalian Zona pellucida. *Cells Tissues Organs* **166,** 148–164.

Primakoff, P., and Myles, D. G. (1983). A map of the guinea pig sperm surface constructed with monoclonal antibodies. *Dev. Biol.* **98,** 417–428.

Primakoff, P., Hyatt, H., and Myles, D. G. (1985). A role for the migrating sperm surface antigen PH-20 in guinea pig sperm binding to the egg zona pellucida. *J. Cell Biol.* **101,** 2239–2244.

Primakoff, P., Lathrop, W., Woolman, L., Cowan, A., and Myles, D. (1988). Fully effective contraception in male and female guinea pigs immunized with the sperm protein PH-20. *Nature* **335,** 543–546.

Rankin, T. L., Tong, Z. B., Castle, P. E., Lee, E., Gore-Langton, R., Nelson, L. M., and Dean, J. (1998). Human ZP3 restores fertility in Zp3 null mice without affecting order-specific sperm binding. *Development* **125,** 2415–2424.

Roseman, S. (1970). The synthesis of complex carbohydrates by multiglycosyltransferase systems and their potential function in intercellular adhesion. *Chem. Phys. Lipids* **5,** 270–297.

Roth, S., McGuire, E. J., and Roseman, S. (1971). Evidence for cell-surface glycosyltransferases. Their potential role in cellular recognition. *J. Cell. Biol.* **51,** 536–547.

Ruggeri, Z. M., and Ware, J. (1992). The structure and function of von Willebrand factor. *Thromb. Haemost.* **67,** 594–599.

Sacco, A. G., Yurewicz, E. C., Subramanian, M. G., and Matzat, P. D. (1989). Porcine zona pellucida: Association of sperm receptor activity with the alpha-glycoprotein component of the Mr = 55,000 family. *Biol. Reprod.* **41,** 523–532.

Saling, P. M. (1981). Involvement of trypsin-like activity in binding of mouse spermatozoa to zonae pellucidae. *Proc. Natl. Acad. Sci. U.S.A.* **78,** 6231–6235.

Saling, P. M., and Storey, B. T. (1979). Mouse gamete interactions during fertilization *in vitro.* Chlortetracycline as a fluorescent probe for the mouse sperm acrosome reaction. *J. Cell Biol.* **83,** 544–555.

Saling, P. M., Sowinski, J., and Storey, B. T. (1979). An ultrastructural study of epididymal mouse spermatozoa binding to zonae pellucidae *in vitro:* Sequential relationship to the acrosome reaction. *J. Exp. Zool.* **209,** 229–238.

Schwoebel, E., Prasad, S., Timmons, T. M., Cook, R., Kimura, H., Niu, E.-M., Cheung, P., Skinner, S., Avery, S., Wilkins, B., and Dunbar, B. S. (1996). Isolation and characterization of a full-length cDNA encoding the 55-kDa rabbit zona pellucida protein. *J. Biol. Chem.* **266,** 7214–7219.

Scully, N. F., Shaper, J. H., and Shur, B. D. (1987). Spatial and temporal expression of cell surface galactosyltransferase during mouse spermatogenesis and epididymal maturation. *Dev. Biol.* **124,** 111–124.

Shalgi, R., Phillips, D. M., and Jones, R. (1989). Status of the rat acrosome during sperm–zona pellucida interactions. *Gamete Res.* **22,** 1–13.

Shur, B. D. (1989). Galactosyltransferase as a recognition molecule during fertilization and development. *In* "The Molecular Biology of Fertilization" (H. Schatten and G. Schatten, eds.), pp. 37–71. Academic Press, San Diego.

Shur, B. D., and Bennett, D. (1979). A specific defect in galactosyltransferase regulation on sperm bearing mutant alleles of the T/t locus. *Dev. Biol.* **71,** 243–259.

Shur, B. D., and Hall, N. G. (1982). A role for mouse sperm surface galactosyltransferase in sperm binding to the egg zona pellucida. *J. Cell Biol.* **95,** 574–579.

Shur, B. D., and Neely, C. A. (1988). Plasma membrane association, purification, and partial characterization of mouse sperm beta 1,4-galactosyltransferase [published erratum appears in J. Biol. Chem. **264**(7), 4264]. *J. Biol. Chem.* **263,** 17706–17714.

Shur, B. D., Evans, S., and Lu, Q. (1998). Cell surface galactosyltransferase: Current issues. *Glycoconj. J.* **15,** 537–548.

Snell, W. J., and White, J. M. (1996). The molecules of mammalian fertilization. *Cell* **85,** 629–637.

Swanson, W. J., and Vacquier, V. D. (1997). The abalone egg vitelline envelope receptor for sperm lysin is a giant multivalent molecule. *Proc. Natl. Acad. Sci. U.S.A.* **94,** 6724–6729.

Swanson, W. J., and Vacquier, V. D. (1998). Concerted evolution in an egg receptor for a rapidly evolving abalone sperm protein. *Science* **281,** 710–712.

Swanson, W. J., Yang, Z., Wolfner, M. F., and Aquadro, C. F. (2001). Positive Darwinian selection drives the evolution of several female reproductive proteins in mammals. *Proc. Natl. Acad. Sci, U.S.A.* **98,** 2509–2514.

Tanphaichitr, N., Moase, C., Taylor, T., Surewicz, K., Hansen, C., Namking, M., Berube, B., Kamolvarin, N., Lingwood, C. A., Sullivan, R., Rattanachaiyanont, M., and White, D. (1998). Isolation of antiSLIP1-reactive boar sperm P68/62 and its binding to mammalian zona pellucida. *Mol. Reprod. Dev.* **49,** 203–216.

Tanphaichitr, N., Smith, J., Mongkolsirikieart, S., Gradil, C., and Lingwood, C. A. (1993). Role of a gamete-specific sulfoglycolipid immobilizing protein on mouse sperm–egg binding. *Dev. Biol.* **156,** 164–175.

Thaler, C. D., and Cardullo, R. A. (1996). The initial molecular interaction between mouse sperm and the zona pellucida is a complex binding event. *J. Biol. Chem.* **271,** 23289–23297.

Thall, A. D., Maly, P., and Lowe, J. B. (1995). Oocyte gal α1,3 gal epitopes implicated in sperm adhesion to the zona pellucida glycoprotein ZP3 are not required for fertilization in the mouse. *J. Biol. Chem.* **270,** 21437–21440.

Topfer-Petersen, E., Romero, A., Varela, P. F., Ekhlasi-Hundrieser, M., Dostalova, Z., Sanz, L., and

Calvete, J. J. (1998). Spermadhesins: A new protein family. Facts, hypotheses and perspectives. *Andrologia* **30,** 217–224.
Turley, E. A., and Roth, S. (1979). Spontaneous glycosylation of glycosaminoglycan substrates by adherent fibroblasts. *Cell* **17,** 109–115.
Urch, U. A., and Patel, H. (1991). The interaction of boar sperm proacrosin with its natural substrate, the zona pellucida, and with polysulfated polysaccharides. *Development* **111,** 1165–1172.
Urch, U. A., Wardrip, N. J., and Hedrick, J. L. (1985). Limited and specific proteolysis of the zona pellucida by acrosin. *J. Exp. Zool.* **233,** 479–483.
Uto, N., Yoshimatsu, N., Lopata, A., and Yanagimachi, R. (1988). Zona-induced acrosome reaction of hamster spermatozoa. *J. Exp. Zool.* **248,** 113–120.
Vacquier, V. D. (1998). Evolution of gamete recognition proteins. *Science* **281,** 1995–1998.
Vacquier, V. D., and Moy, G. W. (1977). Isolation of bindin: The protein responsible for adhesion of sperm to sea urchin eggs. *Proc. Natl. Acad. Sci. U.S.A.* **74,** 2456–2460.
Varki, A. (1994). Selectin ligands. *Proc. Natl. Acad. Sci. U.S.A.* **91,** 7390–7397.
Walensky, L. D., and Snyder, S. H. (1995). Inositol 1,4,5-trisphosphate receptors selectively localized to the acrosomes of mammalian sperm. *J. Cell Biol.* **130,** 857–869.
Ward, C. R., and Kopf, G. S. (1993). Molecular events mediating sperm activation. *Dev. Biol.* **158,** 9–34.
Ward, C. R., Storey, B. T., and Kopf, G. S. (1992). Activation of a Gi protein in mouse sperm membranes by solubilized proteins of the zona pellucida, the egg's extracellular matrix. *J. Biol. Chem.* **267,** 14061–14067.
Wassarman, P. M. (1995). Towards molecular mechanisms for gamete adhesion and fusion during mammalian fertilization. *Curr. Opin. Cell Biol.* **7,** 658–664.
Wassler, M. J., and Shur, B. D. (2000). Clustering of cell surface (beta)1,4-galactosyltransferase I induces transient tyrosine phosphorylation of focal adhesion kinase and loss of stress fibers. *J. Cell Sci.* **113** (Pt. 2), 237–245.
Weerachatyanukul, W., Carmona, E., Soboloff, T., Fluharty, A., White, D., Promdee, L., Ekker, M., Buhr, M. M., and Tanphaichitr, N. (2000). Presence and role of mammalian sperm surface arylsulfatase-A (AS-A) in zonapellucida binding. *Biol. Reprod.* **62** (Suppl. 1), 275a (abstr.).
Williams, R. M., and Jones, R. (1990). Specific binding of sulphated polymers to ram sperm proacrosin. *FEBS Lett.* **270,** 168–172.
Williams, R. M., and Jones, R. (1993). Specificity of binding of zona pellucida glycoproteins to sperm proacrosin and related proteins. *J. Exp. Zool.* **266,** 65–73.
Wolf, D. P., Nishihara, T., West, D. M., Wyrick, R. E., and Hedrick, J. L. (1976). Isolation, physicochemical properties, and the macromolecular composition of the vitelline and fertilization envelopes from *Xenopus* laevis eggs. *Biochemistry* **15,** 3671–3678.
Wyckoff, G. J., Wang, W., and Wu, C. I. (2000). Rapid evolution of male reproductive genes in the descent of man. *Nature* **403,** 304–309.
Yamagata, K., Murayama, K., Okabe, M., Toshimori, K., Nakanishi, T., Kashiwabara, S., and Baba, T. (1998). Acrosin accelerates the dispersal of sperm acrosomal proteins during acrosome reaction. *J. Biol. Chem.* **273,** 10470–10474.
Yamasaki, N., Richardson, R. T., and O'Rand, M. G. (1995). Expression of the rabbit sperm protein Sp17 in COS cells and interaction of recombinant Sp17 with the rabbit zona pellucida. *Mol. Reprod. Dev.* **40,** 48–55.
Yanagimachi, R. (1972). Penetration of guinea-pig spermatozoa into hamster eggs *in vitro*. *J. Reprod. Fertil.* **28,** 477–480.
Yanagimachi, R. (1977). Specificity of sperm-egg interaction. *In* "Immunobiology of Gametes" (M. Edidin and M. H. Johnson, eds.), pp. 255–289. Cambridge University Press, London.
Yanagimachi, R. (1981). Mechanisms of fertilization in mammals. *In* "Fertilization and Embryonic Development *In Vitro*" (L. Mastroianni and J. D. Biggers, eds.), pp. 81–182. Plenum Press, New York.
Yanagimachi, R. (1984). Fertilization in mammals. *Tokai J. Exp. Clin. Med.* **9,** 81–85.
Yanagimachi, R. (1994). Mammalian fertilization. *In* "The Physiology of Reproduction," 2 Ed. (E. Knobil and J. D. Neill, eds.), pp. 189–317. Raven Press, New York.

Yanagimachi, R. and Phillips, D. M. (1984). The status of acrosomal caps of hamster spermatozoa immediately before fertilization *in vivo*. *Gamete Res.* **9,** 1–19.

Yonezawa, N., Aoki, H., Hatanaka, Y., and Nakano, M. (1995). Involvement of N-linked carbohydrate chains of pig zona pellucida in sperm–egg binding. *Eur. J. Biochem.* **233,** 35–41.

Youakim, A., Hathaway, H. J., Miller, D. J., Gong, X., and Shur, B. D. (1994). Overexpressing sperm surface β1,4-galactosyltransferase in transgenic mice affects multiple aspects of sperm–egg interactions. *J. Cell Biol.* **126,** 1573–1583.

Yurewicz, E. C., Hibler, D., Fontenot, G. K., Sacco, A. G., and Harris, J. (1993). Nucleotide sequence of cDNA encoding ZP3a, a sperm-binding glycoprotein from zona pellucida of pig oocyte. *Biochim. Biophys. Acta* **1174,** 211–214.

Yurewicz, E. C., Sacco, A. G., Gupta, S. K., Xu, N., and Gage, D. A. (1998). Hetero-oligomerization-dependent binding of pig oocyte zona pellucida glycoproteins ZPB and ZPC to boar sperm membrane vesicles. *J. Biol. Chem.* **273,** 7488–7494.

6

Signal Transduction Mechanisms Regulating Sperm Acrosomal Exocytosis

Gregory S. Kopf

Center for Research on Reproduction and Women's Health, University of Pennsylvania School of Medicine, Philadelphia

I. Introduction

The acrosome is a secretory vesicle overlying the nucleus of the mature spermatozoon. This organelle is a product of the Golgi complex, and is synthesized and assembled during spermiogenesis. Although there are examples of nonmammalian species in which the sperm do not possess an acrosome, it is clear that in the spermatozoa of all species that do contain this specialized secretory vesicle, the exocytotic release of acrosomal contents (referred to as the "acrosome reaction" or "acrosomal exocytosis") is an important process leading to successful fertilization. The mechanisms by which this exocytosis occurs are of interest to many investigators, not only for the purposes of understanding the normal fertilization process but also for the development of new ways in which to enhance or preclude fertility. An understanding of how acrosomal exocytosis is controlled at the molecular level requires a full knowledge of how this specialized organelle is formed, a detailed analysis of its composition, and an appreciation of how composition, morphology, and function are integrated. There is also an increasing appreciation for how the process of sperm capacitation is intimately tied to events controlling acrosomal exocytosis (see Chapters 3 and 8). It is only when these issues are fully appreciated that we will start to understand how acrosomal exocytosis is regulated and how this regulation dictates function. This review will not be exhaustive, as there are several in-depth reviews of this subject (Kopf and Gerton, 1991; Kopf *et al.,* 1999a; Roldan, 1999; Ward and Kopf, 1993; Yanagimachi, 1994). It is intended to provide an update of some of the latest observations both in and outside the field of fertilization that might ultimately impact on our ability to assemble a unifying set of principles that govern the regulation of sperm acrosomal exocytosis.

II. BIOGENESIS AND MORPHOLOGY OF THE ACROSOME

A complete understanding of the mechanism by which acrosomal exocytosis is regulated must start with a clear understanding of the type of organelle the acrosome actually is. In contrast to the historical assumption that the acrosome is a modified lysosome (Allison and Hartree, 1970), the observations that this organelle (1) is not involved in degradative processes within the sperm itself, (2) contains a dense core of specific acrosomal constituents, (3) is synthesized, assembled and stored for extended periods of time, and (4) undergoes a regulated secretory event in which the constituents of the dense core undergo a time-dependent dissolution (see below) argue more that the acrosome can be considered a type of secretory granule (Burgess and Kelly, 1987; Kelly, 1991) (see Chapter 8). This is an important distinction, because it most likely will dictate the mechanism by which acrosome biogenesis occurs, the mechanism of the signal transduction pathways that lead to acrosomal exocytosis, and the identity of the exocytotic machinery involved in the exocytotic process.

Although the acrosome is present following spermiogenesis, there are several questions pertaining to the formation and maturation of this organelle that remain to be answered. For example, although prominent biogenesis of the acrosome occurs during the Golgi and cap phases of spermiogenesis, it is not clear when during this developmental stage that organelle development actually starts to take place. Furthermore, multiple component proteins comprise the dense core of the acrosome, but little is known regarding whether the synthesis of all of these components occurs at the same time, or whether synthesis is ordered and coordinate. To date, the experimental evidence suggests that synthesis of these components is coordinated in a specific fashion. The mechanism by which these acrosomal components are targeted to this organelle during biogenesis is also not known. Spermatogenic cells possess functional mannose-6-phosphate/insulin-like growth factor II receptors, but it is not clear whether these receptors play a role in the transport of glycoproteins to this secretory granule or whether targeting occurs primarily through the "default" pathway seen in the transport of proteins in other secretory systems. Finally, once these components are packaged into the acrosome, it is not clear what the nature and functional significance are of additional processing of these components (i.e., movement within the organelle; posttranslational modifications) during sperm residence in the testis and/or during residence in the extratesticular male reproductive organs (i.e., epididymis; vas deferens). For example, in the guinea pig, the formation of specific domains within the acrosome has been clearly demonstrated, but the mechanism by which this compartmentalization is established is poorly understood, and an appreciation of the biological role of this compartmentalization is only starting to develop. Answers to all of these questions will no doubt become apparent when a systematic evaluation of the transcription, translation, and posttranslational modifications of the acrosomal proteins are undertaken. An understanding of these processes may greatly further our knowledge of the role of the acrosome in fertilization because it is becoming apparent that this secretory vesicle may have multiple functions (see Section VI). In any event, studies focused on the synthesis and processing of acrosomal components should be considered in the context of the acrosome functioning as a secretory granule and not as a modified lysosome. The reader is directed to a more in-depth review of this particular subject (Kopf and Gerton, 1991) (see also Chapter 8).

Although the gross morphology of the acrosome in spermatozoa of different species varies greatly, the basic molecular morphology of this secretory granule is rather conserved (Yanagimachi, 1994). The contents of the acrosome include structural and nonstructural and nonenzymatic and enzymatic components, and this secretory granule is delimited by both an inner and an outer acrosomal membrane. The granule sits atop and in close apposition to the condensed nucleus. In mammalian spermatozoa, acrosomal exocytosis represents a series of poorly defined events that result in the fusion and vesiculation of the plasma membrane overlying the acrosome with the outer acrosomal membrane, thus creating hybrid membrane vesicles. Events leading to this membrane fusion are controlled by a signal transduction cascade initiated by the binding of the zona pellucida of the egg to specific molecular components of the spermatozoon, a transmembrane signaling event leading to the generation of effector molecules, and the preparation of the participating membranes for membrane fusion. This fusion, as in other systems, likely involves a docking or adhesion step, in which the two membranes adhere, followed by a fusion step, in which the membrane lipid bilayers undergo a destabilization following by a rapid mixing. As described below it is also very possible that changes in membrane dynamics that occur during the capacitation process may play a critical role in preparing these membranes to undergo fusion in response to the appropriate effector.

III. BIOLOGICAL SIGNIFICANCE OF ACROSOMAL EXOCYTOSIS

A. IMPORTANCE OF ACROSOMAL EXOCYTOSIS IN THE CONTEXT OF FERTILIZATION

Although the number of spermatozoa in ejaculates of many species is on the order 10^7–10^8, *in vivo* studies have demonstrated that the number of spermatozoa arriving at the site of fertilization at any particular time is on the order of 1–100 (Yanagimachi, 1994). This indicates that the female tract plays an active role in selecting for the most viable, functionally active spermatozoa from what can be quite a heterogeneous population of cells in the ejaculate. This heterogeneity is quite notable in some species (e.g., human), and occurs as a consequence of cellular age, morphology, motility characteristics, and the ability to undergo capacitation. Migration of spermatozoa through the different regions of the female reproductive tract (vagina, cervix, uterus, oviduct) and the environments that they present, their interactions with the epithelial cells in these regions, and their interactions with the cumulus oophorus-enclosed egg prior to interaction with the zona pellucida provide a very efficient means to select for the most viable, functionally active spermatozoa (*Note:* throughout this review the use of the term "egg" will denote the ovulated oocyte arrested in metaphase II). Acrosomal exocytosis can also be considered a biological selection process designed to ensure the delivery of spermatozoa with optimal fertilization potential to the initial site of spermatozoa–egg interaction proper, i.e., the zona pellucida. In many species, spermatozoa that undergo acrosomal exocytosis prematurely generally display reduced fertilization potential due to a failure to penetrate the cumulus oophorus, an increased propensity for binding to the cells comprising the cumulus oophorus (thus excluding them from interaction with the zona pellucida), and an inability to adhere to the zona pellucida. In many cases those spermatozoa that undergo such premature exocytosis may not represent that population of cells in the ejaculate with optimal attributes; this could be due to morphological and/or cellular defects. Although there are a few exceptions, it is generally accepted that only acrosome-intact spermatozoa can establish high-avidity adhesion to the zona pellucida and thereby participate in the fertilization process. However, it is these exceptions, as well as recent data pertaining to the identity, localization and function of spermatozoa surface-associated zona pellucida binding proteins/receptors that should prompt us to rethink the paradigm of acrosomal status in spermatozoa adhesion to the zona pellucida (see Section VI, and Chapters 5 and 8). This is especially important to consider in light of what is now known about the mechanics of exocytosis in other cellular systems.

The primary function of acrosomal exocytosis following zona pellucida adhesion is to permit the penetration of this unique extracellular matrix so that the sperm cell can gain access to the perivitelline space of the egg. This matrix-induced exocytotic reaction is a regulated event that displays discrete properties addressed later in this chapter, and likely represents the physiologically relevant acrosome reaction. A consequence of zona pellucida penetration is the development of a sharply defined, thin penetration slit that is presumed to result from the limited digestion of the matrix by acrosome-associated proteases. The relative contributions of mechanical forces generated by spermatozoan motility and the aforementioned enzymatic digestion of the zona pellucida by acrosomal proteases in mediating spermatozoan passage through the zona pellucida, however, are not known. As described below, recent data from a number of different laboratories are starting to define new potential and more active roles for the components of the acrosomal matrix in numerous aspects of the fertilization process. Spermatozoa, having undergone zona pellucida-induced acrosomal exocytosis and zona pellucida penetration, arrive in the perivitelline space with the ability to interact and fuse with the plasma membrane of the egg. Because acrosome-intact spermatozoa are unable to adhere to and fuse with the egg plasma membrane, acrosomal exocytosis serves an additional role in fertilization, i.e., the exposure of sperm-associated domains involved in egg plasma membrane binding and fusion (Evans and Kopf, 1998). Acrosomal exocytosis, therefore, serves two biologically important functions. First, in most species studied, it serves as a selection process to ensure the delivery of a subpopulation of viable, acrosome-intact spermatozoa to the zona pellucida, the initial site of sperm–egg interaction proper. Second, zona pellucida-induced acrosomal exocytosis ensures that overtly acrosome-reacted spermatozoa, which are capable of fusion with the egg plasma membrane, are delivered to the perivitelline space, thus enhancing the probability for successful fertilization.

B. RELATIONSHIP OF CAPACITATION TO ACROSOMAL EXOCYTOSIS

Any discussion of the biological significance of acrosomal exocytosis must consider the relationship between this exocytotic event and sperm capacitation. Although capacitation is discussed elsewhere in this volume (Chapter 3), it is addressed at several different points in this chapter because work in several labs has started to integrate these two processes at the molecular level (Kopf *et al.*, 1999b; Visconti *et al.*, 1998; Visconti and Kopf, 1998). From a historical perspective, capacitation has been defined as the time interval between sperm deposition in the female reproductive tract during natural mating and the time at which fertilization occurs. This time interval would, therefore, encompass all of the interactions of spermatozoa with the female tract (Smith, 1998; Suarez, 1998; Verhage *et al.*, 1998). This definition was established following the observation that spermatozoa taken from the female tract immediately following mating did not have the ability to fertilize eggs, and that residency of the spermatozoa in the female tract in some way conferred fertilization capacity. Due, in part, to an increased understanding of sperm biology and to the establishment of more sophisticated assays of sperm function, the definition of capacitation has been narrowed and modified over the years to reflect many investigators' biases regarding the physiological importance of this event. Although fertilization still represents the absolute confirmation that the spermatozoon has undergone capacitation, the ability of these cells to undergo regulated acrosomal exocytosis in response to a biologically relevant ligand (e.g., zona pellucida; progesterone) can be taken as an earlier, upstream endpoint of this extratesticular maturational event. This definition does not take into account acrosomal exocytosis that can occur in response to chemical agents such as A23187, which have been demonstrated to induce this exocytotic event in uncapacitated spermatozoa (Kopf *et al.,* 1999b).

Studies addressing the molecular basis of sperm capacitation have suggested that events occurring during this extratesticular maturational process may be tightly integrated with events that lead to the induction of acrosomal exocytosis (Florman and Babcock, 1991; Kopf *et al.,* 1999b; Visconti *et al.,* 1998; Visconti and Kopf, 1998). For example, dramatic changes in plasma membrane fluidity/dynamics occur during capacitation, and some of these changes lead to and/or are controlled by the initiation of intracellular signaling events that are thought to be key components of capacitation (Gadella and Harrison, 2000; Gadella *et al.,* 1999; Harrison and Miller, 2000; Harrison *et al.,* 1996; Osheroff *et al.,* 1999; Visconti *et al.,* 1999a,b). Such changes in fluidity/dynamics may produce membranes that are now more "fusigenic" in nature. Moreover, Florman and co-workers have provided pharmacological evidence of a T-type Ca^{2+} channel with unique properties in mouse spermatogenic cells (Arnoult *et al.,* 1998). This channel appears to be activated in spermatozoa in response to ZP3, and is responsible for alterations in Ca^{2+} fluxes leading to acrosomal exocytosis. Furthermore, it may be maintained in a low conductance state by protein tyrosine phosphorylation, and then shifted to a high conductance state on dephosphorylation. This shift to the high conductance state could be mediated by a ZP3-induced activation of a protein tyrosine phosphatase activity (Arnoult *et al.,* 1997). Although both of these observations clearly remain working hypotheses at this time, they point to the likelihood that events leading to the capacitated state place the spermatozoon in a "permissive" state such that it can respond to appropriate signals to undergo acrosomal exocytosis. From a biological standpoint, therefore, capacitation and acrosomal exocytosis can be considered a continuum of processes.

IV. PHYSIOLOGICAL SITE OF ACROSOMAL EXOCYTOSIS

A. WHERE DOES IT OCCUR?

A major problem in elucidating the physiological site of acrosomal exocytosis is that many studies of this process have been conducted under conditions *in vitro.* Such studies, although invaluable to our understanding of what we now know about acrosomal exocytosis, do not reflect the actual conditions that the sperm cell encounters at the different levels of the male and female reproductive tracts. For example, many investigators use epididymal spermatozoa for their studies because they are easy to obtain, and they readily undergo capacitation, acrosomal exocytosis, and are able to fertilize eggs *in vitro*. However, such cells are not exposed to the environmental conditions and selection processes that normally occur in the male and female reproductive tracts (e.g., exposure to seminal plasma and/or female tract fluids). Such differences are likely to have significant effects on the numbers and kinetics of spermatozoa undergoing capacitation, which may ultimately affect their ability to undergo acrosomal exocytosis. In addition, typical *in vitro* assays of fertilization utilize very high spermatozoa:egg ratios (e.g., 1000:1; 1,000,000:1) when, as described above, the number of sperm at the site of fertilization *in vivo* is very low (Yanagimachi, 1994). These factors should caution interpretation of *in vitro* results and the extension of those results to the *in vivo* situation

As previously discussed, acrosomal exocytosis may occur in subpopulations of spermatozoa at different regions of the female reproductive tract. These acrosome reactions have no functional significance in the penetration of the zona pellucida per se, but may be important as a selection process to ensure that only the most viable, acrosome-intact spermatozoa with optimal fertilizing potential interact with the zona pellucida.

When spermatozoa are recovered from naturally mated or artificially inseminated females, a majority of the free-swimming spermatozoa in the ampullary region of the oviduct that are not associated with the cumulus–oocyte complex have not undergone acrosomal exocytosis (Bryan, 1974; Cummins and Yanagimachi, 1982; Overstreet and Cooper, 1979; Suarez *et al.*, 1983). In contrast, spermatozoa associated with the cumulus oophorus are either acrosome intact or have initiated acrosomal exocytosis (Bedford, 1972; Cummins and Yanagimachi, 1982; Yanagimachi and Noda, 1970). Spermatozoa found to be associated with the zona pellucida have either initiated acrosomal exocytosis or have completed this process (Cummins and Yanagimachi, 1982; Shalgi *et al.*, 1989). Similar observations have been made under conditions of *in vitro* fertilization (Cherr *et al.*, 1986; Shalgi *et al.*, 1989; Yanagimachi and Phillips, 1984). These reports, plus the observations in several other species that the interaction of spermatozoa with the zona pellucida appears to be initiated by cells with "intact acrosomes" (Cherr *et al.*, 1986; Crozet and Dumont, 1984; Myles *et al.*, 1987; Saling *et al.*, 1979), suggest that physiologically relevant acrosomal exocytosis occurs at the level of the zona pellucida. There are now numerous reports in a variety of mammals demonstrating that the zona pellucida can induce acrosomal exocytosis in a species-specific fashion (Cherr *et al.*, 1986; Cross *et al.*, 1988; Florman and First, 1988a,b; Florman and Storey, 1982b; O'Rand and Fisher, 1987), thus confirming the relevance of the egg's extracellular matrix in regulating this exocytotic event.

B. RELATIONSHIP BETWEEN CAPACITATION AND THE PHYSIOLOGICAL SITE OF ACROSOMAL EXOCYTOSIS

As discussed above, if capacitation is necessary for spermatozoa to undergo physiologically relevant acrosomal exocytosis in response to the zona pellucida, it stands to reason that both the male and female reproductive tracts could influence capacitation. Factors comprising the reproductive tract fluids are thought to bind to and stabilize the sperm plasma membrane, presumably to prevent capacitation from occurring. Other factors remove molecules from the sperm surface, presumably to initiate the capacitation process. Moreover, sperm migration through the different regions of the female reproductive tract (vagina, cervix, uterus, oviduct), their interactions with the epithelial cells in these regions, and their interactions with the cumulus oophorus surrounding the egg have been demonstrated to have significant effects on spermatozoan function, some if not all related to capacitation (Smith, 1998; Suarez, 1998). Although there are numerous reports suggesting that factors present in the fluids of both the male (e.g., epididymal fluids, seminal plasma) and female (e.g., cervical, uterine, and oviductal fluids) reproductive tracts have positive and negative effects on sperm capacitation (Florman and Babcock, 1991; Yanagimachi, 1994), only a limited number of studies have actually examined effects of purified components of these fluids on the capacitation process. Such components include phospholipid and cholesterol-binding proteins of the male tract (Desnoyers and Manjunath, 1992; Moreau *et al.,* 1998; Therien *et al.,* 1995) and glycosaminoglycans of the female tract (Lane *et al.,* 1999; Miller and Ax, 1990). Detailed discussion of these factors is beyond the scope of this review, but they have been discussed in greater depth elsewhere (Florman and Babcock, 1991; Kopf and Visconti, 1998; Kopf *et al.,* 1999b; Smith, 1998; Suarez, 1998; Visconti *et al.,* 1998; Yanagimachi, 1994). Nevertheless, it should be emphasized that as we learn more about the capacitation process, how it is controlled, and how it is integrated with the regulation of acrosomal exocytosis at the molecular level, we should then be in a position to examine the role of specific components of the reproductive tracts on sperm function, as they relate to defining how competence to undergo acrosomal exocytosis is accomplished physiologically.

V. THE ZONA PELLUCIDA AND PROGESTERONE AS PHYSIOLOGICAL INDUCERS OF ACROSOMAL EXOCYTOSIS

Although numerous biological and pharmacological agents have been demonstrated to induce acrosomal exocytosis *in vitro* [summarized in Kopf and Gerton (1991)], the physiological relevance of a majority of these agents in regulating this important exocytotic process is questionable. This review, therefore, focuses on two biological agents, namely, the zona pellucida and progesterone. Although both of these agents can stimulate acrosomal exocytosis, the available evidence to date, when the effects of both agents on acrosomal exocytosis of the same species have been examined, suggests that the mechanisms by which these agents function (e.g., binding proteins/receptor on sperm; signal transduction) are quite different.

A. THE ZONA PELLUCIDA

As described above, from an experimental perspective the zona pellucida has received the most attention as the physiological inducer of acrosomal exocytosis. To date, the preponderance of information regarding the identity of the active component of the zona pellucida, as well as the molecular basis of the spermatozoan–zona pellucida interaction, has been gleaned from studies in the mouse. It is becoming apparent, however, that results of studies in this species have translated to similar results in other species (Bagavant *et al.,* 1993; Moller *et al.,* 1990; Yurewicz *et al.,* 1993), thereby allowing us to conclude with some degree of confidence that the zona pellucida is the universal biological trigger of acrosomal exocytosis in mammals.

In the mouse, the zona pellucida is synthesized and assembled by the growing oocyte during its growth phase; it is composed of three major glycoproteins, designated as ZP1, ZP2, and ZP3 (Greve *et al.,* 1982; Wassarman, 1988). In fact, results of cDNA cloning, as well as characterization of the *zonae pellucidae* from several species, have demonstrated the existence of three major glycoprotein components (Bagavant *et al.,* 1993; Epifano and Dean, 1994; Harris *et al.,* 1994; Moos *et al.,* 1995; Shabanowitz and O'Rand, 1988). The genes encoding each of these proteins are unique and are under the control of zona pellucida-specific promoters that ensure their temporal and tissue-specific expression; the zona pellucida is synthesized only by the oocyte and not by somatic cells (Chamberlain and Dean, 1989; Kinloch and Wassarman, 1989; Lira *et al.,* 1990; Philpot *et al.,* 1987). ZP1, ZP2, and ZP3 are all highly glycosylated, and this posttranslational modification is extremely important for conferring specific biological functions to the zona pellucida (Florman and Wassarman, 1985a; Florman *et al.,* 1984). It should be noted that different designations have been given to the components of the zona pellucida in other species and this has led to some confusion in nomenclature (Harris *et al.,* 1994; Wassarman, 1990; Wassarman and Albertini, 1994); a uniform nomenclature should be carefully considered so as to avoid future confusion. Several studies suggest that the coordinate expression of these individual zona pellucida components is essential for secretion and subsequent assembly of this unique egg-associated extracellular matrix (Epifano *et al.,* 1995; Qi and Wassarman, 1999; Tong *et al.,* 1995; Wassarman *et al.,* 1997, 1998). Following assembly, the zona pellucida is comprised of ZP2/ZP3 heterodimers that are crosslinked in an organized fashion by ZP1 monomers, giving rise to a three-dimensional, relatively insoluble structure that functions biologically as a matrix (Greve and Wassarman, 1985).

The genes encoding the different zona pellucida components have been cloned in several species and the deduced primary polypeptide structures from these other species (including the human) bear remarkable similarity to one another (Harris *et al.,* 1994). These data suggest that the primary protein structure of the zona pellucida components of mammalian eggs are similar to one another and that differences observed in biochemical and biological heterogeneity between species may be specified by the carbohydrate domains. It is notable in this regard that various studies have demonstrated that the primary structure of the egg vitelline envelopes of certain fish, the extracellular coats of these oldest of vertebrates, bear a resemblance to the structure of ZP2 and ZP3 (Wassarman, 1995). These data would suggest that specific domains of egg extracellular coats that are involved in sperm recognition and binding are highly conserved in animal evolution.

In the mouse, the sperm-adhesive and acrosome reaction-inducing activities of the intact zona pellucida are conferred by ZP3, and these biological activities are confined to both the carbohydrate and the protein regions of ZP3 (Bleil and Wassarman, 1980, 1983; Florman *et al.,* 1984; Florman and Wassarman, 1985b; Lopez *et al.,* 1985; Miller *et al.,* 1992; Shur and Hall, 1982; Youakim *et al.,* 1994). Moreover, ZP3 was initially reported to interact exclusively with the plasma membrane overlying the acrosome in acrosome-intact spermatozoa (Bleil and Wassarman, 1986; Cheng *et al.,* 1994), although subsequent data call into question the actual status of the sperm plasma membrane during sperm–ZP3 interaction (Foster *et al.,* 1997; Kim *et al.,* 2001) (see below and Chapters 5 and 8). The fact that ZP3 interacts exclusively with the region of the plasma membrane overlying the acrosome suggests that this region of the spermatozoon possesses specific ZP3 binding proteins and/or receptors, the identity of which is a subject of much controversy and will be considered below. Presently, most is known about the functional domains of mouse ZP3 involved in sperm binding; these domains are encoded by the O-linked oligosaccharide chains of ZP3 (Bleil and Wassarman, 1988; Florman and Wassarman, 1985b; Wassarman, 1995), but there is some controversy regarding the nature of the active oligosaccharides (Johnston *et al.,* 1998; Miller *et al.,* 1993). These functional domains are currently being further refined (Litscher *et al.,* 1995; Rosiere and Wassarman, 1992), and will ultimately be valuable for probing the molecular mechanisms underlying these cell-matrix interactions (Wassarman, 1995). Work in other species has demonstrated that the respective ZP3 equivalent also possesses functional equivalence (Kinloch *et al.,* 1990; Moller *et al.,* 1990; Sacco *et al.,* 1989), although detailed functional analyses have yet to be carried out.

Although the role of ZP2 has not been directly tested, experiments suggest that this zona pellucida component is involved in anchoring the acrosome-reacted spermatozoan on the zona pellucida (Bleil *et al.,* 1988). In fact, it has been suggested that the inner acrosomal membrane, which is retained following acrosomal exocytosis, is the site for binding of ZP2, suggesting that this particular membrane domain possesses specific ZP2 binding proteins/receptors. The demonstration that sp56, a binding protein with high affinity for ZP3 (Bleil and Wassarman, 1990; Bookbinder *et al.,* 1995; Cheng *et al.,* 1994), is present in the acrosomal matrix (Foster *et al.,* 1997; Kim *et al.,* 2001) supports the possibility that ZP3–sp56 interaction may also serve to hold acrosome-reacting/reacted sperm on the zona pellucida. It is likely, therefore, that the ZP2/ZP3 heterodimers function together in an integrated manner to regulate sperm–zona pellucida adhesion and zona pellucida penetration (Chapters 4 and 8). To date, ZP1 is thought to play solely a structural role in the formation and maintenance of the zona pellucida (Greve and Wassarman, 1985).

B. PROGESTERONE

Progesterone was identified as the active component of serum or follicular fluid responsible for the induction of acrosomal exocytosis in human spermatozoa (Baldi *et al.,* 1995; Osman *et al.,* 1989; Revelli *et al.,* 1998). This particular steroid has since been demonstrated to induce acrosomal exocytosis in the spermatozoa of other mammals (Cheng *et al.,* 1998; Melendrez *et al.,* 1994; Purohit *et al.,* 1998; Roldan *et al.,* 1994). Several lines of evidence suggest that this effect is mediated by a sperm surface receptor through a nongenomic pathway (Blackmore *et al.,* 1995; Revelli *et al.,* 1998), although this receptor has still not been unequivocally identified. The signal transduction mechanisms that regulate the progesterone-induced acrosomal exocytosis have also been investigated, and are different than those proposed to regulate zona pellucida-induced acrosomal exocytosis.

There are several questions to be asked regarding the biological relevance of progesterone as an *in vivo* trigger for acrosomal exocytosis. For example, (1) Where does this steroid exert its effect on acrosomal exocytosis in the female reproductive tract? (2) Do the concentrations of this steroid reach levels high enough to induce acrosomal exocytosis? (3) If this steroid functions to induce acrosomal exocytosis prior to the interaction of spermatozoa with the zona pellucida, what is its biological significance? To date, estimates have been made that suggest that the progesterone levels in the oviduct might, under certain conditions, be high enough to induce acrosomal exocytosis. This, however, creates a problem with regard to the biology of fertilization, because it has been demonstrated in many species that acrosome-reacted spermatozoa cannot establish adhesion to the zona pellucida and thus would not fertilize eggs. This conundrum could be resolved, however, if one considered two models of action of this steroid that could be physiologically relevant.

First, as stated above, biological selection of the most functionally competent spermatozoa to fertilize eggs most likely occurs at different regions of the female reproductive tract and is a reasonable mechanism by which a heterogeneous population of spermatozoa give rise to the generation of a vanguard population of fertilizing spermatozoa. As stated earlier, acrosomal exocytosis may represent one mechanism of such a selection process. Progesterone in the female tract may, therefore, induce acrosomal exocytosis of those spermatozoa in the population that are not optimal for fertilization. This hypothesis has not been tested.

Second, it is possible that progesterone could function in conjunction with the zona pellucida to "prime" the sperm so that they undergo efficient acrosomal exocytosis following adhesion to the zona pellucida (see below). In such a scenario, it is not necessary that the concentrations of this steroid be high enough to induce overt occasional oxocytosis. Sub-threshold concentrations could serve as part of the priming mechanism. The idea of a priming event that subsequently leads to overt exocytosis has precedents in other exocytotic systems, and is not inconsistent with more current paradigms of acrosomal exocytosis that have been proposed (see below and Chapter 8). Although this hypothesis has been proposed using *in vitro* assays (Roldan *et al.,* 1994), evidence of its validity can only be approached experimentally once the identity of the sperm surface binding proteins/receptors for progesterone and the zona pellucida are identified (see below).

VI. SPERM-ASSOCIATED RECEPTOR/BINDING PROTEINS FOR THE ZONA PELLUCIDA AND PROGESTERONE

A. THE ZONA PELLUCIDA

One of the most controversial fields in the area of fertilization research has been the elucidation of the identity and function of the spermatozoan-associated zona pellucida/ZP3 binding proteins/receptors that mediate sperm–zona pellucida adhesion and signal transduction leading to acrosomal exocytosis. Clearly, sperm–zona pellucida interactions have several hallmarks of ligand-binding protein/receptor–effector interactions, in which the ligand is a component of a unique extracellular matrix and the subsequent transmembrane signaling leads to an exocytotic event (Kopf and Gerton, 1991; Ward and Kopf, 1993). However, sperm–zona pellucida adhesion also represents the interaction of a motile cell with an extracellular matrix, and models for other cell–cell or cell–matrix interactions should form the basis of our thinking when considering the molecular basis of these adhesion and exocytotic processes, as well as for the design of experiments to test specific aspects of resultant hypotheses (Goldbrunner *et al.,* 1999; Gonzalez-Amaro and Sanchez-Madrid, 1999; Lawrence, 1999; Murphy and Gavrilovic, 1999).

The identification of ZP3 binding proteins/receptors has been hampered by the lack of information pertaining to the precise identity of the active sperm adhesion and acrosomal exocytosis-inducing moieties of the zona pellucida (ZP3 or functional ZP3 equivalents), and the seemingly complex nature of the interactions of these moieties with the sperm surface, of which there is ample experimental evidence (Benau and Storey, 1988; Bleil and Wassarman, 1983; Endo *et al.,* 1987b,c; Florman and Storey, 1982a; Kligman *et al.,* 1991; Leyton and Saling, 1989b; Storey, 1995; Thaler and Cardullo, 1996). Such complex interactions do not preclude the possibility that sperm–zona pellucida interactions likely involve multiple interactions between the zona pellucida and the sperm cell surface, and that acrosomal exocytosis results from the formation of functional complexes containing multiple zona pellucida-associated active domains, as well as multiple, specific sperm-associated zona pellucida binding proteins and signal transducing receptors (Kopf and Gerton, 1991; Ward and Kopf, 1993). Biochemical examination of such functional complexes may reveal that both high- and low-affinity binding interactions comprise these complexes. It is premature and naive, therefore, to conclude that a single component on the sperm surface mediates all of these biological events and that this single component has to be a receptor. In fact, it is these very perceptions that have hindered significant advancement of this field.

In any event, the identity of multiple, seemingly dissimilar ZP3 binding protein/receptor candidates being put forth by numerous laboratories may, at first glance, seem to be confusing because there is no overwhelming evidence to support one candidate over another as being "the molecule." As stated above, it is possible that multiple binding protein/receptor candidates might function to mediate the dual biological functions of ZP3, and that the interaction of these components with ZP3, as well as with one another, must occur in an ordered fashion through high- and low-affinity interactions to first establish adhesion of the spermatozoan to ZP3, which then allows signal transduction to occur in order to initiate acrosomal exocytosis. The concept of multiple, interactive protein domains giving rise to ligand binding and signal transduction is not a novel concept in cell biology, because most cell–cell or cell–substratum interactions occur in this fashion. Moreover, these domains need not necessarily be identical. In the case of sperm–ZP3 interaction these possibilities are based on precedent in the literature pertaining to the interaction between the zona pellucida (or ZP3) and the sperm cell surface to mediate both sperm adhesion and the induction of acrosomal exocytosis. Studies from several laboratories using the mouse model have established the fact that sperm adhesion to the zona pellucida and zona pellucida-induced acrosomal exocytosis are two independent processes (Kopf and Gerton, 1991; Storey and Kopf, 1991), with ZP3-induced acrosomal exocytosis appearing to consist of discrete, independently regulated events (Kligman *et al.*, 1991; Lee and Storey, 1985, 1989; Saling and Storey, 1979; Ward and Storey, 1984). These observations, combined with the experimental evidence that the interaction of ZP3 with the sperm surface may occur in a multivalent or cooperative fashion (Thaler and Cardullo, 1996), and that multiple interactions followed by possible binding protein/receptor aggregation then lead to signal transduction and acrosomal exocytosis, strongly support the hypothesis that sperm–zona pellucida interaction is a multicomponent and highly ordered event.

Several zona pellucida binding proteins/receptors have been studied in detail. Although some can be considered viable binding proteins, to date none of these candidates fulfill all of the properties that one would expect for a specific receptor. When considering the properties of a cell surface receptor, specific criteria should include (1) presence in the appropriate region of the sperm cell involved in sperm adhesion and acrosomal exocytosis, (2) specificity and kinetics of ligand binding, (3) presence in numbers on the sperm surface consistent with ligand binding kinetics and biological effect, (4) ability to couple to specific signal transduction systems, and (5) appropriate cell and tissue expression. It is beyond the scope of this chapter to consider all of the candidates, so only a few will be considered. The reader is encouraged to examine the numerous review articles dealing with this topic (Kopf and Gerton, 1991; Tulsiani *et al.*, 1997; Ward and Kopf, 1993; Wassarman, 1995) (see also Chapter 5, this volume). For the most part, these candidates have the ability to bind to carbohydrates and, in some cases, may possess enzymatic activities that may or may not be involved in their putative zona pellucida binding/receptor activity [see Table 2 in Tulsiani *et al.* (1997)]. The particular candidates that will be discussed have been chosen because they represent diversity with respect to structure and potential function, and because they have received considerable attention. This does not, in any way, diminish the importance of the other candidates that have been studied.

One of the most extensively characterized mouse spermatozoan proteins that possesses specific binding activity for ZP3 is a protein known as "sp56" (Bleil and Wassarman, 1990; Bookbinder *et al.,* 1995; Cheng *et al.,* 1994). This protein binds ZP3 with high affinity, but does not interact with ZP2 or $ZP3_f$ (Bleil and Wassarman, 1990). This protein has lectin-like properties, has properties of an extracellular peripheral membrane protein that is associated with the plasma membrane overlying the sperm head, displays appropriate tissue expression, and bears homology to the family of complement component 4-binding proteins (Bookbinder *et al.,* 1995; Cheng *et al.,* 1994). The domains on sp56 involved in ZP3 binding, and the mechanism by which signal transduction leading to acrosomal exocytosis is integrated with ZP3 binding to this protein, are not known. Although it appears as if sp56 is a ZP3 binding protein, it is not known whether it is a signal-transducing receptor. The guinea pig sperm ortholog of sp56 (AM67) has been cloned and was shown to be present within the acrosomal matrix of the guinea pig sperm (Foster *et al.,* 1997). Moreover, a reevaluation of the localization of mouse sperm sp56 by these investigators demonstrated that it, likewise, was associated with the acrosomal matrix (Kim *et al.,* 2001). This apparent discrepancy in localization (i.e., localization solely to the plasma membrane versus localization to the acrosomal matrix) was based on differences in techniques used for localization of the protein (Cheng *et al.,* 1994; Foster *et al.,* 1997; Suzuki-Toyota *et al.,* 1995). If it is conclusively established that sp56 is a ZP3 binding protein present only within the acrosomal matrix, this does not rule out its function in mediating sperm adhesion to the zona pellucida. It is distinctly possible that "acrosome-intact" spermatozoa that establish adhesion to the zona pellucida via ZP3 (or its functional equivalent in other species) may, in fact, be spermatozoa in which the plasma membrane and outer acrosomal membranes have "docked" with one another and have formed intermediate membrane complexes comprising the vesicular face of the outer acrosomal membrane and the extracellular face of the plasma membrane; such membrane docking and formation of intermediate membrane complexes characterize the events of regulated secretion in other secretory model systems (Burgess and Kelly, 1987; Jahn and Sudhof, 1999; Kelly, 1991; Sudhof, 1995), and it is now becoming clear that the sperm acrosome has associated with it many of the components of the exocytotic machinery (Schulz *et al.,* 1997; Ward *et al.,* 1999; Iida *et al.,* 1999; Michaut *et al.,* 2000; Yunes *et al.,* 2000; Ramalho-Santos *et al.,* 2000). In other model systems of exocytosis, it has been proposed that small fusion pores form that open and close in a dynamic fashion ("flickering pores"), and that these pores form subsequent to the formation of such intermediate membrane complexes and prior to overt exocytosis (Monck and Fernandez, 1996). The formation of these transient pores could be considered a metastable state prior to the complete mixing of the membrane compartments and subsequent exocytosis/secretion. It is entirely possible that stable interactions between an acrosomal matrix-associated sp56 and ZP3 are established during the formation of such a metastable state in which flickering pores (or some other structure) precede the completion of acrosomal exocytosis (i.e., complete membrane vesiculation and lifting of the hybrid membrane vesicles away from the acrosomal matrix). The presence of a metastable state following sperm–zona pellucida/ZP3 interaction and prior to the completion of acrosomal exocytosis has been documented (Kligman *et al.*, 1991), but has not been characterized in great detail. This new model for sperm–zona pellucida interaction is outlined in greater detail in Chapter 8. Elements of this new paradigm should be examined in greater detail, as it could explain apparent "conflicting" data seen by others using different species regarding the acrosomal status of spermatozoa adherent to the zona pellucida (Storey and Kopf, 1991). Moreover, such a model might focus attention on a new role for the acrosomal matrix in sperm–zona pellucida adhesion and zona pellucida penetration. For example, recent data suggest that components of the acrosomal matrix, including the ZP3 binding protein sp56, undergo an organized and time-dependent release following acrosomal exocytosis and this likely plays a key role in sperm–zona pellucida adhesion and penetration (Kim *et al.*, 2001).

A specific form of β-galactosyltransferase (GalTase) has been implicated as a receptor/binding protein for ZP3 in mouse spermatozoa (Dubois and Shur, 1995). This protein has been postulated to mediate sperm–zona pellucida adhesion by interacting with oligosaccharide residues specifically on ZP3, and to induce acrosomal exocytosis through GalTase aggregation on the cell surface, leading to the activation of sperm heterotrimeric G_i proteins (Gong *et al.*, 1995; Miller *et al.*, 1992). Targeted overexpression of this form of the enzyme in spermatozoa, predicted in theory to yield spermatozoa that have an enhanced ability to interact with the zona pellucida, yields sperm that, in fact, display a reduced ability to adhere to the zona pellucida (Youakim *et al.*, 1994). This is apparently due to their hypersensitivity to ZP3, such that they undergo acrosomal exocytosis precociously/spontaneously and, therefore, have a reduced avidity of interaction with the zona pellucida. These experiments were interpreted as demonstrating that successful fertilization requires an optimal, rather than a maximal, concentration of GalTase moieties on the sperm surface. In contrast, targeted mutation of the GalTase gene yields null males that are fertile, yet their spermatozoa interact with less avidity for ZP3 compared to the spermatozoa from wild-type animals and are unable to undergo ZP3-induced acrosomal exocytosis (Lu and Shur, 1997). The authors conclude that although ZP3 binding and induction of acrosomal exocytosis are dispensable for fertilization, these properties impart a physiological advantage to the spermatozoon for fertilization. One can conclude that although the sperm-associated GalTase may be important for certain aspects of sperm–ZP3 interaction, it is not absolutely critical for fertility. These data suggest that although GalTase may play some role in sperm–zona pellucida interaction, it is probably not of primary importance in mediating sperm–ZP3 binding leading to fertilization.

Hardy and Garbers (1994, 1995) utilized pig zonae pellucidae as an affinity matrix to isolate proteins of the pig spermatozoan membranes that display high affinity and specific binding to this extracellular matrix. It should be noted that these membrane proteins were covalently modified (biotinylation) for the purposes of the screening assays, so any binding proteins that might not be efficiently labeled by this technique might go undetected. They identified several classes of proteins that bound to the pig zona pellucida in a species-specific fashion (p105/p45 and p56–62). One of these proteins (p105/p45) displayed high-affinity binding to the zona pellucida, and amino acid sequencing of tryptic peptides from this protein yielded sequences that did not match any sequences in the databases. Subsequent cloning of the cDNA encoding p105/p45 (termed zonadhesin) demonstrated that it encodes a novel protein with a very unique domain structure (Hardy and Garbers, 1995). This protein contains a single putative transmembrane domain, a short basic intracellular C terminal domain, and a very large amino-terminal extracellular domain. The domain sequence of the extracellular region is of great interest, because it represents a mosaic of a unique N-terminal domain, a mucinlike domain, and five tandemly arranged domains that display homology to prepro von Willebrand factor. The protein that is present on spermatozoa is processed at some time postmeiotically, such that the N-terminal and mucin domains are lost. This creates a mature extracellular domain containing the von Willebrand domains. This protein is of interest because von Willebrand factor is a major adhesive glycoprotein that plays a key role in mediating platelet adhesion to the subendothelium under conditions of high shear stress such as occur in small arterioles and arterial capillaries (Ruggeri, 1997; Ruggeri and Ware, 1993). The presence of such a domain on zonadhesin suggests that it may mediate sperm–zona pellucida adhesion, a cell–extracellular matrix interaction that could be considered a cellular adhesion event that occurs under conditions of high shear stress, given the fact that the sperm is motile. The roles of these domains in zona pellucida/ZP3 binding have yet to be confirmed. Likewise, there is no evidence that zonadhesin can function as a signal transducer. Such conclusions will await further experimentation. It is of interest to note that von Willebrand factor forms a complex with several receptor molecules, leading not only to cellular adhesion (platelet adhesion to the subendothelium) but to transmembrane signal transduction, resulting in cellular activation (platelet aggregation) (Falati *et al.,* 1999; Ruggeri, 1997; Ruggeri and Ware, 1993). The morphological, mechanical, and biochemical similarities of these hemostatic events to sperm–zona pellucida/ZP3 adhesion and sperm activation are notable.

Saling and co-workers have described a mouse sperm protein that they designated as "p95," which they postulated could serve as a ZP3 receptor with properties of a receptor tyrosine kinase (Leyton and Saling, 1989a). This protein has some characteristics of a membrane protein and it was postulated that p95 possessed intrinsic tyrosine kinase activity that was modulated by ZP3 (Leyton *et al.,* 1992).

Subsequent purification of this protein demonstrated it to be a germ cell-specific hexokinase (type I) with some unique properties (Kalab *et al.,* 1994; Travis *et al.,* 1998; Visconti *et al.,* 1996). To date, experiments designed to test whether this germ cell-specific hexokinase can function as a ZP3 binding protein and/or receptor have been negative (A. J. Travis, S. B. Moss, and G. S. Kopf, unpublished data). The reported cloning of the human homolog of mouse p95, which was designated as Hu9 or "zona receptor kinase/ZRK" by Burks *et al.* (1995), has revealed that it is c-mer (Bork *et al.,* 1996), a protooncogene member of the axl family of transforming receptor tyrosine kinases (Graham *et al.,* 1994, 1995). The function of Hu9 as a receptor tyrosine kinase for ZP3 in human spermatozoa is presently unclear. It is of interest that targeted gene disruptions of three structurally related receptor tyrosine kinases, namely, Tyro 3, Axl, and Mer, result in viable and fertile mice when a single receptor or any combination of two receptors is targeted (Lu *et al.,* 1999). However, when mice are null for all three receptors the males are sterile as a result of the progressive death of the differentiating germ cells. This phenotype appears to be due to a failure of Sertoli cell support, but the role of these receptors does not appear to be limited to the tissues of the reproductive systems. Whether mouse c-mer is actually expressed in mouse sperm is not known at this time.

In conclusion, although considerable effort by numerous laboratories has been invested in identifying candidate sperm proteins thought to be involved in sperm–zona pellucida/ZP3 adhesion and the induction of acrosomal exocytosis, it is clear that this is a very complex process and that we should be rethinking our current models for how these interactions occur. Part of this thought process should involve a careful reevaluation of the nature of the sperm cell's membranes as they come in contact with the zona pellucida and the possible roles of the acrosomal matrix in mediating sperm–zona pellucida interaction. Consideration of other models of cell–cell adhesion and cell–matrix adhesion should also be kept in mind when approaching this very difficult problem experimentally. It is distinctly possible that some of these aforementioned candidates [as well as others not discussed; see Tulsiani *et al.,* 1997)] may represent only a subset of molecules that mediate the complex nature of interactions between the sperm surface and the zona pellucida.

B. PROGESTERONE

As described above, progesterone has been demonstrated to induce acrosomal exocytosis in the spermatozoa of several different species. Although the exact physiological role of this steroid in regulating sperm cell function as it relates to fertilization is still some matter of debate, this particular paradigm has become of interest to a significant number of scientists working in the field of alternative modes of action of steroid hormones. Clearly, the classical mechanism of action of steroids in specific target tissues is at the level of gene expression. Yet there is an increasing body of literature suggesting that steroids can also function in a nongenomic fashion (Revelli *et al.,* 1998; Schmidt *et al.,* 2000; Wehling, 1997; Falkenstein *et al.,* 2000a,b), and the mechanism by which this occurs is poorly understood. It is now becoming apparent that these nongenomic actions of steroids may be mediated by a new class of cell surface receptors that couple, in some manner, to intracellular signal transduction cascades. For example, aldosterone has been demonstrated to stimulate the Na^+/H^+ exchanger in human leukocytes (Wehling, 1995), and progesterone has been demonstrated to stimulate ion fluxes rapidly in hepatocytes (Waldegger *et al.,* 1995) and human sperm (Blackmore, 1993; Blackmore *et al.,* 1990). One of the earliest reports suggesting the existence of a nongenomic progesterone receptor came from the work of Sadler and Maller (1982), who were interested in defining the biochemical mechanism by which progesterone induces oocyte maturation in *Xenopus* oocytes. Subsequently, numerous reports suggested that various cells/tissues of the reproductive systems possess such nongenomic receptors (Rae *et al.,* 1998a,b; Revelli *et al.,* 1994, 1998). For many years the identity of such receptors had been inferred indirectly by pharmacological studies. Meyer *et al.* (1996) purified and sequenced a high-affinity progesterone binding protein from pig liver microsomal membranes. Purification yielded two major proteins of 28 and 56 kDa that had identical amino-terminal sequences, suggesting that they were related. The amino acid sequence did not match any known sequences in the databases. A full-length cDNA cloning encoding this protein was isolated by reverse transcriptase–polymerase chain reaction (RT–PCR) and screening of a porcine vascular smooth muscle cell cDNA library (Falkenstein *et al.,* 1996). The cDNA sequence predicts a protein of ~22 kDa that contains a single putative transmembrane domain. Database searches revealed some similarity (76% identity at the amino acid level) of this sequence to that of a gene encoding a 25-kDa rat liver protein postulated to be a dioxin-induced protein similar to the cytokine/growth factor/prolactin receptor family (Selmin *et al.,* 1996). The predicted protein does not contain a cysteine-rich domain, or a consensus steroid binding pocket domain, known to be present in classical steroid receptors. N-Terminal antipeptide antiserum to the protein recognized the appropriately sized proteins in the liver (Falkenstein *et al.,* 1999a), and has been subsequently used to demonstrate the existence of a ~45-kDa protein in human sperm extracts that appeared to be present in the sperm head, using indirect immunofluorescence techniques (Buddhikot *et al.,* 1999). It is not known whether the difference in molecular weight of the sperm protein compared to the liver protein is due to posttranslational modifications and/or cross-reaction of antibodies with a related protein. Moreover, both this antipeptide antiserum as well as an antiserum to the recombinant protein blocked progesterone, but not A23187-induced acrosomal exocytosis of human spermatozoa by approximately 50–60% (Buddhikot *et al.,* 1999), and also blocked the progesterone-induced increase in cytosolic Ca^{2+} (Falkenstein *et al.,* 1999b). It should be noted that very high concentrations of antibody were need to observe these inhibitory effects. Clearly, additional experiments will be needed to determine whether, in fact, this protein represents the sperm receptor that mediates the progesterone effects.

Several other proteins with different properties have also been implicated as the functional progesterone receptor on human spermatozoa (Luconi *et al.,* 1998; Sabeur *et al.,* 1996). In these particular studies, antisera against the genomic progesterone receptor were used as a reagent to demonstrate the presence of protein by immunoblotting (Luconi *et al.,* 1998; Sabeur *et al.,* 1996), or to inhibit progesterone-induced acrosomal exocytosis (Sabeur *et al.,* 1996). In both cases, the reagents used recognize the conserved progesterone binding pocket motif that is present in genomic receptors but absent in the putative nongenomic liver microsomal membrane protein receptor. Work by Blackmore *et al.* (1995) using a variety of steroid analogs has demonstrated that the Ca^{2+} influx response of human spermatozoa to progesterone is likely mediated by the interaction of its putative receptor with the β-face of the steroid C and D rings, in contrast to the interaction of the α-face of this steroid with genomic receptors. The data using antisera raised against the genomic receptor must therefore be carefully considered. Finally, the relationship between these putative progesterone receptors and the $GABA_A$ receptor/Cl^- channel, which has been implicated in progesterone-induced acrosomal exocytosis, is not clear at this time (Meizel, 1997). Clearly, further experimentation will be required to resolve the identity of the spermatozoan surface receptor for progesterone and to determine whether the same protein that binds this steroid also mediates signal transduction leading to acrosomal exocytosis. The identity of this receptor is especially important to establish given the recent evidence that supports the idea that the progesterone receptor on the *Xenopus* oocyte surface that functions through a non-genomic pathway to initiate oocyte maturation is, in fact, a conventional progesterone receptor that functions at the genomic level (Bayaa *et al.,* 2000; Tian *et al.,* 2000).

VII. SIGNAL TRANSDUCTION MECHANISMS MEDIATING THE EFFECTS OF THE ZONA PELLUCIDA AND PROGESTERONE

A. INTEGRATION OF SIGNAL TRANSDUCTION PATHWAYS ASSOCIATED WITH CAPACITATION WITH THOSE ASSOCIATED WITH ACROSOMAL EXOCYTOSIS

As described in Section III,B, it is apparent that because sperm capacitation is associated with events such as hyperactivation of motility and the ability to undergo a regulated exocytotic event in response to a physiological agent (e.g., zona pellucida) (Kopf *et al.,* 1999b), analysis of the molecular mechanisms underlying acrosomal exocytosis must consider the integration of membrane and signal transduction events that occur during the capacitation process. If one accepts the hypothesis that elements of acrosomal exocytosis have similarities to other exocytotic systems, it is clear (1) that changes in membrane properties/dynamics are important for the stages of membrane docking, membrane fusion, and membrane vesiculation that ultimately lead to exocytosis, and (2) that intracellular signal transduction in response to a specific initiator (e.g., ligand; membrane potential change) is necessary to initiate and/or complete those molecular events involved in membrane–membrane interaction and fusion. In this regard, sperm capacitation and acrosomal exocytosis can be seen to represent a continuum of events that eventually lead to the exposure of the acrosomal matrix to the extracellular space. This discussion will integrate, where appropriate, those biophysical and/or signaling events that comprise the capacitation process as they relate to those signal transduction events that lead to acrosomal exocytosis.

B. NATURE OF SPERM SIGNAL TRANSDUCTION REGULATING ACROSOMAL EXOCYTOSIS

As with signal transduction processes in most somatic cells, it is clear that signal transduction in sperm represents an integration between changes in ionic conductance and second messenger generation, and that such integration serves to regulate capacitation, motility, and acrosomal exocytosis.

As outlined above, although there are still many questions to be addressed regarding the identity of the binding proteins/receptors on the sperm surface that mediate zona pellucida adhesion and acrosomal exocytosis, experiments focusing on mechanisms of transmembrane signal transduction and activation of intracellular effectors have evolved from several laboratories. If one proposes that ZP (or ZP3) effects on sperm are mediated via specific binding proteins and/or receptors that are coopted into a multimeric functional signaling complex, one could argue that transmembrane signal transduction may have similarities to that seen in somatic cells. Because heterotrimeric G-proteins play a key role in signal transduction in response to the activation of several classes of receptors, experiments have focused on the role of these GTP-binding proteins on ZP- and ZP3-mediated sperm adhesion and acrosomal exocytosis. Sperm-associated heterotrimeric G-proteins of the G_i class appear to play a key role in regulating ZP- or ZP3-mediated signal transduction events leading to the acrosomal exocytosis in mouse, bovine, and human sperm (Florman *et al.,* 1989; Gong *et al.,* 1995; Ward and Kopf, 1993; Ward *et al.,* 1992, 1994; Wilde *et al.,* 1992), and this has been demonstrated using a variety of experimental approaches. In contrast, acrosomal exocytosis induced either spontaneously or with specific agents such as A23187 or progesterone occurs independently of G_i activation, thus demonstrating that coupling to and subsequent activation of this class of heterotrimeric G-proteins is restricted to exocytosis initiated by ZP/ZP3, the physiologically relevant ligand for this secretory event (Endo *et al.,* 1987a; Foresta *et al.,* 1993; Lee *et al.,* 1992; Murase and Roldan, 1996; Tesarik *et al.,* 1993). To date, the only characterized sperm-associated ZP3 binding protein/receptor that has been implicated in coupling to G_i is GalTase (Dubois and Shur, 1995; Gong *et al.,* 1995). Recently, these investigations expressed GalTase in *Xenopus* oocytes and demonstrated that ZP3, but not ZP1 and ZP2, can bind to these oocytes (Shi *et al.,* 2001). ZP3 and GalTase antibodies were able to stimulate GTPγ[^{35}S] binding and increase GTPase activity of membranes from these GalTase expressing oocytes and these agents were also able to trigger cortical granule exocytosis and activation; these events occurred in a pertussin toxin-sensitive fashion. Thus, one of the earliest events following ZP or ZP3 interaction with the sperm surface is transmembrane signaling via sperm-associated G_i. Elucidation of the effector systems upstream and the signaling systems downstream that are entrained and activated to bring about exocytosis (see below) must, therefore, consider the integration with these particular GTP-binding regulatory proteins.

C. IONIC EVENTS MODULATING ACROSOMAL EXOCYTOSIS

The process of capacitation and the signaling events associated with this essential extratesticular maturation process are clearly regulated by transmembrane ionic movements (Kopf *et al.,* 1999b; Visconti *et al.,* 1998). Given the interdependence of capacitation and subsequent signaling events regulating acrosomal exocytosis, integration of ionic events occurring during both of these processes is likely and should be the focus of future investigations.

Changes in sperm Ca^{2+} conductance play essential roles both in capacitation and in acrosomal exocytosis. These changes appear to be modulated by voltage-sensitive Ca^{2+} channels, which may display distinct regionalization in the sperm membrane and may display distinct pharmacological properties (Arnoult *et al.,* 1996a,b, 1998; Beltrán *et al.,* 1994; Darszon *et al.,* 1999; Florman *et al.,* 1998; Florman and Babcock, 1991; Hagiwara and Kawa, 1984; Lievano *et al.,* 1996; Santi *et al.,* 1996; Serrano *et al.,* 1999). It must be emphasized that many, but not all, of these characterizations are based on studies with spermatogenic cells and are extended to sperm, given the restrictions in working with the highly differentiated and motile sperm cell. Although earlier pharmacological studies implicated L-type Ca^{2+} channels in mediating *the zona pellucida*-induced increase in intracellular Ca^{2+} (Florman, 1994; Florman *et al.,* 1992), it is now known that some of the pharmacological agents used can also block T-type Ca^{2+} channels (Arnoult *et al.,* 1998; Lievano *et al.,* 1994). Although there still is some controversy, it is likely that the functional voltage-sensitive Ca^{2+} channels are of the low-voltage-activated T-type variety. Other work suggests that transient elevations of intracellular Ca^{2+} occur following the very early interactions with the zona pellucida (ZP3) and that these transient elevations are entrained as a regulatory system in the sperm by membrane hyperpolarization that occurs during capacitation (Arnoult *et al.,* 1999; Florman *et al.,* 1998). Part of this regulatory constraint during capacitation may involve protein tyrosine phosphorylation such that T channels are maintained in a low or zero conductance state by the tonic protein tyrosine phosphorylation that occurs during capacitation, and are converted to a high conductance state, now activatable by the zona pellucida/ZP3, following dephosphorylation (Arnoult *et al.,* 1997). This hypothesis remains to be tested experimentally. Since calmodulin has also been recently demonstrated to inhibit these T-type Ca^{2+} channels (Lopez-Gonzalez *et al.,* 2001) the involvement of this calcium regulatory protein in this process must be further explored.

It must be emphasized that the activation of such a low-voltage-activated T-type Ca^{2+} channel is not the only mode of ionic regulation of acrosomal exocytosis entrained by the zona pellucida/ZP3. As summarized by Florman *et al.* (1998), zona pellucida/ZP3 sperm adhesion also results in the activation of a cation channel(s) through a G_i-insensitive and voltage-insensitive mechanism; the resultant conductance through this cation channel(s) produces a depolarizing current, leading to the activation of the aforementioned T-type Ca^{2+} channel and transient inward Ca^{2+} currents. In addition, signaling also occurs by the activation of a G_i-sensitive pH regulator, resulting in the transient alkalinization of internal pH. This combination of intracellular alkalinization and transient Ca^{2+} elevations leads to a sustained Ca^{2+} rise, leading to acrosomal exocytosis. Recent work suggests that these sustained elevations of Ca^{2+} are mediated through store operated channels (O'Toole *et al.,* 2000) with characteristics of the transient receptor potential (Trp) protein 2 (Jungnickel *et al.,* 2001). Progesterone has also been reported to induce a significant increase in sperm intracellular Ca^{2+} (Baldi *et al.,* 1991; Blackmore *et al.,* 1990), but uncertainty remains regarding the Ca^{2+} channels recruited under these conditions, as well as their mechanism of regulation.

Other ions (e.g., Na^+, K^+ and HCO_3^-) have been implicated in acrosomal exocytosis (Kopf and Gerton, 1991; Yanagimachi, 1994). A sodium requirement for the acrosome reaction in guinea pig sperm has been demonstrated (Hyne, 1984; Hyne *et al.,* 1984), but these studies did not focus on regulated acrosomal exocytosis. K^+ influx, likewise, has been proposed to regulate the acrosome reaction of hamster sperm through the action of a Na^+,K^+-ATPase (Mrsny and Meizel, 1981); its role in regulated acrosomal exocytosis has not been examined. Moreover, the relationship between such an influx mechanism and the above proposed K^+-dependent hyperpolarization events during capacitation, thought to be required for the control of zona pellucida/ZP3-induced Ca^{2+} channel activation, is not clear. HCO_3^- has been demonstrated to be required for capacitation (Boatman and Robbins, 1991; Visconti *et al.,* 1998) and acrosomal exocytosis induced by the zona pellucida (Lee and Storey, 1986). Whether the HCO_3^- requirement for this exocytotic event is a direct effect or is coupled to other ionic events (e.g., movement through specific antiporters or cotransporters) is also not clear at this time.

D. EFFECTOR ENZYMES AND INTRACELLULAR SECOND MESSENGERS MODULATING ACROSOMAL EXOCYTOSIS

Several different effector/intracellular signaling cascades have been implicated in regulating acrosomal exocytosis in mammalian sperm. These include cyclic nucleotide/phosphorylation cascades, lipid and phospholipid turnover, other phosphorylation cascades, and nitric oxide signaling. However, a vast majority of studies have been pharmacological in nature and do not address the cascades that are specifically activated in response to a biologically relevant ligand such as the zona pellucida, and therefore will not be addressed in this chapter. The reader is encouraged to examine additional reviews concerning these particular subjects (Kopf and Gerton, 1991; Roldan, 1999). Only two areas will be considered below, namely, cyclic nucleotide signaling and lipid/phospholipid turnover.

1. Cyclic Nucleotide Metabolism

Because there is no compelling evidence for the role of cGMP metabolism in regulating zona pellucida-induced acrosomal exocytosis at this time, only cAMP metabolism is considered here. Steady-state concentrations of intracellular cyclic AMP are regulated through both its synthesis (by adenylyl cyclase) and its degradation (by cyclic nucleotide phosphodiesterases), and both of the enzymes involved have been demonstrated to be regulated in a ligand-dependent manner in various cells.

The adenylyl cyclase of sperm cells has properties that set it apart from all other somatic cell adenylyl cyclases. Although initially described and characterized as a soluble enzyme in the testis (Braun and Dods, 1975; Neer, 1978; Neer and Murad, 1979), the enzyme appears to be present in both soluble and particulate forms in germ cells (Adamo *et al.,* 1980), and >90% of the activity appears to be particulate in spermatozoa. The sperm enzyme displays activity that is highly Mn^{2+}-dependent, is not activated by cholera toxin, cannot be activated by G_s, and is only nominally stimulated by forskolin, suggesting that the coupling characteristics of this enzyme and perhaps its molecular structure are different from those of the other members of the adenylyl cyclase family (Garbers and Kopf, 1980; Hildebrandt *et al.,* 1985; Kopf and Gerton, 1991). Although there is no evidence for the regulation of this enzyme by G_s, under very selective conditions this enzyme can be nominally stimulated by GTPγS and G-protein $\beta\gamma$ subunits (Leclerc and Kopf, 1999); these data further support the idea that the mode of regulation of this particular adenylyl cyclase may be unique.

The one property that clearly sets the sperm enzyme apart from all other adenylyl cyclases is its ability to be activated by bicarbonate anion (Garty and Salomon, 1987; Okamura *et al.,* 1985, 1991; Visconti *et al.,* 1995). The partial purification of this enzyme on a 4-acetamido-4′-isothiocyanostilbene-2,2′-disulfonic acid affinity column suggests that the enzyme is either tightly associated with an anion channel and/or possesses a domain(s) that could confer anion channel activity (Okamura *et al.,* 1991). It is also of interest to note that in sea urchin spermatozoa, adenylyl cyclase activity appears to be regulated by membrane potential (Beltran *et al.,* 1996), further supporting the idea that this effector enzyme is exquisitely regulated by the ionic environment. Both Ca^{2+} and HCO_3^- have been implicated in the regulation of sperm cAMP concentrations through their effects to stimulate adenylyl cyclase (Garbers *et al.,* 1982; Garty and Salomon, 1987; Hyne and Garbers, 1979b), but the mechanism by which this occurs is not known. Clearly, this appears to be an important regulatory mechanism governing changes in cAMP metabolism associated with capacitation (Kopf *et al.,* 1999b; Visconti *et al.,* 1995, 1997), as well as the induction of acrosomal exocytosis (Hyne and Garbers,

1979a). In fact, mouse sperm adenylyl cyclase has been demonstrated to be activated by solubilized zona pellucida glycoproteins (Leclerc and Kopf, 1995, 1999), suggesting some potential coupling between sperm–zona pellucida adhesion and events leading to acrosomal exocytosis (see below). However, given all of the unique properties of the sperm enzyme, a complete understanding of its regulation will only be realized following the purification and characterization of the protein.

Buck *et al.* (1999) purified the rat testis soluble adenylyl cyclase (termed "*s*oluble *a*denylyl *c*yclase," or sAC, by these authors) and obtained amino acid sequence from tryptic digests of a 48-kDa protein whose elution properties correlated with enzyme activity. Although the sequences did not match known proteins, subsequent cloning of the cDNA encoding the polypeptide revealed a nucleotide sequence with an open reading frame encoding a putative protein of 187 kDa, suggesting that processing of the protein was likely to occur. Although this is a distinct possibility, recent evidence suggests that alternate splicing might account for these different forms (Jaiswal and Conti, 2001). The contribution of alternative splicing and proteolytic processing to generate multiple forms of this enzyme with different regulatory properties remains an attractive possibility. Comparison of the sequence of the putative protein with other known protein sequences demonstrated a homology to various adenylyl cyclase catalytic domains, the most closely related being those of a number of different cyanobacterial adenylyl cyclases. Comparison with other adenylyl cyclase catalytic domains suggested that these domains in mammalian sAC were likely to have evolved independently of those of the transmembrane forms of adenylyl cyclase (i.e., diverged from an ancestral adenylyl cyclase prior to duplication events that generated the transmembrane adenylyl cyclase family). Of interest was the fact that domains of the protein beyond those regions homologous to adenylyl cyclase did not possess sequence similarity to other known proteins and that there was no consensus transmembrane domains. Although RT–PCR analysis indicates the presence of sAC in almost every tissue examined, its mRNA is clearly expressed at highest levels within the germ cell compartment of the testis, and that expression is first observed in the pachytene spermatocytes and continues throughout spermatogenesis (Buck *et al.,* 1999; Sinclair *et al.,* 2000). Experiments using an antisera directed against both of the catalytic domains of sAC have demonstrated the presence of the 48 k DA protein and higher molecular weight forms on both sperm and testis, as well as in other tissues that are known to regulate bicarbonate concentrations (i.e., kidney, choroid plexus) (Chen *et al,*. 2000) and that have been reported to contain bicarbonate-stimulated adenylyl cyclase activity (Mittag *et al.,* 1993). These antisera were used to immunoprecipitate a bicarbonate-activated adenylyl cyclase activity from testis cytosol. Taken together, these data suggest that sAC might represent a bicarbonate sensor. The mechanism by which bicarbonate anion regulates this enzyme is certainly not clear at this time but will represent a significant step towards understanding the role that this enzyme plays physiologically. Expression of recombinant sAC in HEK 293 cells revealed an enzyme activity that was highly Mn^{2+} dependent, and insensitive to forskolin and GTPγS, consistent with the aforementioned properties of the soluble testis adenylyl cyclase (Buck *et al.*, 1999), and recent data demonstrates that the recombinant protein can be activated by bicarbonate in a concentration-dependent manner (Chen *et al.*, 2000).

These findings are potentially of great significance regarding signal transduction in sperm for a number of reasons. First, the homology of the sAC catalytic domains with those of cyanobacteria could be potentially relevant, given the fact that these eubacteria evolved 2.5 to 3.4 billion years ago in an atmospheric environment rich in H_2O and CO_2. Perhaps the activities of these cyanobacterial adenylyl cyclases were also regulated by CO_2/HCO_3. In fact, the adenylyl activity of purified *Spirulina platensis* CyaC has been recently shown to be stimulated ~2,5-fold by bicarbonate (Chen *et al.*, 2000). Second although sAC protein has been demonstrated in the germ cells (Jaiswal and Conti, 2001) and sperm (Chen *et al.*, 2000), it is of interest that the enzyme appears to be both soluble and particulate in the germ cells, but particulate in the sperm cells. Whether this represents some sort of processing during spermatogenesis, with the subsequent association to a particulate fraction, has yet to be determined. Although the biochemical properties of the sperm adenylyl cyclase are consistent with its being an integral membrane protein (based on solubilization in nonionic detergents), the sequence deduced from the cDNA would not encode a protein with a transmembrane segment. This begs the question as to how the enzyme becomes "particulate." Finally, the soluble nature of this enzyme might be important functionally. As proposed by Buck *et al.* (1999), it is possible that sAC might function intracellularly to produce cAMP locally in different regions of the cell in a manner analogous to the soluble guanylyl cyclases (Wedel and Garbers, 2001). The spatial production of cAMP could be further confined by the ability of sAC to interact with scaffolding proteins, as is the case with protein kinase A and members of the A kinase achoring protein (AKAP) family (Pawson and Scott, 1997). Clearly, many questions remain regarding the structure and function of sAC, and answers to such questions will be key to our understanding of the regulation of germ cell and sperm signal transduction.

As stated above, addition of solubilized zonae pellucidae to membranes of capacitated mouse spermatozoa results in a concentration-dependent activation of the enzyme (Leclerc and Kopf, 1995, 1999). Moreover, the addition of solubilized zonae pellucidae to capacitated mouse sperm cells results in rapid and transient increases in sperm cAMP concentrations over untreated controls, which occur prior to overt acrosomal exocytosis (Noland *et al.*, 1988). In addition, the human sperm acrosome reaction induced by the zona pellucida has been reported to be inhibited by addition of protein kinase A (PKA) inhibitors (Bielfeld *et al.*, 1994). Taken together, these data would be consistent with the notion that changes in cAMP metabolism play a regulatory role in signal transduction leading to this exocytotic event. However, it has yet to be determined which zona pellucida glycoprotein (e.g., ZP1, ZP2, or ZP3) is required for this response, whether zonae pellucidae from fertilized eggs can induce a response, and whether noncapacitated sperm populations can mount a response to zonae pellucidae.

Although there is experimental support for the role of this second messenger in acrosomal exocytosis, it is clear that cAMP is a required second messenger but not the only second messenger involved in this process. For example, addition of dibutyryl cAMP to capacitated spermatozoa will not induce acrosomal exocytosis, even though addition of this membrane-permeable cAMP analog does appear to promote capacitation (Kopf *et al.,* 1999b; Visconti *et al.,* 1995). The mechanisms by which sperm cells integrate changes in cAMP metabolism leading to capacitation, and changes in cAMP metabolism of capacitated cells, leading to acrosomal exocytosis, are not clear at this time.

There is also evidence to suggest that progesterone-induced acrosomal exocytosis may involve a cAMP-mediated pathway. Parinaud and Milhet (1996) have demonstrated that progesterone will increase cAMP concentrations in human sperm in a Ca^{2+}-dependent manner. Using a series of PKA inhibitors and AKAP inhibitors, Harrison *et al.* (2000) have reported a role for PKA and AKAPs in human sperm acrosomal exocytosis initiated by this steroid. A further investigation of the role of this intracellular second messenger system in regulating both the zona pellucida- and progesterone-induced acrosomal exocytosis is warranted to determine the similarity in mechanism of these two ligands.

2. Lipid and Phospholipid Turnover

Given that heterotrimeric G-proteins of the G_i class play an essential role in signal transduction regulating ZP3-mediated acrosomal exocytosis (see above), lipid/phospholipid signaling regulates intracellular Ca^{2+} metabolism and exocytosis in a variety of cells, and G-protein signaling integrates with lipid/phospholipid turnover to regulate cellular function in many different cell types, it is not surprising that many investigators have examined the role of lipid and phospholipid turnover in controlling acrosomal exocytosis.

Incubation of capacitated sperm with either zonae pellucidae or progesterone results in an increase in 1,2-diacylglycerol (O'Toole *et al.,* 1996b; Roldan *et al.,* 1994). Elevations of this lipid messenger are also observed following ionophore A23187 addition to spermatozoa, but are not seen in sperm cells incubated with Ca^{2+} channel blockers [summarized in Roldan (1999)], these observations suggest that ligand-induced increases in intracellular Ca^{2+} occur upstream from the generation of 1,2-diacylglycerol. One source of 1,2-diacylglycerol is likely to result from the hydrolysis of phosphatidylinositol 4,5-bisphosphate (PIP_2) and/or PIP; it has been demonstrated that the hydrolysis of these poly(phosphoinositides) accompanies acrosomal exocytosis in the mouse and human using physiological ligands (Roldan and Murase, 1994; Roldan *et al.,* 1994; Thomas and Meizel, 1989). The hydrolysis of these poly(phosphoinositides) is likely due to phosphatidylinositol 4,5-bisphosphate–phospholipase Cγ, which is present (Feng *et al.,* 1998; Ribbes *et al.,* 1987) and is activated by the zona pellucida in mouse spermatozoa (Tomes *et al.,* 1996). This form is not sensitive to G-protein modulation. Of great interest is the recent report demonstrating that the targeted deletion of the PLCδ4 gene results in a male infertility phenotype in which the sperm of the knockout animals were unable to initiate acrosomal exocytosis in rsponse to the zona pellucida (Fukami *et al.*, 2001). These results strongly support the notion that the activation of this PLCδ isoform in sperm by the zona pellucida plays a critical role in signal transduction leading to acrosomal exocytosis. Diacylglycerol generation in sperm cells responding to the zona pellucida and/or progesterone may also result from the hydrolysis of diacylphosphatidylcholine (O'Toole *et al.*, 1996b; Roldan and Murase, 1994; Roldan *et al.*, 1994) by phosphatidylcholine-specific phospholipase C, which has been localized in bull sperm to the acrosome (Sheikhnejad and Srivastave, 1986).

In somatic cells, 1,2-diacylglycerol functions as an intracellular second messenger to regulate effector enzymes such as protein kinase C (PKC) and phospholipase A_2; these two enzymes have been demonstrated to be activated in sperm by this second messenger under conditions that lead to acrosomal exocytosis (O'Toole *et al.*, 1996a; Roldan and Fragio, 1994). Several isoforms of PKC have been identified in various species of mammalian spermatozoa by immunochemical methods. However, the localization of the different isoforms, although present throughout the various compartments of these cells, varies among species and may represent potential problems with the reagents used (Breitbart *et al.*, 1992; Lax *et al.*, 1997; Rotem *et al.*, 1990). The role of PKC activation in regulating acrosomal exocytosis is, to date, still controversial (Kopf and Gerton, 1991; Roldan, 1999). Most studies have examined the effects of phorbol diesters and PKC inhibitors on nonregulated acrosomal exocytosis and only a few studies have utilized the zona pellucida and/or progesterone. Activation of sperm PKC by these two ligands has not been reported.

The other major product of PIP_2 hydrolysis, namely, inositol 1,4,5-trisphosphate (IP_3), has also been proposed as an intracellular second messenger that in other cell types regulates intracellular Ca^{2+} release through its binding to IP_3 receptors associated with the endoplasmic reticulum. The presence of internal Ca^{2+} stores in sperm has been a question for many years due to the absence of an endoplasmic reticulum (or similar structure) in these highly differentiated cells. However, several observations may point to the possibility of functional IP_3 effects in mature sperm. First, although there are no reports of zona pellucida/progesterone effects on mammalian spermatozoa to increase sperm-associated IP_3, the fucose sulfate glycoconjugate fraction of sea urchin egg jelly that induces acrosomal exocytosis also stimulates IP_3 accumulation in these cells (Domino and Garbers, 1988), suggesting some functional coupling between this egg jelly ligand and sperm phosphoinositide turnover. In addition, IP_3 receptors in the acrosomal region of mammalian sperm cells have been identified by immunochemical methods (Trevino *et al.*, 1998; Walensky and Snyder, 1995; Zapata *et al.*, 1997), but it is unclear whether such receptors are functional. Given the fact that a thapsigargin-sensitive Ca^{2+} pump has been identified in bovine sperm membranes and is thought to localize to the acrosome (Spungin and Breitbart, 1996) and that acrosomal membranes contain a Ca^{2+}-ATPase activity (Gordon, 1973; Gordon *et al.*, 1978), it is possible that spermatozoa do, in fact, have functional intracellular Ca^{2+} stores. The functional role of such stores, however, has yet to be determined. It is tempting to speculate that the thapsigargin-sensitive Ca^{2+} store may represent a store-operated Ca^{2+} channel, and data of O'Toole *et al.* (2000) and Jungnickel *et al.* (2001) suggest that Ca^{2+} entry through such a channel in mouse spermatozoa is initiated by ZP3 and regulates acrosomal exocytosis.

The generation of arachidonic acid and lysophospholipids by phospholipase A_2 has also been proposed as a signaling mechanism regulating acrosomal exocytosis, because lysophospholipids have been implicated in membrane fusion in other systems, and lysophospholipds will induce sperm membrane fusion when added to capacitated sperm cells (Kopf and Gerton, 1991; Roldan, 1999). The enzyme has been reported to be activated in human and boar spermatozoa in response to progesterone (Baldi *et al.,* 1993; Roldan and Vazquez, 1996), but the role of the second messengers generated is still unclear. Further complicating interpretation of these data is the fact that the type of phospholipase A_2 present in spermatozoa and active in response to progesterone is unknown (Roldan, 1999). To date, the effects of zonae pellucidae on the activity of this class of enzymes have not been reported.

In summary, it appears as if lipid and phospholipid turnover may play important roles in acrosomal exocytosis. It is clear from the studies carried out to date that the mechanisms by which this occurs may be complex and that additional studies utilizing biologically relevant inducers of this exocytotic event will be necessary to clarify the roles of these pathways.

E. MOLECULAR MECHANICS OF ACROSOMAL EXOCYTOSIS

As discussed in Section II, one must consider the acrosome more as a secretory granule that can undergo regulated secretion (exocytosis) in response to appropriate ligand–receptor–effector signaling. In this context, the molecular mechanics of plasma/outer acrosomal membrane recognition, docking/adhesion leading to membrane fusion, and overt exocytosis are likely to have some similarity to other well-characterized secretory events (Burgess and Kelly, 1987; Fischer von Mollard *et al.,* 1994; Jahn and Sudhof, 1999; Pfeffer, 1996, 1999; Plattner, 1989; Sudhof, 1995), and the process of sperm capacitation may ready those membranes for their eventual docking/adhesion and subsequent fusion (see Section III). The basic elements of such processes as they relate to ligand-induced acrosomal exocytosis are considered in Chapter 8.

Work from many laboratories studying stimulus–secretion coupling and the mechanics of exocytosis have demonstrated the existence of two classes of proteins that play key roles in protein-mediated membrane fusion. The first class represents a group of membrane-associated proteins that are found to be present on vesicle (v) and target (t) membranes, known collectively as v- and t-SNAREs [soluble NSF-attachment protein (SNAP) receptors], respectively. A highly stable ternary protein complex composed of the t-SNAREs, syntaxin and SNAP-25 (synaptosome-associated protein of 25 kDa), and the v-SNAREs, VAMP (vesicle associated membrane protein)/synaptobrevin, assembles spontaneously and is likely to function as the core machinery for membrane fusion. It is thought that the formation of these stable complexes is a critical step leading to membrane fusion. Botulinum and tetanus neurotoxins have been very valuable in determining the function of these complexes given the fact that specific proteins of these complexes are substrates for the action of these toxins, resulting in a disruption of their normal function. Once membrane fusion has occurred the SNARE complex is then disassembled by the action of N-ethylmaleimide-sensitive fusion protein (NSF), an ATPase, and SNAPs.

The Rab proteins are the second class of proteins that play a key role in this process; these are monomeric GTP-binding proteins belonging to the Ras superfamily. Different members of this family have been demonstrated to localize to different compartments, and the functions of these proteins are mediated by a variety of proteins that facilitate guanine nucleotide exchange and/or hydrolysis (i.e., GTPase-activating proteins, GAPs; guanine nucleotide exchange factors, GEFs; guanine nucleotide dissociation inhibitors, GDIs). It is thought that Rabs could be functioning at several different levels during exocytosis. For example, they might regulate vesicle docking by controlling SNARE complex formation and/or regulate the stability of the SNARE complex. There is also the suggestion that Rabs may control the timing of vesicle fusion.

If one accepts the hypothesis that regulated acrosomal exocytosis may have elements similar to other exocytotic systems, it might be expected that some of the aforementioned components of the fusion engine are present in the membranes overlying the acrosome. Second, one would predict that those proteins might interact with one another in a manner consistent with their putative role(s) in membrane docking and fusion. Finally, it would be predicted that modulating the function of these proteins might result in the appropriate response (i.e., activation or inhibition of specific aspects of spontaneous and/or ligand-regulated exocytosis). However, one must also recognize that unlike the other model systems for the study of exocytosis, whereby secretory granules and protein components are recycled, the acrosome is synthesized and stored for an extended period of time prior to an exocytotic release of its contents that occurs only once. This clearly puts some experimental limitation on the approaches one can take to study this process. Second, because sperm are transcriptionally and translationally inactive, attempts to examine the effects of mutated protein components of the fusion machinery identified (e.g., dominant negative; constitutively active, targeted gene deletions) will require specific germ-line expression of the appropriate construct, which represents a substantial experimental investment.

Work from several laboratories is starting to address the players present in the acrosome and their potential function. Schulz *et al.* (1997) have identified homologs of syntaxin and VAMP in sea urchin spermatozoa. Following exocytosis, both of these proteins are released with the resultant membrane vesicles, indicating that they are both present in the membrane compartment of the acrosome. Immunoprecipitation studies demonstrated that both syntaxin and VAMP interact with one another, and this would be expected given the current models of membrane fusion and exocytosis. Furthermore, syntaxin and VAMP appear to have undergone changes following exocytosis, as demonstrated by changes in their cosedimentation properties. Although these data suggest that these protein constituents are present and can associate with one another, their actual role in membrane fusion was not evaluated. Subsequent work by this group using sea urchin spermatozoa demonstrated that SNAP-25 is also expressed in these cells, that it is released in the shed membrane vesicles following the acrosomal exocytosis, and that it interacts with syntaxin and VAMP both prior to and following exocytosis (Schulz *et al.,* 1998). Recently, several SNARE proteins (e.g., VAMP/synaptobrevin, syntaxin 1, syntaxin 2, and synaptotagmin have been demonstrated to be present in mouse, rat, hamster, monkey, and human sperm (Michaut *et al.,* 2000; Katafuchi *et al.,* 2000; Ramalho-Santos *et al.,* 2000). Michaut *et al.* (2000) have demonstrated, using a streptolysin O-permeabilized human sperm preparation that responds to Ca^{2+} to trigger acrosomal exocytosis, that N-ethylmaleimide-sensitive factor (NSF) is present in the acrosome and that the Ca^{2+} dependent exocytosis in this system requires NSF. Moreover, both Rab 3A (see below) and active NSF appear to be necessary for Ca^{2+} dependent exocytosis in this model system.

Several groups have investigated the identity and role of monomeric GTP-binding regulatory proteins in mammalian sperm acrosomal exocytosis. Rab3A has been demonstrated to be present in and associated with the acrosomal regions of rat (Iida *et al.,* 1999), mouse (Ward *et al.,* 1999), and human (Yunes *et al.,* 2000) spermatozoa, and is lost following acrosomal exocytosis. Iida *et al.* (1999) demonstrated that incubation of mouse sperm cells with a synthetic peptide of the Rab3 effector domain inhibited acrosomal exocytosis induced by A23187, suggesting that Rab3A might function as an inhibitory regulator of the acrosome reaction. These experiments were performed by incubating spermatozoa with the peptide and the A23187 at the same time, and it is not clear how effective these peptides would be in crossing membranes if they had been added prior to the addition of A23187. These results are in contrast to those of Garde *et al.* (1996), who demonstrated that the addition of the effector peptide after the addition of A23187 to ram sperm cells (to make sure that the peptide got in) enhanced acrosomal exocytosis induced by A23187. The results of both of these studies must be interpreted with caution because both authors are using a powerful agent to induce nonregulated exocytosis in an effort to ensure that the peptide is getting to its potential site of action to modulate (either positively or negatively) exocytosis. Yunes *et al.* (2000) have provided the most convincing data to date regarding the role of Rab3A in acrosomal exocytosis. Using a streptolysin O-permeabilized human sperm preparation that could respond to Ca^{2+} and activators of G-proteins (GTPγS) to trigger acrosomal exocytosis, they demonstrated that the Rab3A effector peptide, as well as recombinant Rab3A protein in the GTP-bound form, could cause acrosomal exocytosis; recombinant protein in the inactive GDP-bound state was inactive. In addition, recombinant GDI, which would release Rab proteins from the membrane, inhibited GTPγS-stimulated acrosomal exocytosis. Clearly additional studies will have to be performed with ligands such as the zona pellucida and progesterone.

In summary, we are at a very early stage in understanding the molecular mechanics of acrosomal exocytosis. Some of the general mechanisms will be conserved with other exocytotic systems, but the sperm acrosome clearly has properties that distinguish it from other secretory systems, and it is these differences that will yield exciting results. How the mechanics of membrane fusion and exocytosis integrate at a higher level with upstream signaling pathways will also be an area of future investigation that should provide fruitful results.

ACKNOWLEDGMENTS

I thank all of the present and past members of my laboratory who have contributed to work from my lab cited in this chapter. Their hard work and dedication are gratefully appreciated. I also acknowledge the support of the following funding agencies: the National Institutes of Health, the Fogarty Foundation, the United States Department of Agriculture, the Rockefeller Foundation, and the Lalor Foundation.

REFERENCES

Adamo, S., Conti, M., Geremia, R., and Monesi, V. (1980). Particulate and soluble adenylate cyclase activities of mouse male germ cells. *Biochem. Biophys. Res. Commun.* **97,** 607–613.

Allison, A. C., and Hartree, E. F. (1970). Lysosomal enzymes in the acrosome and their possible role in fertilization. *J. Reprod. Fertil.* **21,** 501–515.

Arnoult, C., Cardullo, R. A., Lemos, J. R., and Florman, H. M. (1996a). Activation of mouse sperm T-type Ca^{2+} channels by adhesion to the egg zona pellucida. *Proc. Natl. Acad. Sci., U.S.A.* **93,** 13004–13009.

Arnoult, C., Kazam, I. G., Visconti, P. E., Kopf, G. S., Villaz, M., and Florman, H. M. (1999). Control of the low voltage-activated calcium channel of mouse sperm by egg ZP3 and by membrane hyperpolarization during capacitation. *Proc. Natl. Acad. Sci. U.S.A.* **96,** 6757–62.

Arnoult, C., Lemos, J. R., and Florman, H. M. (1997). Voltage-dependent modulation of T-type calcium channels by protein tyrosine phosphorylation. *EMBO J.* **16,** 1593–1599.

Arnoult, C., Villaz, M., and Florman, H. M. (1998) Pharmacological properties of the T-type Ca^{2+} current of mouse spermatogenic cells. *Mol. Pharmacol.* **53,** 1104–1111.

Arnoult, C., Zeng, Y., and Florman, H. M. (1996b). ZP3-dependent activation of sperm cation channels regulates acrosomal secretion durin mammalian fertilization. *J. Cell Biol.* **134,** 637–645.

Bagavant, H., Yurewicz, E. C., Sacco, A. G., Talwar G. P., and Gupta, S. K. (1993). Block in porcine gamete interaction by polyclonal antibodies to a pig ZP3β fragment having partial sequence homology to human ZP3. *J. Reprod. Immunol.* **25,** 277–283.

Baldi, E., Casano, R., Falsetti, C., Krausz, C., Maggi, M., and Forti, G. (1991). Intracellular calcium accumulation and responsiveness to progesterone in capacitating human spermatozoa. *J. Androl.* **12,** 323–330.

Baldi, E., Falsetti, C., Krausz, C., Gervasi, G., Carloni, V., Casano, R., and Forti, G. (1993). Stimulation of platelet-activating factor synthesis by progesterone and A23187 in human spermatozoa. *Biochem. J.* **292,** 209–216.

Baldi, E., Krausz, C., Luconi, M., Bonaccorsi, L., Maggi, M., and Forti, G. (1995). Actions of progesterone on human sperm: A model of non-genomic effects of steroids. *J. Steroid Biochem. Mol. Biol.* **53,** 199–203.

Bayaa, M., Booth, R. A., Sheng, Y., and Liu, X. J. (2000). The classical progesterone receptor mediates Xenopus oocyte maturation through a nongenomic mechanism. *Proc. Natl. Acad. Sci. U.S.A.* **97,** 12607–12612.

Bedford, J. M. (1972). An electron microscopic study of sperm penetration into the rabbit egg after natural mating. *Am. J. Anat.* **133,** 213–254.

Beltrán, C., Darszon, A., Labarca, P., and Liévano, A. (1994). A high-conductance voltage-dependent multistate Ca^{2+} channel found in sea urchin and mouse spermatozoa. *Fed. Eur. Biochem. Soc.* **338,** 23–26.

Beltran, C., Zapata, O., and Darszon, A. (1996). Membrane potential regulates sea urchin sperm adenylyl cyclase. *Biochemistry* **35,** 7591–7598.

Benau, D. A., and Storey, B. T. (1988). Relationship between two types of mouse sperm surface sites that mediate binding of sperm to the zona pellucida. *Biol. Reprod.* **39,** 235–240.

Bielfeld, P., Faridi, A., Zaneveld, L. J. D., and De Jonge, C. J. (1994). The zona pellucida-induced acrosome reaction of human spermatozoa is mediated by protein kinases. *Fertil. Steril.* **61,** 536–541.

Blackmore, P. F. (1993). Rapid non-genomic actions of progesterone stimulate Ca^{2+} influx and the acrosome reaction in human sperm. *Cell Signal* **5,** 531–538.

Blackmore, P. F., Bebe, S. J., Danforth, D. R., and Alexander, N. (1990). Progesterone and 17a-hydroxyprogesterone: Novel stimulators of calcium influx in human sperm. *J. Biol. Chem.* **265,** 1376–1380.

Blackmore, P. F., Fisher, J. F., Spilman, C. H., Im, W. B., and Bleasdale, J. E. (1995). "Effect of Steroids on Calcium Fluxes in Human Sperm." John Libbey Eurotext, Ltd., Montrouge, France.

Bleil, J. D., and Wassarman, P. M. (1980). Mammalian sperm-egg interaction: Identification of a glycoprotein in mouse egg zonae pellucidae possessing receptor activity for sperm. *Cell* **20,** 873–882.

Bleil, J. D., and Wassarman, P. M. (1983). Sperm-egg interactions in the mouse: Sequence of events and induction of the acrosome reaction by a zona pellucida glycoprotein. *Dev. Biol.* **95,** 317–324.

Bleil, J. D., and Wassarman, P. M. (1986). Autoradiographic visualization of the mouse egg's sperm receptor bound to sperm. *J. Cell. Biol.* **102,** 1363–1371.

Bleil, J. D., and Wassarman, P. M. (1988). Galactose at the nonreducing terminus of O-linked oligosaccharides of mouse egg zona pellucida glycoprotein ZP3 is essential for the glycoprotein's sperm receptor activity. *Proc. Natl. Acad. Sci. U.S.A.* **85,** 6778–6782.

Bleil, J. D., and Wassarman, P. M. (1990). Identification of a ZP3-binding protein on acrosome-intact mouse sperm by photoaffinity crosslinking. *Proc. Natl. Acad. Sci. U.S.A.* **87,** 5563–5567.

Bleil, J. D., Greve, J. M., and Wassarman, P. M. (1988). Identification of a secondary sperm receptor in the mouse egg zona pellucida: Role in maintenance of binding of acrosome-reacted sperm to eggs. *Dev. Biol.* **128,** 376–385.

Boatman, D. E., and Robbins, R. S. (1991). Bicarbonate: Carbon-dioxide regulation of sperm capacitation, hyperactivated motility, and acrosome reactions. *Biol. Reprod.* **44,** 806–813.

Bookbinder, L. H., Cheng, A., and Bleil, J. D. (1995). Tissue- and species-specific expression of sp56, a mouse sperm fertilization protein. *Science* **269,** 86–89.

Bork, P., Tsai, J. Y., and Silver, L. (1996). Sperm-egg binding protein or proto-oncogene? *Science* **271,** 1431–1435.

Braun, T., and Dods, R., F. (1975). Development of a Mn^{2+}-sensitive, "soluble"adenylate cyclase in rat testis. *Proc. Natl. Acad. Sci. U.S.A.* **72,** 1097–1101.

Breitbart, H., Lax, J., Rotem, R., and Naor, Z. (1992). Role of protein kinase C in the acrosome reaction of mammalian spermatozoa. *Biochem. J.* **281,** 473–476.

Bryan, J. H. D. (1974). Capacitation in the mouse: The response of murine acrosomes to the environment of the female genital tract. *Biol. Reprod.* **10,** 414–420.

Buck, J., Sinclair, M. L., Schapal, L., Cann, M. J., and Levin, L. R. (1999). Cytosolic adenylyl cyclase defines a unique signaling molecule in mammals. *Proc. Natl. Acad. Sci. U.S.A.* **96,** 79–84.

Buddhikot, M., Falkenstein, E., Wehling, M., and Meizel, S. (1999). Recognition of a human sperm surface protein involved in the progesterone-initiated acrosome reaction by antisera against an endomembrane progesterone binding protein from porcine liver. *Mol. Cell. Endocrinol.* **158,** 187–193.

Burgess, T. L., and Kelly, R. B. (1987). Constitutive and regulated secretion of proteins. *Annu. Rev. Cell Biol.* **3,** 243–293.

Burks, D. J., Carballada, R., Moore, H. D. M., and Saling, P. M. (1995). Interaction of a tyrosine kinase from human sperm with the zona pellucida at fertilization. *Science* **269,** 83–86.

Chamberlain, M., and Dean, J. (1989). Genomic organization of a sex specific gene: The primary sperm receptor of the mouse zona pellucida. *Dev. Biol.* **131,** 207–214.

Chen, Y., Cann, M. J., Litvin, T. N., Iourgenko, V., Sinclair, M. L., Levin, L. R., and Buck, J. (2000). Soluble Adenylyl Cyclase as an Evolutionarily Conserved Bicarbonate Sensor. *Science* **289,** 625–628.

Cheng, A., Le, T., Palacios, M., Bookbinder, L. H., Wassarman, P. M., Suzuki, F., and Bleil, J. D. (1994). Sperm-egg recognition in the mouse: Characterization of sp56, a sperm protein having specific affinity for ZP3. *J. Cell. Biol.* **125,** 867–878.

Cheng, F. P., Gadella, B. M., Voorhout, W. F., Fazeli, A., Bevers, M. M., and Colenbrander, B. (1998). Progesterone-induced acrosome reaction in stallion spermatozoa is mediated by a plasma membrane progesterone receptor. *Biol. Reprod.* **59,** 733–742.

Cherr, G. N., Lambert, H., Meizel, S., and Katz, D. F. (1986). *In vitro* studies of the golden hamster sperm acrosome reaction: Completion on the zona pellucida and induction by homologous soluble zonae pellucidae. *Dev. Biol.* **114,** 119–131.

Cross, N. L., Morales, P., Overstreet, J. W., and Hanson, F. W. (1988). Induction of acrosome reactions by the human zona pellucida. *Biol. Reprod.* **38,** 235–244.

Crozet, N., and Dumont, M. (1984). The site of the acrosome reeaction during *in vivo* penetration of the sheep oocyte. *Gamete Res.* **10,** 97–105.

Cummins, J. M., and Yanagimachi, R. (1982). Sperm–egg ratio and the site of the acrosome reaction during *in vivo* fertilization in the hamster. *Gamete Res.* **5,** 239–246.

Darszon, A., Labarca, P., Nishigaki, T., and Espinosa, F. (1999). Ion channels in sperm physiology. *Physiol. Rev.* **79,** 481–510.

Desnoyers, L., and Manjunath, P. (1992). Major proteins of bovine seminal plasma exhibit novel interactions with phospholipid. *J. Biol. Chem.* **267,** 10149–10155.

Domino, S. E., and Garbers, D. L. (1988). The fucose-sulfate glycoconjugate that induces an acrosome reaction in spermatozoa stimulates inositol 1,4,5-trisphosphate accumulation. *J. Biol. Chem.* **263,** 690–695.

Dubois, D. H., and Shur, B. D. (1995). Cell surface β1,4-galactosyltransferase—A signal transducing receptor? *Adv. Exp. Med. Biol.* **376,** 105–114.

Endo, Y., Lee, M. A., and Kopf, G. S. (1987a). Evidence for the role of a guanine nucleotide-binding regulatory protein in the zona pellucida-induced mouse sperm acrosome reaction. *Dev. Biol.* **119,** 210–216.

Endo, Y., Mattei, P., Kopf, G. S., and Schultz, R. M. (1987b). Effects of a phorbol ester on mouse eggs: Dissociation of sperm receptor activity from acrosome reaction-inducing activity of the mouse zona pellucida protein, ZP3. *Dev. Biol.* **123,** 574–577

Endo, Y., Schultz, R. M., and Kopf, G. S. (1987c). Effects of phorbol esters and a diacylglycerol on mouse eggs: Inhibition of fertilization and modification of the zona pellucida. *Dev. Biol.* **119,** 199–209.

Epifano, O., and Dean, J. (1994). Biology and structure of the zona pellucida: A target for immunocontraception. *Reprod. Fertil. Dev.* **6,** 319–330.

Epifano, O., Liang, L., Familari, M., Moos, M. C., Jr., and Dean, J. (1995). Coordinate expression of the three zona pellucida genes during mouse oogenesis. *Development* **121,** 1947–1956.

Evans, J. P., and Kopf, G. S. (1998). Molecular mechanisms of sperm-egg interactions and egg activation. *Andrologia* **30,** 297–307.

Falati, S., Edmead, C. E., and Poole, A. W. (1999) Glycoprotein Ib-V-IX, a receptor for von Willebrand factor, couples physically and functionally to the Fc receptor gamma-chain, Fyn, and Lyn to activate human platelets. *Blood* **94,** 1648–1656

Falkenstein, E., Falkenstein, G. D., and Wehling, M. (1999a). Cloning, expression and characterization of a putative progesterone membrane receptor. *Naunyn Schmiedebergs Arch. Pharmacol.* **(Suppl.) 359,** R13.

Falkenstein, E., Heck, M., Gerdes, D., Grube, D., Christ, M., Weigel, M., Buddhikot, M., Meizel, S., and Wehling, M. (1999b). Specific progesterone binding to a membrane protein and related nongenomic effects on Ca^{2+}-fluxes in sperm. *Endocrinology* **140,** 5999–6002.

Falkenstein, E., Meyer, C., Eisen, C., Scriba, P. C., and Wehling, M. (1996). Full-length cDNA sequence of a progesterone membrane-binding protein from porcine vascular smooth muscle cells. *Biochem. Biophys. Res. Commun.* **229,** 86–89.

Falkenstein, E., Norman, A. W., and Wehling, M. (2000a). Mannheim classification of nongenomically initiated (rapid) steroid action(s). *J. Clin. Endocrinol. Metab.* **85,** 2072–2075.

Falkenstein, E., Tillmann, H. C., Christ, M., Feuring, M., and Wehling, M. (2000b). Multiple actions of steroid hormones—a focus on rapid, nongenomic effects. *Pharmacol. Rev.* **52,** 513–556.

Feng, H., Sandlow, J. I., and Sandra, A. (1998). The c-kit receptor and its possible signal transduction pathway in mouse spermatozoa. *Mol. Reprod. Dev.* **49,** 317–326.

Fischer von Mollard, G., Stahl, B., Li, C., Sudhof, T. C., and Jahn, R. (1994). Rab proteins in regulated exocytosis. *Trends Biochem. Sci.***19,** 164–168.

Florman, H. (1994). Sequential focal and global elevations of sperm intracellular Ca are initiated by the zona pellucida during acrosomal exocytosis. *Dev. Biol.* **165,** 152–164.

Florman, H. M., and Babcock, D. F. (1991). Progress towards understanding the molecular basis of capacitation. *In* "CRC Uniscience—Chemistry of Fertilization" (P. M. Wassarman, ed.), Vol. II, pp. 105–132. CRC Press, Boca Raton.

Florman, H. M., and First, N. L. (1988a). The regulation of acrosomal exocytosis: 1. Sperm capacitation is required for the induction of the acrosome reactions by the bovine zona pellucida *in vitro. Dev. Biol.* **128,** 453–463.

Florman, H. M., and First, N. L. (1988b). Regulation of acrosomal exocytosis: 2. The zona pellucida-induced acrosome reaction of bovine spermatozoa is controlled by extrinsic positive regulatory elements. *Dev. Biol.* **128,** 464–473.

Florman, H. M., and Storey, B. T. (1982a). Characterization of cholinomimetic agents that inhibit *in vitro* fertilization in the mouse. Evidence for a sperm-specific binding site. *J. Androl.* **3,** 157–164.

Florman, H. M., and Storey, B. T. (1982b). Mouse gamete interactions: The zona pellucida is the site of the acrosome reaction leading to fertilization *in vitro. Dev. Biol.* **91,** 121–30.

Florman, H., M., and Wassarman, P., M. (1985a). O-Linked oligosaccharides of mouse egg zp3 account for its sperm receptor activity. *Cell* **41,** 313–324.

Florman, H. M., and Wassarman, P. M. (1985b). O-Linked oligosaccharides of mouse egg ZP3 account for its sperm receptor activity. *Cell* **41,** 313–324.

Florman, H. M., Arnoult, C., Kazam, I. G., Li, C., and O'Toole, C. M. B. (1998). A perspective on the control of mammalian fertilization by egg-activated ion channels in sperm: A tale of two channels. *Biol. Reprod.* **59,** 12–16.

Florman, H. M., Bechtol, K. B., and Wassarman, P. M. (1984). Enzymatic digestion of the functions of the mouse egg's receptor for sperm. *Dev. Biol.* **106,** 243–255.

Florman, H. M., Corron, M. E., Kim, T. D., and Babcock, D. F. (1992). Activation of voltage-dependent calcium channels of mammalian sperm is required for zona pellucida-induced acrosomal exocytosis. *Dev. Biol.* **152,** 304–314.

Florman, H. M., Tombes, R. M., First, N. L., and Babcock, D. F. (1989). An adhesion-associated agonist from the zona pellucida activates G protein-promoted elevations of internal Ca^{2+} and pH that mediate mammalian sperm acrosomal exocytosis. *Dev. Biol.* **135,** 133–146.

Foresta, C., Rossato, M., and Di Virgilio, F. (1993). Ion fluxes through the progesterone-activated channel of the sperm plasma membrane. *Biochem. J.* **294,** 279–283.

Foster, J. A., Friday, B. B., Maulit, M. T., Blobel, C., Winfrey, V. P., Oson, G. E., Kim, K. S., and Gerton, G. L. (1997). AM67, a secretory component of the guinea pig sperm acrosomal matrix, is related to mouse sperm protein sp56 and the complement component 4-binding proteins. *J. Biol. Chem.* **272,** 12714–12722.

Fukami, K., Nakao, K., Inoue, T., Kataoka, Y., Kurokawa, M., Fissore, R. A., Nakamura, K., Katsuki, M., Mikoshiba, K., Yoshida, N., and Takenawa, T. (2001). Requirement of phospholipase Cdelta4 for the zona pellucida-induced acrosome reaction. *Science* **292,** 920–923.

Gadella, B. M., and Harrison, R. A. (2000). The capacitating agent bicarbonate induces protein kinase A-dependent changes in phospholipid transbilayer behavior in the sperm plasma membrane. *Development* **127,** 2407–2420.

Gadella, B. M., Miller, N. G., Colenbrander, B., van Golde, L. M., and Harrison, R. A. (1999). Flow cytometric detection of transbilayer movement of fluorescent phospholipid analogues across the boar sperm plasma membrane: elimination of labeling artifacts. *Mol. Reprod. Dev.* **53,** 108–125.

Garbers, D. L., and Kopf, G. S. (1980). The regulation of spermatozoa by calcium and cyclic nucleotides. *Adv. Cyclic Nucleotide Res.* **13,** 251–306.

Garbers, D. L., Tubb, D. J., and Hyne, R. V. (1982). A requirement of bicarbonate for Ca^{2+}-induced elevations of cyclic AMP in guinea pig spermatozoa. *J. Biol. Chem.* **257,** 8980–8984.

Garde, J., and Roldan, E. R. (1996). rab 3-peptide stimulates exocytosis of the ram sperm acrosome via interaction with cyclic AMP and phospholipase A2 metabolites. *FEBS Lett.* **391,** 263–268.

Garty, N., and Salomon, Y. (1987). Stimulation of partially purified adenylate cyclase from bull sperm by bicarbonate. *Fed. Eur. Biochem. Soc.* **218,** 148–152.

Goldbrunner, R. H., Bernstein, J. J., and Tonn, J. C. (1999). Cell–extracellular matrix interaction in glioma invasion. *Acta Neurochir.* **141,** 295–305.

Gong, X., Dubois, D. H., Miller, D. J., and Shur, B. D. (1995). Activation of a G protein complex by aggregation of β-1,4-galactosyltransferase on the surface of sperm. *Science* **269,** 1718–1721.

Gonzalez-Amaro, R., and Sanchez-Madrid, F. (1999). Cell adhesion molecules: Selectins and integrins. *Crit. Rev. Immunol.* **19,** 389–429.

Gordon, M. (1973). Localization of phosphatase activity on the membranes of the mammalian sperm head. *J. Exp. Zool.* **185,** 111–120.

Gordon, M., Dandekar, P. V., and Eager, P. R. (1978). Identification of phosphatases on the membranes of guinea pig sperm. *Anat. Rec.* **191,** 123–33.

Graham, D. K., Bowman, G. W., Dawson, T. L., Stanford, W. L., Earp, H. S., and Snodgrass, H. R. (1995). Cloning and developmental expression analysis of the murine c-mer tyrosine kinase. *Oncogene* **10,** 2349–2359.

Graham, D. K., Lawson, T. L., Mullaney, D. L., Snodgrass, H. R., and Earp, H. S. (1994). Cloning and mRNA expression analysis of a novel human protooncogene, c-mer. *Cell Growth Differ.* **5,** 647–657.

Greve, J. M., and Wassarman, P. M. (1985). Mouse egg extracellular coat is a matrix of interconnected filaments possessing a structural repeat. *J. Mol. Biol.* **181,** 253–264.

Greve, J. M., Salzmann, G. S., Roller, R. J., and Wassarman, P. M. (1982). Biosynthesis of the major zona pellucida glycoprotein secreted by oocytes during mammalian oogenesis. *Cell* **31,** 749–759.

Hagiwara, S., and Kawa, K. (1984). Calcium and potassium currents in spermatogenic cells dissociated from rat seminiferous tubules. *J. Physiol. (Lond.)* **356,** 135–149.

Hardy, D. M., and Garbers, D. L. (1994). Species-specific binding of sperm proteins to the extracellular matrix (Zona pellucida) of the egg. *J. Biol. Chem.* **269,** 19000–19004.

Hardy, D. M., and Garbers, D. L. (1995). A sperm membrane protein that binds in a species-specific manner to the egg extracellular matrix is homologous to von Willebrand factor. *J. Biol. Chem.* **270,** 26025–26028.

Harris, J. D., Hibler, D. W., Fontenot, G. K., Hsu, K. T., Yurewicz, E. C., and Sacco, A. G. (1994). Cloning and characterization of zona pellucida genes and cDNAs from a variety of mammalian species: The ZPA, ZPB and ZPC gene families. *DNA Seq.* **4,** 361–393.

Harrison, D. A., Carr, D. W., and Meizel, S. (2000). Involvement of protein kinase A and a kinase anchoring protein in the progesterone-initiated human sperm acrosome reaction. *Biol. Reprod.* **62,** 811–820.

Harrison, R. A., and Miller, N. G. (2000). cAMP-dependent protein kinase control of plasma membrane lipid architecture in boar sperm. *Mol. Reprod. Dev.* **55,** 220–228.

Harrison, R. A. P., Ashworth, P. J. C., and Miller, N. G. A. (1996). Bicarbonate/CO_2, an effector of capacitation, induces a rapid and reversible change in the lipid architecture of boar sperm plasma membranes. *Mol. Reprod. Dev.* **45,** 378–391.

Hildebrandt, J. D., Codina, J., Tash, J. S., Kirchick, H. J., Lipschultz, L., Sekura, R. D., and Birnbaumer, L. (1985). The membrane-bound spermatozoal adenylyl cyclase system does not share coupling characteristics with somatic cell adenylyl cyclases. *Endocrinology* **116,** 1357–1366.

Hyne, R. V. (1984). Bicarbonate- and calcium-dependent induction of rapid guinea pig sperm acrosome reactions by monovalent ionophores. *Biol. Reprod.* **31,** 312–323.

Hyne, R. V., and Garbers, D. L. (1979a). Calcium-dependent increase in adenosine 3′,5′-monophosphate and induction of the acrosome reaction in guinea pig spermatozoa. *Proc. Natl. Acad. Sci. U.S.A.* **76,** 5699–5703.

Hyne, R. V., and Garbers, D. L. (1979b). Regulation of guinea pig sperm adenylate cyclase by calcium. *Biol. Reprod.* **21,** 1135–1142.

Hyne, R. V., Higginson, R. E., Kohlman, D., and Lopata, A. (1984). Sodium requirement for capacitation and membrane fusion during the guinea-pig sperm acrosome reaction. *J. Reprod. Fertil.* **70,** 83–94.

Iida, H., Yoshinaga, Y., Tanaka, S., Toshimori, K., and Mori, T. (1999). Identification of Rab3A GTPase as an acrosome-associated small GTP- binding protein in rat sperm. *Dev. Biol.* **211,** 144–155.

Jahn, R., and Sudhof, T. C. (1999). Membrane fusion and exocytosis. *Annu. Rev. Biochem.* **68,** 863–911.

Jaiswal, B. S., and Conti, M. (2001). Identification and functional analysis of splice variants of the germ cell soluble adenylate cyclase. *J. Biol. Chem.,* M011698200.

Johnston, D. S., Wright, W. W., Shaper, J. H., Hokke, C. H., Van den Eijnden, D. H., and Joziasse, D. H. (1998). Murine sperm-zona binding, a fucosyl residue is required for a high affinity sperm-binding ligand; a second site on sperm binds a nonfucosylated, β-galactosyl-capped oligosaccharide. *J. Biol. Chem.* **273,** 1888–1895.

Jungnickel, M. K., Marrero, H., Birnbaumer, L., Lemos, J. R., and Florman, H. M. (2001). Trp2 regulates entry of $Ca2^+$ into mouse sperm triggered by egg ZP3. *Nat. Cell Biol.* **3,** 499–502.

Kalab, P., Visconti, P., Leclerc, P., and Kopf, G. S. (1994). p95, the major phosphotyrosine-containing protein in mouse spermatozoa, is a hexokinase with unique properties. *J. Biol. Chem.* **269,** 3810–3817.

Katafuchi, K., Mori, T., Toshimori, K., and Iida, H. (2000). Localization of a syntaxin isoform, syntaxin 2, to the acrosomal region of rodent spermatozoa. *Mol. Reprod. Dev.* **57,** 375–383.

Kelly, R. B. (1991). Secretory granule and synaptic vesicle formation. *Curr. Opin. Cell Biol.* **3,** 654–660.

Kim, K. S., Cha, M. C., and Gerton, G. L. (2001). Mouse sperm protein sp56 is a component of the acrosomal matrix. *Biol. Reprod.* **64,** 36–43.

Kinloch, R. A., and Wassarman, P. M. (1989). Profile of a mammalian sperm receptor gene. *New Biol.* **1,** 232–238.

Kinloch, R. A., Ruiz-Seiler, B., and Wassarman, P. M. (1990). Genomic organization and polypeptide primary structury of zona pellucida glycoprotein hZP3, the hamster sperm receptor. *Dev. Biol.* **142,** 414–421.

Kligman, I., Glassner, M., Storey, B. T., and Kopf, G. S. (1991). Zona pellucida-mediated acrosomal exocytosis in mouse spermatozoa: Characterization of an intermediate stage prior to the completion of the acrosome reaction. *Dev. Biol.* **145,** 344–355.

Kopf, G. S., and Gerton, G. L. (1991). The mammalian sperm acrosome and the acrosome reaction. *In* "Elements of Mammalian Fertilization" (P. M. Wassarman, ed.), Vol. 1, pp. 153–203. CRC Press, Boca Raton.

Kopf, G. S., Visconti, P. E., and Galantino-Homer, H. (1999b). Capacitation of the mammalian spermatozoon. *Adv. Dev. Biochem.* **5,** 83–107.

Kopf, G. S., Ning, X. P., Visconti, P. E., Purdon, M., Galantino-Homer, H., and Fornes, M. (1999a). Signaling mechanisms controlling mammalian sperm fertilization competence and activation. *In*

"The Male Gamete: From Basic Science to Clinical Applications" (C. Gagnon, ed.), pp. 105–118. Cache River Press, Vienna, IL.

Lane, M. E., Therien, I., Moreau, R., and Manjunath, P. (1999). Heparin and high-density lipoprotein mediates bovine sperm capacitation by different mechanisms. *Biol. Reprod.* **60,** 169–175.

Lawrence, M. B. (1999). Selectin-carbohydrate interactions in shear flow. *Curr. Opin. Chem. Biol.* **3,** 659–664.

Lax, Y., Rubinstein, S., and Breitbart, H. (1997). Subcellular distribution of protein kinase Cα and βI in bovine spermatozoa, and their regulation by calcium and phorbol esters. *Biol. Reprod.* **56,** 454–459.

Leclerc, P., and Kopf, G. S. (1995). Mouse sperm adenylyl cyclase: General properties and regulation by the zona pellucida. *Biol. Reprod.* **52,** 1227–1233.

Leclerc, P., and Kopf, G. S. (1999). Evidence for the role of heterotrimeric guanine nucleotide-binding regulatory proteins in the regulation of the mouse sperm adenylyl cyclase by the egg's zona pellucida. *J. Androl.* **20,** 126–134.

Lee, M. A., and Storey, B. T. (1985). Evidence for plasma membrane impermeability to small ions in acrosome-intact mouse spermatozoa bound to mouse zonae pellucidae, using an aminoacridine fluorescent pH probe: Time course of the zona-induced acrosome reaction monitored by both chlortetracycline and pH probe fluorescence. *Biol. Reprod.* **33,** 235–246.

Lee, M. A., and Storey, B. T. (1986). Bicarbonate is essential for fertilization of mouse eggs; mouse sperm require it to undergo the acrosome reaction. *Biol. Reprod.* **34,** 349–356.

Lee, M. A., and Storey, B. T. (1989). Endpoint of first stage of zona pellucida-induced acrosome reaction in mouse spermatozoa characterized by Acrosomal H and Ca permeability: Population and single cell kinetics. *Gamete Res.* **24,** 303–326.

Lee, M. A., Check, J. H., and Kopf, G. S. (1992). A guanine nucleotide-binding regulatory protein in human sperm mediates acrosomal exocytosis induced by the human zona pellucida. *Mol. Reprod. Dev.* **31,** 78–86.

Leyton, L., and Saling, P. (1989a). 95 kd sperm proteins bind ZP3 and serve as tyrosine kinase substrates in response to zona binding. *Cell* **57,** 1123–1130.

Leyton, L., and Saling, P. M. (1989b). Evidence that aggregation of mouse sperm receptors by ZP3 triggers the acrosome reaction. *J. Cell Biol.* **108,** 2163–2168.

Leyton, L., LeGuen, P., Bunch, D., and Saling, P. (1992). Regulation of mouse gamete interaction by a sperm tyrosine kinase. *Proc. Natl. Acad. Sci. U.S.A.* **89,** 11692–11695.

Lievano, A., Bolden, A., and Horn, R. (1994). Calcium channels in excitable cells: Divergent genotypic and phenotypic expression of alpha 1-subunits. *Am. J. Physiol.* **267,** C411–C424.

Lievano, A., Santi, C. M., Serrano, C. J., Trevino, C. L., Bellve, A. R., Hernandez-Cruz, A., and Darszon, A. (1996). T-type Ca^{2+} channels and alpha1E expression in spermatogenic cells, and their possible relevance to the sperm acrosome reaction. *FEBS Lett.* **388,** 150–154.

Lira, S. A., Kinloch, R. A., Mortillo, S., and Wassarman, P. M. (1990). An upstream region of the mouse ZP3 gene directs expression of firefly luciferase specifically to growing oocytes in transgenic mice. *Proc. Natl. Acad. Sci. U.S.A.* **87,** 7215–7219.

Litscher, E. S., Juntunen, K., Seppo, A., Penttilä, L., Niemelä, R., Renkonen, O., and Wassarman, P. M. (1995). Oligosaccharide constructs with defined structures that inhibit binding of mouse sperm to unfertilized eggs *in vitro*. *Biochemistry* **34,** 4662–4669.

Lopez, L. C., Bayna, E. M., Litoff, D., Shaper, N. L., Shaper, J. H., and Shur, B. D. (1985). Receptor function of mouse sperm surface galactosyltransferase during fertilization. *J. Cell Biol.* **101,** 1501–1510.

Lopez-Gonzalez, I., De La Vega-Beltran, J. L., Santi, C. M., Florman, H. M., Felix, R., and Darszon, A. (2001). Calmodulin antagonists inhibit t-type Ca^{2+} currents in mouse spermatogenic cells and the zona pellucida-induced sperm acrosome reaction. *Dev. Biol.* **236,** 210–309.

Lu, Q., and Shur, B. D. (1997). Sperm from beta 1,4-galactosyltransferase-null mice are refractory to ZP3-induced acrosome reactions and penetrate the zona pellucida poorly. *Development* **124,** 4121–4131.

Lu, Q., Gore, M., Zhang, Q., Camenisch, T., Boast, S., Casagranda, F., Lai, C., Skinner, M. K., Klein, R., Matsushima, G. K., Earp, H. S., Goff, S. P., and Lemke, G. (1999). Tyro-3 family receptors are essential regulators of mammalian spermatogenesis. *Nature* **398,** 723–728.

Luconi, M., Bonaccorsi, L., Maggi, M., Pecchioli, P., Krausz, C., Forti, G., and Baldi, E. (1998). Identification and characterization of functional nongenomic progesterone receptors on human sperm membrane. *J. Clin. Endocrinol. Metab.* **83,** 877–885.

Meizel, S. (1997). Amino acid neurotransmitter receptor/chloride channels of mammalian sperm and the acrosome reaction. *Biol. Reprod.* **56,** 568–574.

Melendrez, C. S., Meizel, S., and Berger, T. (1994). Comparison of the ability of progesterone and heat solubilized porcine zona pellucida to initiate the porcine sperm acrosome reaction *in vitro Mol. Reprod. Dev.* **39,** 433–438.

Meyer, C., Schmid, R., Scriba, P. C., and Wehling, M. (1996). Purification and partial sequencing of high-affinity progesterone-binding site(s) from porcine liver membranes. *Eur. J. Biochem.* **239,** 726–731.

Michaut, M., Tomes, C. N., De Blas, G., Yunes, R., and Mayorga, L. S. (2000). Calcium-triggered acrosomal exocytosis in human spermatozoa requires the coordinated activation of Rab3A and N-ethylmaleimide-sensitive factor. *Proc. Natl. Acad. Sci. U.S.A.* **97,** 9996–10001.

Miller, D. J., and Ax, R. L. (1990). Carbohydrates and fertilization in animals. *Mol. Reprod. Dev.* **26,** 184–198.

Miller, D. J., Gong, X., Decker, G., and Shur, B. D. (1993). Egg cortical granule *N*-acetylglucosaminidase is required for the mouse zona block to polyspermy. *J. Cell Biol.* **123,** 1431–1440.

Miller, D. J., Macek, M. B., and Shur, B. D. (1992). Complementarity between sperm surface β-1,4-galactosyltransferase and egg-coat ZP3 mediates sperm-egg binding. *Nature* **357,** 589–593.

Mittag, T. W., Guo, W. B., and Kobayashi, K. (1993). Bicarbonate-activated adenylyl cyclase in fluid-transporting tissues. *Am. J. Physiol.* **264,** F1060–F1064.

Moller, C. C., Bleil, J. D., Kinloch, R. A., and Wassarman, P. M. (1990). Structural and functional relationships between mouse and hamster zona pellucida glycoproteins. *Dev. Biol.* **137,** 276–286.

Monck, J. R., and Fernandez, J. M. (1996). The fusion pore and mechanisms of biological membrane fusion. *Curr. Opin. Cell Biol.* **8,** 524–533.

Moos, J., Faundes, D., Kopf, G. S., and Schultz, R. M. (1995). Composition of the human zona pellucida and modifications following fertilization. *Hum. Reprod.* **10,** 2467–2471.

Moreau, R., Therien, I., Lazure, C., and Manjunath, P. (1998). Type II domains of BSP-A1/-A2 proteins: Binding properties, lipid efflux, and sperm capacitation potential. *Biochem. Biophys. Res. Commun.* **246,** 148–154.

Mrsny, R. J., and Meizel, S. (1981). Potassium ion influx and Na^+,K^+-ATPase activity are required for the hamster sperm acrosome reaction. *J. Cell Biol.* **91,** 77–82.

Murase, T., and Roldan, E. R. (1996). Progesterone and the zona pellucida activate different transducing pathways in the sequence of events leading to diacylglycerol generation during mouse sperm acrosomal exocytosis. *Biochem. J.* **320,** 1017–1023.

Murphy, G., and Gavrilovic, J. (1999). Proteolysis and cell migration: Creating a path? *Curr. Opin. Cell Biol.* **11,** 614–621.

Myles, D. G., Hyatt, H., and Primakoff, P. (1987). Binding of both acrosome-intact and acrosome-reacted guinea pig sperm to the zona pellucida during *in vitro* fertilization. *Dev. Biol.* **121,** 559–567.

Neer, E. J. (1978). Physical and functional properties of adenylate cyclase from mature rat testis. *J. Biol. Chem.* **253,** 5808–5812.

Neer, E. J., and Murad, F. (1979). Separation of soluble adenylate and guanylate cyclases from the mature rat testis. *Biochim. Biophys. Acta* **583,** 531–534.

Noland, T. D., Garbers, D. L., and Kopf, G. S. (1988). An elevation in cyclic AMP concentration precedes the zona pellucida-induced acrosome reaction of mouse spermatozoa. *Biol. Reprod.* **38,** 74.

Okamura, N., Tajima, Y., Onoe, S., and Sugita, Y. (1991). Purification of bicarbonate-sensitive sperm adenylylcyclase by 4-acetamido-4′-isothiocyanostilbene-2,2′-disulfonic acid-affinty chromatography. *J. Biol. Chem.* **266,** 17754–17759.

Okamura, N., Tajima, Y., Soejima, Y., Madua, H., and Sugita, Y. (1985). Sodium bicarbonate in seminal plasma stimulates the motility of mammalian spermatozoa through direct activation of adenylate cyclase. *J. Biol. Chem.* **260,** 9699–9705.

O'Rand, M. G., and Fisher, S. J. (1987). Localization of zona pellucida binding sites on rabbit spermatozoa and induction of the acrosome reaction by solubilized zonae. *Dev. Biol.* **119,** 551–559.

Osheroff, J. E., Visconti, P. E., Valenzuela, J. P., Travis, A. J., Alvarez, J., and Kopf, G. S. (1999). Regulation of human sperm capacitation by a cholesterol efflux-stimulated signal transduction pathway leading to protein kinase A-mediated up-regulation of protein tyrosine phosphorylation. *Mol. Hum. Reprod.* **5,** 1017–1026.

Osman, R. A., Andria, M. L., Jones, A. D., and Meizel, S. (1989). Steroid induced exocytosis: The human sperm acrosome reaction. *Biochem. Biophys. Res. Commun.* **160,** 828–833.

O'Toole, C. M. B., Arnoult, C., Darszon, A., Steinhardt, R. A., and Florman, H. M. (2000). Ca^{2+} entry through store-operated channels in mouse sperm is initiated by egg ZP3 and drives the acrosome reaction. *Mol. Biol. Cell* **11,** 1571–1584.

O'Toole, C. M. B., Roldan, E. R. S., and Fraser, L. R. (1996a). Protein kinase C activation during progesterone-stimulated acrosomal exocytosis. *Mol. Hum. Reprod.* **2,** 921–927.

O'Toole, C. M. B., Roldan, E. R. S., Hampton, P., and Fraser, L. R. (1996b). A role for diacylglycerol in human sperm acrosomal exocytosis. *Mol. Hum. Reprod.* **2,** 317–326.

Overstreet, J. W., and Cooper, G. W. (1979). The time and location of the acrosome reaction during sperm transport in the female rabbit. *J. Exp. Zool.* **209,** 97–104.

Parinaud, J., and Milhet, P. (1996). Progesterone induces Ca^{2+} dependent 3′,5′-cyclic adenosine monophosphate increase in human sperm. *J. Clin. Endocrinol. Metab.* **81,** 1357–1360.

Pawson, T., and Scott, J. D. (1997). Signaling through scaffold, anchoring, and adaptor proteins. *Science* **278,** 2075–2080.

Pfeffer, S. R. (1996). Transport vesicle docking: SNAREs and associates. *Annu. Rev. Cell Biol.* **12,** 441–461.

Pfeffer, S. R. (1999). Transport-vesicle targeting: Tethers before SNAREs. *Nat. Cell. Biol.* **1,** E17–E22.

Philpot, C., C., Ringuette, M., J., and Dean, J. (1987). Oocyte-specific expression and developmental regulation of zp3, the sperm receptor of the mouse zona pellucida. *Dev. Biol.* **121,** 568–575.

Plattner, H. (1989). Regulation of membrane fusion during exocytosis. *Int. Rev. Cytol.* **119,** 197–286.

Purohit, S. B., Laloraya, M., and Kumar, P. G. (1998). Bicarbonate-dependent lipid ordering and protein aggregation are part of the nongenomic action of progesterone on capacitated spermatozoa. *J. Androl.* **19,** 608–618.

Qi, H., and Wassarman, P. M. (1999). Secretion of zona pellucida glycoprotein mZP2 by growing oocytes from mZP3(+/+) and mZP3(−/−) mice. *Dev. Genet.* **25,** 95–102.

Rae, M. T., Menzies, G. S., and Bramley, T. A. (1998a). Bovine ovarian non-genomic progesterone binding sites: Presence in follicular and luteal cell membranes. *J. Endocrinol.* **159,** 413–27.

Rae, M. T., Menzies, G. S., McNeilly, A. S., Woad, K., Webb, R., and Bramley, T. A. (1998b). Specific non-genomic, membrane-localized binding sites for progesterone in the bovine corpus luteum. *Biol. Reprod,* **58,** 1394–406.

Revelli, A., Massobrio, M., and Tesarik, J. (1998). Nongenomic actions of steroid hormones in reproductive tissues. *Endocr. Rev.* **19,** 3–17.

Revelli, A., Modotti, M., Piffaretti-Yanez, A., Massobrio, M., and Balerna, M. (1994). Steroid receptors in human spermatozoa. *Hum. Reprod.* **9,** 760–766.

Ribbes, H., Plantavid, M., Bennet, P. J., Chap, H., and Douste-Blazy, L. (1987). Phospholipase C from human sperm specific for phosphoinositides. *Biochim. Biophys. Acta* **912,** 245–252.

Roldan, E. R. S. (1999). Signaling for exocytosis: Lipid second messengers, phosphorylation cascades and cross-talk. *In* "The Male Gamete: From Basic Science to Clinical Applications" (C. Gagnon, ed.), pp. 127–138. Cache River Press, Vienna, IL.

Roldan, E. R. S., and Fragio, C. (1994). Diacylglycerols stimulate phospholipase A_2 and subsequent exocytosis in ram spermatozoa. Evidence that the effect is not mediated via protein kinase C. *Biochem. J.* **297,** 225–232.

Roldan, E. R. S., and Murase, T. (1994). Polyphosphoinositide-derived diacylglycerol stimulates the hydrolysis of phosphatidylcholine by phospholipase C during exocytosis of ram sperm acrosome. *J. Biol. Chem.* **269,** 23583–23589.

Roldan, E. R. S., and Vazquez, J. M. (1996). Bicarbonate/CO_2 induces rapid activation of phospholipase A_2 and renders boar spermatozoa capable of undergoing acrosomal exocytosis in response to progesterone. *FEBS Lett.* **396,** 227–232.

Roldan, E. R. S., Murase, T., and Shi, Q. X. (1994). Exocytosis in spermatozoa in response to progesterone and zona pellucida. *Science* **266,** 1578–1581.

Ramalho-Santos, J., Moreno, R. D., Sutovsky, P., Chan, A. W., Hewitson, L., Wessel, G. M., Simerly, C. R., and Schatten, G. (2000). SNAREs in mammalian sperm: possible implications for fertilization. *Dev. Biol.* **223,** 54–69.

Rosiere, T. K., and Wassarman, P. M. (1992). Identification of a region of mouse zona pellucida glycoprotein mZP3 that possesses sperm receptor activity. *Dev. Biol.* **154,** 309–317.

Rotem, R., Paz, G. F., Homonnai, Z. T., Kalina, M., and Naor, Z. (1990). Protein kinase C is present in human sperm: possible role in flagellar motility. *Proc. Natl. Acad. Sci. U.S.A.* **87,** 7305–7308.

Ruggeri, Z. M. (1997). Mechanisms initiating platelet thrombus formation. *Thromb. Haemost.* **78,** 611–616.

Ruggeri, Z. M., and Ware, J. (1993). von Willebrand factor. *FASEB J.* **7,** 308–316.

Sabeur, K., Edwards, D. P., and Meizel, S. (1996). Human sperm plasma membrane progesterone receptor(s) and the acrosome reaction. *Biol. Reprod.* **54,** 993–1001.

Sacco, A. G., Yurewicz, E. C., Subramanian, M. G., and Matzat, P. D. (1989). Porcine zona pellucida: Association of sperm receptor activity with the α-glycoprotein component of the M_r = 55,000 family. *Biol. Reprod.* **41,** 523–532.

Sadler, S. E., and Maller, J. L. (1982). Identification of a steroid receptor on the surface of Xenopus oocytes by photoaffinity labeling. *J. Biol. Chem.* **257,** 355–361.

Saling, P. M., and Storey, B. T. (1979). Mouse gamete interactions during fertilization *in vitro.* Chlortetracycline as a fluorescent probe for the mouse sperm acrosome reaction. *J. Cell Biol.* **83,** 544–555.

Saling, P. M., Sowinski, J., and Storey, B. T. (1979). An ultrastructural study of epididymal mouse spermatozoa binding to zonae pellucidae *in vitro:* Sequential relationship to the acrosome reaction. *J. Exp. Zool.* **209,** 229–238.

Santi, C. M., Darszon, A., and Hernandez-Cruz, A. (1996). A dihydropyridine-sensitive T-type Ca^{2+} current is the main Ca^{2+} current carrier in mouse primary spermatocytes. *Am. J. Physiol.* **271,** C1583–C1593.

Schmidt, B. M., Gerdes, D., Feuring, M., Falkenstein, E., Christ, M., and Wehling, M. (2000). Rapid, nongenomic steroid actions: A new Age? *Front. Neuroendocrinol.* **21,** 57–94.

Schulz, J. R., Sasaki, J. D., and Vacquier, V. D. (1998). Increased association of synaptosome-associated protein of 25 kDa with syntaxin and vesicle-associated membrane protein following acrosomal exocytosis of sea urchin sperm. *J. Biol. Chem.* **273,** 24355–24359.

Schulz, J. R., Wessel, G. M., and Vacquier, V. D. (1997). The exocytosis regulatory proteins syntaxin and VAMP are shed from sea urchin sperm during the acrosome reaction. *Dev. Biol.* **191,** 80–87.

Selmin, O., Lucier, G. W., Clark, G. C., Tritscher, A. M., Vanden Heuvel, J. P., Gastel, J. A., Walker, N. J., Sutter, T. R., and Bell, D. A. (1996). Isolation and characterization of a novel gene induced by 2,3,7,8-tetrachlorodibenzo-*p*-dioxin in rat liver. *Carcinogenesis* **17,** 2609–2615.

Serrano, C. J., Trevino, C. L., Felix, R., and Darszon, A. (1999). Voltage-dependent Ca(2+) channel subunit expression and immunolocalization in mouse spermatogenic cells and sperm. *FEBS Lett.* **462,** 171–176.

Shabanowitz, R. B., and O'Rand, M. G. (1988). Characterization of the human zona pellucida from fertilized and unfertilized eggs. *J. Reprod. Fertil.* **82,** 151–161.

Shalgi, R., Phillips, D. M., and Jones, R. (1989). Status of the rat acrosome during sperm-zona pellucida interactions. *Gamete Res.* **22,** 1–13.

Sheikhnejad, R. G., and Srivastave, P. N. (1986). Isolation and properties of a phosphatidylcholine-specific phopsholipase C from bull seminal plasma. *J. Biol. Chem.* **261,** 7544–7549.

Shi, X., Amindari, S., Paruchuru, K., Skalla, D., Burkin, H., Shur, B. D., and Miller, D. J. (2001). Cell surface beta-1,4-galactosyltransferase-I activates G protein-dependent exocytotic signaling. *Development* **128,** 645–654.

Shur, B. D., and Hall, N. G. (1982). A role for mouse sperm surface galactosyltransferase in sperm binding to the egg zona pellucida. *J.Cell Biol.* **95,** 574–579.

Sinclair, M. L., Wang, X. Y., Mattia, M., Conti, M., Buck, J., Wolgemuth, D. J., and Levin, L. R. (2000). Specific expression of soluble adenylyl cyclase in male germ cells. *Mol. Reprod. Dev.* **56,** 6–11.

Smith, T. T. (1998). The modulation of sperm function by the oviductal epithelium. *Biol. Reprod.* **58,** 1102–1104.

Spungin, B., and Breitbart, H. (1996). Calcium mobilization and influx during sperm exocytosis. *J. Cell Sci.* **109,** 1947–1955.

Storey, B. T. (1995). Interactions between gametes leading to fertilization: The sperm's eye view. *Reprod. Fertil. Dev.* **7,** 927–942.

Storey, B. T., and Kopf, G. S. (1991). *In* Spermatozoa. "Fertilization in the Mouse.", pp. 167–216. Plenum Press, New York.

Suarez, S. S. (1998). The oviductal sperm reservoir in mammals: Mechanisms of formation. *Biol. Reprod.* **58,** 1105–1107.

Suarez, S. S., Katz, D. F., and Overstreet, J. W. (1983). Movement characteristics and acrosomal status of rabbit spermatozoa recovered at the site and time of fertilization. *Biol. Reprod.* **29,** 1277–1287.

Sudhof, T. C. (1995). The synaptic vesicle cycle: A cascade of protein–protein interactions. *Nature* **375,** 645–653.

Suzuki-Toyota, F., Maekawa, M., Cheng, A., and Bleil, J. D. (1995). Immuno-colloidal gold labeled surface replica, and its application to detect sp56, the egg recognition and binding protein, on the mouse spermatozoon. *J. Electron. Microsc. (Tokyo)* **44,** 135–139.

Tesarik, J., Carreras, A., and Mendoza, C. (1993). Differential sensitivity of progesterone and zona pellucida-induced acrosome reactions to pertussis toxin. *Mol. Reprod. Dev.* **34,** 183–189.

Thaler, C. D., and Cardullo, R. A. (1996). The initial molecular interaction between mouse sperm and the zona pellucida is a complex binding event. *J. Biol. Chem.* **271,** 23289–23297.

Therien, I., Bleau, G., and Manjunath, P. (1995). Phosphatidylcholine-binding proteins of bovine seminal plasma modulate capacitation of spermatozoa by heparin. *Biol. Reprod.* **52,** 1372–1379.

Thomas, P., and Meizel, S. (1989). Phosphatidylinositol 4,5-bisphosphate hydrolysis in human sperm stimulated with follicular fluid or progesterone is dependent upon Ca^{2+} influx. *Biochem. J.* **264,** 539–546.

Tian, J., Kim, S., Heilig, E., and Ruderman, J. V. (2000). Identification of XPR-1, a progesterone receptor required for Xenopus oocyte activation. *Proc. Natl. Acad. Sci. U.S.A.* **97,** 14358–14363.

Tomes, C., McMaster, C. R., and Saling, P. M. (1995). Activation of mouse sperm phosphatidylinositol-4,5 bisphosphate-phospholipase C by zona pellucida is modulated by tyrosine phosphorylation. *Mol. Reprod. Dev.* **43,** 196–204.

Tong, Z. B., Nelson, L. M., and Dean, J. (1995). Inhibition of zona pellucida gene expression by antisense oligonucleotides injected into mouse oocytes. *J. Biol. Chem.* **270,** 849–853.

Travis, A. J., Foster, J. A., Rosenbaum, N. A., Visconti, P. E., Gerton, G. L., Kopf, G. S., and Moss, S. B. (1998). Targeting of a germ cell-specific type 1 hexokinase lacking a porin-binding domain to the mitochondria as well as to the head and fibrous sheath of murine spermatozoa. *Mol. Biol. Cell* **9,** 263–276.

Trevino, C. L., Santi, C. M., Beltran, C., Hernandez-Cruz, A., Darszon, A., and Lomeli, H. (1998). Localisation of inositol trisphosphate and ryanodine receptors during mouse spermatogenesis: Possible functional implications. *Zygote* **6,** 159–172.

Tulsiani, D. R. P., Yoshida-Komiya, H., and Araki, Y. (1997). Mammalian fertilization: A carbohydrate-mediated event. *Biol. Reprod.* **57,** 487–494.

Verhage, H. G., Mavrogianis, P. A., O'Day-Bowman, M. B., Schmidt, A., Arias, E. B., Donnelly, K. M., Boomsma, R. A., Thibodeaux, J. K., Fazleabas, A. T., and Jaffe, R. C. (1998). Characteristics of an oviductal glycoprotein and its potential role in the fertilization process. *Biol. Reprod.* **58,** 1098–1101.

Visconti, P. E., and Kopf, G. S. (1998). Regulation of protein phosphorylation during sperm capacitation. *Biol. Reprod.* **59,** 1–6.

Visconti, P. E., Galantino-Homer, H., Moore, G. D., Bailey, J. L., Ning, X. P., Fornes, M., and Kopf, G. S. (1998). The molecular basis of sperm capacitation. *J. Androl.* **19,** 242–248.

Visconti, P. E., Galantino-Homer, H., Ning, X., Moore, G. D., Valenzuela, J. P., Jorgez, C. J., Alvarez, J. G., and Kopf, G. S. (1999a). Cholesterol efflux-mediated signal transduction in mammalian sperm. Beta-cyclodextrins initiate transmembrane signaling leading to an increase in protein tyrosine phosphorylation and capacitation. *J. Biol. Chem.* **274,** 3235–3242.

Visconti, P. E., Johnson, L. R., Oyaski, M., Fornes, M., Moss, S. B., Gerton, G. L., and Kopf, G. S. (1997). Regulation, localization, and anchoring of protein kinase A subunits during mouse sperm capacitation. *Dev. Biol.* **192,** 351–363.

Visconti, P. E., Moore, G. D., Bailey, J. L., Leclerc, P., Connors, S. A., Pan, D., Olds-Clarke, P., and Kopf, G. S. (1995). Capacitation of mouse spermatozoa. II. Protein tyrosine phosphorylation and capacitation are regulated by a cAMP-dependent pathway. *Development* **121,** 1139–1150.

Visconti, P. E., Ning, X., Fornes, M. W., Alvarez, J. G., Stein, P., Connors, S. A., and Kopf, G. S. (1999b). Cholesterol efflux-mediated signal transduction in mammalian sperm: Cholesterol release signals an increase in protein tyrosine phosphorylation during mouse sperm capacitation. *Dev. Biol.* **214,** 429–443.

Visconti, P. E., Olds-Clarke, P., Moss, S. B., Kalab, P., Travis, A. J., de las Heras, M., and Kopf, G. S. (1996). Properties and localization of a tyrosine phosphorylated form of hexokinase in mouse sperm. *Mol. Reprod. Dev.* **43,** 82–93.

Waldegger, S., Beisse, F., Apfel, H., Breit, S., Kolb, H. A., Haussinger, D., and Lang, F. (1995). Electrophysiological effects of progesterone on hepatocytes. *Biochim. Biophys. Acta* **1266,** 186–190.

Walensky, L., D., and Snyder, S., H. (1995). Inositol 1,4,5-trisphosphate receptors selectively localized to the acrosome of mammalian sperm. *J. Cell Biol.* **130,** 857–869.

Ward, C. R., and Kopf, G. S. (1993). Molecular events mediating sperm activation. *Dev. Biol.* **158,** 9–34.

Ward, C. R., and Storey, B. T. (1984). Determination of the time course of capacitation in mouse spermatozoa using a chlortetracycline fluorescence assay. *Dev. Biol.* **104,** 287–296.

Ward, C. R., Faundes, D., and Foster, J. A. (1999). The monomeric GTP binding protein, rab3A, is associated with the acrosome in mouse sperm. *Mol. Reprod. Dev.* **53,** 413–421.

Ward, C. R., Storey, B. T., and Kopf, G. S. (1992). Activation of a G_i protein in mouse sperm membranes by solubilized proteins of the zona pellucida, the egg's extracellular matrix. *J. Biol. Chem.* **267,** 14061–14067.

Ward, C. R., Storey, B. T., and Kopf, G. S. (1994). Selective activation of G_{i1} and G_{i2} in mouse sperm by the zona pellucida, the egg's extracellular matrix. *J. Biol. Chem.* **269,** 13254–13258.

Wassarman, P. M. (1988). Zona pellucida glycoproteins. *Annu. Rev. Biochem.* **57,** 415–442.

Wassarman, P. M. (1990). Profile of a mammalian sperm receptor. *Development* **108,** 1–17.

Wassarman, P. M. (1995). Towards molecular mechanisms for gamete adhesion and fusion during mammalian fertilization. *Curr. Top. Cell Biol.* **7,** 658–664.

Wassarman, P. M., and Albertini, D. F. (1994). The mammalian ovum. *In* "The Physiology of Reproduction" (E. Knobil and J. D. Neill, eds.), pp. 79–122. Raven Press, New York.

Wassarman, P. M., Liu, C., Chen, J., Qi, H., and Litscher, E. S. (1998). Ovarian development in mice bearing homozygous or heterozygous null mutations in zona pellucida glycoprotein gene mZP3. *Histol. Histopathol.* **13,** 293–300.

Wassarman, P. M., Qi, H., and Litscher, E. S. (1997). Mutant female mice carrying a single mZP3 allele produce eggs with a thin zona pellucida, but reproduce normally. *Proc. R. Soc. Lond. B. Biol. Sci.* **264,** 323–328.

Wedel, B. J., and Garbers, D. L. (2001). The guanylyl cyclase family at Y2K. *Ann. Rev. Physiol.* **63,** 215–233.

Wehling, M. (1995). Looking beyond the dogma of genomic steroid action: Insights and facts of the 1990s. *J. Mol. Med.* **73,** 439–447.

Wehling, M. (1997). Specific, nongenomic actions of steroid hormones. *Annu. Rev. Physiol.* **59,** 365–393.

Wilde, M. W., Ward, C. R., and Kopf, G. S. (1992). Activation of a G protein in mouse sperm by the zona pellucida, an egg-associated extracellular matrix. *Mol. Reprod. Dev.* **31,** 297–306.

Yanagimachi, R. (1994). Mammalian fertilization. *In* "The Physiology of Reproduction" (E. Knobil and J. D. Neill, eds.), pp. 189–317. Raven Press, New York.

Yanagimachi, R., and Noda, Y. D. (1970). Ultrastructural changes in the hamster sperm head during fertilization. *J. Ultrastruct. Res.* **31,** 465–485.

Yanagimachi, R., and Phillips, D. M. (1984). The status of acrosomal caps of hamster spermatozoa immediately before fertilization *in vitro. Gamete Res.* **9,** 1–9.

Youakim, A., Hathaway, H. J., Miller, D. J., Gong, X., and Shur, B. D. (1994). Overexpressing sperm surface β1,4-galactosyltransferase in transgenic mice affects multiple aspects of sperm-egg interactions. *J. Cell Biol.* **126,** 1573–1583.

Yunes, R., Michaut, M., Tomes, C., and Mayorga, L. S. (2000). Rab3A triggers the acrosome reaction in permeabilized human spermatozoa. *Biol. Reprod.* **62,** 1084–1089.

Yurewicz, E. C., Hibler, D., Fontenot, G. K., Sacco, A. G., and Harris, J. (1993). Nucleotide sequence of cDNA encoding ZP3 alpha, a sperm-binding glycoprotein from zona pellucida of pig oocyte. *Biochim. Biophys. Acta* **1174,** 211–214.

Zapata, O., Ralston, J., Beltran, C., Parys, J. B., Chen, J. L., Longo, F. J., and Darszon, A. (1997). Inositol triphosphate receptors in sea urchin sperm. *Zygote* **5,** 355–364.

7

Regulation of Sperm Ion Currents

Alberto Darszon, Felipe Espinosa, Blanca Galindo, Daniel Sánchez, and Carmen Beltrán

Departamento de Genética y Fisiología Molecular, Instituto de Biotecnología, Universidad Nacional Autónoma de México, Cuernavaca, Morelos

I. Importance of Ion Channels

In recent years it has been shown that ion channels are essential elements in cell signaling. This has stimulated their study enormously (Hille, 1992; Jan and Jan, 1997). The generation of a new individual involves the fusion of a spermatozoon and an egg, a process called fertilization. This process requires fully mature and competent male and female gametes, and the appropriate communication between them. The egg emits long- and short-range signals that influence sperm function and determine proper gamete interaction, which leads to fertilization. Much additional work is needed to reveal the detailed molecular mechanisms that mediate the sperm–egg dialogue. There is, however, mounting evidence that ion channels are deeply involved in gamete signaling. For example, the sperm acrosome reaction (AR), an exocytotic process necessary for fertilization in many species, is inhibited by ion channel blockers [reviewed in Florman *et al.* (1998); Darszon *et al.* (1999)].

This chapter focuses on how sperm ion channels participate in the information exchange between gametes and between gametes and their environment. Species for which there is available information on ion channel participation in specific sperm functions will be discussed. The authors' limitations will no doubt reflect biases, and some important contributions will be left out; an apology is offered together with reference to several excellent reviews on general aspects of gamete interaction and function that may cover these omissions (Garbers, 1989; Hoshi *et al.*, 1994; Suzuki, 1995; Florman *et al.*, 1998; Benoff, 1998; Vacquier, 1998; Darszon *et al.*, 1999, 2001; Publicover and Barrat, 1999; Flesch and Gadella, 2000; Baldi *et al.*, 2000; Wassarman *et al.*, 2001).

Spermatozoa are very small. This has limited their electrophysiological characterization and enhanced the use of complementary strategies to learn how their ion transport systems participate in gamete communication. *In vivo* measurements of intracellular Ca^{2+} ($[Ca^{2+}]_i$), intracellular pH (pH_i) (Babcock, 1983; Schackmann and Chock, 1986; Guerrero and Darszon, 1989a,b; Florman *et al.*, 1989, 1992; Zeng *et al.*, 1996), membrane potential (E_M) (Schackmann *et al.*, 1981; González-Martínez and Darszon, 1987; García-Soto *et al.*, 1987; Babcock *et al.*, 1992; Arnoult *et al.*, 1996a), and patch-clamp techniques (Guerrero *et al.*, 1987; Babcock *et al.*, 1992; Weyand *et al.*, 1994; Espinosa *et al.*, 1997), together with reconstitution in planar and spherical bilayers [reviewed in Darszon *et al.* (1994, 1996); see also Cox and Peterson (1989) and Chan *et al.* (1997)], have revealed the presence of Ca^{2+}, K^+, cation, and Cl^- channels in spermatozoa. Researchers are now combining these strategies to explore how ion channels participate in the sperm responses to the egg coats, including the regulation of their activities.

Four alternatives have emerged to circumvent the sperm size limitation: (1) Sea urchin spermatozoa have been swollen in diluted seawater. The swollen cells are spherical (~4 μm in diameter) and they regulate their E_M, pH_i, and $[Ca^{2+}]_i$. Their main virtue is that they can be patch-clamped (Babcock *et al.*, 1992), a difficult endeavor with normal spermatozoa (Guerrero *et al.*, 1987). (2) Ion channels have been transferred to lipid bilayers directly using mouse and sea urchin spermatozoa (Beltrán *et al.*, 1994). This strategy opens new avenues to explore cell–cell interactions, such as sperm–egg fusion, at the single-channel level. (3) The sequences of many physiologically relevant ion channels have been obtained (Hille, 1992), opening the possibility of exploring testicular libraries with probes designed for specific channels. This approach has resulted in the cloning and heterologous expression of cyclic nucleotide-gated channels present in mouse and sea urchin sperm cells (Weyand *et al.*, 1994; Gauss *et al.*, 1998) and a peculiar member of the family of Ca^{2+} activated K^+ channels expressed in mouse spermatogenic cells

(Schreiber *et al.,* 1998). (4) Spermatozoa are terminal cells lacking protein synthesis machinery and most RNAs; therefore the study of gene expression and protein assembly requires the use of the progenitor spermatogenic cells. In addition, spermatogenic cells, particularly pachytene spermatocytes and round spermatids, which are at the later stages of differentiation, are much larger than mature spermatozoa and thus are easier to patch-clamp (Liévano *et al.*, 1996; Santi *et al.*, 1996; Arnoult *et al.,* 1996b, 1997).

II. SPERM ION TRANSPORT AND ENVIRONMENTAL SENSING

In their journey toward the egg, spermatozoa experience important alterations in their ionic milieu that influence their functional state. For instance, they acquire the potential for motility only after leaving the testis of organisms that fertilize externally (e.g., sea urchins and teleost fishes), and as they pass through the vas deferens in internal fertilizers such as reptiles, birds, and mammals. Many factors determine the acquisition of motility potential, but among the most important are concentration changes in external HCO_3^-, H^+, and Ca^{2+}, and exposure to glycoproteins. Motility is initiated when sperm cells are spawned into the reproductive ground or ejaculated into the female reproductive tract. Activation is triggered by ionic or osmotic changes. It is believed that sperm ion channels are involved in these transduction events (Morisawa, 1994; Darszon *et al.,* 1999; Krasznai *et al.,* 2000).

A. SEA URCHINS

In the male gonads sea urchin spermatozoa are immotile because the high CO_2 tension in semen keeps pH_i acid (~7.2) with respect to seawater (Johnson *et al.,* 1983). Motility and respiration are repressed because dynein, the ATPase that drives the flagellum, is inactive below pH 7.3 (Schackmann *et al.,* 1981; Christen *et al.,* 1982; Lee *et al.,* 1983). The concentration of CO_2 decreases when spermatozoa are spawned, contributing to H^+ release, a pH_i increase to ~7.4, and motility initiation (Nishioka and Cross, 1978; Christen *et al.,* 1982; Johnson *et al.,* 1983). At this pH_i dynein hydrolyzes ATP to ADP, activating mitochondrial respiration 50-fold. The energy produced in the mitochondria reaches the flagellum through a phosphocreatine shuttle (Tombes and Shapiro, 1985).

The pH_i increase that occurs with sperm dilution in seawater is Na^+ dependent. Activation is inhibited in Na^+-free seawater, and can be restored by adding Na^+ or NH_4^+ (Schackmann *et al.,* 1981; Christen *et al.,* 1982, 1983; Johnson *et al.,* 1983; Lee *et al.,* 1983; Bibring *et al.,* 1984). An unusual amiloride-insensitive, Mg^{2+}- and voltage-dependent, Na^+/H^+ exchange is responsible for this alkalinization. It has been studied in isolated sperm flagella and in vesicles derived from them (Lee, 1984a,b, 1985). This Na^+/H^+ exchange is, in addition, modulated by Zn^{2+} (Clapper and Eppel, 1985). The Na^+,K^+-ATPase maintains low intracellular Na^+ and participates in regulating pH_i (Gatti and Christen, 1985).

The pH_i rise that accompanies sperm activation also depends on the concentration of external K^+ ($[K^+]_e$), which is higher in semen than in seawater. Although the sea urchin sperm's resting membrane potential (−36 to −56 mV) (Schackman *et al.,* 1981; García-Soto *et al.,* 1987) is not very sensitive to $[K^+]_e$, increasing it to 100 m*M* inhibits activation. Thus, spawning could hyperpolarize these spermatozoa and stimulate the voltage-dependent Na^+/H^+ exchange, and adenylyl cyclase (AC) (Cook and Babcock, 1993b; Beltrán *et al.,* 1996). The rise in cAMP may activate a protein kinase A (PKA) that phosphorylates axonemal proteins, contributing to sperm motility (Garbers, 1989). This hypothesis is consistent with the presence of K^+ channels in the plasma membrane of these cells (Liévano *et al.,* 1985; Guerrero *et al.,* 1987).

B. FISH

High $[K^+]_e$ is responsible for keeping trout spermatozoa immotile in the seminal tract (Morisawa and Suzuki, 1980; Morisawa *et al.,* 1983). A decrease in $[K^+]_e$ initiates sperm motility (Morisawa and Okuno, 1982) and causes an immediate transient increase in cAMP (Morisawa and Ishida, 1987). Motility activation requires a cAMP-dependent phosphorylation of axonemal proteins (Morisawa and Hayashi, 1985). A decrease in $[K^+]_e$ leads to hyperpolarization and to activation of motility, in a pH_i-independent fashion, whereas depolarization results in inactivation (Boitano and Omoto, 1991).

K^+ channel blockers such as tetraethylammonium ion (TEA^+) and Ba^{2+} suppress K^+ efflux and sperm motility in salmonoid fish. Verapamil, a Ca^{2+} channel blocker, inhibits trout sperm motility. Transient increases in $[Ca^{2+}]_i$ coming from intracellular stores may also mediate motility activation (Boitano and Omoto, 1992). In marine teleosts (puffer and flounder) motility activation ensues on hypertonic dilution in nonelectrolyte solutions apparently involving an increase in pH_i and $[Ca^{2+}]_i$ (Oda and Morisawa, 1993). In fish, a K^+-dependent hyperpolarization and the subsequent increase in [cAMP] are essential for motility initiation (Morisawa, 1994; Krasznai *et al.,* 2000). These findings point to the importance of ion channels in motility.

C. MAMMALS

Ion concentrations vary significantly in the course of spermatozoa passage through the epididymis. Na^+ decreases from more than 100 m*M* in the caput to less than 50 m*M* in the cauda (Jenkins *et al.,* 1980), and K^+ increases from ~20 to ~40 m*M* in this transition. Because mouse sperm membrane resting potential is driven principally by K^+ (Espinosa and Darszon, 1995), increasing $[K^+]_e$ can depolarize and open voltage-dependent Ca^{2+} channels [reviewed in Florman *et al.* (1998), Darszon *et al.* (2001)] present in mouse spermatozoa, possibly triggering premature ARs. However, the low $[Ca^{2+}]$ in epididymal fluids (Jenkins *et al.,* 1980) and the decrease in $[Na^+]_e$, which may acidify pH_i (Zeng *et al.,* 1996), would counterbalance the tendency to open Ca^{2+} channels, in this manner preventing spontaneous ARs.

$[Ca^{2+}]_i$ is important for spermatozoa to acquire the capacity to fertilize the egg. $[Ca^{2+}]_i$ progressively increases during sperm maturation, and in some species leads to hyperactivated motility (White and Aitken, 1989) and spontaneous ARs (Langlais and Roberts, 1985; Bavister, 1986; Yanagimachi, 1988). This is the case for capacitation, a time-dependent acquisition of fertilizing ability that regulates the efficiency of acrosome exocytosis in spermatozoa and coordinates it with egg contact to ensure fertilization (Chang, 1951; Austin, 1951). This poorly understood process normally occurs in the female reproductive tract and involves changes in $[Ca^{2+}]_i$ and other ions, plasma membrane lipid remodeling as well as modifications in protein phosphorylation (Fraser, 1987; Visconti *et al.*, 1999; Baldi *et al.*, 2000). $[Ca^{2+}]_i$ steadily elevates during sperm capacitation *in vitro* and reaches a plateau in ~100 minutes (Baldi *et al.*,1991; DasGupta *et al.*, 1993). Sperm capacitation is regulated by seminal plasma decapacitating factors (Okamura *et al.*, 1990; Boettger-Tong *et al.*, 1993) and by factors present in the female fluids *in vivo* or added to the capacitating media *in vitro* (Lakoski *et al.*, 1988; Yanagimachi, 1988). Some of these factors modulate Ca^{2+} plasma membrane permeability. Caltrin, a seminal plasma protein, was shown to inhibit $^{45}Ca^{2+}$ uptake by spermatozoa (Rufo *et al.*, 1982; Lardy and San Agustin, 1989; Clark *et al.*, 1993). Heparin, essential for bovine sperm *in vitro* capacitation, regulates $[Ca^{2+}]_i$ by modulating voltage-dependent Ca^{2+} channels (VDCC), possibly binding to specific plasma membrane receptors (Parrish *et al.*, 1989; Calvette *et al.*, 1996; Córdoba *et al.*, 1997). Compounds that mediate Ca^{2+} release from internal Ca^{2+} stores, such as thapsigargin, seem to accelerate capacitation (Mendoza and Tesarik, 1993). Calreticulin, a Ca^{2+} binding protein (Nakamura *et al.*, 1993), and the inositol 1,4,5-trisphosphate receptor ($InsP_3R$) (Walensky and Snyder, 1995; Treviño *et al.*, 1998) have been localized to the acrosome of several mammalian species, indicating Ca^{2+} may be stored and released from this organelle.

Ca^{2+}, $NaHCO_3$ and serum albumin are three key components necessary for capacitation in mouse spermatozoa (Visconti and Kopf, 1998). $[Ca^{2+}]_e$ as well as $[HCO_3^-]_e$, have been shown to modulate protein phosphorylation during capacitation of mouse (Visconti *et al.*, 1995a,b) and human spermatozoa (Baldi *et al.*, 1996; Luconi *et al.*, 1996; Naz, 1996; Emiliozzi and Fenichel, 1997). HCO_3^- is necessary in capacitating media for protein tyrosine phosphorylation and for hyperactivated motility. These effects have been proposed to result from [cAMP] increases mediated by AC (Boatman and Robbins, 1991; Shi and Roldán, 1995; Visconti *et al.*, 1995a,b; Luconi *et al.*, 1996). It is thought albumin is required to remove cholesterol from the membrane. It has been proposed that the decrease in cholesterol content alters membrane architecture and somehow leads to an elevation of cAMP levels in spermatozoa (Arnoult *et al.*, 1999).

Maturation and capacitation are influenced by pH_i (White and Aitken, 1989; Gatti *et al.*, 1993; Hammamah *et al.*, 1996; Zeng *et al.*, 1996). During capacitation pH_i increases in mouse spermatozoa mainly through a Na^-, Cl^-, and HCO_3^--dependent mechanism (Zeng *et al.,* 1996). Because pH_i may modulate sperm Ca^{2+} permeability (Babcock and Pfeiffer, 1987; Darszon *et al.,* 1999), an acidic pH_i could contribute to maintain membrane potential (Calzada *et al.,* 1988), and $[Ca^{2+}]_i$ low, thus preventing untimely AR.

During capacitation of bovine and mouse spermatozoa, K^+ permeability increases, hyperpolarizing the cells from around −30 mV to −60 mV (Zeng *et al.,* 1995). A pH_i-dependent inwardly rectifying K^+ channel recently identified in spermatogenic cells could link the pH_i and E_M changes associated to capacitation (Muñoz-Garay *et al.,* 2001).

III. MODULATION OF SPERM ION TRANSPORT BY DIFFUSIBLE EGG COMPONENTS

After spawning, the gametes of external fertilizers undergo an immense dilution. Their encounter requires long-range information as to their whereabouts. On the other hand, internal fertilizers, whose gametes interact in the female reproductive tract, need long-range signaling to prepare them for fertilization and to promote preferential interactions of the egg with the fittest subpopulations of spermatozoa. Among these signals, some stimulate the directed movement of spermatozoa toward the egg (chemotaxis) and/or enhance their motility and metabolism (chemokinesis). Secretions from the egg or from the female reproductive organs have been described to cause chemotaxis in spermatozoa from plant and animal species (Miller, 1985; Morisawa, 1994). Distinguishing between chemotaxis and chemokinesis can be difficult, so both will be considered as long-range gamete signaling processes (Ward and Kopf, 1993).

A. SEA URCHINS

The jelly surrounding the sea urchin egg contains small peptides (~10–14 amino acids) that alter the metabolic state and motility of sperm species specifically (with restrictions). It has been proposed that these peptides may also promote ARs by acting in concert with the main egg jelly-derived inducer of this process (Yamaguchi *et al.,* 1988; Shimizu *et al.,* 1990; but see Schulz *et al.,* 1997). Cooperativity between egg factors may contribute to the success of fertilization.

Speract (or SAP-1) is a decapeptide (Gly-Phe-Asp-Leu-Asn-Gly-Gly-Gly-Val-Gly) that has been isolated from *Strongylocentrotus purpuratus* and *Hemicentrotus pulcherrimus* egg jelly (Hansbrough and Garbers, 1981a; Suzuki *et al.,* 1980, 1981). At concentrations as low as picomolar, speract can stimulate sperm phospholipid metabolism, respiration, and motility (Hansbrough *et al.,* 1980; Suzuki and Yoshino, 1992). Furthermore, this peptide profoundly alters the plasma membrane permeability of sea urchin spermatozoa. At nanomolar concentrations speract and resact (Cys-Val-Thr-Gly-Ala-Pro-Gly-Cys-Val-Gly-Gly-Gly-Arg-Leu-NH2), a similar peptide isolated from *Arbacia punctulata* (Suzuki *et al.,* 1984), stimulate uptake of $^{22}Na^+$ and $^{45}Ca^{2+}$, and release of H^+ and K^+ (Hansbrough and Garbers, 1981b; Repaske and Garbers, 1983; Lee and Garbers, 1986). These permeability changes lead to increases in $[Ca^{2+}]_i$ and pH_i (Schackmann and Chock, 1986). In addition, these peptides increase the levels of cGMP and cAMP (Kopf *et al.,* 1979; Yoshino *et al.,* 1989).

In *S. purpuratus* sperm flagella and flagellar plasma membrane vesicles, speract triggers a K^+-dependent hyperpolarization, probably mediated by the opening of K^+ channels (Lee and Garbers, 1986; Garbers, 1989). In these membranes guanosine 5′-*O*-(3-thiotriphosphate) (GTPγS) stimulates the speract-induced hyperpolarization, suggesting the participation of a G-protein (Lee, 1988). Indeed, G_i (Kopf *et al.,* 1986; Bentley *et al.,* 1986a), G_s, and three low-molecular-weight G-proteins (Cuéllar-Mata *et al.,* 1995; Castellano *et al.,* 1997) have been found in sea urchin spermatozoa. However, experiments with flagellar vesicles containing guanosine 5′-*O*-(2-thiodiphosphate) (GDPβS), where the speract response is not inhibited, raise doubts as to the participation of Gα in this response (Lee, 1988; Galindo *et al.,* 2000).

Determinations of $^{22}Na^+$ influx and pH using fluorescent dyes revealed that speract induces a 1:1 Na^+/H^+ exchange in sea urchin spermatozoa (Lee, 1984a,b; Schackmann and Chock, 1986). Curiously, although this exchange is electroneutral, it is activated by the hyperpolarization triggered by speract (Lee, 1984a). It will be worthwhile to reexamine this point using methods that have equal time resolution to measure changes in pH_i and intracellular Na^+.

Cross-linking experiments have indicated that functional speract analogs bind to a 77-kDa transmembrane polypeptide (Dangott and Garbers, 1984). This putative speract receptor was purified, sequenced, and cloned from *S. purpuratus* (Dangott *et al.,* 1989; Dangott, 1991). The present hypothesis is that the speract–receptor complex transiently activates the sperm membrane guanylyl cyclase (Garbers, 1989). In *A. punctulata,* nanomolar resact binds directly to a membrane guanylyl cyclase (Shimomura *et al.,* 1986), stimulating it. Thereafter the enzyme is dephosphorylated, changing its apparent molecular mass from 160 to 150 kDa (Ward and Vacquier, 1983; Suzuki *et al.,* 1984; Ward *et al.,* 1985b). The phosphorylated enzyme is more active (Ramaro and Garbers, 1985; Ward *et al.,* 1985b), and alkaline pH enhances its dephosphorylation (Ward *et al.,* 1986; Bentley *et al.,* 1986b). The sea urchin resact receptor was the first cloned and sequenced member of a family of guanylyl cyclases that are surface receptors participating in a new signal transduction pathway (Singh *et al.,* 1988).

In swollen sea urchin spermatozoa, picomolar concentrations of speract provoke a long-lasting, K^+-selective permeability increase, mediated by K^+ channels, as indicated by patch-clamp experiments (Babcock *et al.,* 1992) (see Figure 7.1). The increase in [cGMP] induced by this peptide (>100 p*M*) opens TEA^+-insensitive K^+-selective channels that hyperpolarize sperm cells by activating Na^+/H^+ exchange (Babcock *et al.,* 1992; Reynaud *et al.,* 1993; Cook and Babcock, 1993a). The resulting increase in pH_i inhibits guanylyl cyclase (Suzuki *et al.,* 1984; Ward *et al.,* 1986; Bentley *et al.,* 1986b) and stimulates AC, which is sensitive to pH_i (Cook and Babcock, 1993a,b), membrane potential (Beltrán *et al.,* 1995), and $[Ca^{2+}]$ (Garbers, 1989). The decrease in [cGMP] would diminish K^+ permeability and repolarize spermatozoa (Cook and Babcock, 1993a).

Speract (>100 pM) transiently increases $[Ca^{2+}]_i$ and, after the hyperpolarization, induces a Ca^{2+}-dependent net depolarization in swollen (Babcock *et al.*, 1992; Reynaud *et al.*, 1993; Cook and Babcock 1993a) and in nonswollen spermatozoa (Beltrán *et al.*, 1996; Labarca *et al.*, 1997). These results indicate that speract opens a Ca^{2+}-permeable channel. This channel allows Mn^{2+} through, and appears to be regulated by cAMP (Cook and Babcock, 1993b). A Na^+/Ca^{2+} exchanger also contributes to the speract-induced $[Ca^{2+}]_i$, and to $[Ca^{2+}]_i$ regulation (Schackmann and Chock, 1986). Thus, in normal sperm two (or more) ion channels with distinct selectivity and pharmacology may contribute to the depolarization triggered by nanomolar speract: a cAMP- and/or pH_i-regulated Ca^{2+} channel (Darszon *et al.*, 1990; Babcock *et al.*, 1992; Cook and Babcock, 1993b) and a cAMP-regulated K^+ channel that allows Na^+ influx (Labarca *et al.*, 1995). A poorly selective K^+ channel ($PK^+/PNa^+ = 5$), directly modulated by cAMP, was studied in planar lipid bilayers with incorporated flagellar membranes. This channel is blocked by TEA^+ (30 mM) and Ba^{2+}; its opening in seawater would depolarize the cells (Labarca *et al.*, 1995). The participation of this channel in the speract-induced repolarization may explain its Na^+ dependence (Reynaud *et al.*, 1993; Labarca *et al.*, 1995, 1997).

A cAMP-regulated K^+ channel cloned recently from sea urchin testis and functionally expressed in HEK 293 cells has properties similar to those of the channel described above in planar bilayers (Gauss *et al.*, 1998). The channel, named *SPIH,* is a 767-amino acid polypeptide ($M_r \sim 88$) with significant sequence similarity to cyclic nucleotide-gated (CNG) and ether-a-gogo (EAG)/HERG channels. The channel is only about four times more selective for K^+ than for Na^+ and is much more sensitive to cAMP than to cGMP. It was immunodetected in the sperm flagella (Gauss *et al.*, 1998). This channel is a member of a growing family of hyperpolarizing potential- and cyclic nucleotide-gated (HCN) channels that are acti-

FIGURE 7.1 Working hypothesis of the mechanisms involved in the resact and speract responses of sea urchin spermatozoa. In *Arbacia punctulata* flagellar membranes, resact directly activates guanylyl cyclase (1); in flagellar membranes of *Strongylocentrotus purpuratus,* speract indirectly activates guanylyl cyclase by binding to its receptor (2). The transient increase in [cGMP] could directly (most likely), or after X_n steps, open a K^+ channel (3) responsible for the initial transient hyperpolarization. This hyperpolarization can stimulate adenylyl cyclase (4), initiate other important alterations in membrane potential (ΔE_M), and possibly activate Na^+/Ca^{2+} exchange (5). Suitable concentrations of speract hyperpolarize spermatozoa enough to activate Na^+/H^+ exchange (6) and increase intracellular pH (ΔpH_i). The pH_i changes may directly or indirectly modulate guanylyl (1) and adenylyl (4) cyclases and possibly also some kinases, phosphatases, and phosphodiesterases. The increase in [cAMP] activates a cAMP-dependent poorly K^+-selective channel possibly involved in the speract-induced depolarization (7). Concomitant changes in pH_i and [cAMP] may modulate a Ca^{2+} channel (8). E_M (I), pH_i (II), and $[Ca^{2+}]_i$ (III) changes induced by speract (100 nM) in *S. purpuratus* spermatozoa in alkaline seawater are shown on the right side. Upward deflections indicate depolarization (I), alkalinization (II), and $[Ca^{2+}]_i$ increase (III). E_M was measured with the fluorescent probe Dis-C_3-(5), pH_i with 2′,7′-bis(2-carboxyethyl)-5(6)-carboxyfluorescein, and $[Ca^{2+}]_i$ with fura-2 [for details see Darszon *et al.* (1994)]. The cell recordings shown correspond to a cGMP-activated conductance in patched swollen spermatozoa (3) (D. Sánchez and A. Darszon, unpublished), the cAMP-regulated K^+ channel (7) in planar lipid bilayers, and a K^+ channel from swollen spermatozoa (9).

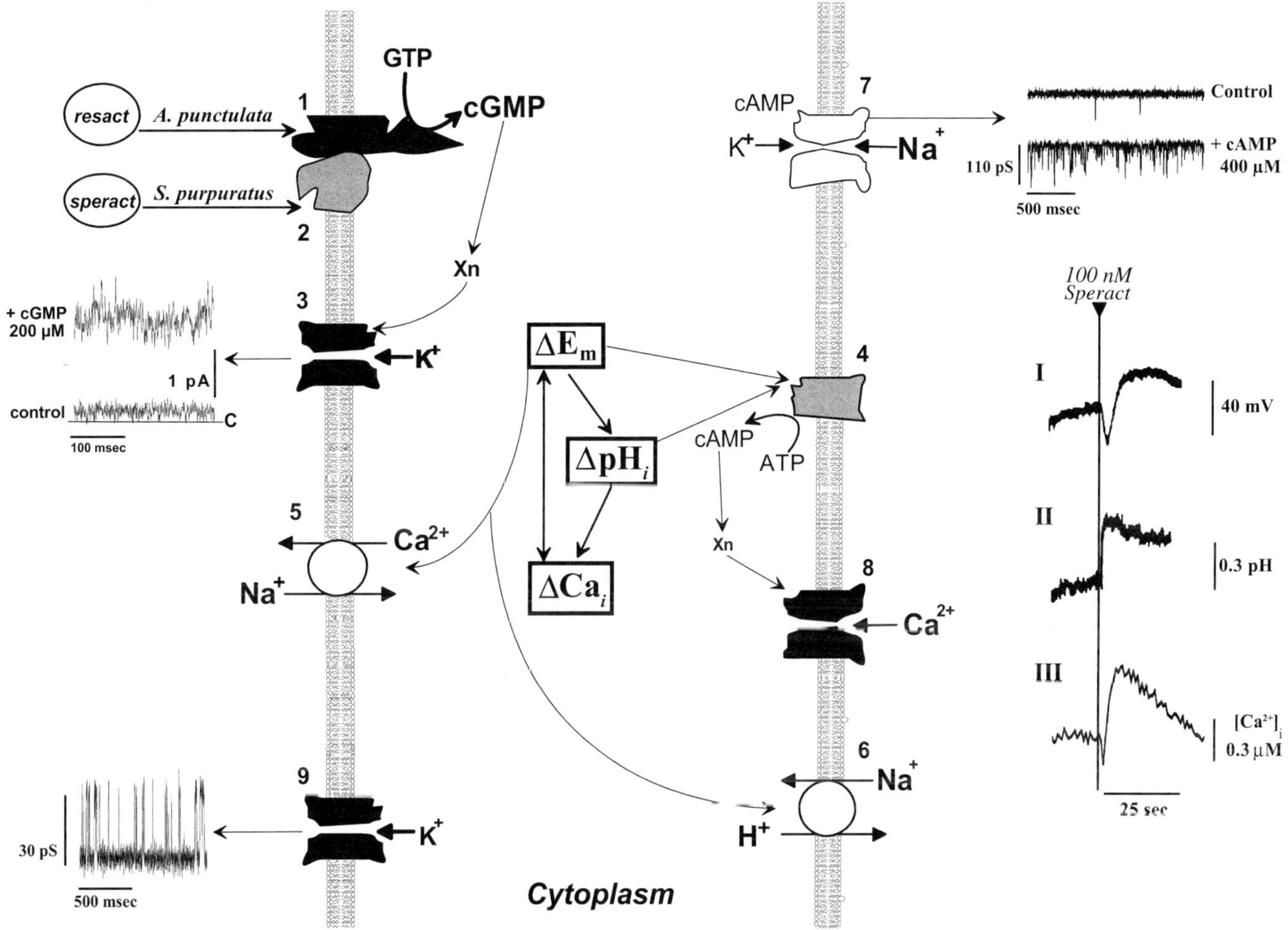

resact
A. punctulata
speract
S. purpuratus
GTP
cGMP
Xn
+ cGMP
200 μM
1 pA
control
C
100 msec
Ca2+
Na+
30 pS
500 msec
ΔEm
ΔpHi
ΔCai
Cytoplasm
cAMP
K+
Na+
Control
+ cAMP
400 μM
110 pS
500 msec
cAMP
ATP
Xn
Ca2+
Na+
H+
100 nM
Speract
I
40 mV
II
0.3 pH
III
[Ca2+]i
0.3 μM
25 sec
1
2
3
4
5
6
7
8
9

vated both by cyclic nucleotides and by hyperpolarizing potentials (Clapham, 1998). The members of this new family are important in shaping the autonomous rhythmic activity of single neurons and the periodicity of network oscillations. *SPIH* could participate in modulating sperm motility (Kaupp and Seifert, 2001), though it has not been recorded in sperm channels activated by cAMP and hyperpolarizing potentials, sharing selectivity properties with *SPIH,* have been recently reported (Sánchez *et al.,* 2001). It remains to be seen if they are *SPIH.*

Even though these egg peptides have chemokinetic effects, chemotaxis has been clearly demonstrated only in *A. punctulata.* Spermatozoa from this species are attracted by nanomolar concentrations of resact, which changes their swimming pattern from a circular to a straighter trajectory. External Ca^{2+} is required for this response (Ward *et al.,* 1985a). Adding 50 μM speract together with 100 μM isobutyl methyl xantine (IBMX), a phosphodiesterase inhibitor, produces asymmetric flagellar movements in *S. purpuratus* spermatozoa (Cook *et al.,* 1994). Although AR is induced under this condition (Schackmann and Chock, 1986), these results were used to derive an interesting model to explain how spermatozoa may detect an increasing egg peptide gradient over a broad concentration range (Cook *et al.,* 1994; Darszon *et al.,* 1996). New binding studies with fluorescent speract analogs indicate that peptide binding is cooperative and modulated by pH_i. Speract may affect sperm trajectory only for a few seconds in the vicinity of the egg (0.5 mm) (Nishigaki *et al.,* 2000, 2001). Using the available information, a working model for the action mechanism of speract is delineated in Figure 7.1.

1. Adenylyl Cyclase

The adenylyl cyclases catalyze the formation of cAMP, an intracellular second messenger in almost all animal cells (Antoni, 1997). In *Paramecium,* an AC not modulated by G-proteins and directly stimulated by hyperpolarization was described and was proposed to be associated with an ion channel (Schultz *et al.,* 1992). The sea urchin sperm AC is modulated by $[Ca^{2+}]_i$ and pH_i (Garbers, 1989; Cook and Babcock, 1993a,b) and is insensitive to G-proteins (Hildebrandt *et al.,* 1985; Garbers, 1989). This AC is also stimulated by hyperpolarizing sea urchin spermatozoa, independently of $[Ca^{2+}]$ and pH_i (Beltrán *et al.,* 1996). Sperm hyperpolarization is triggered by components of the egg outer envelope (Darszon *et al.,* 1999; 2001). Therefore, membrane potential activation of AC could modulate sperm motility, chemotaxis, and AR. It seems worthwhile to explore further the interplay between sperm membrane potential and AC activity since mammalian somatic cell voltage-dependent ACs have been shown to exist (Reddy *et al.,* 1995).

A mammalian soluble form of AC (sAC) preferentially expressed in testis and closely related to cyanobacterial ACs was purified, cloned, and functionally expressed. Though the full-length cDNA predicts a 187-kDa protein, the catalytically active purified form of the enzyme is 48 kDa (Buck *et al.,* 1999). This AC is directly modulated by bicarbonate and not by G proteins or pH_i (Okamura *et al.,* 1985; Chen *et al.,* 2000), and is present in mature sperm. This AC could participate in sperm maturation, capacitation, motility, and AR (Sinclair *et al.,* 2000). It is likely that sea urchin and other marine sperm possess a similar AC.

B. MAMMALS

Even though mammalian spermatozoa in the female reproductive tract follow an arranged pathway toward the egg, long-range gamete communication may be important. A significant fraction of ejaculated cells from various mammalian species seem to have reduced motility when stored in the caudal isthmus of the oviduct (Hunter and Nichol, 1986). Spermatozoa that leave their storage sites minutes after ovulation reach the ampullary region (Flechon and Hunter, 1981). These observations have led to the belief that eggs or follicle cells release factors that activate motility and guide sperm cells toward the ovulated egg. Such factors may enhance productive encounters among the fittest gametes, specially considering that the sperm:egg ratio is low (1:1 to 1:10) at the fertilization site (Yanagimachi, 1988; Ward and Kopf, 1993).

In vivo, human spermatozoa are attracted by follicular factors (Villanueva-Diaz *et al.,* 1990). Diluted human follicular fluid may contain a chemoattractant that can change the swimming pattern of human spermatozoa (Ralt *et al.,* 1991). Reportedly, only a small fraction (2–12%) of human spermatozoa undergoes chemoattraction by follicular factors. They appear to acquire their chemotactic responsiveness as they capacitate, a state proposed to be transient (see Chapter 3, this volume). These results suggest that sperm chemotaxis to follicular factors *in vivo* may selectively recruit capacitated spermatozoa for egg fertilization (Cohen-Dayag *et al.,* 1995). In mammals chemotaxis might be needed to recruit a selected population of capacitated spermatozoa to fertilize the egg (Eisenbach, 1999). Much remains to be done in mammals to fully understand the involvement of ion channels in motility regulation and chemotaxis.

In sea urchin spermatozoa cGMP plays a key role in chemotaxis (Garbers, 1989; Darszon *et al.,* 1999). Though this has not been shown in mammalian spermatozoa, it turns out that the first sperm ion channel cloned was a cyclic nucleotide-gated channel from mouse. At least two subunits (α and β) form these channels. The α subunit displays the channel activity, but β alone is not functionally active. Channel species with properties different from those of homooligomeric channels result from the coexpression of α and β subunits (Kaupp, 1995). Initially, the α subunit was cloned from bovine testis (Weyand *et al.,* 1994). It is 78% homologous in terms of amino acid sequence to CNG channels in chicken photoreceptors and contains the cyclic nucleotide binding site, pore sequence, transmembrane segments, and S4-voltage sensor motif characteristic of the CNG channel family. The channel expressed in *Xenopus* oocytes has a single channel conductance of 20 pS. It is permeable to Ca^{2+}, selects poorly between Na^+ and K^+, is blocked by Mg^{2+}, and has a much higher affinity for cGMP (>100-fold) than for cAMP. Small cGMP-induced currents thought to arise from single-channel transitions of <10 pS were detected in vesicles derived probably from sperm cytoplasmic droplets. Similar small currents were recorded from inside-out patches from human and bovine spermatozoa exposed to cGMP (Weyand *et al.,* 1994).

In bovine testis one short and several long, less abundant transcripts of CNG channel β subunits were identified (Wiesner *et al.,* 1998). Immunolocalization revealed that the α subunit is present along the entire sperm flagellum, whereas the short β subunit is found only in the principal piece of the flagellum. Various combinations of α and β subunits have different permeability to Ca^{2+}. Ca^{2+} microdomains may exist due to the distinct localization of the β subunits in the flagellum, and may be the basis for flagellar bending control (Wiesner *et al.,* 1998).

IV. MODULATION OF SPERM ION TRANSPORT DURING THE ACROSOME REACTION

In all species whose sperm cells possess an acrosome, successful fertilization requires the acrosome reaction. This reaction allows spermatozoa to penetrate through the outer vestments of the egg, and to recognize and fuse with the egg plasma membrane (Yanagimachi, 1994; Vacquier, 1998; Wassarman *et al.,* 2001). Induction of this fundamental process involves short-range interactions with components from the egg's outer layers, and also with other components of the female reproductive tract for internal fertilizers.

A. SEA URCHIN

When spermatozoa encounter the jelly layer surrounding the sea urchin egg, the AR is triggered (Dan, 1952; Tilney, 1985). A fucose sulfate polymer (FSP) is the egg jelly component responsible for inducing the AR (SeGall and Lennarz, 1979; Garbers *et al.,* 1983; Alves *et al.,* 1997; Vacquier and Moy, 1997). This reaction encompasses acrosomal content release (Dan, 1952; Summers and Hylander, 1975), exposure of material required for sperm–egg binding (Vacquier and Moy, 1977; Glabe and Lennarz, 1979), and extension of the acrosomal tubule with its surrounding membrane destined to fuse with the egg (Trimmer and Vacquier, 1986).

In normal seawater, external Ca^{2+} and Na^{+} are needed for the AR (Dan, 1954; Collins and Epel, 1977; Schackmann and Shapiro, 1981). Na^{+} and Ca^{2+} entry, and H^{+} and K^{+} efflux, are activated within seconds of FSP binding to spermatozoa (Schackmann *et al.,* 1978; Garbers and Kopf, 1980; Schackmann and Shapiro, 1981; Garbers, 1989; Schackmann, 1989). As a consequence E_M (Schackmann *et al.,* 1984; González-Martínez and Darszon, 1987; García-Soto *et al.,* 1987), $[Ca^{2+}]_i$ (Trimmer *et al.,* 1986; Guerrero and Darszon, 1989a,b), and pH_i (Lee *et al.,* 1983; Guerrero and Darszon, 1989b) change. In addition, FSP increases [cAMP] (Garbers and Kopf, 1980), PKA (Garbers *et al.,* 1980; García-Soto *et al.,* 1991), nitric oxide synthase (Kuo *et al.,* 2000) and phospholipase D activity (Domino and Garbers, 1989), and inositol 1,4,5-trisphosphate levels. The increase in [cAMP] results from AC stimulation and was reported to precede the AR (Garbers and Kopf, 1980; Garbers, 1981), occurs in the head when triggered by A23187 or nigericin. It does, however, also depend on Ca^{2+} uptake (Watkins *et al.,* 1978). The relationship between the FSP-induced changes in permeability and second messenger levels is still unclear.

The receptor for the egg jelly factor that triggers the AR is considered to be a 210-kDa sperm plasma membrane protein (REJ) that has been cloned (Moy *et al.,* 1997). Purified REJ binds to FSP, and a monoclonal antibody (mAb) to it induces a sperm AR that can be blocked by the purified protein. The mAbs to REJ bind to a narrow collar of plasma membrane over the acrosome and also to the entire flagellum (Trimmer *et al.,* 1985). REJ has an epidermal growth factor (EGF) module, two C-type lectin carbohydrate-recognition modules and a novel 700-residue module, named REJ, that displays extensive homology with the human polycystic kidney disease protein, polycystin-1 (Moy *et al.,* 1996). This protein, encoded by the PKD1 gene, co-assembles at the plasma membrane with polycystin-2, the gene product of PKD2, to form a channel that is proposed to regulate renal tubular morphology and function (Hanaoka *et al.,* 2000). Autosomal dominant polycystic kidney disease is associated with mutations in either PKD1 or PKD2 (Arnaout, 2001). Polycystin-2 has sequence similarity to voltage-gated Ca^{2+} and other cation channels, especially within domains that form the pore (Nomura *et al.,* 1998). This protein itself may form Ca^{2+}-permeable nonselective channels (Hanaoka *et al.,* 2000; González-Perrett, *et al.,* 2001).

FSP induces a transient hyperpolarization followed by a depolarization in *Lytechinus pictus* spermatozoa. The K^+ dependence of the hyperpolarization suggests it is mediated by K^+ channels (González-Martínez and Darszon, 1987). Increasing $[K^+]$ in seawater from 10 to 40 m*M* also blocks Ca^{2+} uptake, the AR (Schackmann *et al.,* 1978), and the pH_i increase that accompanies the reaction (Guerrero and Darszon, 1989b). These results suggest that FSP increases pH_i, at least in part, by activating a Na^+/H^+ exchange stimulated by the hyperpolarization (González-Martínez *et al.,* 1992).

Ca^{2+} uptake and the AR are inhibited by antagonists of Ca^{2+} channels—verapamil and dihydropyridines (DHPs) (Schackmann *et al.,* 1978; Garbers and Kopf, 1980; Kazazoglou *et al.,* 1985; García-Soto and Darszon, 1985) and K^+ channels (TEA^+) (Shackmann *et al.,* 1978). These results demonstrate the crucial participation of ion channels in triggering ARs. The first K^+ single-channel recordings were achieved in bilayers made at the tip of patch pipettes with incorporated, purified plasma membranes from sperm flagella. Three types of monovalent cation channels were identified, with conductances of 22, 46, and 88 pS. Because two of them conducted K^+ and were blocked by TEA^+, they were classified as K^+ channels. Infrequently, a Cl^- channel of 148 pS was also detected (Liévano *et al.,* 1985). Despite of the size limitation, the presence of K^+ channels in spermatozoa was confirmed in patch-clamp recordings made directly on sea urchin sperm heads. Single-channel transitions of 40, 60, and 180 pS were observed (Guerrero *et al.,* 1987).

Experiments performed with fluorescent Ca^{2+}-sensitive dyes indicated the participation of two different Ca^{2+} channels in the sea urchin sperm AR (Guerrero and Darszon, 1989a,b; Schackmann, 1989). Binding of egg jelly to its receptor opens a Ca^{2+}-selective channel that inactivates; this channel can be blocked by trifluoperazine (TFP), verapamil, and DHPs. A second channel, insensitive to the later blockers, opens with a 5-second delay, does not inactivate, and allows Mn^{2+}

to permeate. The second channel and the AR are both inhibited by conditions that prevent the egg jelly-induced pH_i increase, leaving a transient rise in $[Ca^{2+}]_i$ due to the opening of the first channel. The operation of the two Ca^{2+} channels is linked; blocking the first channel inhibits the second. The pH_i increase induced by FSP is Ca^{2+} dependent (Guerrero *et al.,* 1998), and this could be the connection between the two Ca^{2+} channels, because the second one is regulated by pH_i (Guerrero and Darszon, 1989b). Another possible link between the two channels is Ca^{2+}-induced Ca^{2+} release (Berridge, 1993).

In many cells the emptying of Ca^{2+}-stores is coupled to the opening of store operated channels (SOCs) in the plasma membrane (Parekh and Penner, 1997; Barritt, 1999). Current results suggest that the second channel belongs to the family of store operated Ca^{2+} channels (SOCs) (González-Martínez *et al.,* 2001; Parekh and Penner, 1997), and may be important in the AR of many species (Santi *et al.,* 1998: O'Toole *et al.,* 2000). Since sperm do not have endoplasmic reticulum, the internal store is likely to be the acrosome. *S. purpuratus* sperm produce IP3 (Domino *et al.,* 1989) during AR and possess IP3 receptors (Zapata *et al.,* 1997).

Two types of Ca^{2+} channels have been dectected in planar lipid bilayers with fused plasma membranes isolated from *S. purpuratus* spermatozoa (10 m*M* $CaCl_2$) (Liévano *et al.,* 1990; Darszon *et al.,* 1994): The first is a 50 pS voltage-dependent channel that has not been characterized; the second is a voltage-dependent channel having a high main conducting state of 172 pS and several subconductive states. The latter channel discriminates poorly between divalent and monovalent cations ($PCa^{2+}/PNa^{+} = 5.9$), is insensitive to verapamil and nisoldipine, and is blocked by Cd^{2+} and Co^{2+} at concentrations similar to those required to inhibit AR.

Ca^{2+}-selective ($PBa^{2+}/PK^{+} = \sim 4$), high conductance, multi-state, voltage-dependent channels similar to the one just described in *S. purpuratus* sperm membranes have been transferred to planar lipid bilayers directly using spermatozoa from sea urchin and mouse. Interestingly, the properties of these channels resemble some of those displayed by the PKDL and PKD1-PKD2 channels (Beltrán *et al.,* 1994).

A 150-pS anion-selective channel ($NO^{2-} > SCN^{-} > Br^{-} > Cl^{-}$) has also been recorded in planar lipid bilayers with incorporated sperm plasma membranes or vesicles formed from an enriched preparation of REJ. The channel has a high open probability at the holding potentials tested, often displays substates, and is partially blocked by 4,4′-diisothiocyanatostilbene-2,2′-disulfonic acid (DIDS) (see Figure 7.2). DIDS blocks the AR in *S. purpuratus* sea urchin spermatozoa, indicating the possible involvement of Cl^{-} channels in the AR (Morales *et al.,* 1993).

Acrosome reaction inactivation (ARI) is a process that renders spermatozoa irreversibly refractory to the egg jelly. This process, triggered by the egg jelly, is associated with an increase in $[Ca^{2+}]_i$. However, a rise in $[Ca^{2+}]_i$ alone is not sufficient to induce ARI, because artificially increasing $[Ca^{2+}]_i$ with an ionophore or by rising pH_e does not trigger ARI. In contrast to the AR, which strictly requires Ca^{2+}, ARI can be triggered almost equally well by Sr^{2+}. On the other hand, Mn^{2+} inhibits ARI, although it does not affect the AR. Thus the mechanisms involved in ARI differ from those leading to the AR. High pH_e can trigger the AR in previously inactivated spermatozoa by opening the same Ca^{2+} channels activated by the egg jelly. ARI requires egg jelly receptor activation, and originates from the uncoupling between the egg jelly receptor and the Ca^{2+} channels (Guerrero *et al.,* 1998).

Many important questions about the sea urchin sperm ARs remain unanswered. How does the egg jelly receptor orchestrate ion transport? Although G-proteins are present in spermatozoa, there is no evidence of their participation in the AR. Why is Na^+/H^+ exchange stimulated by a hyperpolarization, and what are the identities and characteristics of the proteins that perform this transport?. Why must pH_i increase for the AR to occur, to open the high-conductance pH-sensitive Ca^{2+} channel, to hyperpolarize, or to activate proteases (Farach *et al.,* 1987; Matsumara and Aketa, 1990), AC, some kinases, or phosphatase? Are the increases in cAMP and $InsP_3$ related to the permeability changes that occur during the AR? The cAMP-regulated channel described earlier may participate in the AR (Labarca *et al.,* 1995). Because the sea urchin sperm AC is modulated by $[Ca^{2+}]_i$, pH_i, and membrane potential, it could be a coincidence detector involved in the AR (Beltrán *et al.,* 1995). A summary of the events and possible mechanisms is presented in Figure 7.2.

Recently, low-density lipid rafts were isolated from sea urchin sperm. SuREJ1, the speract receptor, a 63 kDa glycosylphosphatidylinositol (GPI)-anchored protein, Gs, AC, GC and PKA were found in them. Only the speract receptor, the GPI-anchored protein and Gs, co-immunoprecipitated, suggesting their functional association in the speract response (Ohta *et al.,* 2000). The characterization of signal transduction microdomains will hopefully shed light on sperm function.

B. STARFISH

The starfish egg jelly contains three biologically active compounds necessary to trigger the AR: (1) acrosome reaction-inducing substance (ARIS), a high-molecular-weight fucose-sulfated glycoprotein, (2) Co-ARIS, a diffusible, non-species-specific sulfated steroid (Matsui *et al.,* 1986b; Nishiyama *et al.,* 1987a,b), and (3) asterosaps, a 34 amino acid peptide containing an intramolecular disulfate bond essential for function (Nishigaki *et al.,* 1996). Crude jelly stimulates Ca^{2+} influx, modulates [cAMP], increases pH_i in a Na^+-dependent fashion (Tubb *et al.,* 1979; Matsui *et al.,* 1986a,c; Hoshi *et al.,* 1990, 1991), and leads to sperm histone degradation (Amano *et al.,* 1992a,b). In seawater containing >10 m*M* $CaCl_2$ and at $pH > 8.0$, ARIS induces AR species specifically (Ikadai and Hoshi 1981a,b; Matsui *et al.,* 1986a). Its carbohydrate and sulfate moieties are believed to be responsible for the biological activity and species specificity (Matsui *et al.,* 1986a,b; Okinaga *et al.,* 1992). In *Asterias amurensis,* a pentasaccharide repeat containing xylose, sufated fucose and galactose is the AR inducer (Koyota *et al.,* 1997). Variations in the fine structure of sulfated polysaccharides in EJ contribute to species specificity of fertilization in marine animals. Starfish spermatozoa have species-specific receptors for ARIS on their head region (Ushiyama *et al.,* 1993; Longo *et al.,* 1995).

In normal seawater ARIS and Co-ARIS or ARIS and asterosap are required for induction of the AR. A sperm chemoattractant from the starfish *Pycnopodia helianthoides* shows high homology with asterosap (Miller and Vogt, 1996), suggest-

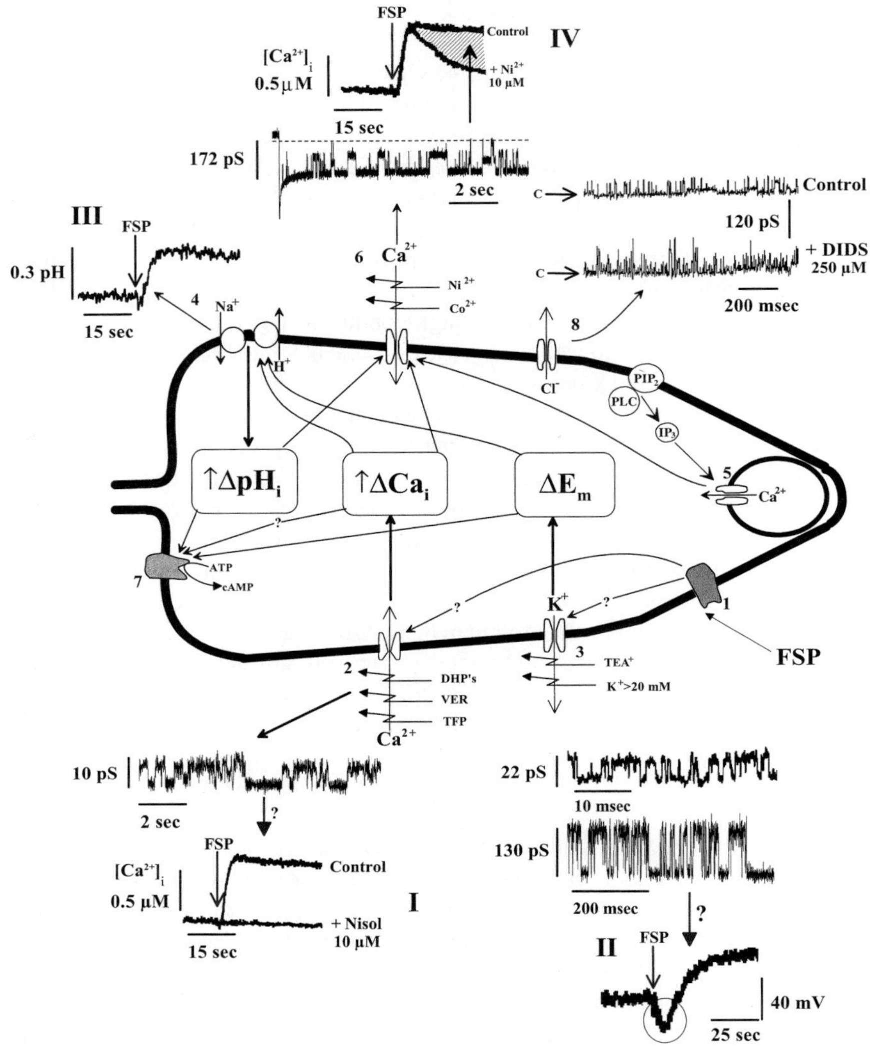

FIGURE 7.2 Possible mechanisms involved in the sea urchin sperm acrosome reaction. It is not known how binding of egg factor fucose sulfate polymer (FSP) to the sperm receptor (1) leads to the opening of a Ca^{2+} channel (2), which inactivates. This channel is sensitive to dihydropyridines (DHPs), verapamil (VER), and trifluoperazine (TFP). Activation of this channel initiates $[Ca^{2+}]_i$ (I) elevation, sensitive to nisoldipine (Nisol). Simultaneously or immediately after the first Ca^{2+} channel (2) opens, a K^+ channel (3) is activated, hyperpolarizing *L. pictus* spermatozoa (II, circle) and stimulating a voltage- and Ca^{2+}-dependent Na^+/H^+ exchange (4) that increases pH_i (III). FSP also increases the sperm levels of inositol triphosphate (IP_3), which may release Ca^{2+} from the acrosome (5) and signal a store-operated Ca^{2+} channel (SOC) (6) (González-Martínez *et al.,* 2001). The alkalinization modulates the opening of the second Ca^{2+} channel (6), and further depolarizes the cell. The FSP-induced hyperpolarization ($-\Delta E_M$) and the increases in $[Ca^{2+}]_i$ and pH_i activate the sperm adenylyl cyclase (7), mainly found in the flagella but also present in the head, where cAMP elevation may regulate various channels. DIDS-sensitive Cl^- channels (8) may participate in setting resting E_M. Although the shown single-channel records from sperm membranes incorporated into planar lipid bilayers may arise from channels responsible for increasing $[Ca^{2+}]_i$ (I, IV) and changing E_M (II) during the acrosome reaction, this has not been demonstrated.

ing it may be a potent chemoattractant. Under physiological conditions, antiasterosap rabbit IgG neutralizes the ability of egg jelly to induce the AR, thus asterosap is important for this reaction. The asterosap receptor is believed to be a 130-kDa guanylyl cyclase. There are $\sim 10^5$ receptors/sperm, mostly localized in the tail (Nishigaki *et al.,* 1996).

ARIS and Co-ARIS cannot cause detectable changes in sperm $[Ca^{2+}]_i$ or pH_i by themselves. However, a mixture of the compounds increases $[Ca^{2+}]_i$, induces AR, and results in a pH_i decrease (Matsui *et al.,* 1986a,c; Hoshi *et al.,* 1990). Apparently starfish ARs do not require a pH_i increase, whereas sea urchin ARs do.

Maitotoxin, a marine toxin that may activate Ca^{2+} channels, induces the AR in starfish *Asterina pectinifera* spermatozoa, suggesting the participation of these channels in the reaction (Amano *et al.,* 1993). This effect depends on external Ca^{2+} and is blocked by the Ca^{2+} channel blocker verapamil. On the other hand, increasing K^+ to 30 m*M* KCl in seawater inhibits ARs, indicating the possible participation of K^+ channels in the starfish AR.

C. MAMMALS

The zona pellucida (ZP) is believed to be the main mediator of the mammalian sperm AR. The ZP consists mainly of two to four sulfated glycoproteins, depending on the species. In the mouse, ZP3 (83 kDa) displays most of the sperm-binding and AR-inducing activity of unfertilized eggs (Bleil, 1991; Wassarman *et al.,* 2001) (see Chapters 9 and 10, this volume).

Species-specific gamete interactions are required to attain ARs, suggesting the presence of specialized sperm receptors. Numerous candidate proteins for ZP3 have been proposed: in mouse spermatozoa, a 95-kDa tyrosine kinase (Leyton *et al.,* 1992), a (β1–4)galactosyltransferase (GalTase-R) (Gong *et al.,* 1995), and a lectin sp56 (Bookbinder *et al.,* 1995); and in guinea pig spermatozoa, a hyaluronidase (Gmachl and Kreil, 1993). Also trypsinlike proteins (Boettger-Tong *et al.,* 1993) and spermadhesins (Hardy and Garbers, 1995; Gao and Garbers, 1998) have been proposed as receptors (Storey, 1995; McLeskey *et al.,* 1998). An active discussion of the physiological relevance of many of these candidates has been undertaken (McLeskey *et al.,* 1998; Wassarman *et al.,* 2001).

Why would several ZP3 sperm receptors be needed to induce ARs? Multiple concerted and cooperative interactions between ZP3 and sperm surface components, possibly involving receptor aggregation, may be required for transduction leading to ARs (Bleil and Wassarman, 1983; Kopf *et al.,* 1989; Leyton and Saling, 1989; Boettger-Tong *et al.,* 1992; Aarons *et al.,* 1992; Gong *et al.,* 1995).

The ZP-induced AR requires external Ca^{2+} (Yanagimachi, 1988). In mature spermatozoa solubilized ZP raises pH_i and $[Ca^{2+}]_i$, and leads to acrosomal exocytosis (Florman and First, 1988; Florman *et al.,* 1989). In a single spermatozoon loaded with fluorescent ion indicator dyes, ZP increases $[Ca^{2+}]_i$ before exocytosis occurs (Florman *et al.,* 1989; Storey *et al.,* 1992; Florman, 1994). Multiple G-proteins, such as G_i, and G_z, have been detected in mammalian spermatozoa (Glassner *et al.,* 1991). The ZP-induced AR and its associated ion fluxes are inhibited by pertussis toxin (PTX), a specific inactivator of the G_i class of heterotrimeric G-proteins, in mouse, bovine, and human spermatozoa (Endo *et al.,* 1987, 1988; Florman *et al.,* 1989; Lee *et al.,* 1992). ZP activates G_{i1} and G_{i2} in mouse spermatozoa (Ward *et al.,* 1994), and ion channels are regulated by G-proteins (Wikman and Clapham, 1995). It has been shown that the pH_i increase necessary for ZP-induced AR is the PTX-sensitive step (Arnoult *et al.,* 1996a). It is now necessary to determine which plasma membrane proteins interact with G_i to learn how pH_i is regulated during the AR. It is noteworthy that GalTase-R has been shown to interact with G_i (Gong *et al.,* 1995).

Besides ZP3, other agonists, including progesterone (Thomas and Meizel, 1989; Blackmore *et al.,* 1990; Baldi *et al.,* 1991; Meizel *et al.,* 1997), γ-aminobutyric acid (GABA) (Wistrom and Meizel, 1993; Roldán *et al.,* 1994; Shi *et al.,* 1997), glycine (Meléndrez and Meizel, 1995), EGF (Lax *et al.,* 1994), ATP (Foresta *et al.,* 1996), and the platelet-activating factor (PAF) (Sengoku *et al.,* 1996), have been shown to induce ARs. What is the physiological relevance of these "alternative" pathways? As has been proposed for progesterone, they may enhance capacitation (DasGupta *et al.,* 1994; Barboni *et al.,* 1995), potentiate the ZP-induced AR (Roldan *et al.,* 1994), promote sperm hyperactivation (a motility state important for fertilization), and/or induce chemotaxis (Villanueva-Diaz *et al.,* 1995; Eisenbach, 1999). On the other hand, some of these transduction mechanisms could be vestiges from previous differentiation stages.

In the central nervous system progesterone metabolites enhance the interaction of GABA with the GABA receptor (GABA-R). This receptor is a multisubunit protein containing a Cl^- channel (Rabow *et al.,* 1995). The GABA-R has been detected in boar and ram spermatozoa (Erdo and Werkele, 1990). The human sperm responses triggered by progestins appear to involve steroid interaction with a sperm steroid receptor/Cl^- channel complex, similar to, but distinct from, the $GABA_A$/Cl^- channel complex (Wistrom and Meizel, 1993). GABA or glycine induces the AR in human, mouse, and porcine spermatozoa, and antagonists to their putative ligand-gated Cl^- channels inhibit it (Meizel, 1997). The ZP-induced AR is inhibited in sperm from mice defective in the glycine receptor/Cl^- channel (Sato *et al.,* 2000).

Residue phosphorylation is a ubiquitous mechanism used by cells to regulate protein function, and thus ion channel activity (Hille, 1992). During capacitation and ZP- or progesterone-induced ARs, a set of proteins between 20 and 220 kDa are phosphorylated (Visconti *et al.,* 1995a; Baldi *et al.,* 1996, 2000; Naz, 1996). In human spermatozoa the progesterone-induced AR is accompanied by a large extracellular-dependent increase in $[Ca^{2+}]_i$ (Thomas and Meizel, 1989; Blackmore *et al.,* 1990; Baldi *et al.,* 1991), and by Cl^- efflux (Turner and Meizel, 1995; Meizel and Turner, 1996; Sabeur *et al.,* 1996). The progesterone-induced $[Ca^{2+}]_i$ increase has a rapid rising phase and a long-lasting plateau phase, and is affected by tyrosine kinase inhibitors (Bonaccorsi *et al.,* 1995; Meizel and Turner, 1996; Kirkman-Brown *et al.,* 2000). These inhibitors affect the plateau phase and not the rising one (Bonaccorsi *et al.,* 1995). Nevertheless, Mendoza *et al.* (1995) found that genistein, a tyrosine kinase inhibitor, had no effect on $[Ca^{2+}]_i$ changes induced by progesterone. The progesterone-induced $[Ca^{2+}]_i$ rise and the AR are not sensitive to

PTX, implying a different transduction path and Ca^{2+} uptake mechanism than those involved in the ZP-triggered AR (Tesarik *et al.*, 1993a; Murase and Roldán, 1996).

Alterations in phospholipid and cAMP metabolism have been shown to occur during the induction of the AR by the ZP, by progesterone, and by nonphysiological agents such as the Ca^{2+} ionophore A23187 (Thomas and Meizel, 1989; Kopf and Gerton, 1990; Fraser and Monks, 1990; Florman and Babcock, 1991; Roldan *et al.*, 1994; Baldi *et al.*, 2000). These pathways may regulate PKC and PKA activity, leading to phosphorylation of several proteins during the AR (Mendoza *et al.*, 1995; Doherty *et al.*, 1995; Baldi *et al.*, 1996). PKCα and PKCβII have been immunolocalized in the equatorial segment of human spermatozoa (Rotem *et al.*, 1992). The time courses of the ZP-triggered AR and the cell distribution of PKC are modified by biologically active phorbol diesters and diacylglycerols (Lee *et al.*, 1987; Endo *et al.*, 1991; Lax *et al.*, 1997). Moreover, PKC translocation from cytosol to the plasma membrane is dependent on $[Ca^{2+}]_e$ (Lax *et al.*, 1997). Like the AR, the progesterone- and the ZP-induced Ca^{2+} permeability changes are also sensitive either to PKC inhibitors in human spermatozoa (Foresta *et al.*, 1995) or to PKC and PKA inhibitors in plasma membrane vesicles, and in isolated acrosomes of bovine spermatozoa (Spungin and Breitbart, 1996; Breitbart and Spungin, 1997). Agonists for these kinases bypass the need for $[Ca^{2+}]_e$ in the induction of the AR, especially when combined (Mendoza *et al.*, 1995; Doherty *et al.*, 1995). These results may indicate that $[Ca^{2+}]_i$ rises stimulate these kinases during intermediate steps of the phyiologically relevant AR. Artificial activation of the kinases removes the $[Ca^{2+}]_e$ requirements for the final stages, when membrane fusion occurs.

Planar bilayer experiments with incorporated mammalian sperm plasma membranes have indicated the presence of several types of divalent permeable channels (10–20 and 50–60 pS) (Cox and Peterson, 1989; Cox *et al.*, 1991; Chan *et al.*, 1997). As indicated before, ion channels can be transferred to planar lipid bilayers directly using mouse spermatozoa. A high-conductance, voltage-dependent poorly Ca^{2+}-selective channel, similar to the one described from sea urchin sperm membranes, was detected using this approach. This channel must be important because it is present in such diverse species, and could participate in the AR (Beltrán *et al.*, 1994).

Tiwari-Woodruff *et al.* (1994) characterized an interesting dihydropyridine-sensitive, 10-pS Ca^{2+} channel from boar sperm membranes. This channel may be involved in the AR, although it did not display the expected voltage dependence of VDCCs. Similar experiments with mouse sperm plasma membranes revealed the presence of (1) an 80-pS anion channel, (2) a cation channel (P_{Na^+}/P_{K^+} = 2.5) with two modes of gating, and (3) the high-conductance Ca^{2+} channel described above (Figure 7.3) (Labarca *et al.* 1995). This latter channel was blocked by micromolar concentrations of ruthenium red, which inhibits the AR in sea urchin spermatozoa (Labarca *et al.*, 1995).

Voltage-dependent Ca^{2+} channels have been shown to be present in the plasma membrane of mammalian spermatozoa. Bull (Florman and Babcock, 1991) and ram (Babcock and Pfeiffer, 1987) spermatozoa undergo dihydropyridine-, benzothiazepine-, and phenylalkylamine-sensitive $[Ca^{2+}]_i$ increases, which depend

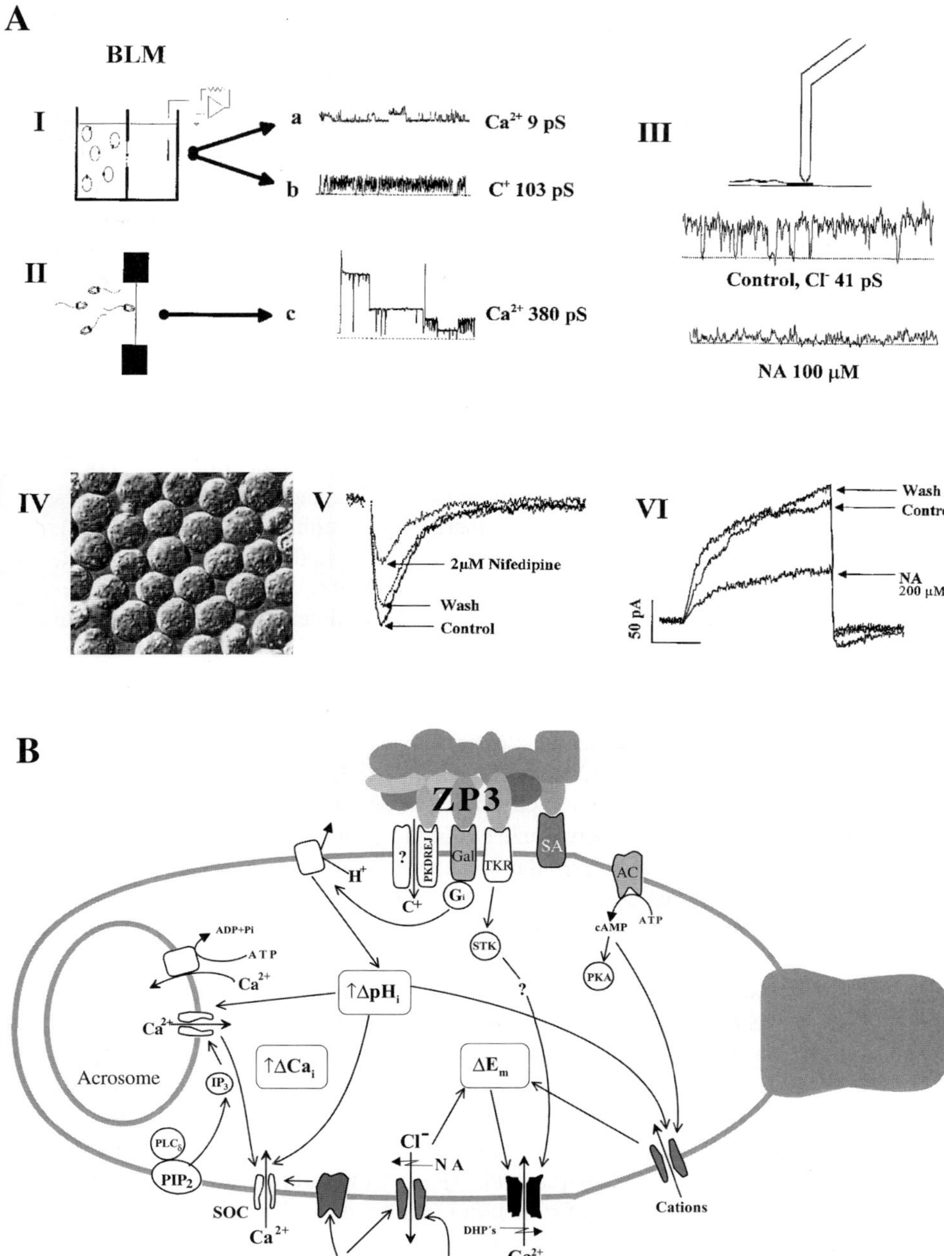

FIGURE 7.3 Mammalian sperm ion channels (A) and their possible relation to the acrosome reaction (B). (A) Ion channels can be transferred from either purified sperm plasma membranes (I) or directly from intact spermatozoa (II) under fusogenic conditions. In the recordings, dotted baselines indicate the closed level of the channel. (Ia) A small-conductance Ca^{2+} channel from boar sperm plasma

on $[Ca^{2+}]_e$ and pH_e, when depolarized by $[K^+]_e$. Both species have modestly high affinity binding sites for PN200-110 ($K_D \sim 0.4\ \mu M$), an L-type VDCC antagonist. Experiments in which the AR is induced by a combined elevation of pH_i and a K^+-induced depolarization (Florman *et al.*, 1992; Arnoult *et al.*, 1996a) are consistent with the participation of VDCCs during ARs, and suggest the presence of K^+ channels in the mammalian sperm plasma membrane. K^+-selective and TEA^+-sensitive channels have been observed in bilayers containing rat sperm plasma membranes (Chan *et al.*, 1997).

The molecular identity of the Ca^{2+} channel involved in the AR is still unknown (Florman *et al.*, 1998; Benoff, 1998; Darszon *et al.*, 1999). Though submicromolar concentrations of DHPs block typical L-type Ca^{2+} channels, micromolar concentrations of them are required to inhibit the AR and the increase in $[Ca^{2+}]_i$. As described in Section V, mouse pachytene spermatocytes and round spermatids display basically only T-type Ca^{2+} currents (Hagiwara and Kawa, 1984; Liévano *et al.*, 1996; Santi *et al.*, 1996; Arnoult *et al.*, 1996b). Similar micromolar concentrations of DHPs, pimozide, and Ni^{2+} block the T-type channels from spermatogenic cells and inhibit both the AR and the increase in $[Ca^{2+}]_i$ associated with this reaction (Arnoult *et al.*, 1996a,b). Thus, it is likely that T-type Ca^{2+} channels participate in the ZP3-induced increase in mouse sperm $[Ca^{2+}]_i$, leading to the AR (Florman *et al.*, 1998; Darszon *et al.*, 2001). New evidence suggests that N-type Ca^{2+} channels may be present in mouse spermatogenic cells and mature sperm, making a minor contribution to the Ca^{2+} currents in these cells (Wennemuth *et al.*, 2000).

Ion-selective fluorescent probes have allowed the dissection of two phases of the ZP3-induced increase in $[Ca^{2+}]_i$. The first phase is a fast, transient elevation of

---

membranes (modified from Tiwari-Woodruff *et al.*, 1994). (Ib) A nonselective cationic channel from mouse sperm plasma membranes (modified from Labarca *et al.*, 1995). (IIc) A large-conductance Ca^{2+} channel from mouse spermatozoa (modified from Beltrán *et al.*, 1994). (III) Mouse sperm patch-clamp recordings (top diagram) of a niflumic acid (NA)-sensitive Cl^- channel. Whole-cell recordings from pachytene spermatocytes (~16 μm diameter, IV) showing nifedipine-sensitive T-type Ca^{2+} currents (V) (modified from Liévano *et al.*, 1996), and niflumic acid-sensitive outward Cl^- currents (VI) (Espinosa *et al.*, 1997). The (B) sperm acrosome reaction initiates when egg ZP3 binds to sperm receptors, which may have to aggregate. Four receptor candidates are illustrated: GalTase (Gal), PKREJ, a 95-kDa tyrosine kinase receptor (TKR), and a spermadhesin (SA). In addition, specific receptors for progesterone and GABA, which may mediate the AR, are considered. The stimulated receptors can apparently activate several targets: (1) G_i proteins sensitive to PTX that regulate pH_i acting directly or indirectly through a H^+ transporter; (2) open channels that depolarize the cell—both ZP3 and progesterone have been shown to induce sperm depolarization [Cl^- fluxes via $GABA_A$-type receptors or other Cl^- channels may contribute to this membrane potential (E_M) change; the ZP3-induced depolarization would open a T-type VDCC responsible for a fast transient (50–100 msec) increase in $[Ca^{2+}]_i$]; (3) phospholipase Cδ (PLCδ), resulting in IP_3 production and release of Ca^{2+} from the acrosome (Fukami *et al.*, 2001). Ca^{2+} depletion from the acrosome would open a SOC, necessary to maintain elevated $[Ca^{2+}]_i$ and achieve AR. The increase in pH_i might enhance the IP_3 sensitivity of its receptor and also modulate SOCs. Changes in $[Ca^{2+}]_i$, pH_i and E_M may also coordinately regulate adenylyl cyclase (AC). The elevated levels of second messengers such as $[Ca^{2+}]_i$ and cAMP may modulate plasma membrane and acrosome ion channels directly, or through cAMP-dependent protein kinases, tyrosine kinases, and/or protein kinase C. The interplay between $[Ca^{2+}]_i$, pH_i, kinases, phosphatases, and ion channels may be anticipated to regulate membrane fusion, culminating in the AR.

$[Ca^{2+}]_i$ that occurs within 40–50 msec, to values of approximately 10 μM, relaxing to resting values within the next 200 msec (Arnoult *et al.,* 1999). The kinetics of activation and inactivation and pharmacology of this transient elevation are consistent with the properties of low voltage-activated (LVA), T-type Ca^{2+} channels (Arnoult *et al.,* 1999). The second phase, a much slower and sustained elevation in $[Ca^{2+}]_i$, follows the ZP3-induced fast transitory response. Many seconds to minutes are needed to develop this slow response and the AR occurs only after a high, sustained $[Ca^{2+}]_i$ level is reached (Arnoult *et al.,* 1996a,b). The kinetics of the slow, sustained elevation in $[Ca^{2+}]_i$ are incompatible with the properties of T-type Ca^{2+} channels (Bean and McDonough, 1998); therefore, at least another pathway for Ca^{2+} is necessary to trigger the AR. LVA Ca^{2+} channel blockers added before ZP3 also inhibit the sustained elevation in $[Ca^{2+}]_i$ (Arnoult *et al.,* 1996a). These results indicate that the transient increase in $[Ca^{2+}]_i$ due to the ZP-induced activation of T-type channels is necessary to open a second Ca^{2+} pathway that keeps $[Ca^{2+}]_i$ elevated enough to allow the AR (Florman *et al.,* 1998; Darszon *et al.,* 1999).

The T-type Ca^{2+} channels are likely to be inactivated at the resting potential of capacitated sperm cells (−55 mV) (Zeng *et al.,* 1995), thus it is not clear if they could be opened by a depolarization. As stated above, mouse, bull, and human spermatozoa do not undergo ARs when depolarized with K^+ unless external or internal pH_i is raised (Florman *et al.,* 1992; Arnoult *et al.,* 1996a). Actually, the fact that human cells do not undergo dihydropyridine-sensitive $[Ca^{2+}]_i$ increases in response to K^+, or using agonists that induce a Na^+-dependent depolarization, has misled researchers to conclude that the cells lack VDCCs (Foresta and Rossato, 1997). Possibly a hyperpolarization is required, as in sea urchin spermatozoa (González-Martínez *et al.,* 1992), to remove Ca^{2+} channel inactivation (Liévano *et al.,* 1996). This hyperpolarization has not yet been detected, however. In addition to K^+ channels, mouse spermatozoa may also posses H^+ channels (Zeng *et al.,* 1995) that could contribute to hyperpolarization as ZP increases pH_i. Other regulatory mechanisms could also be involved.

Once VDCCs are ready to open, how does ZP depolarize spermatozoa? Arnoult *et al.* (1996a) have shown that homologous ZP or ZP3 will induce a 30-mV depolarization mediated by a nonselective cation channel in bovine or mouse sperm cells. This latter depolarization seems too slow to activate T-type Ca^{2+} channels. Planar bilayer (Labarca *et al.,* 1995; Chan *et al.,* 1997) (see Figure 7.3) and patch-clamp studies (Espinosa *et al.,* 1997) have revealed the presence of poorly selective cation channels in mammalian spermatozoa, which could participate in this depolarization. A homologue of REJ, PKDREJ, is express only in mouse and human testis, in a pattern that coincides with sperm maturation (Hughes *et al.,* 1999). PKDREJ has sequence similarity with the membrane-associated region of polycystin-1 which forms ion channels with polycystin-2 (Hanaoka *et al.,* 2000). Thus, the PKDREJ protein could be part of a ligand (ZP3) gated channel that initiates AR by depolarizing sperm. Activation of anion channels would be an alternative since the Cl^- equilibrium potential is ~−17−−30 mV (Garcia and Meizel, 1999). Another possibility, if present in mature spermatozoa, two newly cloned channels (mSlo3 and hHCN4) from testis mRNA could be candidates to accomplish a ZP3-induced depolarization. mSlo3, a channel with extensive sequence similarity to large-conductance K^+ channels activated by Ca^{2+} and voltage, is sensitive to pH and is not modulated by Ca^{2+} (Schreiber *et al.*, 1998). It is possible that the ZP-induced pH_i increase opens mSlo3, if its voltage dependence in mature spermatozoa is different from the one it displays when heterologously expressed (Schreiber *et al.*, 1998). Nevertheless, Arnoult *et al.* (1996a) reported that even though the ZP-induced pH_i increase is inhibited by PTX, the depolarization is not. hHCN4 is a homolog of sea urchin sperm SPIH. This channel belongs to the pacemaker channel family, members of which are activated by hyperpolarization and regulated directly by cyclic nucleotides (Seifert *et al.*, 1999).

The question then is, after the T-type Ca^{2+} channel opens transiently (Santi *et al.*, 1996; Bean and McDonough, 1998), which channel is responsible for the sustained $[Ca^{2+}]_i$ elevation? It has been reported that a pH_i-sensitive Ca^{2+} channel is present in the plasma membrane of mouse spermatogenic cells, as well as in immotile testicular spermatozoa. These cells also posses a store-operated Ca^{2+} channel, possibly the pH_i-sensitive channel (Santi *et al.*, 1998). This channel could be responsible for the sustained $[Ca]_i$ elevation necessary for the AR (Darszon *et al.*, 1999; O'Toole *et al.*, 2000). Figure 7.3B illustrates a working model of the induction of mammalian sperm ARs.

Up to now the only means to elevate $[Ca^{2+}]_i$ during the sperm AR that has been discussed is influx of external Ca^{2+} through plasma membrane channels. Nevertheless, compounds known to release Ca^{2+} from internal stores, such as thapsigargin, induce ARs in human, mouse, and hamster spermatozoa (Meizel and Turner, 1993; Blackmore, 1993; Walensky and Snyder, 1995; Llanos, 1998). Interestingly this response is dependent on $[Ca^{2+}]_e$, suggesting cross-talk between internal and external Ca^{2+} pathways. Alkaline pH_i favors Ca^{2+} release through $InsP_3R$, adding to the possible modulation pathways of $[Ca^{2+}]_i$ rise in spermatozoa (Berridge, 1993). Thapsigargin and $InsP_3$ (but not caffeine) preclude ATP-dependent $^{45}Ca^{2+}$ uptake in permeablized spermatozoa, as well as in isolated acrosomes (Walensky and Snyder, 1995; Spungin and Breitbart; 1996). The putative $InsP_3$-induced Ca^{2+} release from isolated acrosomes is sensitive to H89, a PKA inhibitor, suggesting that the $InsP_3R$ is regulated by that kinase (Spungin and Breitbart, 1996; Breitbart and Spungin, 1997). As previously mentioned, $InsP_3$ receptors have been selectively immunolocalized to the acrosomal cap of mature mouse and human spermatozoa (Walensky and Snyder, 1995; Treviño *et al.*, 1998), and in the acrosome, postacrosome, and along the tail in bull and ram spermatozoa (Dragileva *et al.*, 1999).

Mature mouse sperm do undergo a thapsigargin stimulated Ca^{2+} uptake that has similar kinetics and sensitivity to Ni^{2+} and DHPs as the second phase of Ca^{2+} influx induced by ZP3 (O'Toole *et al.*, 2000). Thus, the sustained Ca^{2+} uptake required for the AR occurs through SOCs Some of the transient receptor potential (*trp*) gene products code for SOCs (Harteneck *et al.*, 2000). The seven *trp* mamalian homologues so far identified are present in spermatogenic cells (Vannier *et al.,* 1999; Treviño, unpublished). Furthermore, new results indicate that *trp2* regulates the ZP3-induced Ca^{2+} influx in mouse sperm (Jungnickel *et al.,* 2001).

V. SPERMATOGENIC CELLS, A NEW TOOL TO STUDY SPERM ION CHANNELS

Spermatozoa are very small differentiated terminal cells unable to make proteins. Because of this their ion channels must be synthesized during spermatogenesis. Being easier to patch-clamp than sperm cells, spermatocytes are also a promising model to investigate ion channel function during spermatogenesis and to define the pharmacology and regulation of sperm ion channels. Ca^{2+} currents change during rat spermatogenesis (Hagiwara and Kawa, 1984), but little is known about the role of ion channels in this differentiation process.

Because Ca^{2+} channels are fundamental in sperm physiology, their genotypic and phenotypic expression is being studied in the late differentiation stages using the mouse model. Initial experiments with oligonucleotide probes to the subunit (α_1), which contains the pore and the voltage sensor of the various voltage-dependent Ca^{2+} channels (L, T, P/Q, and R), indicated that only α_{1E}, and to a much lesser extent α_{1A}, transcripts were present in pachytene spermatocytes and in round and condensed spermatids (Liévano *et al.,* 1996). Using similar strategies additional studies revealed that transcripts for α_{1C} (Benoff, 1998; Espinosa *et al.,* 1999), α_{1G} and α_{1H} (Espinosa *et al.,* 1999) are also present in spermatogenic cells. α_{1A}, α_{1C}, and α_{1E} were immunodetected in mature spermatozoa (Benoff, 1998; Westenbroeck and Babcock, 1999; Serrano *et al.,* 1999). In addition to α_1, Ca^{2+} channels are formed by auxiliary subunits $\alpha_{2\delta}$ and β (Felix, 1999). The presence of the four known genes encoding the β subunits in mouse spermatogenic cells was demonstrated using reverse transcriptase–polymerase chain reaction (RT–PCR). Immunolocalization studies detected $\beta1$, $\beta2$, and $\beta3$ in these cells and in mature spermatozoa (Serrano *et al.,* 1999).

Mainly low voltage-activated Ca^{2+} channels of the T-type have been found in spermatogenic cells (Liévano *et al.,* 1996; Santi *et al.,* 1996; Arnoult *et al.,* 1996b, 1997, 1999). These Ca^{2+} currents are sensitive to micromolar concentrations of nifedipine, Ni^{2+}, amiloride, and pimozide (see Figure 7.3). Because the mouse sperm AR and the uptake of Ca^{2+} that triggers it are also inhibited by these blockers, it is likely that a T-type Ca^{2+} channel is involved in inducing this reaction (Arnoult *et al.,* 1996b, 1999). In dissociated mouse pachytene spermatocytes and round spermatids, the T-type Ca^{2+} currents have been reported to be positively modulated by dephosphorylation and albumin and negatively modulated by tyrosine-dependent phosphorylation (Arnoult *et al.,* 1997; Espinosa *et al.,* 2000; see Figure 7.3). Recent patch-clamp recordings in spermatogenic cells suggest that N-type Ca^{2+} channels may also contribute to Ca^{2+} influx in these cells. In agreement, α_{1B} subunits were immunodetected in rodent sperm membranes (Wennemuth *et al.,* 2000). Detection of the messenger is still pending.

Rises in $[Ca^{2+}]_i$ are key signals for cell division, differentiation, and maturation. Similarly, they appear to be important for the unique processes of meiosis and spermatogenesis, carried out exclusively by male germ cells. Intracellular alkalinization and rises of $[Ca^{2+}]_i$ are also important during capacitation and the AR in mammalian spermatozoa. Experiments in individual spermatogenic cells loaded with Ca^{2+} and pH_i-sensitive fluorescent dyes have indicated progressively higher resting $[Ca^{2+}]_i$ in cells at more advanced stages of maturation. Brief applications of 25 m*M* NH_4Cl that increase pH_i produced an initial drop in $[Ca^{2+}]_i$ followed by a two- to threefold transient rise (Santi *et al.,* 1998). These $[Ca^{2+}]_i$ rises are due to Ca^{2+} influx because they are basically abolished in the absence of external Ca^{2+}. No significant contribution of Ca^{2+} release from intracellular stores was detected. Alkalinization-induced Ca^{2+} influx was inhibited by 0.2 and 1 m*M* Ni^{2+} but was insensitive to nifedipine at concentrations up to 20 μ*M*. This Ca^{2+} influx pathway was also permeable to Sr^{2+}, Ba^{2+}, and Mn^{2+}. Ca^{2+} transients potentiated with repeated NH_4Cl applications. Experiments with thapsigargin and cyclopiazonic acid suggest that this novel pH-dependent Ca^{2+} permeation pathway may correspond to a SOC. This channel is also present in testicular spermatozoa (Santi *et al.,* 1998).

Considering the possible role of SOCs in the sperm AR, the presence of Ca^{2+} release channels from intracellular stores has been analyzed in different stages of spermatogenic cell differentiation (Treviño *et al.,* 1998). Messenger RNAs for the three $InsP_3R$ subtypes (I, II, and III) were detected in spermatogonia as well as in all subsequent stages of spermatogenesis. Immunolocalization studies revealed that $InsP_3$ receptors are homogeneously distributed throughout the cytoplasm at early stages and become selectively localized to the Golgi complex as differentiation proceeds. Consistent with this distribution pattern, spermatogonia underwent a large intracellular Ca^{2+} release in response to Ca^{2+}-ATPase inhibitor thapsigargin, whereas smaller responses were detected in late spermatocytes and spermatids (Treviño *et al.,* 1998).

The three known genes (I, II, and III) encoding ryanodine receptors were also found to be expressed at all stages of spermatogenesis. However, experiments with specific antibodies for each of the RyR subtypes indicate that only types I and III are present in spermatogenic cells. RyRs remain homogeneously scattered in the cytoplasm at all stages of differentiation, in contrast to $InsP_3$ receptors, which undergo a dramatic subcellular redistribution. Caffeine and ryanodine did not induce any responses in spermatogenic cells, indicating that $InsP_3$ receptors may participate more significantly than RyRs in spermatogenesis, particularly during cell proliferation (Treviño *et al.,* 1998).

Spermatogenic cells are also endowed with voltage-gated Cl^- currents, blocked by niflumic acid (IC_{50} = 100 μ*M*) (Figure 7.3) (Espinosa *et al.,* 1998). In addition, several K^+ currents have been identified in these cells, including a TEA^+-sensitive, non-inactivating outward current (Hagiwara and Kawa, 1984) and a rapidly activating and sustained inwardly rectifying current (Muñoz-Garay *et al,* 2001). Interestingly, mice deficient in plasma membrane Cl^- channels undergo massive degeneration of male sperm cells (Bösl *et al.,* 2001). Correlating the presence and cellular distribution of various ion channels with their functional status at different stages of spermatogenesis, will allow a better understanding of their participation in differentiation, and in the meticulously choreographed signaling process required for the AR in mature spermatozoa.

VI. CONCLUDING REMARKS

Cell signaling is fundamental in determining the behavior of organisms. The propagation of life in many species depends on the dialogue between gametes, ion channels being elementary tools of cell communication. At the present time there is background information about some of the ion channels present in spermatozoa. Future study will determine the molecular mechanisms that regulate these channels in the cell. Combining molecular biological strategies and electrophysiology in spermatogenic cells, and the transfer of ion channels directly from spermatozoa to planar bilayers, opens new avenues to explore how ion channels participate in spermatogenesis, and how they are regulated in mature spermatozoa cells. It is hoped that this will allow a deeper understanding of the finely orchestrated events that lead to spermatozoa activation, induction of the acrosome reaction, and in the end to the generation of a new individual.

ACKNOWLEDGMENTS

This work was supported by grants from CONACyT (27707-N to A.D. and 32052-N to C.B.), DGAPA, the Howard Hughes Medical Institute, and the International Centre for Genetic Engineering and Biotechnology, to A.D. The authors thank Otilia Zapata, José Luis de la Vega, Ignacio López-González, and Shirley Ainsworth for technical help, and Claudia Treviño for discussions and help with the manuscript.

REFERENCES

Aarons, P., Battle, T., Boettger-Tong, H., and Poirier, G. R. (1992). Role of monoclonal antibody J-23 in inducing acrosome reactions in capacitated mouse spermatozoa. *J. Reprod. Fertil.* **96,** 49–59.

Alves, A.-P., Mulloy, B., Diniz, J. A., and Mourau, P. A. S. (1997). Sulfated polysaccharides from the egg jelly layer are species-specific inducers of the acrosomal reaction in sperm of sea urchins. *J. Biol. Chem.* **272,** 6965–6971.

Amano, T., Okita, Y., and Hoshi, M. (1992a). Treatment of starfish sperm with egg jelly induces the degradation of histones. *Dev. Growth Diff.* **34,** 99–106.

Amano, T., Okita, Y., Okinaga, T., Matsui, T., Nishiyama, I., and Hoshi, M. (1992b). Egg jelly components responsible for histone degradation and acrosome reaction in the starfish, *Asterina pectinifera. Biophys. Biochem. Res. Commun.* **187,** 274–278.

Amano, T., Okita, Y., Yasumoto, T., and Hoshi, M. (1993). Maitotoxin induces acrosome reaction and histone degradation of starfish *Asterina pectinifera* sperm. *Zool. Sci.* **10,** 307–312.

Antoni, F. A. (1997). Calcium regulation of adenylyl cyclase: Relevance for endocrine control. *Trends Endocrinol. Metab.* **8,** 7–14.

Arnaout, M. A. (2001). Molecular genetics and pathogenesis of autosomal dominant polycystic kidney disease. *Annu. Rev. Med.* **52,** 93–123.

Arnoult, C., Cardullo, R. A., Lemos, J. R., and Florman, H. M. (1996b). Activation of mouse sperm T-type Ca^{2+} channels by adhesion to the egg zona pellucida. *Proc. Natl. Acad. Sci. U.S.A.* **93,** 13004–13009.

Arnoult, C., Kazam, I. G., Visconti, P. E., Kopf, G. S., Villaz, M., and Florman, H. M. (1999). Control of the low voltage-activated calcium channel of mouse sperm by egg ZP3 and by membrane hyperpolarization during capacitation. *Proc. Natl. Acad. Sci. U.S.A.* **96,** 6757–6762.

Arnoult, C., Lemos, J. R., and Florman, H. M. (1997). Voltage-dependent modulation of T-type calcium channel by protein tyrosine phosphorylation. *EMBO J.* **16,** 1593–1599.

Arnoult, C., Zeng, Y., and Florman, H. (1996a). ZP3-dependent activation of sperm cation channels regulates acrosomal secretion during mammalian fertilization. *J. Cell Biol.* **134,** 637–645.

Austin, C. R. (1951). Observations on the penetration of the sperm into the mammalian egg. *Aust. J. Sci. Res.* [*B*] **4,** 581–596.

Babcock, D. F. (1983). Examination of the intracellular ionic environment and of ionophore action by null point measurements employing the fluorescein chromophore. *J. Biol Chem.* **258,** 6380–6389.

Babcock, D. F., and Pfeiffer, D. R. (1987). Independent elevation of cytosolic [Ca^{2+}] and pH of mammalian sperm by voltage-dependent and pH-sensitive mechanisms. *J. Biol. Chem.* **262,** 15041–15047.

Babcock, D. F., Bosma, M. M., Battaglia, D. E., and Darszon, A. (1992). Early persistent activation of sperm K^+ channels by the egg peptide speract. *Proc. Natl. Acad. Sci. U.S.A.* **89,** 6001–6005.

Baldi, E., Casano, R., Falsetti, C., Krausz, C., Maggi, M., and Forti, G. (1991). Intracellular calcium accumulation and responsiveness to progesterone in capacitating human spermatozoa. *J. Androl.* **5,** 323–330.

Baldi, E., Luconi, M., Bonaccorsi, L., Krausz, C., and Forti, G. (1996). Human sperm activation during capacitation and acrosome reaction: Role of calcium, protein phosphorylation and lipid remodelling pathways. *Front. Biosci.* **1,** d189–d205.

Baldi, E., Luconi, M., Bonaccorsi, L., Muratori, M., and Forti, G. (2000). Intracellular events and signaling pathways involved in sperm acquisition of fertilizing capacity and acrosome reaction. *Front. Biosci.* **5,** E110-E123

Barboni, E., Mattioli, M., and Seren, E. (1995). Influence of progesterone on boar sperm capacitation. *J. Endocrinol.* **144,** 13–18.

Barritt, G. J. (1999). Receptor-activated Ca^{2+} inflow in animal cells: A variety of pathways tailored to meet different intracellular Ca^{2+} signalling requirements. *Biochem. J.* **337,** 153–169.

Bavister, B. D. (1986). Animal *in vitro* fertilization and embryo development. *In* "Developmental Biology" (L. W. Browder, ed.), Vol. 4, pp. 81–148. Plenum Press, New York.

Bean, B. P., and McDonough, S. I. (1998). Two for T. *Neuron* **20,** 825–828.

Beltrán, C., Darszon, A., Labarca, P., and Liévano, A. (1994). A high-conductance multistate Ca^{2+} channel found in sea urchin and mouse spermatozoa. *FEBS Lett.* **338,** 23–26.

Beltrán, C., Zapata, O., and Darszon, A. (1995). Membrane potential regulates sea urchin sperm adenylyl cyclase [Annual Meeting, San Francisco, California]. *Biophys. J.,* p. 39.

Beltrán, C., Zapata, O., and Darszon, A. (1996). Membrane potential regulates sea urchin sperm adenylylcyclase. *Biochemistry* **35,** 7591–7598.

Benoff, S. (1998). Voltage dependent calcium channels in mammalian spermatozoa. *Front. Biosci.* **3,** d1220–d1240.

Bentley, J. K., Garbers, D. L., Domino, S. E., Nolan, T. D., and Van Dop, C. (1986a). Spermatozoa contain a guanine nucleotide binding protein ADP ribosylated by pertussis toxin. *Biochem. Biophys. Res. Commun.* **138,** 728–734.

Bentley, J. K., Tubb, D. J., and Garbers, D. L. (1986b). Receptor-mediated activation of spermatozoa guanylate cyclase. *J. Biol. Chem.* **261,** 14859–14862.

Berridge, M. J. (1993). Inositol trisphosphate and calcium signalling. *Nature* **361,** 315–325.

Bibring, T., Baxandall, J., and Harter, C. C. (1984). Sodium-dependent pH regulation in active sea urchin sperm. *Dev. Biol.* **101,** 425–435.

Blackmore, P. F. (1993). Thapsigargin elevates and potentiates the ability of progesterone to increase intracellular free calcium in human sperm: Possible role of perinuclear calcium. *Cell. Calcium* **14,** 53–60.

Blackmore, P. F., Beebe, S. J., Danforth, D. R., and Alexander, N. (1990). Progesterone an 17α-hydroxyprogesterone. Novel stimulators of calcium influx in human sperm. *J. Biol. Chem.* **265,** 1376–1380.

Bleil, J. D. (1991). Sperm receptors of mammalian eggs. *In* "Elements of Mammalian Fertilization" (P. M. Wassarman, ed.), Vol. I, pp. 133–151. CRC Press, Boca Raton.

Bleil, J. D., and Wassarman, P. M. (1983). Sperm–egg interactions in the mouse: Sequence of events and induction of the acrosome reaction by a zona pellucida glycoprotein. *Dev. Biol.* **95,** 317–324.

Boatman, D. E., and Robbins, R. S. (1991). Bicarbonate:carbon-dioxide regulation of sperm capacitation, hyperactivated motility and acrosome reactions. *Biol. Reprod.* **44,** 806–813.

Boettger-Tong, H., Aaron, D. J., Biegler, B. E., George, B., and Poirier, G. R. (1993). Binding of a murine proteinase inhibitor to the acrosome region of the human sperm head. *Mol. Reprod. Dev.* **36,** 346–353.

Boettger-Tong, H., Aarons, D., Biegler, B., Lee, T., and Poirier, G. R. (1992). Competition between zonae pellucidae and a proteinase inhibitor for sperm binding. *Biol. Reprod.* **47,** 716–722.

Boitano, S., and Omoto, C. K. (1991). Membrane hyperpolarization activates trout sperm without an increase in intracellular pH. *J. Cell Sci.* **98,** 343–349.

Boitano, S., and Omoto, C. K. (1992). Trout sperm swimming patterns and role of intracellular Ca^{2+}. *Cell Motil. Cytoskeleton* **21,** 74–82.

Bonaccorsi, L., Luconi, M., Forti, G., and Baldi, E. (1995). Tyrosine kinase inhibition reduces the plateau phase of the calcium increase in response to progesterone in human sperm. *FEBS Lett.* **364,** 83–86.

Bookbinder, L. H., Cheng, A., and Bleil, J. D. (1995). Tissue- and species-specific expression of sp56, a mouse sperm fertilization protein. *Science* **269,** 86–89.

Bösl, M. R., Stein, V., Hubner, C., Zdebik, A. A., Jordt, S. E., Mukhopadhyay, A. K., Davidoff, M. S., Holstein, A. F., and Jentsch, T. J. (2001). Male germ cells and photoreceptors, both dependent on close cell-cell interactions, degenerate upon ClC-2 Cl(-) channel disruption. *EMBO J.* **20,** 1289–1299.

Breitbart, H., and Spungin, B. (1997). The biochemistry of the acrosome reaction. *Mol. Hum. Reprod.* **3,** 195–202.

Buck, J., Sinclair, M. L., Schapal, L., Cann, M. J., and Levin L. R. (1999). Cytosolic adenylyl cyclase defines a unique signaling molecule in mammals. *Proc. Natl. Acad. Sci. U.S.A.* **96,** 79–84.

Calvete, J. J., Mann, K., Sanz, L., Raida, M., and Topfer-Petersen, E. (1996). The primary structure of BSP-30K, a major lipid-, gelatin-, and heparin-binding glycoprotein of bovine seminal plasma. *FEBS Lett.* **399,** 147–152.

Calzada, L., Bernal, A., and Loustaunau, E. (1988). Effect of steroid hormones and capacitation on membrane potential of human spermatozoa. *Arch. Androl.* **21,** 121–128.

Castellano, L. E., Martínez-Cadena, G., López-Godinez, J., Obregon, A. and García-Soto, J. (1997). Subcellular localization of the GTP-binding protein Rho in the sea urchin sperm. *Eur. J. Cell Biol.* **74,** 329–335.

Chan, H. C., Zhou, T. S., Fu, W. O., Wang, W. P., Shi, Y. L., and Wong, P. Y. D. (1997). Cation and anion channels in rat and human spermatozoa. *Biochim. Biophys. Acta* **1323,** 117–129.

Chang, M. C. (1951). Fertilizing capacity of spermatozoa deposited into the fallopian tubes. *Nature* **168,** 697–698.

Chen, Y., Cann, M. J., Litvin, T. N., Iourgenko, V., Sinclair, M. L., Levin, L. R., and Buck, J. (2000). Soluble adenylyl cyclase as an evolutionarily conserved bicarbonate sensor. *Science* **289,** 625–628.

Christen, R. W., Schackmann, R. W., and Shapiro, B. M. (1982). Elevation of intracellular pH activates respiration and motility of sperm of the sea urchin *Strongylocentrotus purpuratus*. *J. Biol. Chem.* **257,** 14881–14890.

Christen, R., Schackmann, R. W., and Shapiro, B. M. (1983). Interaction between sperm and sea urchin egg jelly. *Dev. Biol.* **98,** 1–14.

Clapham, D. E. (1998). Not so funny anymore: Pacing channels are cloned. *Neuron* **21,** 5–7.

Clapper, D. L., and Epel, D. (1985). The *limulus* sperm motility initiating peptide initiates acrosome reactions in sea water lacking potassium. *J. Exp. Zool.* **236,** 211–217.

Clark, E. N., Corron, M. E., and Florman, H. M. (1993). Caltrin, the calcium transport regulatory peptide of spermatozoa, modulates acrosome exocytosis in the response to the egg's zona pellucida. *J. Biol. Chem.* **268,** 5309–5316.

Cohen-Dayag, A., Tur-Kaspa, I., Dor, J., Mashiash, S., and Eisenbach, M. (1995). Sperm capacitation in humans is transient and correlates with chemotactic responsiveness to follicular factors. *Proc. Natl. Acad. Sci. U.S.A.* **92,** 11039–11043.

Collins, F., and Epel, D. (1977). The role of calcium ions in the acrosome reaction of sea urchin sperm. Regulation of exocytosis. *Exp. Cell Res.* **106,** 211–222.

Cook, S. P., and Babcock, D. F. (1993a). Selective modulation by cGMP of the K^+ channel activated by speract. *J. Biol. Chem.* **268,** 22402–22407.

Cook, S. P., and Babcock, D. F. (1993b). Activation of Ca^{2+} permeability by cAMP is coordinated through the pH_i increase induced by speract. *J. Biol. Chem.* **268,** 22408–224013.

Cook, S. P., Brokaw, C. J., Muller, C. H., and Babcock, D. F. (1994). Sperm chemotaxis: Egg peptides control cytosolic calcium to regulate flegellar responsen. *Dev. Biol.* **165,** 10–10.

Córdoba, M., Santa-Coloma, T. A., Beorlegui, N. B., and Beconi, M. T. (1997). Intracellular calcium variation in heparin-capacitated bovine sperm. *Biochem. Mol. Biol. Int.* **41,** 725–733.

Cox, T., and Peterson, R. N., (1989). Identification of calcium conducting channels in isolated boar sperm plasma membranes. *Biochem. Biophys. Res. Commun.* **161,** 162–168.

Cox, T., Campbell, P., and Peterson, R. N. (1991). Ion channels in boar sperm plasma membranes: Characterization of a cation selective channel. *Mol. Reprod. Dev.* **30,** 135–147.

Cuellar-Mata, P., Martinez-Cadena, G., Castellano, L. E., Aldana-Veloz, G. Novoa-Martinez, G., Vargas, I., Darszon, A., and García-Soto, J. (1995). Multiple GTP-binding proteins in sea urchin sperm: Evidence for Gs and small G-proteins. *Dev. Growth Differ.* **37,** 173–181.

Dan, J. C. (1952). Studies on the acrosome. I. Reaction to egg-water and other stimuli. *Biol. Bull.* **103,** 54–66.

Dan, J. C. (1954). Studies on the acrosome. III. Effect of Ca^{2+} deficiency *Biol. Bull.* **107,** 335–349.

Dangott, L. J. (1991). Isolation and characterization of a cDNA clone for a speract receptor homologue in the sea urchin *Arbacia punctulata. FASEB J.* **5,** A1227.

Dangott, L. J., and Garbers, D. L. (1984). Identification and partial characterization of the receptor for speract. *J. Biol. Chem.* **259,** 13712–13716.

Dangott, L. J., Jordan, J. E., Bellet, R. A., and Garbers, D. L. (1989). Cloning of the mRNA of the protein that crosslinks to the egg peptide speract. *Proc. Natl. Acad. Sci. U.S.A.* **86,** 2128–2132.

Darszon, A., Beltrán, C., Felix, R., Nishigaki, T., and Treviño, C. L. (2001). Ion transport in sperm signaling. *Dev. Biol.* In press.

Darszon, A., Bosma, M., Battaglia, D. E., and Babcock, D. F. (1990). Volume expanded sperm: Models for study of early responses required for fertilization. *J. Cell Biol.* **111,** 115a.

Darszon, A., Labarca, P., Beltrán, C., García-Soto, J., and Liévano, A. (1994). Sea urchin sperm: An ion channel reconstitution study case. *Methods: Methods Enzymol.* **6,** 37–50.

Darszon, A., Liévano, A., and Beltrán, C. (1996). Ion channels: Key elements in gamete signaling. *In* "Current Topics in Develomental Biology" (R. A. Pedersen, and G. P. Schatten, eds.). Vol. 34, pp. 117–167. Academic Press, San Diego.

Darszon, S., Labarca, P., Nishigaki, T., and Espinosa, F. (1999). Ion channels in sperm physiology. *Physiol. Rev.* **79,** 481–510.

DasGupta, S., Mills, C. L., and Fraser, L. R. (1993). Ca^{2+}-related changes in the capacitation state of human spermatozoa assessed by a chlortetracycline fluorescence assay. *J. Reprod. Fertil.* **99,** 135–143.

DasGupta, S., O'Toole, C., Mills, C. L., and Fraser, L. R. (1994). Effect of pentoxifylline and progesterone on human sperm capacitation and acrosomal exocytosis. *Hum. Reprod.* **9,** 2103–2109.

Doherty, C. M., Tarchala, S. M., Radwanska, E., and Jonge, C. J. (1995). Characterization of two second messenger pathways and their interactions in eliciting the human sperm acrosome reaction. *J. Androl.* **16,** 36–46.

Domino, S. T., and Garbers, D. L. (1989). Activation of phospholipase D by the fucose-sulfate glycoconjugate that induces an acrosome reaction in spermatozoa. *J. Biol. Chem.* **264,** 9412–9419.

Domino, S. E., Bocckino, S. B., and Garbers, D. L. (1989). Activation of phospholipase D by the fucose-sulfate glycoconjugate that induces an acrosome reaction in spermatozoa. *J. Biol. Chem.* **264,** 9412–9419.

Dragileva, E., Rubinstein, S., and Breitbart, H. (1999). Intracellular Ca^{2+}–Mg^{2+}-ATPase regulates calcium influx and acrosomal exocytosis in bull and ram spermatozoa. *Biol. Reprod.* **61,** 1226–1234.

Eisenbach, M. (1999). Mammalian sperm chemotaxis and its association with capacitation. *Dev. Genet.* **25,** 87–94.

Emiliozzi, C., and Fenichel, P. (1997). Protein tyrosine phosphorylation is associated with capacitation of human sperm *in vitro* but is not sufficient for its completion. *Biol. Reprod.* **56,** 674–679.

Endo, Y., Komatsu, S., Hirai, M., Suzuki, S., and Shimizu, N. (1991). Protein kinase C activity and protein phosphorylation in mouse sperm. *Nippon Sanka Fujinka Gakkai Zasshi* **43,** 109–116.

Endo, Y., Lee, M. A., and Kopf, G. S. (1987). Evidence for a role for a guanine nucleotide binding regulatory protein in the zona pellucida induced sperm acrosome reaction. *Dev. Biol.* **119,** 210–216.

Endo, Y., Lee, M. A., and Kopf, G. S. (1988). Characterization of an islet activating protein-sensitive site in mouse sperm that is involved in the zona pellucida-induced acrosome reaction. *Dev. Biol.* **129,** 12–24.

Erdo, S. L., and Werkele, L. (1990). $GABA_A$ type binding sites on membranes of spermatozoa. *Life Sci.* **47,** 1147–1151.

Espinosa, F., and Darszon, A. (1995). Mouse sperm membrane potential: Changes induced by Ca^{2+}. *FEBS Lett.* **372,** 119–125.

Espinosa, F., De la Vega-Beltrán, J. L., López-González, I., Delgado, R., Labarca, P., and Darszon, A. (1998). Mouse sperm patch-clamp recordings reveal single Cl^- channels sensitive to niflumic acid, a blocker of the sperm acrosome reaction. *FEBS Lett.* **426,** 47–51.

Espinosa, F., Delgado, R., de la Vega, J. L., Labarca, P., and Darszon, A. (1997). Single channel recording in mouse sperm. *Biophys. J.* **72,** A154.

Espinosa, F., López-González, I., Muñoz-Garay, C., Felix, R., de la Vega-Beltrán, J. L., Kopf, G. S., Visconti, P. E., and Darszon, A. (2000). Dual regulation of the T-type Ca^{2+} current by serum albumin and beta-estradiol in mammalian spermatogenic cells. *FEBS Lett.* **475,** 251–256.

Espinosa, F., Lopéz-González, I., Serrano, C. J., Gasque, G., De La Vega-Beltrán, J. L., Treviño, C. L., and Darszon, A. (1999). Anion channel blockers differentially affect T-type Ca^{2+} currents of mouse spermatogenic cells, α1E currents expressed in *Xenopus* oocytes and the sperm acrosome reaction. *Dev. Genet.* **25,** 103–114.

Farach, H. A., Jr., Mundy, D. I., Strittmatter, W. J., and Lennarz, W. J. (1987). Evidence for the involvement of metalloendoproteases in the acrosome reaction in sea urchin sperm. *J. Biol. Chem.* **262,** 5483–5487.

Félix, R. (1999). Voltage-dependent Ca^{2+} channel $\alpha_2\delta$ auxiliary subunit: Structure, function and regulation. *Receptors Channels* **6,** 351–362.

Flechon, J.-E., and Hunter, R. H. F. (1981). Distribution of spermatozoa in the utero-tubal junction and isthmus of pigs, and their relationship with the luminal epithelium after mating: A scanning electron microscope study. *Tissue Cell* **13,** 127–139.

Flesch, F. M., and Gadella, B. M. (2000). Dynamics of the mammalian sperm plasma membrane in the process of fertilization. *Biochim. Biophys. Acta* **1469,** 197–235.

Florman, H. M. (1994). Sequential focal and global elevations of sperm intracellular Ca^{2+} are initiated by the zona pellucida during acrosomal exocytosis. *Dev. Biol.* **165,** 152–164.

Florman, H. M., and Babcock, D. F. (1991). Progress toward understanding the molecular basis of capacitation. *In* "Elements of Mammalian Fertilization. I. Basic concepts" (P. M. Wasssarman, ed.), pp. 105–132. CRC Press, Boca Raton.

Florman, H. M., and First, N. L. (1988). The regulation of acrosomal exocytosis II. The zona pellucida-induced acrosome reaction of bovine spermatozoa is controlled by extrinsic positive regulatory elements. *Dev. Biol.* **128,** 464–473.

Florman, H. M., Arnoult, C., Kazam, I. G., Li, C., and O'Toole, C. M. B. (1998). A perspective on the control of mammalian fertilization by egg-activated ion channels in sperm: A tale of two channels. *Biol. Reprod.* **59,** 12–16.

Florman, H., Corron, M. E., Kim, T. D.-H., and Babcock, D. F. (1992). Activation of voltage-dependent calcium channels of mammalian sperm is required for zona pellucida-induced acrosomal exocytosis. *Dev. Biol.* **152,** 304–314.

Florman, H., Tombes, R. M., First, N. L., and Babcock, D. F. (1989). An adhesion associated agonist from the zona pellucida activates G protein promoted elevations of internal Ca^{2+} and pH that mediate mammalian sperm acrosomal exocytosis. *Dev. Biol.* **135,** 133–147.

Foresta, C., and Rossato, M. (1997). Calcium influx pathways in human spermatozoa. *Mol. Hum. Reprod.* **3,** 1–4.

Foresta, C., Rossato, M., Chiozzi, P., and Di Virgilio, F. (1996). Mechanism of human sperm activation by extracellular ATP. *Am. Physiol. Soc.* **270,** C1709–C1714.

Foresta, C., Rossato, M., and Di Virgilio, F. (1995). Differential modulation by protein kinase C of progesterone-activated responses in human sperm. *Biochem. Biophys. Res. Commun.* **206,** 408–413.

Fraser, L. R. (1987). Minimum and maximum extracellular Ca^{2+} requirements during mouse sperm capacitation and fertilization *in vitro*. *J. Reprod. Fertil.* **81,** 77–89.

Fraser, L. R., and Monks, N. J. (1990). Cyclic nucleotides and mammalian sperm capacitation. *J. Reprod. Fertil.* (Suppl.) **42,** 9–21.

Fukami, K., Nakao, K., Inoue, T., Kataoka, Y., Kurokawa, M., Fissore, R. A., Nakamura, K., Katsuki, M., Mikoshiba, K., Yoshida, N., and Takenawa, T. (2001). Requirement of phospholipase Cδ4 for the zona pellucida-induced acrosome reaction. *Science* **292,** 920–923.

Galindo, B. E., Beltrán, C., Cragoe, E. J. J., and Darszon, A. (2000). Participation of a K^+ channel modulated directly by cGMP in the speract-induced signaling cascade of *Strongylocentrotus purpuratus* sea urchin sperm. *Dev. Biol.* **221,** 285–294.

Gao, Z., and Garbers, D. L. (1998). Species diversity in the structure of zonadhesin, a sperm-specific membrane protein containing multiple cell adhesion molecule-like domains. *J. Biol. Chem.* **273,** 3415–3421.

Garbers, D. L. (1981). The elevation in cyclic AMP concentrations in flagella-less sea urchin sperm heads. *J. Biol. Chem.* **256,** 620–624.

Garbers, D. L. (1989). Molecular basis of fertilization. *Annu. Rev. Biochem.* **58,** 719–742.

Garbers, D. L., and Kopf, G. S. (1980). The regulation of spermatozoa by calcium and cyclic nucleotides. *Adv. Cyclic Nucleotides Res.* **13,** 251–306.

Garbers, D. L., Kopf, G. S., Tubb, D. J., and Olson, G. (1983). The elevation of sperm cyclic AMP concentrations by a fucose-sulfate rich complex associated with eggs. I. Structural characterization. *Biol. Reprod.* **29,** 1211–1220.

Garbers, D. L., Tubb, D. J., and Kopf, G. S. (1980). Regulation of sea urchin sperm cyclic AMP-dependent protein kinases by an egg associated factor. *Biol. Reprod.* **22,** 526–532.

García-Soto, J., and Darszon, A. (1985). High pH-induced acrosome reaction and Ca^{2+} uptake in sea urchin sperm suspended in Na^+-free sea water. *Dev. Biol.* **110,** 338–345.

García-Soto, J., González-Martínez, M., de Delatorre, L., and Darszon, A. (1987). Internal pH can regulate Ca^{2+} uptake and the acrosome reaction in sea urchin sperm. *Dev. Biol.* **120,** 112–120.

García-Soto, J., Araiza, L. M., Barrios, M., Darszon, A., and Luna-Arias, J. P. (1991). Endogenous activity of cyclic nucleotide-dependent protein kinase in plasma membranes isolated from *Strongylocentrotus purpuratus* sea urchin sperm. *Biochem. Biophys. Res. Commun.* **180,** 1436–1445.

García, M. A., and Meizel, S. (1999). Determination of the steady-state intracellular chloride concentration in capacitated human spermatozoa. *J. Androl.* **20,** 88–93.

Gatti, J.-L., and Christen, R. (1985). Regulation of internal pH of sea urchin sperm. *J. Biol. Chem.* **260,** 7599–7602.

Gatti, J.-L., Chevrier, C., Paquignon, M., and Dracheux, J. L. (1993). External ionic conditions, internal pH and motility of ram and boar spermatozoa. *J. Reprod. Fertil.* **98,** 439–449.

Gauss, R., Seiffert, R., and Kaupp, U. B. (1998). Molecular identification of a hyperpolarization-activated channel in sea urchin sperm. *Nature* **393,** 583–587.

Glabe, C. G., and Lennarz, W. (1979). Species-specific sperm adhesion in sea urchins: A quantitative investigation of binding mediated egg agglutination. *J. Cell Biol.* **83,** 595–604.

Glassner, M., Jones, J., Kligman, I., Woolkalis, M. J., Gerton, G. L., and Kopf, G. S. (1991). Immunocytochemical and biochemical characterization of guanine nucleotide-binding regulatory proteins in mammalian spermatozoa. *Dev. Biol.* **146,** 438–450.

Gmachl, M., and Kreil, G. (1993). Bee venom hyaluronidase to a membrane protein of mammalian sperm. *Proc. Natl. Acad. Sci. U.S.A.* **90,** 3569–3573.

Gong, X., Dubois, D. H., Miller, D. J., and Shur, B. (1995). Activation of a G protein complex aggregation of β-1,4-galactosyltransferase on the surface of sperm. *Science* **269,** 1718–1720.

González-Martínez, M., and Darszon, A. (1987). A fast transient hyperpolarization occurs during the sea urchin sperm acrosome reaction induced by egg jelly. *FEBS Lett.* **218,** 247–250.

González-Martínez, M., Galindo, B. E., de De La Torre, L., Zapata, O., Rodíguez, E., Florman, H. M., and Darszon, A. (2001). A sustained increase in intracellular Ca^{2+} is required for the acrosome reaction in sea urchin sperm. *Dev. Biol.* **236,** 220–229.

González-Martínez, M. T., Guerrero, A., Morales, E., de Delatorre, L., and Darszon, A. (1992). A depolarization can trigger Ca^{2+} uptake and the acrosome reaction when preceded by a hyperpolarization in *L. pictus* sea urchin sperm. *Dev. Biol.* **150,** 193–202.

Gonzalez-Perret, S., Kim, K., Ibarra, C., Damiano, A. E., Zotta, E., Batelli, M., Harris, P. C., Reisin, I. L., Arnaout, M. A., and Cantiello, H. F. (2001). Polycystin-2, the protein mutated in autosomal dominant polycystic kidney disease (ADPKD), is a Ca^{2+}-permeable nonselective cation channel. *Proc. Natl. Acad. Sci. U.S.A* **98,** 1182–1187.

Guerrero, A., and Darszon, A. (1989a). Egg jelly triggers a calcium influx which inactivates and is inhibited by calmodulin antagonists in the sea urchin sperm. *Biochem. Biophys. Acta* **980,** 109–116.

Guerrero, A., and Darszon, A. (1989b). Evidence for the activation of two different Ca^{2+} channels during the egg jelly induced acrosome reaction of sea urchin sperm. *J. Biol. Chem.* **264,** 19593–19599.

Guerrero, A., García, L., Zapata, O., Rodriguez, E., and Darszon, A. (1998). Acrosome reaction inactivation in sea urchin sperm. *Biochim. Biophys. Acta* (in press).

Guerrero, A., Sánchez, J. A., and Darszon, A. (1987). Single-channel activity in sea urchin sperm revealed by the patch-clamp technique. *FEBS Lett.* **220,** 295–298.

Hagiwara, N., and Kawa, K. (1984). Calcium and potassium currents in spermatogenic cells dissociated from rat seminiferous tubules. *J. Physiol. (London)* **356,** 135–149.

Hammamah, S., Maginoux, E., Royere, D., Bartheleny, C., Dacheux, J. L., and Gatti, J. L. (1996). Internal pH of human spermatozoa: efect of ions, human folicular gluid and progesterone. *Mol. Hum. Reprod.* **2,** 219–224.

Hanaoka, K., Qian, F., Boletta, A., Bhunia, A. K., Piontek, K., Tsiokas, L., Sukhatme, V. P., Guggino, W. B., and Germino, G. G. (2000). Co-assembly of polycystin-1 and -2 produces unique cation-permeable currents. *Nature* **408,** 990–994.

Hansbrough, J. R., and Garbers, D. L. (1981a). Speract. Purification and characterization of a peptide associated with eggs that activate spermatozoa. *J. Biol. Chem.* **256,** 1477–1542.

Hansbrough, J. R., and Garbers, D. L. (1981b). Sodium dependent activation of sea urchin spermatozoa by speract and monencin. *J. Biol. Chem.* **256,** 2235–2241.

Hansbrough, J. R., Kopf, G. S., and Garbers, D. L. (1980). The stimulation of sperm metabolism by a factor associated with eggs and by 8-bromoguanosine 3′,5′-monophosphate. *Biochim. Biophys. Acta* **620,** 82–91.

Hardy, D. M., and Garbers, D. L. (1995). A sperm membrane protein that binds in a species-specific manner to the egg extracellular matrix is homologous to von Willebrand factor. *J. Biol. Chem.* **270,** 26025–26028.

Harteneck, C., Plant, T. D., and Schultz, G. (2000). From worm to man: Three subfamilies of TRP channels. *Trends Neurosci.* **23,** 159–166.

Hildebrandt, J. D., Codina, J., Tash, J. S., Kirchick, H. J., Lipschultz, L., Sekura, R. D., and Birnbaumer, L. (1985). The membrane-bound spermatozoal adenylyl cyclase system does not share coupling characteristics with somatic cell adenylyl cyclases. *Endocrinology* **116,** 1357–1366.

Hille, B. (1992). "Ionic Channels of Excitable Membranes," 2nd Ed. Sinauer Associates Inc., Sunderland, Massachusetts.

Hoshi, M., Amano, T., Okita, Y., Okinaga, T., and Matsui, T. (1990). Egg signals for triggering the acrosome reaction in starfish spermatozoa. *J. Reprod. Fertil.* (Suppl.) **42,** 23–31.

Hoshi, M., Nishigaki, T., Ushiyama, A., Okinaga, T., Chiba, K., and Matsumoto, M. (1994). Egg-jelly signal molecules for triggering the acrosome reaction in starfish spermatozoa. *Int. J. Dev. Biol.* **38,** 167–174.

Hoshi, M., Okinaga, T. M., Kontani, K., Araki, T., and Chiba, K. (1991). Acrosome reaction-inducing glycoconjugate in the jelly coat of star fish eggs. *In* "Comparative Spermatology 20 Years After" (B. Baccetti, ed.), pp. 175–180. Raven Press, New York.

Hughes, J., Ward, C. J., Aspinwall, R., Butler, R., and Harris, P. C. (1999). Identification of a human homologue of the sea urchin receptor for egg jelly: A polycystic kidney disease-like protein. *Hum. Mol. Genet.* **8,** 543–549.

Hunter, R. H. F., and Nichol, R. (1986). A preovulatory temperature gradient between the isthmus and the ampulla of pig oviducts during the phase of sperm storage. *J. Reprod. Fertil.* **77,** 599–606.

Ikadai, H., and Hoshi, M. (1981a). Biochemical studies on the acrosome reaction of starfish, *Asterias amurensis.* I. Factors participating in the acrosome reaction. *Dev. Growth. Differ.* **23,** 73–80.

Ikadai, H., and Hoshi, M. (1981b). Biochemical studies on the acrosome reaction of starfish, *Asterias amurensis.* II. Purification and characterization of the acrosome reaction-inducing substance. *Dev. Growth. Differ.* **23,** 81–88.

Jan, Y. N., and Jan, L. Y. (1997). Cloned potassium channels from eukaryotes and prokaryotes. *Annu. Rev. Neurosci.* **20,** 91–123.

Jenkins, A. D., Lechene, C. P., and Howard, S. S. (1980). Concentrations of seven elements in the intraluminal fluids of the rat seminiferous tubules, rat testis, and epididymis. *Biol. Reprod.* **23,** 981–987.

Johnson, C. H., Clapper, D. L., Winkler, M., Lee, H. C., and Epel, D. (1983). A volatile inhibitor immobilizes sea urchin sperm in semen by depressing intracellular pH. *Dev. Biol.* **98,** 493–501.

Jungnickel, M. K., Marrero, H., Birnbaumer, L., Lemos, J. R., and Florman, H. M. (2001). Trp2 regulates entry of Ca^{2+} into mouse sperm triggered by egg ZP3. *Nat. Cell. Biol.* **3,** 499–502.

Kaupp, U. B. (1995). Family of cyclic nucleotide gated ion channels. *Curr. Opin. Neurobiol.* **5,** 434–442.

Kaupp, U., and Seifert, R. (2001). Molecular diversity of pacemaker ion channels. *Annu. Rev. Physiol.* **63,** 235–257.

Kazazoglou, T., Schackmann, R. W., Fossett, M., and Shapiro, B. M. (1985). Calcium channel antagonists inhibit the acrosome reaction and bind to plasma membranes of sea urchin sperm. *Proc. Natl. Acad. Sci. U.S.A.* **82,** 1460–1464.

Kirkman-Brown, J. C., Bray, C., Stewart, P. M., Barratt, C. L., and Publicover, S. J. (2000). Biphasic elevation of $[Ca^{2+}]_i$ in individual human spermatozoa exposed to progesterone. *Dev. Biol.* **222,** 326–335.

Kopf, G. S., and Gerton, G. L. (1990). The mammalian sperm acrosome and the acrosome reaction. *In* "Elements of Mammalian Fertilization. I. Basic Concepts" (P. M. Wassarman, ed.), pp. 153–203. CRC Uniscience Series, CRC Press, Boca Raton.

Kopf, G. S., Endo, Y., Mattei, P., Kurasawa, S., and Schultz, R. M. (1989). Egg-induced modifications of the murine zona pellucida. *In* "Mechanisms of Egg Activation" (R. L. Nuccitelli, H. W. Clark, and G. N. Cher, eds.), pp. 249–272. Plenum, New York.

Kopf, G. S., Tubb, D. J., and Garbers, D. L. (1979). Activation of sperm respiration by a low molecular weight egg factor and by 8-bromoguanosine 3′, 5′-monophosphate. *J. Biol. Chem.* **254,** 8554–8560.

Kopf, G. S., Woolkalis, M. J., and Gerton, G. L. (1986). Evidence for a guanine nucleotide-binding regulatory protein in invertebrate an mammalian sperm: Identification by islet activating protein catalyzed ADP rybosylation and immunochemical methods. *J. Biol. Chem.* **261,** 7327–7331.

Koyota, S., Wimalasiri, K. M., and Hoshi, M. (1997). Structure of the main saccharide chain in the acrosome reaction-inducing substance of the starfish, *Asterias amurensis. J. Biol. Chem.* **272,** 10372–10376.

Krasznai, Z., Marian, T., Izumi, H., Damjanovich, S., Balkay, L., Tron, L., and Morisawa, M. (2000). Membrane hyperpolarization removes inactivation of Ca^{2+} channels, leading to Ca^{2+} influx and subsequent initiation of sperm motility in the common carp. *Proc. Natl. Acad. Sci. U.S.A.* **97,** 2052–2057.

Kuo, R. C., Baxter, G. T., Thompson, S. H., Stricker, S. A., Patton, C., Bonaventura, J., and Epel, D. (2000). NO is necessary and sufficient for egg activation at fertilization. *Nature* **406,** 633–636.

Labarca, P., Santi, C., Morales, E., Zapata, O., Beltrán, C., Liévano, A., and Darszon, A. (1995). Possible participation of a cAMP regulated K^+ channel detected in planar bilayers in the speract-induced depolarization of sea urchin spermatozoa. *Biophys. J.* **68,** A224.

Labarca, P., Santi, C., Zapata, O., Beltrán, C., Liévano, A., Sandoval, Y., and Darszon, A. (1997). Possible participation of a cAMP regulated K^+ channel from the sea urchin sperm in the speract response. *In* "From Ion Channels to Cell-to-Cell Conversations" (R. Latorre, ed.), Plenum, New York.

Labarca, P., Santi, C., Zapata, O., Morales, E., Beltrán, C., Liévano A., and Darszon A. (1996). cAMP regulated K^+-selective channel from the sea urchin sperm plasma membrane. *Dev. Biol.* **174** (2), 271–280.

Lakoski, K. A., Carron, C. P., Cabot, C. L., and Saling, P. M. (1988). Epididymal maturation and the acrosome reaction in mouse sperm: Response to zona pellucida develops coincident with modification of M42 antigen. *Biol. Reprod.* **38,** 221–233.

Langlais, J., and Roberts, K. D. (1985). A molecular membrane model of sperm capacitation and the acrosome reaction of mammalian spermatozoa. *Gamete Res.* **12,** 183–224.

Lardy, H. A., and San Agustin, J. (1989). Caltrin and calcium regulation of sperm activity. *In* "The Cell Biology of Fertilization" (H. Schatten, and G. Schatten, eds.), pp. 29–39. Academic Press, New York.

Lax, Y., Rubinstein, S., and Breitbart, H. (1994). Epidermal growth factor induces acrosomal exocytosis in bovine sperm. *FEBS Lett.* **339,** 234–238.

Lax, Y., Rubinstein, S., and Breitbart, H. (1997). Subcellular distribution of PKCα and β1 in bovine spermatozoa and their regulation by calcium and phorbol esters. *Biol. Reprod.* **56,** 454–459.

Lee, H. C. (1984a). Sodium and proton transport in flagella isolated from sea urchin spermatozoa. *J. Biol. Chem.* **259,** 4957–4963.

Lee, H. C. (1984b). A membrane potential-sensitive Na^+/H^+ exchange system in flagella isolated from sea urchin spermatozoa. *J. Biol. Chem.* **259,** 15315–15319.

Lee, H. C. (1985). The voltage-sensitive Na^+/H^+ exchange in flagella isolated from sea urchin sperm. *J. Biol. Chem.* **260,** 10794–10799.

Lee, H. C. (1988). Internal GTP stimulates the speract receptor mediated voltage changes in sea urchin spermatozoa membrane vesicles. *Dev. Biol.* **126,** 91–97.

Lee, H. C., and Garbers, D. L. (1986). Modulation of the voltage sensitive Na^+/H^+ exchange in sea urchin spermatozoa through membrane potential changes induced by the egg peptide speract. *J. Biol. Chem.* **261,** 16026–16032.

Lee, H. C., Johnson, C., and Epel, D. (1983). Changes in internal pH associated with the initiation of motility and the acrosome reaction of sea urchin sperm. *Dev. Biol.* **95,** 31–45.

Lee, M. A., Kopf, G. S., and Storey, B. T. (1987). Effects of phorbol esters and diacyglycerol on the mouse sperm acrosome reaction induced by the zona pellucida. *Biol. Reprod.* **36,** 617–627.

Lee, M. A., Check, J. H., and Kopf, G. S. (1992). A guanine nucleotide-binding regulatory protein in human sperm mediates acrosomal exocytosis induced by the human zona pellucida. *Mol. Reprod. Dev.* **31,** 78–86.

Leyton, L., and Saling, P. (1989). Evidence that aggregation of mouse sperm receptors by ZP3 triggers the acrosome reaction. *J. Cell Biol.* **108,** 2163–2168.

Leyton, L., LeGuen, P., Bunch, D., and Saling, P. M. (1992). Regulation of mouse gamete interaction by a sperm tyrosine kinase. *Proc. Natl. Acad. Sci. U.S.A.* **89,** 11692–11695.

Liévano, A., Sánchez, J., and Darszon, A. (1985). Single channel activity of bilayers derived from sea urchin sperm plasma membranes at the tip of a patch-clamp electrode. *Dev. Biol.* **112,** 235–295.

Liévano, A., Santi, C. M., Serrano, C. J., Treviño, C. L., Bellvé, A. R., Hernández-Cruz, A., and Darszon, A. (1996). T-type Ca^{2+} channels and α_{1E} expression in spermatogenic cells, and their possible relevance to the sperm acrosome reaction. *FEBS Lett.* **388,** 150–154.

Liévano, A., Vega Saenz de Miera, E. C., and Darszon, A., (1990). Ca^{2+} channels from the sea urchin sperm plasma membrane. *J. Gen. Physiol.* **95,** 273–296.

Llanos, M. N. (1998). Thapsigargin stimulates acrosomal exocytosis in hamster spermatozoa. *Mol. Reprod. Dev.* **51,** 84–91.

Longo, F. J., Ushiyama, A., Chiba, K., and Hoshi, M. (1995). Ultrastructural localization of acrosome reaction-inducing substance (ARIS) on sperm of the starfish *Asterias amurensis. Mol. Reprod. Dev.* **41,** 91–99.

Luconi, M., Krausz, C., Forti G., and Baldi, E. (1996). Extracellular calcium negatively modulates tyrosine phosphorylation and tyrosine kinase activity during capacitation of human spermatozoa. *Biol. Reprod.* **55,** 207–216.

Matsui, T., Nishiyama, I., Hino, A., and Hoshi, M. (1986a). Induction of the acrosome reaction in starfish. *Dev. Growth Differ.* **28,** 339–348.

Matsui, T., Nishiyama, I., Hino, A., and Hoshi, M. (1986b). Acrosome reaction-inducing substance purified from the egg jelly inhibits the jelly-induced acrosome reaction in starfish: An apparent contradiction. *Dev. Growth Differ.* **28,** 349–357.

Matsui, T., Nishiyama, I., Hino, A., and Hoshi, M. (1986c). Intracellular pH changes of starfish sperm upon the acrosome reaction. *Dev. Growth Differ.* **28,** 443–448.

Matsumara, K., and Aketa, K. (1990). Proteasome (multicatalytic protease) of sea urchin sperm and its possible participation in the acrosome reaction. *Mol. Reprod. Dev.* **29,** 189–199.

McDonald, R. L., and Olsen, R. W. (1994). $GABA_A$ receptor channels. *Annu. Rev. Neurosci.* **17,** 596–602.

McLeskey, S. B., Dowds, C., Carballada, R., White, R. R., and Saling, P. M. (1998). Molecules involved in mammalian sperm-egg interaction. *Int. Rev. Cytol.* **177,** 57–113.

Meizel, S. (1997). Amino acid neurotransmitter receptor/chloride channels of mammalian sperm and the acrosome reaction. *Biol. Reprod.* **56,** 569–574.

Meizel, S., and Turner, K. O. (1993). Initiation of the human sperm acrosome reaction by thapsigargin. *J. Exp. Zool.* **267,** 350–355.

Meizel, S., and Turner, K. O. (1996). Chloride efflux during the progesterone-initiated human sperm acrosome reaction is inhibited by lavenductin A, a tyrosine kinase inhibitor. *J. Androl.* **17,** 327–330.

Meizel, S., Turner, K. O., and Nuccitelli, R. (1997). Progesterone triggers a wave of increased free calcium during the human sperm acrosome reaction *Dev. Biol.* **182,** 67–75.

Melendrez, C., and Meizel, S. (1995). Studies of porcine and human sperm suggesting a role for a sperm glycine receptor/Cl^- channel in the zona pellucida-initated acrosome reaction. *Biol. Reprod.* **53,** 676–683.

Mendoza, C., and Tesarik, J. (1993). A plasma-membrane progesterone receptor in human sperm is switched on by increasing intracellular free calcium. *FEBS Lett.* **330,** 57–60.

Mendoza, C., Soler, A., and Tesarik, J. (1995). Nongenomic steroid action Independent targeting of a plamsa membrane calcium channel and a tyrosine kinase. *Biochem. Biophys. Res. Commun.* **210,** 518–523.

Miller, R. L. (1985). Sperm chemo-orientation in the metazoa. *In* "Biology of Fertilization. Biology of the Sperm" (C. B. Metz, and A. Monroy, eds.), Vol. 2, pp. 275–337. Academic Press, New York.

Miller, R. L., and Vogt, R. (1996). An terminal partial sequence of the 13kDa *Picnopodia helianthoides* sperm chemoattractant 'startrak' possesses sperm-attracting activity. *J Exp. Biol.* **199,** 311–318.

Morales, E., de De la Torre, L., Moy, G., Vacquier, V. D., and Darszon, A. (1993). Anion channels in the sea urchin sperm plasma membrane. *Mol. Reprod. Dev.* **36,** 174–182.

Morisawa, M. (1994) Cell signaling mechanisms for sperm motiity. *Zool. Sci.* **11,** 647–662.

Morisawa, M., and Hayashi, H. (1985). Phosphorylation of a 15 K axonemal protein is the trigger initiating trout sperm motility. *Biomed. Res.* **6,** 181–184.

Morisawa, M., and Ishida, K. (1987). Short-term changes in levels of cyclic AMP, adenylate cyclase, and phosphodiesterase during the initiation of sperm motility in rainbow trout. *J. Exp. Zool.* **242,** 199–204.

Morisawa, M., and Okuno, M. (1982). Cyclic AMP induces maturation of trout sperm axoneme to initiate motility. *Nature (London)* **295,** 703–704.

Morisawa, M., and Suzuki, K. (1980). Osmolarity and potassium ion: Their roles in initiation of sperm motility in teleosts. *Science* **210,** 1145–1147.

Morisawa, M., Suzuki, K., and Morisawa, S. (1983). Effects of potassium and osmolality on spermatozoan motility of salmonid fishes. *J. Exp. Biol.* **107,** 105–113.

Moy, G. W., Mendoza, L. M., Schulz, J. R., Swanson, W. J., Glabe, C. G., and Vacquier, V. D. (1996). The sea urchin sperm receptor for egg jelly is a modular protein with extensive homology to the human polycystic kidney disease protein, PKD1. *J. Cell. Biol.* **133,** 809–817.

Muñoz-Garay, C., de la Vega-Beltrán, J. L., Delgado, R., Labarca, P., Felix, R., and Darszon, A. (2001). Inwardly rectifying K^+ channels in spermatogenic cells: Functional expression and implication in sperm capacitation. *Dev. Biol.* **234,** 261–274.

Murase, T., and Roldán E. R. S. (1996). Progesterone and the zona pellucida activate different transducing pathways in the sequence of events leading to diacylglycerol generation during mouse sperm acrosome reaction. *Biochem. J.* **320,** 1017–1023.

Nakamura, M., Moriya, M., Baba, T., Michikawa, Y., Yamanobe, T., Arai, K., Okinaga, S., and Kobayashi, T. (1993). An endoplasmic reticulum protein, calreticulin, is transported into the acrosome of rat sperm. *Exp. Cell Res.* **205,** 101–110.

Naz, R. K. (1996). Protein tyrosine phosphorylation and signal transduction during capacitation-acrosome reaction and zona pellucida binding in human sperm. *Arch. Androl.* **37,** 47–55.

Nishigaki, T., and Darszon, A. (2000). Real-time measurements of the interactions between fluorescent speract and its sperm receptor. *Dev. Biol.* **223,** 17–26.

Nishigaki, T., Chiba, K., Miki, W., and Hoshi, M. (1996). Structure and function of asterosaps sperm activating peptides from the jelly coat of starfish eggs. *Zygote* **4,** 237–245.

Nishigaki, T., Zamudio, F. Z., Possani, L. D., and Darszon, A. (2001). Time-resolved sperm responses to an egg peptide measured by stopped-flow fluorometry. *Biochem Biophys Res Commun.* **284,** 531–535.

Nishioka, D., and Cross, N. (1978). The role of external sodium in sea urchin fertilization. *In* "Cell Reproduction" (E. R. Dirksen, D. Prescott, and C. F. Fox, eds.), pp. 403–413. Academic Press, New York.

Nishiyama, I., Matsui, T., and Hoshi, M. (1987a). Purification of CoARIS, a cofactor for acrosome reaction-inducing substance, from the egg jelly of starfish. *Dev. Growth Differ.* **29,** 161–169.

Nishiyama, I., Matsui, T., Fujimoto, Y., Ikekawa, N., and Hoshi, M. (1987b). Correlation between the molecular structure and the biological activity of CoARIS, a cofactor for acrosome reaction-inducing substance. *Dev. Growth Differ.* **29,** 171–176.

Nomura, H., Turco, A. E., Pei, Y., Kalaydjieva, L., Schiavello, T., Weremowicz, S., Ji, W., Morton, C. C., Meisler, M., Reeders, S. T., and Zhou, J. (1998). Identification of PKDL, a novel polycystic kidney disease 2-like gene whose murine homologue is deleted in mice with kidney and retinal defects. *J. Biol. Chem.* **273,** 25967–25973.

Oda, S., and Morisawa, M. (1993). Rises of intracellular Ca^{2+} and pH mediate the initiation of sperm motility by hyperosmolality in marine teleosts. *Cell Motil. Cytoskel.* **25,** 171–178.

Ohta, K., Sato, C., Matsuda, T., Toriyama, M., Vacquier, V. D., Lennarz, W. J., and Kitajima, K. (2000). Co-localization of receptor and transducer proteins in the glycosphingolipid-enriched, low density, detergent-insoluble membrane fraction of sea urchin sperm. *Glycoconj. J.* **17,** 205–214.

Okamura, N., Onoe, S., Kawakura, K., Tajima, Y., and Sugita, Y. (1990). Effects of a membrane-bound trypsin-like proteinase and seminal proteinase inhibitors on the bicarbonate-sensitive adenylate cyclase in porcine sperm plasma membranes. *Biochim. Biophys. Acta* **1035,** 83–89.

Okamura, N., Tajima, Y., Soejima, A., Masuda, H., and Sugita, Y. (1985). Sodium bicarbonate in seminal plasma stimulates the motility of mammalian spermatozoa through direct activation of adenylate cyclase. *J. Biol. Chem.* **260,** 9699–9705.

Okinaga, T., Ohashi, Y., and Hoshi, M. (1992). A novel saccharide, Xyl-3Gall-(SO_3^-)3,4Fuc-, is present in acrosome reaction-inducing substance (ARIS) of the starfish, *Asterias amurensis. Biophys. Biochem. Res. Commun.* **186,** 405–410.

O'Toole, C. M., Arnoult, C., Darszon, A., Steinhardt, R. A., and Florman, H. M. (2000). Ca^{2+} entry through store-operated channels in mouse sperm is initiated by egg ZP3 and drives the acrosome reaction. *Mol. Biol. Cell.* **11,** 1571–1584.

Parekh, A. B., and Penner, R. (1997). Store depletion and calcium influx. *Physiol. Rev.* **77,** 901–930.

Parrish, J. J., Susko-Parrish, J. L., and First, N. L. (1989). Capacitation of bovine sperm by heparin: Inhibitory effect of glucose and role of intracellular pH. *Biol. Reprod.* **41,** 683–699.

Publicover, S. J., and Barratt, C. L. (1999). Voltage-operated Ca^{2+} channels and the acrosome reaction: which channels are present and what do they do? *Hum. Reprod.* **14,** 873–879.

Rabow, L. E., Russek, S. J., and Farb, D. H. (1995). From ion currents to genomic analysis: Recent advances in $GABA_A$ receptor research. *Synapse* **21,** 189–274.

Ralt, D., Goldenberg, M., Fetterolf, P., Thompson, D., Dor, J., Mashiach, S., Garbers, D. L., and Eisenbach, M. (1991). Sperm attraction to a follicular factor(s) correlates with human egg fertilizability. *Proc. Natl. Acad. Sci. U.S.A.* **88,** 2840–2844.

Ramaro, C. S., and Garbers, D. L. (1985). Receptor-mediated regulation of guanylate cyclase activity in spermatozoa. *J. Biol. Chem.* **260,** 8390–8396.

Reddy, R., Smith, D., Wayman, G., Wu, Z., Villacres E. C., and Storm, D. R. (1995). Voltage-sensitive adenylyl cyclase activity in cultured neurons. A calcium-independent phenomenon. *J. Biol. Chem.* **270,** 14340–14346.

Repaske, D. R., and Garbers, D. L. (1983). A hydrogen ion flux mediates stimulation of respiratory activity by speract in sea urchin spermatozoa. *J. Biol Chem.* **258,** 6025–6029.

Reynaud, E., de Delatorre, L., Zapata, O., Liévano, A., and Darszon, A. (1993). Ionic bases of the membrane potential changes induced by speract in swollen sea urchin sperm. *FEBS Lett.* **329,** 210–214.

Roldán, E. R. S., Murase, T., and Shi Q. (1994). Exocytosis in spermatozoa in response to progesterone and zona pellucida. *Science* **266,** 1578–1581.

Rotem, R., Paz, G. F., Homonnai, Z. T., Kalina, M., Lax, J., Breitbart, H., and Naor, Z. (1992). Ca^{2+}-independent induction of acrosome reaction by protein kinase C in human sperm. *Endocrinology* **131,** 2235–2243.

Rufo, G. A., Jr., Singh, J. P., Babcock, D. F., and Lardy, H. A. (1982). Purification and characterization of a calcium transport inhibitor protein from bovine seminal plasma. *J. Biol. Chem.* **257,** 4627–4632.

Sabeur, K., Edwards, D. P., and Meizel, S. (1996). Human sperm plasma membrane progesterone receptor(s) and the acrosome reaction. *Biol. Reprod.* **54,** 993–1001.

Sánchez, D., Labarca, P., and Darszon, A. (2001). Sea urchin sperm cation-selective channels directly modulated by cAMP. *FEBS Lett.* **503,** 111–115.

Santi, C. M., Darszon, A., and Hernández-Cruz, A. (1996). A dihydropyridine-sensitive T-type Ca^{2+} current is the main Ca^{2+} current carrier in mouse primary spermatocytes. *Am. J. Physiol.* **271,** C1583–C1593.

Santi, C. M., Santos, T., Hernández, A., and Darszon, A. (1998). Properties of a novel pH-dependent Ca^{2+} permeation pathway present in male germ cells with possible roles in spermatogenesis and mouse sperm function. *J. Gen. Physiol.* **112,** 33–53.

Sato, Y., Son, J. H., Tucker, R. P., and Meizel, S. (2000). The zona pellucida-initiated acrosome reaction: Defect due to mutations in the sperm glycine receptor/Cl^- channel. *Dev. Biol.* **227,** 211–218.

Schackmann, R. W. (1989). Ionic regulation of the sea urchin sperm acrosome reaction and stimulation by egg-derived peptides. *In* "The Cell Biology of Fertilization" (H. Schatte and G. Schatten, eds.), pp. 3–28. Academic Press, San Diego.

Schackmann, R. W., and Chock, P. B. (1986). Alteration of intracellular $[Ca^{2+}]$ in sea urchin sperm by the egg peptide speract. *J. Biol. Chem.* **261,** 8719–8728.

Schackmann, R. W., and Shapiro, B. M. (1981). A partial sequence of ionic changes associated with the acrosome reaction of *Strongylocentrotus purpuratus. Dev. Biol.* **81,** 145–154.

Schackmann, R. W., Christen, R., and Shapiro, B. M. (1981). Membrane potential depolarization and increased intracellular pH accompany the acrosome reactions of sea urchin sperm. *Proc. Natl. Acad. Sci. U.S.A.* **78,** 6066–6070.

Schackmann, R. W., Christen, R., and Shapiro, B. M. (1984). Measurement of plasma membrane and mitochondrial membrane potentials in sea urchin sperm. *J. Biol. Chem.* **259,** 13914–13922.

Schackmann, R. W., Eddy, E. M., and Shapiro, B. M. (1978). The acrosome reaction and activation of *Strongylocentrotus purpuratus* sperm: Ion requirements and movements. *Dev. Biol.* **65**(2), 483–495.

Schreiber, M., Wei, A., Yuan, A., Gaut, J., Saito, M., and Salkoff, L. (1998). Slo3, a novel pH-sensitive K^+ channel from mammalian spermatocytes. *J. Biol. Chem.* **273,** 3509–3516.

Schultz, J. E., Klumpp, S., Benz, R., Schürhoff-Goeters, W. J. C., and Schmid, A. (1992). Regulation of adenylyl cyclase from *Paramecium* by an intrinsic potassium conductance. *Science* **255,** 322–326.

Schulz, J. R., Wessel, G. M., and Vacquier, V. D. (1997). The exocytosis regulatory proteins syntaxin and VAMP are shed from sea urchin during the acrosome reaction. *Dev. Biol.* **191,** 80–87.

SeGall, G. K., and Lennarz, W. J. (1979). Chemical characterization of the component of the jelly coat from sea urchin eggs responsible for induction of the acrosome reaction. *Dev. Biol.* **71,** 33–48.

Seifert, R., Scholten, A., Gauss, R., Micheva, A., Lichter, P., and Kaupp, U. B. (1999). Molecular characterization of a slowly gating human hyperpolarization-activated channel predominantly expressed in thalamus, heart and testis. *Proc. Natl. Acad. Sci. U.S.A.* **96,** 9391–9396.

Sengoku, K., Tamate, K., Takuma, N., Takaoka, Y., Yoshida, T., Nishiwaki, K., and Ishikawa, M. (1996). Involvement of protein kinases in platelet activating factor-induced acrosome reaction of human spermatozoa. *Mol. Hum. Reprod.* **2,** 401–404.

Serrano, C. J., Treviño, C. L., Felix, R., and Darszon, A. (1999). Votage-dependent Ca^{2+} channel subunit expression and immunolocalization in mouse spermatogenic cells and sperm. *FEBS Lett.* **462,** 171–176.

Shi, Q., and Roldán, E. R. S. (1995). Evidence that a $GABA_A$-like receptor is involved in progesterone-induced acrosomal exocytosis in mouse spermatozoa. *Biol. Reprod.* **52,** 373–381.

Shi, Q., Yuan, Y., and Roldán, E. R. S. (1997). γ-Aminobutiric acid (GABA) induces the acrosome reaction in human spermatozoa. *Mol. Hum. Reprod.* **3,** 677–683.

Shimizu, T., Kinoh, H., Yamaguchi, M., and Suzuki, N. (1990). Purification and characterization of the egg jelly macromolecules, sialoglycoprotein and fucose sulfate glycoconjugate, of the sea urchin *Hemicentrotus pulcherrimus. Dev. Growth Differ.* **32,** 473–487.

Shimomura, H. L., Dangott, L. J., and Garbers, D. L. (1986). Covalent coupling of a resact analog to guanylate cyclase. *J. Biol. Chem.* **261,** 15778–15782.

Sinclair, M. L., Wang, X. Y., Mattia, M., Conti, M., Buck, J., Wolgemuth, D. J., and Levin, L. R. (2000). Specific expression of soluble adenylyl cyclase in male germ cells. *Mol. Reprod. Dev.* **56,** 6–11.

Singh, S., Lowe, D. G., Thorpe, D. S., Rodriguez, H., Kuang, W. J., Dangott, L. J., Chinkers, M., Goeddel, D. V., and Garbers, D. L. (1988). Membrane guanylate cyclase is a cell-surface receptor with homology to protein kinase. *Nature (London)* **334,** 708–712.

Spungin, B., and Breitbart, H. (1996). Calcium mobilization and influx during sperm exocytosis. *J. Cell Sci.* **109,** 1947–1955.

Storey, B. (1995). Interactions between gametes leading to fertilization: The sperm's eye view. *Reprod. Fertil. Dev.* **7,** 927–942.

Storey, B. T., Hourani, C. L., and Kim, J. B. (1992). A transient rise in intracellular Ca^{2+} is a precursor reaction to the zona pellucida-induced acrosome reaction in mouse sperm and is blocked by the induced acrosome reaction inhibitor 3-quinuclidinyl benzilate. *Mol. Reprod. Dev.* **32**(1), 41–50

Summers, R. G., and Hylander, B. L. (1975). Species specific acrosome reaction and primary gamete binding in echinoids. *Exp. Cell Res.* **96,** 63–68.

Suzuki, N. (1995). Structure, function and biosynthesis of sperm-activating peptides and fucose sulfate glycoconjugate in the extracellular coat of sea urchin eggs. *Zoolog. Sci.* **12,** 13–27.

Suzuki, N., and Yoshino, K. (1992). The relationship between amino acid sequences of sperm-activating peptides and the taxonomy of equinoids. *Comp. Biochem. Physiol.* **102B,** 679–690.

Suzuki, N., Nomura, K., and Ohtake, H. (1980). Sperm activating peptides obtained from jelly coat of sea urchin eggs. *Zool. Mag.* **89,** 350.

Suzuki, N., Nomura, K., Ohtake, H., and Isaka, S. (1981). Purification and primary structure of sperm-activating peptides from the jelly coat of sea urchin eggs. *Biochem. Biophys. Res. Commun.* **99,** 1238–1244.

Suzuki, N., Shimomura, H., Radany, E. W., Ramarao, C. S., Ward, G. E., Bentley, J. K., and Garbers, D. L. (1984). A peptide associated with eggs causes a mobility shift in a major plasma membrane in a major plasma membrane protein of spermatozoa. *J. Biol. Chem.* **259,** 14874–14879.

Tesarik, J., Carreras, A., and Mendoza, C. (1993a). Differential sensitivity of progesterone- and zona pellucida-induced acrosome reactions to pertussis toxin. *Mol. Reprod. Dev.* **34,** 183–189.

Tesarik, J., Moos, J., and Mendoza, C. (1993b). Stimulation of a protein tyrosine phosphorylation by progesterone receptor on the cell surface of human sperm. *Endocrinology* **133,** 328–133.

Thomas, P., and Meizel, S. (1989). Phosphatidyl inositol 4,5-bisphosphate hydrolysis in human sperm stimulated with follicular fluid or progesterone is dependent upon Ca^{2+} influx. *Biochem. J.* **264,** 539–546.

Tilney, L. G. (1985). The acrosomal reaction. *In* "Biology of Fertilization" (C. B. Metz and A. Monroy, eds.), Vol. 2, pp. 157–213. Academic Press, Orlando.

Tiwari-Woodruf, S., Peterson, R. N., and Cox, T. (1994). Calcium channels from boar sperm plasma membranes with L-type characteristics: Sensitivity to Bay-K enantiomers. *Biophys J.* **66,** A423.

Tombes, R. M., and Shapiro, B. M. (1985). Metabolite channeling: A phosphorylcreatine shuttle to mediate high energy phosphate transport between sperm mitochondria and tail. *Cell* **41,** 325–334.

Treviño, C. L., Santi, C. M., Beltrán, C., Hernández-Cruz, A., Darszon, A., and Lomelí, H. (1998). Localization of IP_3 and ryanodine receptors during mouse spermatogenesis: Possible funcional implications. *Zygote* **6,** 159–172.

Trimmer, J. S., and Vacquier, V. D. (1986). Activation of sea urchin gametes. *Annu. Rev. Cell Biol.* **2,** 1–26.

Trimmer, J. S., Schackmann, R. W., and Vacquier, V. D. (1986). Monoclonal antibodies increase intracellular Ca in sea urchin spermatozoa. *Proc. Natl. Acad. Sci. U.S.A.* **83,** 9055–9059.

Trimmer, J. S., Trowbridge, I. S., and Vacquier, V. D. (1985). Monoclonal antibody to a membrane glycoprotein inhibits the acrosome reaction and associated Ca^{2+} and H^+ fluxes of sea urchin sperm. *Cell* **40,** 697–703.

Tubb, D. J., Kopf, G. S., and Garbers, D. L. (1979). Starfish and horseshoe crab egg factors cause elevations of cyclic nucleotide concentrations in spermatozoa from starfish and horseshoe crabs. *J. Reprod. Fertil.* **56,** 539–542.

Turner, K. O., and Meizel, S. (1995). Progesterone-mediated efflux of cytosolic chloride during the humane sperm acrosome reaction. *Biochem. Biophys. Res. Commun.* **213,** 774–780.

Ushiyama, A., Araki, T., Chiba, K., and Hoshi, M. (1993). Specific binding of acrosome-reaction-inducing substance to the head of starfish spermatozoa. *Zygote* **1,** 121–127.

Vacquier, V. D. (1998). Evolution of gamete recognition proteins. *Science* **281,** 1995–1998.

Vacquier, V. D., and Moy, G. W. (1977). Isolation of bindin: The protein responsible for adhesion of sperm to sea urchin eggs. *Proc. Natl. Acad. Sci. U.S.A.* **74,** 2456–2460.

Vacquier, V. D., and Moy, G. W. (1997). The fucose sulfate polymer of egg jelly binds to sperm REJ and is the inducer of the sea urchin sperm acrosome reaction. *Dev. Biol.* **192,** 125–135.

Vannier, B., Peyton, M., Boulay, G., Brown, D., Qin N., Jiang, M., Zhu, X., and Birnbaumer, L. (1999). Mouse trp2, the homologue of the human trpc2 pseudogene, encodes mTrp2, a store depletion-activated capacitative Ca^{2+} entry channel. *Proc. Natl. Acad. Sci. U.S.A* **96,** 2060–2064

Villanueva-Diaz, C., Arias-Martinez, J., Bermejo-Martinez, L., and Vadillo-Ortega F. (1995). Progesterone induces human sperm chemotaxis. *Fertil. Steril.* **64,** 1183–1188.

Villanueva-Diaz, C., Vadillo-Ortega, F., Kably-Ambe, A., Diaz-Perez, M., and Krivitzky, S. K. (1990). Evidence that human follicular fluid contains a chemoattractant for spermatozoa. *Fertil. Steril.* **54,** 1180–1182.

Visconti, P. E., and Kopf, G. S. (1998). Regulation of protein phosphorylation during sperm capacitation. *Biol. Reprod.* **59,** 1–6.

Visconti, P. E., Bailey, J. L., Moore, G. D., Pan, D., Olds-Clarke, P., and Kopf, G. S. (1995a). Capacitation of mouse spermatozoa. I. Correlation between the capacitation state and protein tyrosine phosphorylation. *Development* **121,** 1129–1137.

Visconti, P. E., Galantino-Homer, H., Ning, X., Moore, G. D., Valenzuela, J. P., Jorgez, C. J., Alvarez, J. G., and Kopf, G. S. (1999). Cholesterol efflux-mediated signal transduction in mammalian sperm. beta-cyclodextrins initiate transmembrane signaling leading to an increase in protein tyrosine phosphorylation and capacitation. *J. Biol. Chem.* **274,** 3235–3242.

Visconti, P. E., Moore, G. D., Bailey, J. L., Lecler P., Connors, S. A., Pan, D., Olds-Clarke P., and Kopf, G. S. (1995b). Capacitation of mouse spermatozoa. II. Protein tyrosine phosphorylation and capacitation are regulated by a cAMP-dependent pathway. *Development* **121,** 1139–1150.

Walensky, L. D., and Snyder, S. H. (1995). Inositol 1,4,5-trisphosphate receptors selectively localized to the acrosomes of mammalian sperm. *J. Cell Biol.* **130,** 857–859.

Ward, G. E., and Vacquier, V. D. (1983). Dephosphorylation of a major sperm membrane protein is induced by egg jelly during sea urchin fertilization. *Proc. Natl. Acad. Sci. U.S.A.* **30,** 5578–5582.

Ward, G. R., and Kopf, G. (1993). Molecular events mediating sperm activation. *Dev. Biol.* **158,** 9–34.

Ward, G. E., Brokaw, C. J., Garbers, D. L., and Vacquier, V. D. (1985a). Chemotaxis of *Arbacia punctulata* spermatozoa to resact, a peptide from the egg jelly layer. *J. Cell Biol.* **101,** 2324–2329.

Ward, G. E., Garbers, D. L., and Vacquier, V. D. (1985b). Effects of extracellular egg factors on sperm guanylate cyclase. *Science* **227,** 768–770.

Ward, G. E., Moy, G. W., and Vacquier, V. D. (1986). Phosphorylation of membrane-bound guanylate cyclase of sea urchin spermatozoa. *J. Cell Biol.* **103,** 95–101.

Ward, C. R., Storey, B. T., and Kopf, G. S. (1994). Selective activation of Gi1 and Gi2 in mouse sperm by the zona pellucida, the egg's extracellular matrix. *J. Biol. Chem.* **269,** 13254–13258.

Wassarman, P. M., Jovine, L., and Litscher, E. S. (2001). A profile of fertilization in mammals. *Nat. Cell Biol.* **3,** E59-E64.

Watkins, H. D., Kopf, G. S., and Garbers, D. L. (1978). Activation of sperm adenylate cyclase by factors associated with eggs. *Biol. Reprod.* **19,** 890–894.

Wennemuth, G., Westenbroek, R. E., Xu, T., Hille, B., and Babcock, D. F. (2000). CaV2.2 and CaV2.3 (N- and R-type) Ca^{2+} channels in depolarization-evoked entry of Ca^{2+} into mouse sperm. *J. Biol. Chem.* **275,** 21210–21217.

Westenbroek, R. E., and Babcock, D. F. (1999). Discrete regional disributions suggest diverse functional roles of calcium channel α_1 subunits in sperm. *Dev. Biol.* **207,** 457–469.

Weyand, I., Godde, M., Frings, S., Welner, J., Muller, F, Altenhofen, W., Hatt, H., and Kaupp, B. (1994). Cloning and functional expression of a cyclic-nucleotide-gated channel from mammalian sperm. *Nature (London)* **368,** 859–863.

White, D. R., and Aitken, R. J. (1989). Relationship between calcium, cyclic AMP, ATP and intracellular pH and the capacity of hamster spermatozoa to express hyperactivated motility. *Gamete Res.* **22,** 163–177.

Wickman, K. D., and Clapham, D. E. (1995). G-protein regulation of ion channels. *Curr. Opin. Neurobiol.* **5,** 278–285.

Wiesner, B., Weiner, J., Middendorff, R., Hagen, V., Kaupp, U. B., and Weyand, I. (1998). Cyclic nucleotide-gated channels on the flagellum control Ca^{2+} entry into sperm. *J. Cell Biol.* **142,** 473–484.

Wistrom, C. A., and Meizel, S. (1993). Evidence supporting involvement of a unique human steroid receptor/Cl^- channel complex in the progesterone-initiated acrosome reaction. *Dev. Biol.* **159,** 679–690.

Yamaguchi, M., Niwa, T., Kurita, M., and Suzuki, N. (1988). The participation of speract in the acrosome reaction of *Hemocentrotus pulcherrimus* spermatozoa by the egg jelly molecules, fucose-rich glycoconjugate and sperm-activating peptide I. *Dev. Growth Differ.* **30,** 159–167.

Yanagimachi, R. (1988). Mammalian fertilization. *In* "The Physiology of Reproduction" (E. Knobol and J. D. Neil, eds.), Vol. 1, pp. 135–185. Raven Press, New York.

Yanagimachi, R. (1994). Mammalian fertilization. *In* "The Physiology of Reproduction" (E. Knobil and J. D. Neil, eds.), pp. 189. Raven Press, New York.

Yoshino, K., Kajura, H., Nomura, K., Takao, T., Shimonishi, Y., Kurita, M., Yamaguchi, M., and Suzuki, N. (1989). A halogenated amino acid-containing sperm activating peptide and its related peptides isolated from the egg jelly of sea urchins, *Tripneuste gratilla, Pseudoboletia maculata, Strongylocentrotus nudus, Echinometra mathei* and *Heterocentrotus mammillatus. Comp. Biochem. Physiol.* **94B,** 739–751.

Zapata, O., Ralston, J., Beltrán, C., Parys, J. B., Chen, J. L., Longo, F. J., and Darszon, A. (1997). Inositol triphosphate receptors in sea urchin sperm. *Zygote* **5,** 355–364.

Zeng, Y., Clark, E. N., and Florman, H. M. (1995). Sperm membrane potential: Hyperpolarization during capacitation regulates zona pellucida-dependent acrosomal secretion. *Dev. Biol.* **171,** 554–563.

Zeng, Y., Oberdorf, J. A., and Florman H. M. (1996). pH regulation in mouse sperm: Identification of Na^+-, Cl^--, and HCO_3^--dependent and arylaminobenzoate-dependent regulatory mechanisms and characterization of their roles in sperm capacitation. *Dev. Biol.* **173,** 510–520.

8

Function of the Sperm Acrosome

George L. Gerton

Center for Research on Reproduction and Women's Health, University of Pennsylvania Medical Center, Philadelphia

I. Introduction

A. Focus of Review

The sperm acrosome, an exocytotic vesicle on the anterior aspect of the sperm head, is essential for fertilization (Kopf and Gerton, 1991). Males whose spermatozoa have poorly formed acrosomes or lack acrosomes altogether are infertile (Baccetti *et al.,* 1991; Escalier *et al.,* 1992; Schill, 1991; Sotomayor and Handel, 1986) and cannot naturally reproduce without intervention by assisted reproductive technologies such as intracytoplasmic sperm injection (Hamberger *et al.,* 1998). In addition, several studies have indicated that permature loss of the acrosome can be a symptom of subfertility (Bartoov *et al.,* 1994; Benoff *et al.,* 1993; Fenichel *et al.,* 1991; Marshburn *et al.,* 1991; Mundy *et al.,* 1994). Despite the

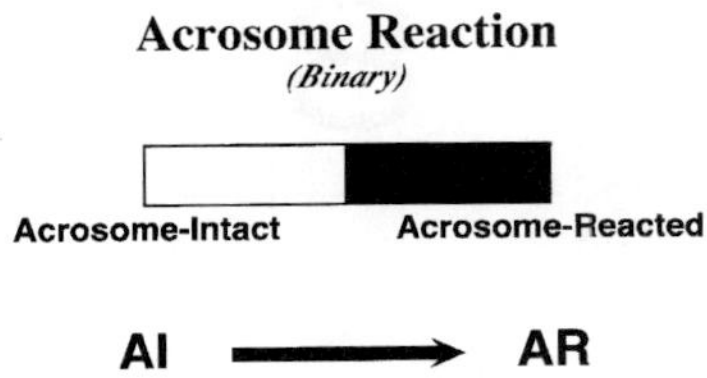

FIGURE 8.1 Depiction of the acrosome reaction model for explaining the function of the acrosome in fertilization. This model emphasizes two states for spermatozoa: acrosome-intact and acrosome-reacted. Using a computing metaphor, this is a binary system; the acrosome is either "on" or "off." Any intermediates between the two extremes are inconsequential. Likewise, the term *reaction* implies a one-step process as depicted in the equation.

well-documented importance of the acrosome, its actual role in mammalian fertilization is poorly understood. Potential functions for the acrosome involve issues of sperm adhesion to the zona pellucida, zona pellucida penetration, and gamete fusion.

In reviewing the function of the acrosome for this chapter, I will assess the currently accepted model for the states of the acrosome during fertilization. This model posits that acrosomal dynamics are represented by two states: acrosome-intact and acrosome-reacted (Figure 8.1). To use computer parlance, this is a binary or digital system; the acrosome is either "on" or "off." Any intermediates between these two states do not exist or are insignificant; the major concern is whether the acrosome is present or not. However, the prevalence of controversies in the field of fertilization suggests that the biology of acrosomal dynamics may be more subtle (and complicated). To return to the computer jargon, the states of the acrosome may be more akin to an analog system, one where the status of the acrosome is continuously variable. Rather than an on/off or "black or white" model, the analog paradigm postulates that there are transitional intermediates between the two extremes (acrosome intact and acrosome reacted) and that the transitional intermediates represent important functional states during the fertilization process.

During the course of this treatise, the function of the acrosome will be addressed in four sections. First, I discuss some salient features about the acrosome, including a description of this interesting cellular organelle, morphological points, the nature of the acrosome, and the biogenesis and maturation of this structure. Next, I address the current "binary" model for acrosomal dynamics, which, for historical and semantic reasons in this review, I call the acrosome reaction model. Then, I outline the basis of the "analog" paradigm for acrosomal dynamics, which I have termed the acrosomal exocytosis model. As a further semantic clarification, the term "acrosome reaction" is used only in reference to the acrosome reaction model because it connotes a two-step process, similar to a chemical reaction of the sort depicted in Figure 8.1. It is my contention that the term "acrosomal exocytosis" is a more accurate term to describe acrosomal dynamics because it implies the secretory nature of the acrosome and infers that the acrosome utilizes exocytotic machinery similar to that in other secretory cells. As discussed below, the homology of acrosomal exocytosis to other secretory systems has growing experimental support.

B. WHAT IS THE ACROSOME?

1. Features of the Acrosome

The features of acrosomes vary from species to species, leading some to consider that the functional importance of the acrosome may differ between species. However, there are several common morphological, structural, and compositional properties of the acrosome worth noting for this review. Morphologically, the spermatozoa of some nonmammalian species do not have acrosomes (Baccetti *et al.,* 1989), but in those spermatozoa that do, the acrosome lies on the anterior aspect of the sperm head (Eddy and O'Brien, 1994). Acrosome size varies remarkably from one species to the next. For example, the mouse sperm acrosome is very difficult to detect without specific histochemical staining methods whereas the acrosome of the guinea pig spermatozoon is quite large and has a prominent, apical segment that can be readily seen by standard phase-contrast or Nomarski differential interference contrast microscopy.

The acrosome can be considered a compartmentalized structure. In all cases, the contents of the acrosome are enclosed by a single, continuous acrosomal membrane that can be further delineated into two subdomains. The inner acrosomal membrane is closely apposed to the nuclear membranes whereas the outer acrosomal membrane is present just under the plasma membrane overlying the acrosome (Figure 8.2). During the course of acrosomal secretion, the outer acrosomal membrane fuses with the plasma membrane to form hybrid membrane vesicles that are eventually released from the spermatozoon. Some of the vesicles may be released with acrosomal matrix material in a complex known as the acrosomal shroud (VandeVoort *et al.,* 1997). The inner acrosomal membrane then becomes the de facto plasma membrane in the acrosomal region. In addition to the membranes of the acrosome, the acrosome as a whole may be considered to be regionalized. The apical segment is the section of the acrosome that extends beyond the tip of the sperm nucleus. In some species such as the guinea pig, the apical segment is quite prominent. The principal segment is that region of the acrosome that is in contact with the anterior region of the sperm nucleus. Finally, the equatorial segment delineates the posterior margin of the acrosome and, after acrosomal secretion, is frequently is demarcated by a "lip" formed by the residual outer acrosomal membrane/plasma membrane junction. Because of the acidic pH within the acrosome, the microenvironments near the outer acrosomal membrane and inner acrosomal membrane may be quite distinct, especially during the course of acrosomal secretion when the external medium with a neutral pH begins to enter the acrosome and mix with its contents.

Within the acrosomes of the spermatozoa of some species (Olson and Winfrey, 1985a; Olson and Winfrey, 1994; Olson *et al.,* 1988; Westbrook-Case *et al.,* 1994),

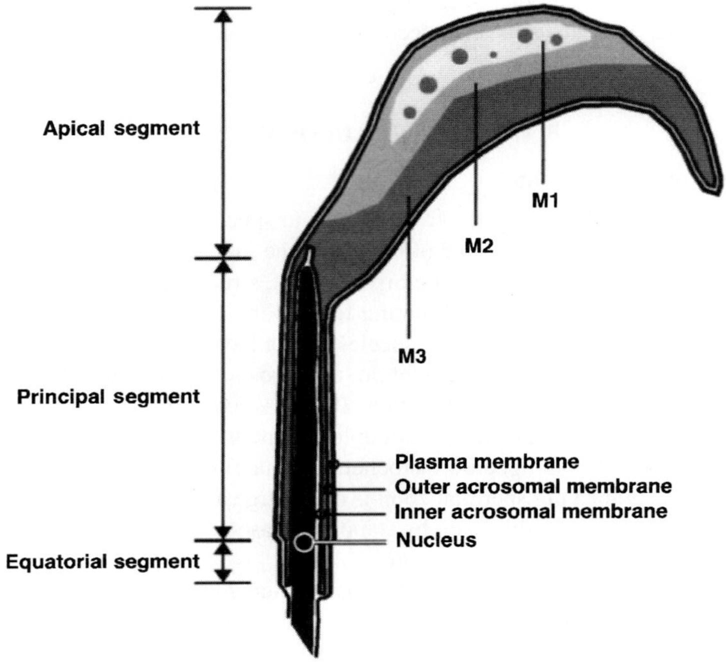

FIGURE 8.2 Graphical illustration of the domain structure of the guinea pig sperm acrosome. The acrosome of this species' sperm is quite large and is partitioned into clear morphological domains (M1, M2, M3). As cited in the text and listed in Table 8.1, spermatozoa of other mammals, including mouse, rat, human, bull, and hamster, have detectable domains within the acrosomes.

the contents can be seen to form morphologically distinct domains that can generally be distinguished by transmission electron microscopy as varying degrees of electron dense material (Figure 8.2). In the guinea pig sperm acrosome, the domains are designated as M1, M2, and M3 (Westbrook-Case *et al.,* 1994); different acrosomal components have distinctly different distributions among the different morphological domains (Table 8.1). Although not widely recognized, the relatively smaller acrosomes of mouse spermatozoa also contain distinct domains that can be identified immunochemically; for instance, the 155,000 M_r protein recognized by monoclonal antibody mMC101 is specifically localized in the cortex of the anterior region of the mouse sperm acrosome (Toshimori *et al.,* 1995).

The domain concept can be extended to biochemical properties as well as morphological features. By fractionating the spermatozoa of some species with the proper buffering conditions and a nonionic detergent such as Triton X-100, it is possible to isolate a particulate, membrane-free component of the acrosome, known as the acrosomal matrix (Hardy *et al.,* 1991; Huang *et al.,* 1985; Hyatt and

TABLE 8.1 Examples of Proteins Found within Specific Acrosomal Domains

Species	Protein name	Domain	Reference
Guinea pig	Proacrosin	M2, M3	Hardy *et al.* (1991)
	AM50	M3	Westbrook-Case *et al.* (1994)
	AM67 (sp 56 ortholog)	M1 (absent from spherical zones)	Foster *et al.* (1997)
	CRISP-2 (autoantigen 1)	M1, M2, M3	Hardy *et al.* (1991)
	Dipeptidyl peptidase II	M1 (absent from spherical zones)	Talbot and DiCarlantonio (1985)
	Acrin 1 (MN7 antigen)	M1 spherical zones and outer acrosomal matrix-associated materials posterior to the dorsal bulge but not on the ventral outer acrosomal membrane	Saxena *et al.* (1999); Yoshinaga *et al.* (1998)
Mouse	mMC101 antigen	Anterior acrosome	Toshimori *et al.* (1995)
Mouse, rat, human, hamster	Equatorin	Equatorial segment	Toshimori *et al.* (1992, 1998)
Hamster	AM29 and AM22 antigens	M1 and M2 (excluded from the equatorial segment)	Olson *et al.* (1998)
Bovine	OMC32 (SP-10 homolog)	Outer acrosomal membrane-associated matrix complex and the inner acrosomal membrane or the equatorial segment	Olson *et al.* (1997)
Human	SP-10	Principal segment and posterior bulb of equatorial segment	Foster *et al.* (1994)

Gwatkin, 1988). In some cases, it is possible to isolate an acrosomal particle with its associated outer acrosomal membrane and plasma membrane (Olson and Winfrey, 1985a,b; Olson *et al.,* 1987). Other acrosomal proteins can be recovered in a soluble form from these fractionated sperm preparations. The recognition of the acrosomal matrix and the soluble compartment as discrete entities is key to an understanding the acrosomal exocytosis model.

2. Lysosome or Secretory Granule?

As indicated in the commentary above, the acrosome should be considered to be an exocytotic organelle. However, this has not always been the case. Earlier studies classified the acrosome as a "specialized lysosome" (Allison and Hartree, 1970). This classification was based on two principal findings: (1) vital staining with euchrysine 3R demonstrated that acrosomes have an acidic internal pH, as do lysosomes; (2) the acrosome was reported to contain several hydrolytic activities similar to do those found within lysosomes (acid phosphatase, arylsulfatase, β-*N*-acetylglucosaminidinase, phospholipase A, and proteases). The attribution of these lysosomal features to the acrosome has frequently led to the incorrect assumption that the acrosome is a bag of readily soluble digestive or degradative enzymes.

As the study of cellular secretion has evolved over the past 30 years, it has become clear that it is more accurate to refer to the acrosome as a regulated secretory vesicle. As reviewed by Burgess and Kelly (1987), the hallmarks of regulated secretion, as contrasted to the constitutive process, are (1) that secretion is coupled to an extracellular stimulus (e.g., the zona pellucida-stimulated acrosomal secretion in mouse spermatozoa), (2) that secretory products are concentrated and condensed, and (3) that secretory granules are stored for long periods of time. In addition, regulated secretory granules have an electron-opaque content known as "dense cores." These are very stable structures that can often be seen after secretion. In fact, the membranes surrounding regulatory secretory granules can frequently be removed without disturbing the condensed core. As discussed above, this is the case when the acrosomal matrices are isolated following detergent treatment of spermatozoa (Hardy *et al.,* 1991). All of these properties apply to the sperm acrosome. Finally, components of the secretory machinery found in other cells have begun to be identified in association with the acrosomal membranes (Katafuchi *et al.,* 2000; Michaut *et al.,* 2000, 2001; Ramalho-Santos *et al.,* 2000; Schulz *et al.,* 1997, 1998).

C. BIOGENESIS AND MATURATION

3. Biogenesis during Spermatogenesis

Many advances have been made during the past decade concerning the pathways involved in intracellular protein transport, especially as it relates to secretion. These pathways appear to be operative in the biogenesis of the acrosome; however, space does not permit here a detailed discusson of the field, and several reviews are available (Allan and Balch, 1999; Gerst, 1999). Briefly, secretory protein synthesis begins on cytoplasmic ribosomes, which are then targeted to the rough endoplasmic reticulum via the signal sequence of the nascent polypeptide. As translation continues, the nascent polypeptide is extruded through the endoplasmic reticulum membrane and into the lumen, where initial steps of glycosylation, protein folding, and multimerization occur. Additional posttranslational modification steps (glycosylation reactions, proteolytic processing, disulfide bonding, etc.) take place as the protein moves from the endoplasmic reticulum through the Golgi into the trans-Golgi network (TGN).

The selective aggregation of regulated secretory proteins in the TGN is thought to be a key step in their sorting to secretory granules (Burgess and Kelly, 1987; Chanat and Huttner, 1991; Seethaler and Huttner, 1991; Tooze, 1991; Tooze *et al.,* 1993). Many studies have indicated that secretory granule proteins, such as the chromogranins and secretogranins that are widespread in endocine cells and neurons but absent in spermatogenic cells, have the ability to aggregate regulated secretory proteins selectively while excluding constitutively secreted proteins (Huttner *et al.,* 1991; Tooze, 1991, 1992; Tooze and Huttner, 1990; Yoo, 1993). In addition, disulfide cross-linking of chromogranin B is required for proper sorting to the regulated pathway; if these bonds are unable to form, this protein is constitutively secreted (Chanat *et al.,* 1993). These data emphasize the importance of secondary and tertiary structure information in targeting to the regulated secretory granules. Some of these processes involve chaperonin proteins such as calreticulin and heat-shock proteins. Calmegin, a testis-specific homolog of the chaperonin calnexin, may be involved in this process because calmegin-null mice are virtually infertile as a consequence of poor adhesion of spermatozoa to the zona pellucida (Ikawa *et al.,* 1997). Similar to chromogranin and secretogranins, the acrosomal matrix may include aggregating factors that interact with other components to sort and maintain acrosomal proteins in the developing acrosome.

After successful aggregation, the regulated secretory proteins interact with the membranes of the TGN and bud off to form the immature secretory granule. Not all of the constitutive proteins are segregated from the regulated secretory proteins following passage from the TGN into the immature secretory granule (Bauerfeind and Huttner, 1993). In many secretory cells, the immature granules then fuse to form the mature secretory granule. This is the case for the acrosome as well. Small vesicles containing acrosomal proteins can first be detected in late pachytene spermatocytes; following meiosis, the vesicles are presumably distributed to the daughter spermatids (Anakwe and Gerton, 1990). The acrosomal vesicles then fuse early during spermiogenesis, and the resulting single acrosomal vesicle associates with the nucleus. The site of interaction between the acrosomal vesicle membrane and the nuclear membrane regionalizes the acrosomal membrane into the presumptive inner and outer acrosomal membrane domains.

In all regulated secretory cells, condensation of the contents of the secretory vesicles occurs when vesicles lacking regulated secretory product bud from the maturing secretory granule, recycling the excess membrane and unneeded lumenal material from the fused immature secretory granules back to the TGN. The removal of excess material by vesicles apparently occurs from the developing spermatid acrosome, because carbohydrate-containing vesicles are associated with the developing acrosomes of ram spermatids at a time when morphometry detects a decrease in the volume of the acrosome (Courtens, 1978). Similar periacrosomal vesicles have also been seen by others (Griffiths *et al.,* 1981; Pelletier and Friend, 1983; Sandoz, 1970) and it may be that clathrin-coated vesicles remove material from the developing acrosome, as is suspected in other secretory systems (Tooze *et al.,* 1993). In addition, evidence indicates that the tubulobulbar complexes, cytoplasmic projections from the heads of late spermatids into the invaginations of Sertoli cell plasma membranes, may be involved in eliminating excess acrosomal contents prior to spermiation (Tanii *et al.,* 1999). Thus, some constitutive proteins or other nonessential components may be transiently packaged into the immature acrosome. The changes described by Courtens (1978) in spermatids and observed by others in somatic cells could allow the recycling of sorting receptors to the TGN, could increase the fidelity of segregation of acrosomal proteins from constitutive secretory proteins and lysosomal proteins, and/or could remove certain products resulting from the processing of regulated secretory components.

Studies of secretory granule biogenesis have also demonstrated several roles for GTP-binding proteins in the protein trafficking pathway (Ferro-Novick and Novick, 1993). GTP-binding proteins have been shown to be involved in the transport of vesicles from the endoplasmic reticulum, through the Golgi and to secretory granules. Heterotrimeric G-proteins may also be involved in acrosome biogenesis, because some subtypes have been localized to developing secretory granules of many cells, including the developing and mature acrosomes of mouse spermatogenic cells and spermatozoa as well as guinea pig spermatozoa (Ahnert-Hilger *et al.,* 1994; Garty *et al.,* 1988; Glassner *et al.,* 1991; Hinsch *et al.,* 1992; Karnik *et al.,* 1992; Konrad *et al.,* 1995).

4. Epididymal Maturation

Following spermiation, the acrosome is altered as the spermatozoa traverse the epididymis. For example, Fawcett and Hollenberg (1963) demonstrated that there is a progressive morphological differentiation of the structure of the large acrosome of guinea pig spermatozoa as they travel down the epididymis. The acrosome of testicular spermatozoa is relatively coplanar with the nucleus and it terminates distally as a rounded, blunt-ended tip. However, after transit through the epididymis, the morphology is modified such that the distal boundary of the acrosome now tapers to a slender edge that, in histological sections, can be seen to be curled. In addition, the apical segment of the acrosome inclines ventrally and is no longer coplanar with the nucleus. These studies have been extended by the examination of the localizaton of antigen MN7 during epididymal maturation in the guinea pig (Yoshinaga *et al.,* 1998). This antigen was initially distributed throughout the electron-lucent dorsal matrix in immature spermatozoa but became more restricted to spherical bodies within the electron-lucent area of the M1 acrosomal domain as the spermatozoa matured in the epididymis. Whether these morphological and structural changes are related to the acquisition of fertility as the spermatozoa traverse the epididymis is not known.

Although the actual biochemical basis underlying the morphological alterations in the acrosome has yet to be explained, internal components of the acrosome, such as the protease zymogen proacrosin, do become modified during the course of epididymal transit. In extracts of guinea pig testis, caput epididymis, and corpus epididymis the major band of proacrosin has an apparent molecular mass of 55,000 kDa, although a M_r 50,000 minor form begins to appear in the corpus epididymis. By contrast, proacrosin of cauda epididymis and vas deferens spermatozoa is M_r 50,000. Further examination demonstrated that the oligosaccharides of proacrosin are altered during epididymal transit and that this modification occurs in the corpus epididymis. Proacrosins of other species have also been shown to be altered as a function of epididymal transit (Baba *et al.,* 1994b; Mukerji and Meizel, 1979; NagDas *et al.,* 1992). The ability to detect MC101, an antigen of the cortex of the apical region of mouse sperm acrosomes, increases as the spermatozoa progress through the epididymis (Toshimori *et al.,* 1995). The regulation of these processes is not understood but may be controlled, in part, by secreted proteins from the epididymis that bind to spermatozoa. Several epididymal proteins are known to bind to the surface of spermatozoa in the periacrosomal region (Cooper, 1998; Klinefelter and Hamilton, 1985; Moore *et al.,* 1994) and it is conceivable that alterations to the intraacrosomal environment may result from the interactions of specific ligands with signal-transducing receptors on the sperm surface.

5. Capacitation

The function of the acrosome is also affected by the state of capacitation of the spermatozoa. As defined by Yanagimachi (1994), capacitation consists of the physiological (functional) changes that render spermatozoa competent to fertilize an egg.[1] Yanagimachi also notes that many investigators surmise that a major aspect of capacitation is the removal or alteration of a stabilizer or protective coat from the sperm plasma membrane, sensitizing the spermatozoa to fertilization conditions and promoting their ability to interact with eggs. One such proposed substance is the acrosome stabilizing factor, which has been termed a "decapacitation" factor because of the reversible nature of its action (Thomas *et al.,* 1986; Wilson and Oliphant, 1987). Thus, any consideration of the function of the acrosome must take into account the influence of capacitation.

II. THE PREVAILING VIEW: THE ACROSOME REACTION MODEL

A. ACROSOMAL DYNAMICS AS A TWO-STATE OR BINARY REACTION

As mentioned previously, the prevailing view of acrosomal dynamics, the acrosome reaction model, emphasizes the acrosome-intact and acrosome-reacted states of spermatozoa. This paradigm deemphasizes the importance of intermediates and does not promote a role for acrosomal matrix proteins in sperm–zona interactions. This perspective of acrosomal dynamics is explained very thoroughly in the review by Yanagimachi (1994). In this model, the outer acrosomal membrane and the plasma membrane fuse in multiple places, allowing for the rapid release or exposure of the acrosomal components thought, principally, to be enzymes (Figure 8.3). The vesiculated intermediate (B in Figure 8.3) is considered to be short-lived; the acrosomal matrix components either dissipate or are shed with vesiculated

[1] For the purposes of this review, I define *egg* as the female gamete capable of being fertilized. In mammals, this is an MII metaphase-arrested oocyte. In some other species, meiosis may be completed prior to fertilization by the spermatozoon.

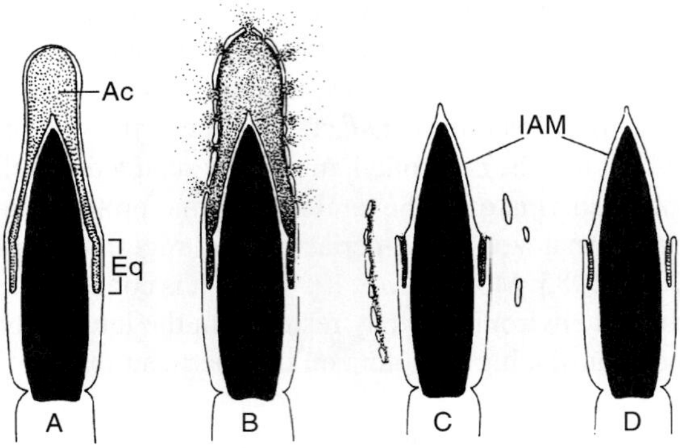

FIGURE 8.3 The acrosome reaction model. In this view, spermatozoa are stimulated to release their contents by a biological stimulus such as the zona pellucida. Hybrid membrane vesicles resulting from the fusion of the outer acrosomal membrane with the plasma membrane are shed from the sperm cell. Some vesicles may be seen as a complex known as the acrosomal shroud. The acrosomal contents (Ac) are immediately lost from the sperm cell by rapid diffusion. Advocates of this model generally discount or ignore the presence of a residual acrosomal matrix. Acrosome-reacted spermatozoa are frequently depicted with completely bare inner acrosomal membranes (IAM), as shown in panel D; Eq, equatorial segment. Reproduced with permission from Yanagimachi (1994).

membranes from the sperm surface, leaving the bare inner acrosomal membrane. Thus, in the acrosome reaction model, the primary consideration governing the role of the acrosomal components is whether the acrosomal membranes are intact or completely vesiculated.

B. SPONTANEOUS SECRETION

How does the acrosome reaction model explain the loss of acrosomes? In this view, acrosomes can be lost from spermatozoa via normal physiological events, such as those occurring during fertilization (true acrosome reactions), or they may become detached through mechanical shearing or other processes such as occurs when moribund or dead spermatozoa degenerate (false acrosome reactions) (Bedford, 1970). In addition, the process of acrosomal secretion can occur adventitiously, but the acrosome reaction model categorizes these spontaneous acrosome reactions as false, nonphysiological, or spurious because this model assumes that physiologically important acrosome reactions take place when the spermatozoa encounter the zona pellucida, not in the medium. However, spontaneous acrosome reactions do not occur accidentially. For example, capacitation greatly increases the occurrence of spontaneous acrosome reactions. Furthermore, the incidence of spontaneous acrosome reactions is dependent on the species, the animal strain, medium composition, state of epididymal storage, pre- and postejaculation conditions, and immunological condition of the spermatozoa (Yanagimachi, 1994). Importantly, although spontaneous acrosome reactions are viewed as nonphysiological, Yanagimachi points out that spermatozoa that have undergone spontaneous acrosome reactions are still capable of fertilizing zona-free eggs.

C. SPERM–ZONA PELLUCIDA INTERACTIONS

1. Acrosomal Status

As mentioned above and discussed in other chapters of this book, the acrosome reaction model assumes that sperm–zona pellucida interactions are governed by the acrosomal status of the sperm cell. Building on the pioneering mouse sperm studies of Saling *et al.* (1979) and others examining the cells from various species such as rat, hamster, rabbit, cattle, pig, sheep, monkey, and human [see Yanagimachi (1994) for references], the concept developed that the acrosomes must be intact for sperm–zona pellucida adhesion. Over the years, the concept of a requirement for spermatozoa to have an intact acrosome to initiate this adhesion has driven the search in this field for a "receptor" on the sperm plasma membrane that, after binding to a ligand in the zona, transduces an acrosome-reaction-inducing signal. However, there are several reasons to step back and reevaluate these data.

One motive to revisit the acrosome reaction model is the body of literature suggesting that there are exceptions to the "acrosome-intact" requirement. For example, Myles *et al.* (1987) convincingly demonstrated that guinea pig spermatozoa are capable of adhesion to the zona pellucida in both acrosome-intact and acrosome-reacted states. Video recordings of the interactions between capacitated guinea pig sperm and cumulus-invested guinia pig oocytes have shed new light on these events (Schroer *et al.*, 2000). Although the resolution of the recording was not adequate to identify sperm with swollen acrosomes, acrosome-reacted sperm could easily be identified and were never observed to penetrate the cumulus. Acrosome-intact sperm did penetrate the cumulus and were observed on the zona but they were not tightly bound. Sperm that were fully acrosome-reacted were adherent, suggesting that zona adhesion is acrosomal status-dependent. Similar results have also been obtained with human sperm cells (Morales *et al.*, 1989). Acrosome-reacted spermatozoa were recovered from the perivitelline space of fertilized rabbit eggs and used to reinseminate the eggs *in vitro*. Over 20% of the eggs challenged by these spermatozoa were fertilized, indicating that the spermatozoa did not need intact acrosomes to be competent for zona pellucida adhesion, zona penetration, and fertilization (Valdivia *et al.*, 1999). Furthermore, the acrosome reaction model, which emphasizes the role of the acrosome-intact state in the adhesion of spermatozoa to the extracellular matrix (zona pellucida) investing an egg, is principally applicable to mammals. In species such as the sea urchin, the spermatozoa must undergo acrosome reactions before they can adhere to the vitelline layer, the sea urchin equivalent of a zona pellucida. Fusion of the outer acrosomal membrane with the spermatozoa plasma membrane and extension of the acrosomal process exposes bindin, the acrosomal matrix protein material responsible for tethering the sea urchin spermatozoa to the vitelline layer (Vacquier and Moy, 1977).

Another challenge to understanding the role of the acrosome in fertilization is the difficulty determining the true acrosomal status of sperm cells that adhere to extracellular coats surrounding eggs. The spermatozoa of some species, such as the guinea pig, have very large acrosomes that are quite easy to visualize by phase-contrast or differential interference microscopy. In many other species, it is more difficult to identify the acrosome-intact and acrosome-reacted states, so special assays have been developed for this purpose. The problem, however, is that each assay measures a different parameter of the acrosome. For example, protocols using protein stains such as Coomassie blue are primarily assaying for the presence of acrosomal (contents) protein (Larson and Miller, 1999); this assay says nothing about the integrity of the membranes overlying the acrosome. Lectin or antibody binding assays detect the presence of specific components inside the acrosome or on the outer acrosomal membrane or plasma membrane (Cross and Meizel, 1989), but again these approaches do not address membrane integrity. Fluorescent reporter dyes may represent various parameters, such as pH or ionic gradients (Lee and Storey, 1985), but these assays may be compromised by nonspecific binding to acrosomal components or by difficulties with dye loading. Even inspection of spermatozoa via light or electron microscopy has pitfalls. Because of the underlying matrix inside the acrosome, an apparently "intact" acrosome may actually possess points of membrane fusion or rupture that cannot be readily detected. Another major concern is the inconstancy of membranes; once a sperm sample has been processed for an acrosomal status assay, does the assay result truly represent the state of the acrosome at the time of intervention? Finally, one must be critical when comparing the data from different experiments that use alternative assays. Specifically, one assay may report a spermatozoon as "acrosome intact" (e.g., a positive Coomassie blue staining pattern) when another method would categorize the same spermatozoon as "acrosome reacted" (e.g., lack of pH gradient as measured by 9-aminoacridine).

2. Adhesion to the Zona Pellucida

The proposal that the sperm acrosomes must be intact for the cells to interact with the zona pellucida led to the presumption that the plasma membrane overlying the acrosome contains a binding protein or receptor-like molecule that recognizes and binds a ligand in the zona pellucida. At the same time that Saling *et al.* (1979) concluded that the acrosome must be intact for zona adhesion, Bleil and Wassarman (1980) demonstrated that the mouse zona pellucida or one of its constituents, the ZP3 glycoprotein, could block the adhesion of spermatozoa to unfertilized mouse eggs. Subsequently, Bleil *et al.* (1988) demonstrated that another zona pellucida glycoprotein, ZP2, did not affect the initial adhesion of spermatozoa to zonae, but did interfere with maintenance of adhesion when the acrosome reaction occurred on the zona pellucida. Bleil and co-workers also found that soybean trypsin inhibitor blocked adhesion of spermatozoa to the zonae, but not in the initial steps of this process; rather, this agent interfered with the continued adhesion of acrosome-reacted cells. This finding contrasts with the previous work of Saling (1981), who concluded that trypsin inhibitors blocked the interaction between spermatozoa and the zona surface but affected neither penetration through zonae nor gamete fusion. Recently, soybean trypsin inhibitor has been used as a probe for the acrosome reaction in motile cynomolgus macaque sperm (Tollner *et al.*, 2000). In contrast with the mouse, soybean trypsin inhibitor does not interfere with sperm-zona binding in this species (Yudin *et al.*, 1999). Although this issue has not been fully resolved, these results and the findings of others led to the concept that there are two types of adhesion: an initial (primary) attachment of acrosome-intact spermatozoa to the zona and a subsequent (secondary) adhesion of the acrosome-reacted spermatozoa. Thus, in the acrosome reaction model, the consequences of sperm–zona adhesion include the stimulation of the acrosome reaction to enable the release of proteins (hydrolases) necessary for penetration of the zona and the unmasking of some mechanism for the continued attachment to the zona as the sperm cell penetrates this extracellular egg coat. Some investigators have proposed that molecules on the inner acrosomal membrane may mediate the secondary adhesion but the acrosome reaction model does not adequately address how a sperm cell can efficiently adhere to and yet simultaneously pass through the zona to the reach the oolemma.

3. Zona Recognition Proteins

If spermatozoa adhere to the zona pellucida via ZP3 (and ZP2), then what are the sperm molecules that interact with the zona ligands? Over the years many candidates have been proposed to act as binding proteins or signaling receptors on sperm plasma membrane. Some proteins are still very attractive and viable possibilities, but others lack adequate experimental support. Although it is not my intention to review that literature, I would like focus on a few interesting proteins that have an acrosomal association and have been proposed to be involved in zona binding, because I feel there is much to be learned from looking at the available data and trying to develop alternative interpretations that eliminate the controversies.

Sea urchin bindin was probably the first acrosomal protein demonstrated to have a definitive function in fertilization (Vacquier and Moy, 1977). When sea urchin spermatozoa encounter the jelly coats surrounding eggs, substances in the jelly induce the spermatozoa to undergo acrosomal secretion (Vacquier and Moy, 1997). As part of this event, an actin store lying just under the inner acrosomal membrane at the tip of the sperm head polymerizes into a filamentous projection called the acrosomal process. As the acrosomal process elongates, substances from the interior of the acrosome coat the acrosomal process and act as the glue to attach the spermatozoon to the vitelline layer, the counterpart in sea urchins of the mammalian zona pellucida. To isolate and characterize the protein constituents of this acrosomal material, Vacquier and colleagues took advantage of the fact that, like other regulated secretory granules, the dense core of intact acrosomes from sea urchin sperm cells can be isolated. These acrosomal matrix particles bind to the vitelline layer, as demonstrated by the species-specific agglutination of unfertilized sea urchin eggs, and essentially contain large amounts of a single protein, bindin (Glabe and Vacquier, 1977; Vacquier and Moy, 1977). Thus, adhesion of spermatozoa to the vitelline layer of sea urchin eggs is accomplished via an acrosomal matrix protein.

The identity of a protein(s) that enables the adhesion of mammalian sperm–zona pellucida seems less clear because the acrosomal reaction model holds that primary attachment (acrosome-intact cells) and secondary adhesion (acrosome-reacted cells) exist. In cases such as the guinea pig, the acrosomal status of spermatozoa that bind to zonae pellucidae is somewhat debatable. The previously cited study of Myles *et al.* (1987) demonstrated that guinea pig spermatozoa are capable of adhering to the zona pellucida in both acrosome-intact and acrosome-reacted states. An earlier study of Huang *et al.* (1981) had concluded that only acrosome-reacted spermatozoa could bind to the zona. Furthermore, this adhesion could be blocked by fucoidan, which acts by binding to the inner acrosomal membrane and equatorial domains (Huang and Yanagimachi, 1984). Work of Jones and Williams (1990) indicated that fucoidan bound to several proteins, three of which were proacrosin (48,000 M_r) and two forms of acrosin (34,000 and 32,000 M_r). These authors suggested that acrosome-reacted guinea pig spermatozoa retain sufficient proacrosin/acrosin in association with the inner acrosomal membrane to mediate binding to the zona pellucida in a manner analogous to that of bindin. Many of these cells still contained proteolytic activity (presumably, derived from proacrosin) to mediate the binding to and/or penetration of the zona pellucida. Studies in other systems have also proposed a role for acrosin and other acrosomal proteins in sperm–zona interactions (Mori *et al.,* 1995).

As mentioned above, the notable studies of Bleil and Wassarman (1980) defined the importance of sperm–ZP3 interactions in the mouse. In these studies, ZP3 purified from unfertilized eggs not only inhibited sperm adhesion to zonae but it also induced an acrosome reaction. A major undertaking by Bleil and co-workers identified mouse sperm sp56 as a zona-binding protein and led to its consideration as an egg recognition molecule (Bleil and Wassarman, 1990). The amino acid sequence of sp56, deduced from its cDNA, demonstrated that this protein was a member of the complement regulatory protein family (Bookbinder *et al.,* 1995). A novel method to visualize immunocolloidal gold particles on surface replicas was used to localize sp56 on whole mounts of capacitated spermatozoa (Suzuki-Toyota *et al.,* 1995). As a result, these researchers concluded that sp56 is an extracellular sperm surface protein, in agreement with its proposed function as an egg recognition protein. However, other workers identified the guinea pig ortholog of sp56, termed AM67, as a component of the intracellular, acrosomal matrix (Foster *et al.,* 1997). To resolve the conflict between the different reported locations of these closely related, homologous sperm proteins, conventional immunoelectron microscopic procedures were used to reexamine the location of sp56 in mouse spermatozoa. These studies concluded that mouse sp56, like guinea pig AM67, was an intracellular, acrosomal protein. Surface labeling was observed only when the spermatozoa were lightly fixed prior to antibody treatment or when the membranes over the acrosome had vesiculated or ruptured. Mouse sperm sp56 is also part of an acrosomal matrix in mouse sperm (Kim *et al.,* 2001b). As viewed from the acrosome reaction model, these findings seem to discount sp56 as a plasma membrane protein important in the initial adhesion of spermatozoa to the zona pellucida. However, as described later, the acrosomal exocytosis model envisions an important functional role for this acrosomal protein.

Another component of spermatozoa that has been implicated in zona pellucida adhesion is hyaluronidase, an enzyme that hydrolyzes the endo-*N*-acetylhexosaminic bonds of hyaluronate and chondroitin sulfates A and C. Originally studied as the protein identified by monoclonal antibody PH-20, sperm hyaluronidase has been implicated to have a required function in sperm–zona pellucida adhesion (Primakoff *et al.,* 1985). Interestingly, the PH-20 antigen has a dual localization in acrosome-intact cells. Some of the protein is found on the posterior head of guinea pig, and the rest is found on the inner acrosomal membrane. After vesiculation of the outer acrosomal membrane and the plasma membrane, the PH-20 antigen moves from the posterior head and accumulates exclusively on the inner acrosomal membrane with the other preexisting PH-20 antigen. These findings have also been independently confirmed by Jones and his associates (Hou *et al.,* 1996; Shalgi *et al.,* 1990). Thus, hyaluronidase could be involved in adhesion of acrosome-intact as well as acrosome-reacted spermatozoa to the zona pellucida because it is present on the sperm surface before and after the loss of the membranes overlying the acrosome.

4. Zona Pellucida-Stimulated Secretion

After the initial demonstration that ZP3 can block sperm–zona adhesion, Bleil and Wassarman (1983) demonstrated that ZP3 could also induce acrosome reactions. These results are extremely significant and demonstrate that ZP3 can act not only as a ligand for a binding protein on the sperm surface, but that this zona subunit can also transmit a signal through a molecule on the sperm surface to stimualte acrosomal secretion. These studies in the mouse have guided experimentation in other mammalian systems to confirm that similar processes occur in nonrodents. In contrast to mammals, the concept has developed that spermatozoa from nonmammalian species, to elicit acrosome reactions, use a different type of signaling mechanism that does not involve direct interaction with the vitelline coat or layer. For example, in the case of sea urchins, the acrosome reaction-inducing factor is present in the jelly coats surrounding the eggs (Vacquier and Moy, 1997). However, I will not discuss signaling processes inducing the fusion of the plasma and outer acrosomal membranes as it relates to acrosomal secretion, because this issue is addressed by others in this volume.

5. Zona Penetration

The proteinaceous zona pellucida acts as a barrier to spermatozoa and other particles while allowing the permeation of variously sized molecules, either from the perivitelline space to the external medium or from the outside milieu to the oolemma. The spaces between the fibers of the zona are not large enough to enable spermatozoa to infiltrate, so a mechanism must exist to facilitate pass-through. Several penetration mechanisms have been proposed. Very simply, a biophysical model has been suggested whereby the spermatozoa push their way through the zona by virtue of the motive force exerted by the flagellum. Measurements of the mechanical force needed for a sperm cell to push directly through the zona do not support this proposal (Green, 1987). Bedford (1998) revisited this issue and offered the "hypothesis of oscillating thrust," which proposes that the characteristic sperm head shape and movements deform the zona to create a narrow penetration slit through the zona pellucida. A second possible mechanism could involve the use of proteins that disassemble the zona in a regional area by a noncatalytic process. Although this mechanism has not gained favor with those studying mammalian sperm penetration of the zona pellucida, it has very strong experimental support in the abalone (Lewis *et al.,* 1982). The abalone vitelline coat lysin is an acrosomal protein that functions, without cleaving any covalent bonds, to break down interactions operative between the vitelline coat macromolecules. In mammals, however, the most popularly held belief is that zona pellucida penetration by spermatozoa is accomplished via proteolysis of the zona.

Several lines of evidence have indicated a role for proteolysis or a proteolytic activity in the penetration of the zona pellucida. As mentioned above, acrosomes were thought to be specialized lysosomes (Allison and Hartree, 1970), so it was not a large stretch of the imagination to envision the acrosome reaction as a wholesale dumping of varous lytic enzymes that would digest a hole in the zona, through which spermatozoa could easily pass. The zona proteolysis model gained further support with the demonstration that spermatozoa contain a trypsin-like protease zymogen, proacrosin, that could be activated on exposure to a neutral pH milieu (Srivastava *et al.,* 1965; Stambaugh and Buckley, 1968). Other studies demonstrated that trypsin antagonists could have strong inhibitory actions on fertilization (Bleil *et al.,* 1988; Fraser, 1982; Liu and Baker, 1993; Saling, 1981). However, it was not always clear that the effects of the trypsin inhibitors were on zona penetration. Indeed, as mentioned above, studies by Saling and others demonstrated that one effect of trypsin inhibitors is not on zona penetration, but is, in fact, on zona adhesion (Benau and Storey, 1987; Bleil *et al.,* 1988; Fraser, 1982; Saling, 1981). Urch (1991) reviewed the literature and noted that acrosin does not fully obey all the criteria for a zona lysin as defined by Hoshi (1985). Nevertheless, for about a quarter of a century, it seemed well accepted (but not formally proved) that acrosin is the zona pellucida lysin and is essential for mammalian fertilization.

With the develoment of methods for creating targeted mutations in genes encoding specific proteins, it became possible to test the function of the proacrosin/acrosin system in fertilization. Baba and his colleagues and, later, Engel and his associates successfully created mutant mice null for the proacrosin gene (Adham *et al.,* 1997; Baba *et al.,* 1994a). Surprisingly, the homologous mutant mice were completely fertile. Although, fertility rates for the male mutant mice and the resulting litter sizes were completely normal, there was a distinct phenotype. When wild-type eggs were fertilized *in vitro* by spermatozoa from homozygous mutant males, the kinetics of fertilization were delayed relative to spermatozoa from wild-type males (Baba *et al.,* 1994a). Furthermore, in a head-to-head competition between mutant and wild-type spermatozoa, the wild-type cells were always the successful spermatozoa (Adham *et al.,* 1997). Thus, the spermatozoa from the proacrosin-null mutant mice could still fertilize in the absence of acrosin but they were at a competitive disadvantage when compared to cells from wild-type males (Adham *et al.,* 1997; Baba *et al.,* 1994a).

If the proacrosin/acrosin system is not essential, then how do spermatozoa penetrate the zona? What effect do trypsin-like protease inhibitors have on the fertilization of mouse eggs by proacrosin-null mice? Additional studies from the Baba laboratory have begun to provide answers to these questions. Although the acrosin-deficient spermatozoa penetrate the zona pellucida, the addition of *p*-aminobenzamidine to the medium still causes a significant inhibition of fertilization *in vitro* (Yamagata *et al.,* 1998a). This suggests that there is a *p*-aminobenzamidine-sensitive protease(s) other than acrosin participating in the zona penetration step. Subsequently, a nonacrosin protease with a size of 42 kDa was identified in the supernatant of the acrosome-reacted sperm cell suspension. The enzyme is inhibited by typical inhibitors of trypsin-like proteases such as *p*-aminobenzamidine, diisopropyl fluorophosphate, and *N*-α-tosyl-L-lysine chloromethyl ketone. Following up on this finding, Baba and colleagues identified several new homologs of serine proteases in spermatozoa that are implicated in zona pellucida penetration (Kohno *et al.,* 1998; Ohmura *et al.,* 1999). However, further work has questioned whether the acrosomal serine protease system is similar among mouse, rat, and hamster (Yamagata *et al.,* 1999).

III. AN ALTERNATIVE PARADIGM: THE ACROSOMAL EXOCYTOSIS MODEL

In reexamining acrosomal dynamics in a contemporary context, another paradigm for acrosomal dynamics is developing. Many of the ideas behind the acrosomal exocytosis model are not necessarily new or novel. However, a fresh interpretation of the currently available data and a synthesis of these ideas into a coherent paradigm may help to provide a more encompassing way to understand the role of the acrosome in fertilization. Again, as part of this discussion, I want to draw attention to the change in nomenclature from the term *acrosome reaction* to the term *acrosomal exocytosis,* to reflect the paradigm shift.

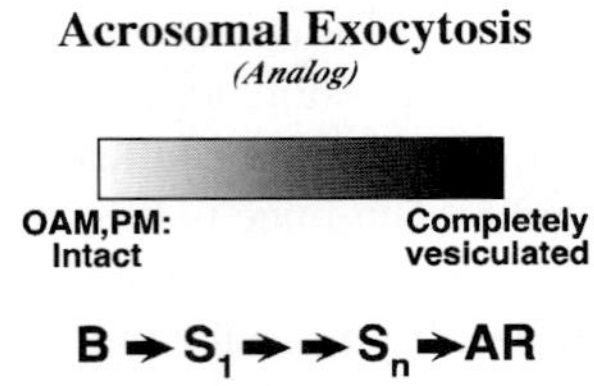

FIGURE 8.4 Depiction of the acrosomal exocytosis model as an analog system. In this paradigm, acrosomal dynamics are viewed as continuously variable, not the single-step process symbolized by the acrosome reaction model (Figure 8.1). As a consequence of capacitation, spermatozoa become capable of acrosomal secretion. As illustrated here, between the initial intact acrosome state and the point where the outer acrosomal and plasma membranes have fully vesiculated and are shed exist transient intermediates (B, banded; S, spotty; AR, acrosome reacted). The process of acrosomal exocytosis can also be expressed as an equation where B represents the capacitated state and AR represents the vesiculated state [as defined by the CTC assay (Lee and Storey, 1985)]. S_1 and S_n represent the serial transitional intermediates. As viewed from the perspective of the acrosomal exocytosis model, the transitional states (gray zone in the diagram or S_1 and S_n in the equation) function in sperm–zona adhesion and penetration. OAM, Outer acrosomal membrane; PM, plasma membrane.

Relative to the acrosome reaction model, a major departure in the acrosomal exocytosis model is to recognize and emphasize that transitional intermediates of acrosomal exocytosis exist ephemerally and that these dynamic states are functionally important in the fertilization process. Thus, in contrast to the binary acrosome reaction model, whereby the acrosome is viewed as either intact or reacted, this alternative model views acrosomal dynamics as an analog system, whereby the status of the acrosome is continuously variable (Figure 8.4). A second important concept is that sperm capacitation promotes and initiates this process but that specific ligands (e.g., the zona pellucida) or pharmacological agents (e.g., ionophores, progesterone) can greatly accelerate acrosomal exocytosis by stimulating the fusion of the plasma and outer acrosomal membranes, thus imparting a competitive advantage to spermatozoa that respond at the right time and place. A third departure from the acrosome reaction model is the acceptance that spontaneous acrosomal exocytosis is physiologically relevant and represents a slower, but mechanistically similar, version of the ligand-accelerated process. A fourth critical point is the recognition of the compartmental nature of the acrosome, including the existence of soluble and particulate (i.e., acrosomal matrix) components as well as specific physical domains within the acrosome.

The acrosomal exocytosis model is summarized diagrammatically in Figure 8.5. At least five hypotheses can be developed to test this model. Each will be described in greater detail below along with supporting data. The model proposes that there are transitional intermediates of exocytosis that represent capacitated spermatozoa whose outer acrosomal and plasma membranes have partially fused in limited areas, exposing the acrosomal contents at the sperm surface. Some of the exposed components on the outer perimeter of the acrosomal matrix come in contact with the zona pellucida and mediate gamete adhesion. Some acrosomal

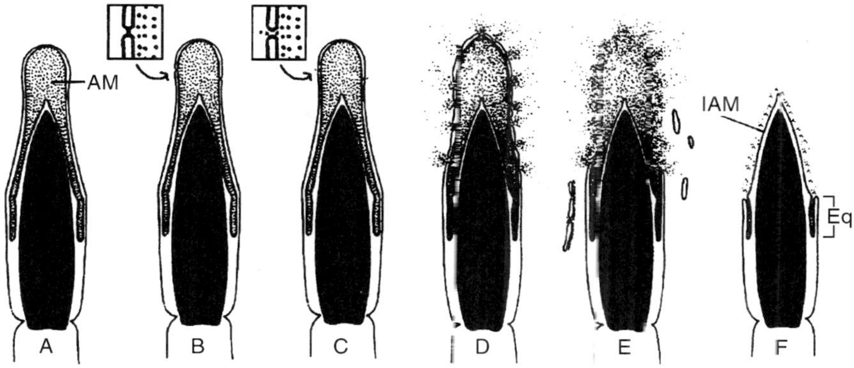

FIGURE 8.5 The acrosomal exocytosis model. In this paradigm, continuously variable states (B–E) of acrosomal secretion are recognized as functionally important intermediates. Dynamic fusion pores (insets, B and C) are hypothesized to precede complete outer acrosomal and plasma membrane fusion (D). Some soluble components rapidly diffuse from the sperm cell but proteins of the acrosomal matrix (AM) remain particulate and stay associated with the sperm head for a prolonged period (E). These components gradually dissipate, leaving an inner acrosomal membrane (IAM) that may maintain a layer of acrosomal matrix material (F). This layer is generally not detectable by standard microscopic procedures but requires immunohistochemistry of specific acrosomal proteins to be visualized. Eq, Equatorial segment. Adapted with permission from Yanagimachi (1994).

components mediate the penetration of the spermatozoa through the zona by the restricted disassembling of this structure either enzymatically or stoichiometrically. As a consequence of exposure to the external milieu, acrosomal components are gradually dispersed as a result of their inherent solubility properties or are released following proteolytic processing of the acrosomal matrix. In the microenvironment at the periphery of the exposed acrosomal matrix, the pH approaches the neutrality of the surrounding milieu, leading to the localized activation of acrosomal proteases (e.g., acrosin), which act to process and disperse the acrosomal matrix. Meanwhile, the (proximal) perinuclear acrosomal matrix is processed more slowly, perhaps as a result of the localized concentration of protease inhibitors that have yet to diffuse away (Figure 8.6). Thus, the acrosomal matrix dissolves from the outer zone to the inner recesses. Following the dissolution of the acrosomal matrix from the outer margins, the freshly exposed, underlying acrosomal matrix materials can then reinitiate the zona adhesion and start the cycle over again. In this continuously variable (analog) manner, the sperm cell can then ratchet its way through the zona pellucida (Figures 8.6 and 8.7).

A. TRANSITIONAL STATES HYPOTHESIS

The first hypothesis is termed the *transitional states hypothesis* and states that acrosomal exocytosis occurs via a continuum of events, passing through transi-

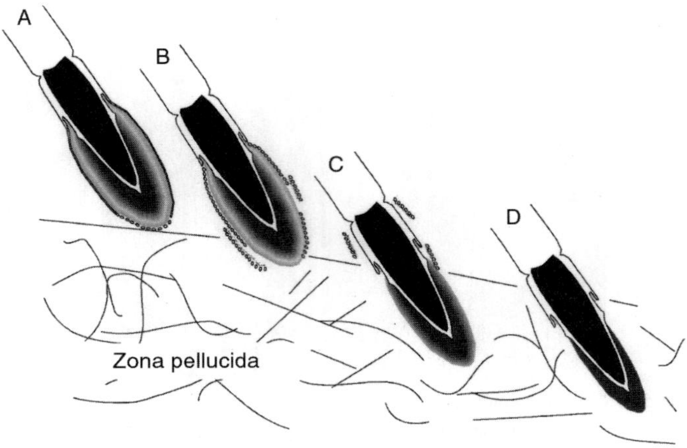

FIGURE 8.6 Role of the acrosomal matrix components in sperm–zona pellucida interactions. As a consequence of capacitation, the outer acrosomal and plasma membranes begin to fuse in limited areas and the sperm cell comes in contact with the zona pellucida (A). Some exposed acrosomal contents mediate gamete adhesion and the penetration of the spermatozoa through the zona by the restricted disassembly of this structure, either enzymatically or stoichiometrically (B, C). Exposed acrosomal components are gradually dispersed as a result of their inherent solubility properties or are released following proteolytic processing of the acrosomal matrix. Hybrid vesicles of the outer acrosomal and plasma membranes may be shed as the acrosomal shroud (B, C). The acrosomal matrix dissolves from the outer zone to the inner recesses as the spermatozoon penetrates the zona pellucida (B–D). For clarity, only the sperm heads are shown.

tional states whereby the plasma membrane and outer acrosomal membrane interactions lead to the progressive exposure of the acrosomal components prior to the time that the plasma membrane and outer acrosomal membrane completely fuse into hybrid membrane vesicles. This hypothesis arises from a rethinking of sperm–zona pellucida interactions that considers paradigms for secretion in other cell systems. When acrosomal exocytosis is examined in perspective with current models for regulated secretion, e.g., the "flickering pore" hypothesis for mast cells (Monck and Fernandez, 1996), a starting point is provided for developing a paradigm to explain the events involved in capacitation, egg recognition, zona pellucida adhesion, and zona pellucida penetration. In this view, the secretion is believed to commence with the formation of small, dynamic fusion pores by hemifusion of the apposed leaflets of the vesicular and plasma membranes; these pores appear to close after release of minute amounts of secretory products.

At the outset, however, the concept of "flickering pores" may be inappropriate for a paradigm addressing acrosomal dynamics. In contrast to other secretory systems, spermatozoa are terminally differentiated cells with a singular purpose: fertilize an egg. As such, there is not a need for the existence of a system to recycle the secretory machinery; once the spermatozoon has initiated acrosomal exocytosis, it has made a commitment that cannot be rescinded. Furthermore, there is lit-

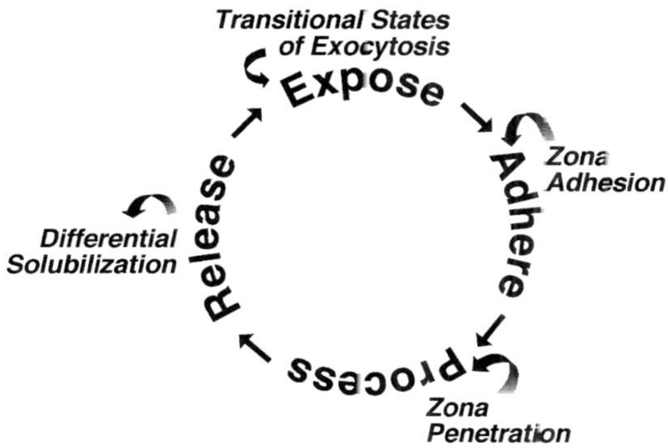

FIGURE 8.7 Integration of the hypotheses of the acrosomal exocytosis model to form the "ratchet" mechanism for sperm cell adhesion and penetration of the zona pellucida. The process starts with capacitation, leading to the exposure of acrosomal components that mediate the adhesion of the sperm cells to the zona pellucida. The exposed surface acrosomal matrix proteins then aid in the penetration of the zona. As illustrated in Figure 8.6, dissolution of the acrosomal matrix takes place from the outer margins to the inner recesses of the matrix, releasing the contact of the sperm cell with the zona and simultaneously exposing fresh matrix to reinitiate the ratchet cycle. This provides a mechanism for a spermatozoon to adhere to and, concurrently, penetrate the zona pellucida.

tle or no experimental support for the presence of "flickering" fusion pores in spermatozoa, and it would be extremely difficult to make a convincing demonstration of such pores in spermatozoa, because the size of the sperm head and regionalization of the sperm plasma membrane preclude the use of membrane capacitance measurements. Therefore, the term *dynamic fusion pore* will be used to refer to the initial points of fusion of the outer acrosomal and plasma membranes because this term connotes pores or points of fusion that are somewhat fluid in nature, position, and size. The "transient states" hypothesis is applied to spermatozoa by postulating that capacitation represents the development of progressive membrane states whereby the outer acrosomal membrane and plasma membrane from transient, dynamic fusion points, leading to the incremental exposure and, eventually, to the release of acrosomal components.

Artificial membranes have been shown to form fusion pores comparable with initial exocytotic pores in the absence of proteinaceous channels (Chanturiya *et al.,* 1997). In cells such as spermatozoa, there is likely to be a mechanism to regulate the complete fusion of cellular membranes and the subsequent release of materials. This control could be at the level of the proteins that regulate the membrane fluidity by managing the lipid composition of the membranes, including cholesterol (Cross, 1996; Nolan and Hammerstedt, 1997; Visconti *et al.,* 1999). Flaherty and Olson (1988) studied guinea pig spermatozoa induced to undergo synchronous acrosome reactions by preincubation in a Ca^{2+}-free medium containing lysolecithin. They assessed the acrosomal status following the addition of Ca^{2+} and found that fusion between the outer acrosomal membrane and plasma membrane occurred at the boundaries but not within periacrosomal plasma membrane junctional zones over the apical segment. Stable, nonfusigenic domains were noted in both the plasma membrane and outer acrosomal membrane of the apical segment. Membrane-associated assemblies were proposed to maintain these domains and to control the membrane fusion events. Thus, it is conceivable that regulation of the membrane fusion would also coordinate the exposure and release of acrosomal contents.

In addition, signal transduction pathways such as those involving heterotrimeric guanine nucleotide regulatory proteins (G-proteins) also control exocytosis in somatic cells (Gasman *et al.,* 1997; Lagriffoul *et al.,* 1996; Ohnishi *et al.,* 1997). Acrosomal exocytosis, as assessed by the chlortetracycline (CTC) assay, is also regulated by G-proteins, because the ability of mouse spermatozoa to undergo exocytosis is inhibited in a concentration-dependent manner by the G_i inactivator, pertussis toxin, but the ability of the cells to adhere to structurally intact zona pellucida is not inhibited by the toxin (Endo *et al.,* 1987, 1988). $G_{i\alpha}$ proteins are present in the acrosomal region of mammalian spermatozoa, are lost from the spermatozoa as a result of acrosomal exocytosis, and are recoverable in the hybrid membrane vesicles released from the cells (Glassner *et al.,* 1991). Furthermore, in exocytotic systems, docking of the vesicle membrane and the plasma membrane are thought to be mediated via specific membrane protein interactions. The SNARE (soluble NSF attachment protein receptor) hypothesis states that every transport vesicle contains on its surface proteins that interact with cognate partners on the target membranes, leading to the subsequent fusion of the vesicular and plasma membranes. Similar to the studies of G-proteins in mammalian spermatozoa, Schulz *et al.* (1997) demonstrated that the hybrid membrane vesicles shed from sea urchin spermatozoa during acrosomal exocytosis contained proteins implicated in the SNARE hypothesis of exocytosis, providing support that these proteins play a role in acrosomal exocytosis.

Once membrane fusion has been initiated, the expansion of the dense core of the secretory vesicle, a so-called smart polymer, may be controlled by hypotonic stress, the ionic environment, proteolysis, or pH (Monck *et al.,* 1991). Hypotonic stress, causing tension within the vesicle membrane, promotes complete fusion of granule membrane with the plasma membranes. In capacitated spermatozoa, the acrosomal material closest to a developing or transient pore may be exposed to the external milieu of the cell. If this happens, a capacitated spermatozoon encountering an egg could actually adhere to the zona pellucida via the exposed acrosomal proteins. Such adhesion might stabilize a dynamic pore. Under these conditions, the ability of acrosomal ion or proton pumps to maintain a gradient between the acrosomal lumen and the external milieu could break down. The loss of such a gradient would lead to an increase in the pH and a change in the ionic environment within the acrosome. The acrosomal material, acting as a smart polymer, could then begin expanding, and acrosomal exocytosis would be driven to completion.

Is there strong evidence for transitional states prior to the completion of acrosomal exocytosis? One hint of such comes from the work of Storey, Kopf, and colleagues. Based on different distribution patterns of fluorescence with the dye chlortetracycline, three basic stages of acrosomal exocytosis were defined (Lee and Storey, 1985). Capacitated, "acrosome-intact" spermatozoa display a banded (B) pattern, progress to an intermediate spotty (S) pattern, and then proceed to the fully acrosome-reacted (AR) pattern (the B → S → AR transition, Figure 8.4). In view of the acrosomal exocytosis model, the S pattern may represent one or more transitional states when hemifused domains or dynamic fusion pores between the outer acrosomal membrane and plasma membrane have formed. Furthermore, spermatozoa can be "trapped" in the S phase by exposing them to the zonae pellucidae of eggs treated with 12-*O*-tetradecanoyl phorbol-13-acetate (TPA). In this situation, the spermatozoa undergo a B → S transition, but do not complete acrosomal exocytosis (i.e., these cells are arrested in the S pattern) (Kligman *et al.*, 1991). Loss of the transmembrane pH gradient in the anterior portion of the sperm head, monitored by the fluorescent pH probe 9-*N*-dodecyl aminoacridine, follows the B → S transition in spermatozoa incubated with zonae pellucidae from untreated, unfertilized eggs, indicating the presence of stable, open pores. However, no loss of the transmembrane pH gradient is observed when the B → S transition is induced using zonae pellucidae from TPA-treated eggs, indicating either a lack of pores or an ability to maintain a gradient when small, dynamic pores are formed. If S pattern-arrested spermatozoa are treated with solubilized zonae pellucidae from unfertilized egges or with the ionophore A23187, the transmembrane pH gradient is rapidly lost and the spermatozoa undergo the S → AR transition. Furthermore, if nigericin is added to B pattern spermatozoa, the pH gradient is discharged but the B → S transition is not induced. On the other hand, if nigericin is added to S pattern-arrested spermatozoa adhering to zonae pellucidae from TPA-treated eggs, the transmembrane pH gradient is discharged and the S → AR transition immediately ensues. Electron microscopic evaluation of spermatozoa arrested in the S pattern with zonae pellucidae from TPA-treated eggs reveals that many of the spermatozoa display the acrosome-intact morphology but that the plasma membrane and outer acrosomal membrane have started to lift away from most of the sperm heads (Kligman *et al.*, 1991). This morphology may be, in part, artifactual if the acrosome contents have begun to expand as a smart polymer (see discussion above) but, during the dehydration steps for electron microscopy, have become condensed, leaving a plasma membrane that appears to be lifting away. These results suggest that zona pellucidae from TPA-treated and fertilized eggs are modified such that the zona ligands inducing the S → AR transition are lost or are inactivated. Thus, in a test of the transitional states hypothesis, these spermatozoa may be found to bind to the zonae pellucidae from TPA-treated eggs via exposed acrosomal proteins but may not be capable of completing the acrosomal exocytosis at the accelerated rate induced by zonae pellucidae from unfertilized eggs.

B. ZONA PELLUCIDA BINDING HYPOTHESIS

The acrosomal exocytosis model builds on the observation that mouse spermatozoa acrosomes contain zona pellucida-binding proteins such as sp56 and other possible attachment proteins such as proacrosin. The zona pellucida-binding hypothesis states that acrosomal matrix components possess ligand-binding properties that mediate binding to the zonae pellucidae of unfertilized eggs. There are two corollaries to the zona pellucida-binding hypothesis: (1) Acrosomal components should bind poorly or not at all to the zonae pellucidae of fertilized eggs. (2) Released (soluble) acrosomal components that have been secreted from spermatozoa should bind poorly or not at all to the zonae pellucidae of unfertilized eggs.

In an application of models from other systems of cell biology, sperm–egg interactions may be considered in terms of an interaction between the extracellular matrices of two cells in contact. The zona pellucida is clearly an extracellular matrix secreted by the oocyte during its growth (Liang and Dean, 1993; Wassarman, 1990). By the same token, the material released from the spermatozoa may be considered an extracellular matrix that coats the sperm head during and immediately following acrosomal exocytosis. Furthermore, the gradual dispersion of matrix components from spermatozoa undergoing acrosomal exocytosis can be thought in terms of the remodeling of the extracellular matrix that is caused by other motile cells that must pass through the extracellular matrix of stromal tissues (Ashkenas *et al.,* 1996).

The evidence demonstrating that the mouse sperm zona pellucida-binding protein sp56 and its guinea pig ortholog, AM67, are components of the acrosomal matrix has been discussed above. There is very strong evidence to support sp56 as a zona pellucida-binding protein (Bleil and Wassarman, 1990; Cheng *et al.,* 1994; Cohen and Wassarman, 2001). The finding that sp56 is within the acrosome and not on the plasma membrane (Foster *et al.,* 1997) does not negate a role for sp56 in the initial phases of sperm–zona interactions. In fact, from the perspective of the acrosomal exocytosis model, zona-binding proteins within the acrosome enable the spermatozoa to attach to the zona pellucida in the first place.

The role of the acrosomal matrix in binding to the zona pellucida is also supported by studies in the macaque (VandeVoort *et al.,* 1997). In these experiments, the investigators examined the number of spermatozoa adhering to the zona pellucida under a variety of conditions. These investigators concluded that macaque spermatozoa that undergo acrosomal exocytosis on the zona surface are bound by the acrosomal shroud before zona penetration. They also found that the capacity of spermatozoa that had "acrosome reacted" prior to interacting with the oocyte zone was significantly reduced. On the other hand, the afffinity of spermatozoa for the zona increased when the acrosomal shroud was removed and the inner acrosomal membrane was exposed. The authors conclude that this sequence of events occurs naturally during the transition from "primary" to "secondary" adhesion on the zona surface. Although these findings seem to be consistent with the acrosome reaction model rather than the acrosomal exocytosis model, one must be cautious in interpretation. Is it really the inner acrosomal membrane that is important here, or is the increased adhesion mediated by acrosomal matrix material coating the inner acrosomal membrane?

The existence of an acrosomal shroud has intrigued investigators for years. The structure appears to be formed from the hybrid membrane vesicles formed by the fusion of the outer acrosomal and plasma membranes that are held together by some matrix-like material underlying the former outer acrosomal membrane region. Circumstantial evidence suggests that acrosomal matrix components, such as AM50 and AM67 (sp56), may interact with plasma membrane proteins in a physiologically significant manner. Guinea pig sperm sp50, a protein of the same size as the acrosomal pentraxin AM50 (Noland *et al.,* 1994; Reid and Blobel, 1994; Westbrook-Case *et al.,* 1994), was found to bind to erythrocyte ghosts in a calcium-dependent manner (Hernandez *et al.,* 1996). An antibody that recognized sp50 also detected a protein of M_r ~42,000, similar to that of $AM50_{AR}$, the form of AM50 released as a result of acrosomal exocytosis (Westbrook-Case *et al.,* 1994). Comparable to the Ca^{2+}-dependent association of AM50 with apical segments [a complex of plasma membrane, outer acrosomal membrane, and acrosomal matrix (Noland *et al.,* 1994)], sp50 was also able to bind to plasma membrane fragments and the outer acrosomal membrane of guinea pig spermatozoa (Hernandez *et al.,* 1996). If sp50 is AM50, the ability of this protein to bind to the plasma membrane and outer acrosomal membrane of guinea pig spermatozoa is consistent with the acrosomal exocytosis model. Furthermore, both AM50 and AM67 were initially discovered independently in the Gerton, Olson, and Blobel laboratories (Foster *et al.,* 1997; Noland *et al.,* 1994; Reid and Blobel, 1994; Westbrook-Case *et al.,* 1994). Interestingly, Reid and Blobel identified these proteins because they copurified with fertilin by affinity chromatography on a column of monoclonal antibody PH-1. Fertilin is a transmembrane protein of the sperm surface implicated in sperm–egg plasma membrane fusion. One explanation for these results is that AM50, AM67, and fertilin share an antigenic epitope. Alternatively, these proteins may form a complex *in vivo* or as a result of cell lysis. Cho *et al.* (1998) created male mice with a targeted mutation in the fertilin-β gene and found that the fertility rate of the homozygous mutant mice is greatly diminished, resulting in part from a defect in sperm–zona pellucida adhesion. Whether this defect demonstrates a role for fertilin in the exposure of acrosomal matrix components or indicates that plasma membrane constituents and components of the acrosomal matrix interact with each other needs to be addressed.

C. ZONA PENETRATION HYPOTHESIS

The zona penetration hypothesis proposes that acrosomal components mediate the penetration of the zona pellucida by spermatozoa. The actual mechanism for accomplishing this is still to be elucidated. The proacrosin knockout experiments eliminated acrosin as an essential zona penetrating agent in the mouse (Adham *et al.,* 1997; Baba *et al.,* 1994a). However, as discussed previously, other serine proteases may be involved, individually or in combination with proacrosin/acrosin (Kohno *et al.,* 1998; Ohmura *et al.,* 1999; Yamagata *et al.,* 1998a). Other hydrolases may also be involved, including glycosidases (Tulsiani *et al.,* 1998). For example, hyaluronidase (PH-20) has been implicated in the penetration of the macaque zona pellucida (Yudin *et al.,* 1999). In addition, one should not neglect the concept that there could be a noncatalytic mechanism for zona penetration, as is the case for the abalone vitelline envelope lysin (Lewis *et al.,* 1982). Bedford (1998) has recently reviewed the penetration of the zona pellucida by eutherian spermatozoa and concluded enzymatic lysis is not involved in this process. Instead, he favors the hypothesis of oscillating physical thrust described above. This model discounts soluble acrosomal components or inner acrosomal membrane-bound lysins in the zona penetration process.

D. DIFFERENTIAL RELEASE HYPOTHESIS

The differential time-release hypothesis for acrosomal material dispersion states that each specific acrosomal protein has a different rate of release from spermatozoa that is dependent on its intrinsic properties and interactions with other components. Corollaries of the differential time-release hypothesis are that soluble components are quickly released from spermatozoa following induction of acrosomal exocytosis and that unprocessed acrosomal matrix components remain insoluble for a prolonged period of time but may undergo posttranslational modifications coincident with their release from the acrosomal matrix.

The differential release hypothesis is a formalization of the concept articulated by Hardy *et al.* (1991). These authors proposed that the compartmentalization of the acrosome provides a mechanism for the differential release of acrosomal enzymes during acrosomal exocytosis. Their model is that soluble proteins such as CRISP-2 [also known as autoantigen 1 and Tpx-1 (Foster and Gerton, 1996; Hardy *et al.,* 1988; Kasahara *et al.,* 1989)] and enzymes such as dipeptidyl peptidase (DiCarlantonio and Talbot, 1988) would be free to diffuse from the acrosome at the outset of exocytosis, whereas release of acrosin would require proteolysis of the acrosomal matrix, with the consequence that acrosin complexed with the acrosomal matrix would remain with the spermatozoa for a relatively much longer period of time. The presence of specific proteins retained in association with the sperm acrosomes and released into the medium surrounding guinea pig sperm induced to undergo exocytosis with the ionophore A23187 has been studied (Kim *et al.,* 2001a). CRISP-2, a soluble component of the acrosome, was rapidly lost from the acrosome soon after ionophore treatment. On the other hand, acrosomal matrix components remained associated with the sperm for longer periods of time. AM67 was released at a slower rate than CRISP-2 but at a faster rate than two other matrix proteins, AM50 and proacrosin. Further support for this concept comes from the analysis of boar and guinea pig proacrosin-binding protein, a component of the acrosomal matrix (Baba *et al.,* 1994b), as well the studies of others on the hydrolase-binding activities in the acrosomal matrix of hamster spermatozoa (NagDas *et al.,* 1996a). Experiments using soluble, recombinant green fluorescent protein expressed in the mouse sperm acrosome provide additional support for this hypothesis (Yamagata *et al.,* 1998b).

Do acrosomal matrix components undergo posttranslational modifications coincident with their release? Studies of the alterations and activation of proacrosin support this corollary (Baba *et al.*, 1989a,b; NagDas *et al.*, 1996b; Noland *et al.*, 1989). Furthermore, AM50, the acrosomal matrix pentraxin, is converted into a 42- to 43-kDA doublet protein ($AM50_{AR}$) and is released into the incubation medium during acrosomal exocytosis, suggesting that proteolysis of matrix components affects their solubility (Westbrook-Case *et al.*, 1994). Thus, if acrosomal matrix components are involved in sperm–egg adhesion but the release of matrix molecules is inhibited by reagents such as trypsin inhibitors, the penetration of the zona pellucida will also be inhibited because spermatozoa will be incapable of releasing their initial points of contact and establishing new interactions. Likewise, if the acrosomal matrix component acrosin is primarily involved in processing other acrosomal matrix material (and not zona pellucida penetration), an absence of acrosin in null mutants could lead to a delay in fertilization because the spermatozoa have a more difficult time releasing the zona contact mediated by acrosomal matrix components. Such a delay in fertilization for the spermatozoa from acrosin-null mice has been correlated with the dispersion of acrosomal proteins (Adham *et al.*, 1997; Baba *et al.*, 1994a; Yamagata *et al.*, 1998b). A similar conclusion was reached by Fraser (1982), who examined the effects of high concentrations of the serine protease inhibitor, *p*-aminobenzamidine, on spermatozoa. As determined by membrane vesiculation, acrosomal exocytosis is initiated but acrosomal matrix dispersal is inhibited. These results indicate that a major role of acrosin is to regulate the release of acrosomal matrix proteins from spermatozoa undergoing exocytosis.

E. CONSERVATION OF MECHANISM HYPOTHESIS

The evolutionary conservation of mechanism hypothesis states that the acrosomes of spermatozoa from most species will function in a similar manner. On the protein level, this implies that the functional components of the mammalian sperm acrosome are conserved among species. Proacrosin has been studied in many species and the homologies have been noted in the catalytic domains of the zymogen (Adham *et al.*, 1990; Baba *et al.*, 1989b,c; Kashiwabara *et al.*, 1990). Hyaluronidase is also well conserved (Hou *et al.*, 1996; Lathrop *et al.*, 1990; Lin *et al.*, 1993; ten Have *et al.*, 1998). CRISP-2 has highly conserved homologs in the mouse, human, rat, and guinea pig (Foster and Gerton, 1996; Maeda *et al.*, 1998; Mizuki *et al.*, 1992). Guinea pig AM67 and mouse sp56 are orthologs (Bookbinder *et al.*, 1995; Foster *et al.*, 1997). Proacrosin-binding protein is also conserved among species (Baba *et al.*, 1994b). Homologs of guinea pig AM50 (i.e., apexin or p50) have been identified in rat, human, and mouse brains and it is likely that these homologs are present in spermatozoa because the mRNAs encoding these proteins are highly expressed in testes of these species (Hsu and Perin, 1995; Tsui *et al.*, 1996). If the primary structures of these components are highly conserved, it is likely that their functions are also conserved.

IV. OTHER CONSIDERATIONS OF ACROSOMAL PROTEINS

Although we do not completely understand the roles of the acrosomal components in fertilization, we stand on the brink of a new wave of functional studies concerning this interesting organelle. The progress being made in understanding signal transduction processes, the pathways controlling secretion, and the basis of cell–cell adhesion will be extremely useful to unlocking the function of the acrosome. The advances in genome projects and functional proteomics will also help in this regard as we learn from other systems the functions of specific proteins with homologs in the acrosome. There are several themes that may provide clues to the functions of some of the acrosomal proteins. For example, several of the proteins in the acrosome and membranes of spermatozoa are related to complement regulatory proteins (e.g., the relationship between sp56/AM67 and complement 4-binding protein). Some proteins (e.g., AM50) are related to ligand-binding proteins that are known to interact with carbohydrates, membranes, or other proteins (Kolb-Bachofen, 1991; O'Brien *et al.,* 1999). CRISP-2 is similar to epididymal CRISP-1 (also known as acidic epididymal glycoprotein and protein D/E) and is part of a larger family of proteins that include proteins from insect and reptile venoms (Foster and Gerton, 1996; Haendler *et al.,* 1993; Lu *et al.,* 1993; Mochca-Morales *et al.,* 1990). Interestingly, PH-20 was originally identified as sperm hyaluronidase on the basis of its amino acid sequence homology with the bee venom enzyme (Gmachl and Kreil, 1993). The functional properties of some acrosomal proteins such as sp10 (Foster *et al.,* 1994) will be problematic until homologies to other proteins with known activities have been identified.

V. FUTURE DIRECTIONS

To illustrate how we are on the threshold of a new appreciation concerning the role of acrosomal components in fertilization, we need only to look at recent results with null mutants for various sperm proteins. I have already discussed how the proacrosin knockout mice have given us a new appreciation of the role of acrosin in acrosomal matrix dispersal rather than zona pellucida penetration. Several examples are present in the literature, and although the phenotypes were somewhat unexpected, I believe we can learn much about the role of acrosomal proteins in zona pellucida interactions if we examine these results with open minds. In the case of the null mice for the chaperonin calmegin, the spermatozoa do not adhere to the zona pellucida very effectively, but several acrosomal proteins were examined and found to be present (Ikawa *et al.,* 1997). However, we do not know about the states of capacitation of these cells [e.g., the CTC staining characteristics or phosphotyrosine-containing proteins (Lee and Storey, 1985; Visconti *et al.,* 1995)]. As mentioned above, spermatozoa of mutant mice deficient in fertilin-β are defective in migration from the uterus into the oviduct, adhesion to the zona pellucida, sperm–egg plasma membrane adhesion, and gamete fusion (Cho *et al.,* 1998). These results suggest that fertilin-β might have a direct role in spermatogenic development, oviduct migration, or sperm–zona adhesion. Male mice lacking both the somatic and germ cell variants of angiotensin-converting enzyme show lower fertility (in addition to lowered blood pressure, thickened arteries in the kidneys, and atrophy of the renal cortex (Krege *et al.,* 1995). If the germ cell isotype of the angiotensin-converting enzyme is specifically eliminated and the somatic form left intact, the males show decreased fertility resulting from a sperm defect in oviductal transport and adhesion to the zona pellucida (Hagaman *et al.,* 1998). The fertility of male mice with the mutation affecting the somatic and germ cell variants can be rescued by a transgene encoding the rabbit germ cell variant (Ramaraj *et al.,* 1998). Although the sperm angiotensin-converting enzyme has been reported to be localized on the cytoplasmic face of the periacrosomal plasma membrane of equine and macaque spermatozoa (Dobrinski *et al.,* 1997), this bears a reexamination. Angiotensin-converting enzyme is found on the extracellular face of somatic cell plasma membranes or as a secreted protein in blood; it would not be expected to be found in the cytoplasmic compartment of spermatozoa.

VI. SUMMARY

Currently there are many questions concerning the role of the acrosome in fertilization. To start, we still do not really have a clear understanding of what are the true characteristics of capacitation at the cellular and molecular levels. Does capacitation represent a stage of readiness for acrosomal exocytosis? Is it, perhaps, the initial stages of a continuum of exocytotic steps? Can capacitation really be reversed, and, if so, what is the mechanism? Second, we have much to learn concerning the role that acrosomal components play in adhesion to the zona pellucida. Are the components of the acrosome the actual agents that mediate the meaningful adhesion of the spermatozoa to the zona pelucida, or is one of the candidate plasma membrane proteins the authentic "receptor" or "recognition protein" for the zona pellucida? A third major interest is the mechanism that the spermatozoon uses to penetrate the zona pellucida. Does this cell use a hydrolytic method to create a passaageway through the zona? If so, then what enzyme or enzymes are involved? It may turn out that proteases, glycosidases, and noncatalytic mechanisms all contribute to the breeching of the zona barrier by the spermatozoon.

Clearly, the story regarding the function of the acrosome in fertilization is far from complete. What I hope to have accomplished through this reassessment of acrosomal dynamics is to challenge the current binary view of acrosomal dynamics and the role this secretory organelle plays in capacitation and zona pellucida interactions. Although many of the ideas presented here are not new, I hope that the synthesis of these concepts into the acrosomal exocytosis model will provide another perspective and a set of hypotheses to enable us to address the function of the acrosome. To return to the computer metaphor, the acrosome is not something that is here one moment (on/acrosome-intact) and gone the next (off/acrosome-reacted). Between these two extremes are continuously variable intermediates that should not be ignored or discounted. These intermediate states are inherently difficult to study, but they do exist, even only fleetingly, and probably serve important functions for zona adhesion and penetration.

ACKNOWLEDGMENTS

This review integrates the ideas and results of many people, too numerous to list. However, I thank Dan Hardy for encouraging me to lay these thoughts out on paper. I also thank all workers who have studied the dynamics of the acrosome. It is a fascinating organelle, but we still have much to learn about this important organelle and its function in fertilization. I have been fortunate to collaborate with several people from other institutions whose work has influenced my own tremendously. In particular, Gary Olson's work on the acrosomal matrix, Tadashi Baba's studies of acrosomal proteases, and Dan Hardy's paper on a mechanism for the differential release of acrosomal enzymes have been quite illuminating. I also thank my colleagues in the Center for Research on Reproduction and Women's Health of the University of Pennsylvania for their comments and suggestions. It has been particularly fun to kick these ideas around with former and present members of my laboratory, such as Jim Foster and Kye-Seong Kim. Finally, I have the good fortune of having Bayard Storey and Greg Kopf as colleagues; their studies and our discussions have helped to shape the concepts put forth in this review. I acknowledge the National Institutes of Health (HD-22899) for supporting my laboratory's studies on the sperm acrosome.

REFERENCES

Adham, I. M., Klemm, U., Maier, W.-M., and Engel, W. (1990). Molecular cloning of human preproacrosin cDNA. *Hum. Genet.* **84,** 125–128.

Adham, I. M., Nayernia, K., and Engel, W. (1997). Spermatozoa lacking acrosin protein show delayed fertilization. *Mol. eprod. Dev.* **46,** 370–376.

Ahnert-Hilger, G., Schäfer, T., Spicher, K., Grund, C., Schultz, G., and Wiedenmann, B. (1994). Detection of G-protein heterotrimers on large dense core and small synaptic vesicles of neuroendocrine and neuronal cells. *Eur. J. Cell Biol.* **65,** 26–38.

Allan, B. B., and Balch, W. E. (1999). Protein sorting by directed maturation of Golgi compartments. *Science* **285,** 63–66.

Allison, A. C., and Hartree, E. F. (1970). Lysosomal enzymes in the acrosome and their possible role in fertilization. *J. Reprod. Fertil.* **21,** 501–515.

Anakwe, O. O., and Gerton, G. L. (1990). Acrosome biogenesis begins during meiosis: Evidence from the synthesis and distribution of an acrosomal glycoprotein, acrogranin, during guinea pig spermatogenesis. *Biol. Reprod.* **42,** 317–328.

Ashkenas, J., Muschler, J., and Bissell, M. J. (1996). The extracellular matrix in epithelial biology: Shared molecules and common themes in distant phyla. *Dev. Biol.* **180,** 433–444.

Baba, T., Azuma, S., Kashiwabara, S.-I., and Toyoda, Y. (1994a). Sperm from mice carrying a targeted mutation of the acrosin gene can penetrate the oocyte zona pellucida and effect fertilization. *J. Biol. Chem.* **269,** 31845.

Baba, T., Kashiwabara, S.-I., Watanabe, K., Itoh, H., Michikawa, Y., Kimura, K., Takada, M., Fukamizu, A., and Arai, Y. (1989a). Activation and maturation mechanisms of boar acrosin zymogen based on the deduced primary structure. *J. Biol. Chem.* **264,** 11920–11927.

Baba, T., Michikawa, Y., Kawakura, K., and Arai, Y. (1989b). Activation of boar proacrosin is effected by processing at both N- and C-terminal portions of the zymogen molecule. *FEBS Lett.* **244,** 132–136.

Baba, T., Niida, Y., Michikawa, Y., Kashiwabara, S.-I., Kodaira, K., Takenaka, M., Kohno, N., Gerton, G. L., and Arai, Y. (1994b). An acrosomal protein, sp32, in mammalian sperm is a binding protein specific for two proacrosins and an acrosin intermediate. *J. Biol. Chem.* **269,** 10133–10140.

Baba, T., Watanabe, K., Kashiwabara, S.-I., and Arai, Y. (1989c). Primary structure of human proacrosin deduced from its cDNA sequence. *FEBS Lett.* **244,** 296–300.

Baccetti, B., Burrini, A. G., Collodel, G., Piomboni, P., Renieri, T., and Sensini, C. (1989) Localization of acrosomal enzymes in Arthropoda, Echinodermata and Vertebrata. *J. Submicrosc. Cytol. Pathol.* **21,** 385–389.

Baccetti, B., Burrini, A. G., Collodel, G., Piomboni, P., and Renieri, T. (1991). A "miniacrosome" sperm defect causing infertility in two brothers. *J. Androl.* **12,** 104–111.

Bartoov, B., Eltes, F., Lunenfeld, E., Har-Even, D., Lederman, H., and Lunenfeld, B. (1994). Sperm quality of subfertile males before and after treatment with human follicle-stimulating hormone. *Fertil. Steril.* **61,** 727–734.

Bauerfeind, R., and Huttner, W. B. (1993). Biogenesis of constitutive secretory vesicles, secretory granules and synaptic vesicles. *Curr. Opin. Cell Biol.* **5,** 628–635.

Bedford, J. M. (1970). Sperm capacitation and fertilization in mammals. *Biol. Reprod.* **2** (Suppl. 2), 128–158.

Bedford, J. M. (1998). Mammalian fertilization misread? Sperm penetration of the eutherian zona pellucida is unlikely to be a lytic event. *Biol. Reprod.* **59,** 1275–1287.

Benau, D. A., and Storey, B. T. (1987). Characterization of the mouse sperm plasma membrane zona-binding site sensitive to trypsin inhibitors. *Biol. Reprod.* **36,** 282–292.

Benoff, S., Hurley, I., Cooper, G. W., Mandel, F. S., Hershlag, A., Scholl, G. M., and Rosenfeld, D. L. (1993). Fertilization potential *in vitro* is correlated with head-specific mannose-ligand receptor expression, acrosome status and membrane cholesterol content. *Hum. Reprod.* **8,** 2155–2166.

Bleil, J. D., and Wassarman, P. M. (1980). Mammalian sperm-egg interaction: Identification of a glycoprotein in mouse egg zonae pellucidae possessing receptor activity for sperm. *Cell* **20,** 873–882.

Bleil, J. D., and Wassarman, P. M. (1983). Sperm-egg interactions in the mouse: Sequence of events and induction of the acrosome reaction by a zona pellucida glycoprotein. *Dev. Biol.* **95,** 317–324.

Bleil, J. D., and Wassarman, P. M. (1990). Identification of a ZP3 binding protein on acrosome-intact mouse sperm by photoaffinity crosslinking. *Proc. Natl. Acad. Sci. U.S.A.* **87,** 5563–5567.

Bleil, J. D., Greve, J. M., and Wassarman, P. M. (1988). Identification of a secondary sperm receptor in the mouse egg zona pellucida: Role in maintenance of binding of acrosome-reacted sperm to eggs. *Dev. Biol.* **128,** 376–85.

Bookbinder, L. H., Cheng, A., and Bleil, J. D. (1995). Tissue- and species-specific expression of sp56, a mouse sperm fertilization protein. *Science* **269,** 86–89.

Burgess, T. L., and Kelly, R. B. (1987). Constitutive and regulated secretion of proteins. *Annu. Rev. Cell Biol.* **3,** 243–293.

Chanat, E., and Huttner, W. B. (1991). Milieu-induced, selective aggregation of regulated secretory proteins in the trans-Golgi network. *J. Cell Biol.* **115,** 1505–1519.

Chanat, E., Weib, U., Huttner, W. B., and Tooze, S. A. (1993). Reduction of the disulfide bond of chromogranin B (secretogranin I) in the trans-Golgi network causes its missorting to the constitutive secretory pathway. *EMBO J.* **12,** 2159–2168.

Chanturiya, A., Chernomordik, L. V., and Zimmerberg, J. (1997). Flickering fusion pores comparable with initial exocytotic pores occur in protein-free phospholipid bilayers. *Proc. Natl. Acad. Sci. U.S.A.* **94,** 14423–14428.

Cheng, A., Le, T., Palacios, M., Bookbinder, L. H., Wassarman, P. M., Suzuki, F., and Bleil, J. D. (1994). Sperm-egg recognition in the mouse: Characterization of sp56, a sperm protein having specific affinity for ZP3. *J. Cell Biol.* **125,** 867–878.

Cho, C., Bunch, D. O., Faure, J. E., Goulding, E. H., Eddy, E. M., Primakoff, P., and Myles, D. G. (1998). Fertilization defects in sperm from mice lacking fertilin beta. *Science* **281,** 1857–1859.

Cohen, N., and Wassarman, P. M. (2001). Association of egg zona pellucida glycoprotein mZP3 with sperm protein sp56 during fertilization in mice. *Int. J. Dev. Biol.* **45,** 569–576.

Cooper, T. G. (1998). Interactions between epididymal secretions and spermatozoa. *J. Reprod. Fertil.* (Suppl.) **53,** 119–136.

Courtens, J. L. (1978). Release of carbohydrate-containing vesicles by the developing acrosomes of ram spermatids—Relation to morphometric changes of the acrosome. *J. Ultrastruct. Res.* **65,** 182–189.

Cross, N. L. (1996). Effect of cholesterol and other sterols on human sperm acrosomal responsiveness. *Mol. Reprod. Dev.* **45,** 212–217.

Cross, N. L., and Meizel, S. (1989). Methods for evaluating the acrosomal status of mammalian sperm. *Biol. Reprod.* **41,** 635–641.

DiCarlantonio, G., and Talbot, P. (1988). Evidence for the sequential deployment of secretory enzymes during the normal acrosome reaction of guinea pig sperm *in vitro. Gamete Res.* **21,** 425–438.

Dobrinski, I., Ignotz, G. G., Fagnan, M. S., Yudin, A. I., and Ball, B. A. (1997). Isolation and characterization of a protein with homology to angiotensin converting enzyme from the periacrosomal plasma membrane of equine spermatozoa. *Mol. Reprod. Dev.* **48,** 251–260.

Eddy, E. M., and O'Brien, D. A. (1994). The spermatozoon. *In* "The Physiology of Reproduction," 2nd Ed. (E. Knobil and J. D. NEill, eds.), Vol. 1, pp. 29–77. Raven Press, New York.

Endo, Y., Lee, M. A., and Kopf, G. S. (1987). Evidence for the role of a guanine nucleotide-binding regulatory protein in the zona pellucida-induced mouse sperm acrosome reaction. *Dev. Biol.* **119,** 210–216.

Endo, Y., Lee, M. A., and Kopf, G. S. (1988). Characterization of an islet-activating protein-sensitive site in mouse sperm that is involved in the zona pellucida-induced acrosome reaction. *Dev. Biol.* **129,** 12–24.

Escalier, D., Bermùdez, D., Gallo, J. M., Viellefond, A., and Schrèvel, J. (1992). Cytoplasmic events in human meiotic arrest as revealed by immunolabelling of spermatocyte proacrosin. *Differentiation* **51,** 233–243.

Fawcett, D. W., and Hollenberg, R. D. (1963). Changes in the acrosome of guinea pig spermatozoa during passage through the epididymis. *Z. Zellforsch.* **60,** 276–292.

Fenichel, P., Donzeau, M., Farahifar, D., Basteris, B., Ayraud, N., and Hsi, B. L. (1991). Dynamics of human sperm acrosome reaction: Relation with *in vitro* fertilization. *Fertil. Steril.* **55,** 944–999.

Ferro-Novick, S., and Novick, P. (1993). The role of GTP-binding proteins in transport along the exocytic pathway. *Annu. Rev. Cell Biol.* **9,** 575–599.

Flaherty, S. P., and Olson, G. E. (1988). Membrane domains in guinea pig sperm and their role in the membrane fusion events of the acrosome reaction. *Anat. Rec.* **220,** 267–280.

Foster, J. A., and Gerton, G. L. (1996). Autoantigen 1 of the guinea pig sperm acrosome is the homologue of mouse TPX-1 and human TPX1 and is a member of the cysteine-rich secretory protein (CRISP) family. *Mol. Reprod. Dev.* **44,** 221–229.

Foster, J. A., Friday, B. B., Maulit, M. T., Blobel, C., Winfrey, V. P., Olson, G. E., Kim, K.-S., and Gerton, G. L. (1997). AM67, a secretory component of the guinea pig sperm acrosomal matrix, is related to mouse sperm protein sp56 and the complement component 4-binding proteins. *J. Biol. Chem.* **272,** 12714–12722.

Foster, J. A., Klotz, K. L., Flickinger, C. J., Thomas, T. S., Wright, R. M., Castillo, J. R., and Herr, J. C. (1994). Human SP-10: Acrosomal distribution, processing, and fate after the acrosome reaction. *Biol. Reprod.* **51,** 1222–1231.

Fraser, L. R. (1982). *p*-Aminobenzamidine, an acrosin inhibitor, inhibits mouse sperm penetration of the zona pellucida but not the acrosome reaction. *J. Reprod. Fertil.* **65,** 185–194.

Garty, N. B., Galiani, D., Aharonheim, A., Ho, Y. K., Phillips, D. M., Dekel, N., and Salomon, Y. (1988). G-proteins in mammalian gametes: An immunocytochemical study. *J. Cell Sci.* **91,** 21–31.

Gasman, S., Chasserot-Golaz, S., Popoff, M. R., Aunis, D., and Bader, M.-F. (1997). Trimeric G proteins control exocytosis in chromaffin cells. G_o regulates the peripheral actin network and catecholamine secretion by a mechanism involving the small GTP-binding protein Rho. *J. Biol. Chem.* **272,** 20564–20571.

Gerst, J. E. (1999). SNAREs and SNARE regulators in membrane fusion and exocytosis. *Cell. Mol. Life Sci.* **55,** 707–734.
Glabe, C. G., and Vacquier, V. D. (1977). Species specific agglutination of eggs by bindin isolated from sea urchin sperm. *Nature* **267,** 836–838.
Glassner, M., Jones, J., Kligman, I., Woolkalis, M. J., Gerton, G. L., and Kopf, G. S. (1991). Immunocytochemical and biochemical characterization of guanine nucleotide-binding regulatory proteins in mammalian spermatozoa. *Dev. Biol.* **146,** 438–450.
Gmachl, M., and Kreil, G. (1993). Bee venom hyaluronidase is homologous to a membrane protein of mammalian sperm. *Proc. Natl. Acad. Sci. U.S.A.* **90,** 3569–3573.
Green, D. P. (1987). Mammalian sperm cannot penetrate the zona pellucida solely by force. *Exp. Cell Res.* **169,** 31–38.
Griffiths, G., Warren, G. Stuhlfauth, I., and Jockusch, B. M. (1981). The role of clathrin-coated vesicles in acrosome formation. *Eur. J. Cell Biol.* **26,** 52–60.
Haendler, B., Krätzschmar, J., Theuring, F., and Schleuning, W.-D. (1993). Transcripts of cysteine-rich secretory protein-1 (CRISP-1; DE/AEG) and the novel related CRISP-3 are expressed under androgen control in the mouse salivary gland. *Endocrinology* **133,** 192–198.
Hagaman, J. R., Moyeer, J. S., Bachman, E. S., Sibony, M., Magyar, P. L., elch, J. E., Smithies, O., Krege, J. H., and O'Brien, D. A. (1998). Angiotensin-converting enzyme and male fertility. *Proc. Natl. Acad. Sci. U.S.A.* **95,** 2552–2557.
Hamberger, L., Lundin, K., Sjogren, A., and Soderlund, B. (1998). Indications for intracytoplasmic sperm injection. *Hum. Reprod.* **13** (Suppl. 1), 128–133.
Hardy, D. M., Huang, T. T., Driscoll, W. J., Tung, K. K., and Wild, G. C. (1988). Purification and characterization of the primary acrosomal autoantigen of guinea pig epididymal spermatozoa. *Biol. Reprod.* **38,** 423–437.
Hardy, D. M., Oda, M. N., Friend, D. S., and Huang, T. T. F., Jr. (1991). A mechanism for differential release of acrosomal enzymes during the acrosome reaction. *Biochem. J.* **275,** 759–766.
Hernandez, E. O., Roa-Espitia, A. L., Cruz Trejo, J., and Mujica, A. (1996). Sperm protein (sp50) binds to acrosome and plasma membranes in a Ca(2^+)-dependent manner: Possible role in acrosome reaction. *Mol. Reprod. Dev.* **43,** 366–375.
Hinsch, K. D., Hinsch, E., Aumüller, G., Tychowiecka, I., Schultz, G., and Schill, W. B. (1992). Immunological identification of G protein alpha- and beta-subunits in tail membranes of bovine spermatozoa. *Biol. Reprod.* **47,** 337–346.
Hoshi, M. (1985). Lysins. *In* "Biology of Fertilization" (C. B. Metz and A. Monroy, eds.), Vol. 2, pp. 431. Aacademic Press, New York.
Hou, S. T., Ma, A., Jones, R., and Hall, L. (1996). Molecular cloning and characterization of rat sperm surface antigen 2B1, a glycoprotein implicated in sperm-zona binding. *Mol. Reprod. Dev.* **45,** 193–203.
Hsu, Y.-C., and Perin, M. S. (1995). Human neuronal pentraxin II: Conservation, genomic structure and chromosomal localization. *Genomics* **28,** 220–227.
Huang, T. T., and Yanagimachi, R. (1984). Fucoidin inhibits attachment of guinea pig spermatozoa to the zona pellucida through binding to the inner acrosomal membrane and equatorial domains. *Exp. Cell Res.* **153,** 363–373.
Huang, T. T., Fleming, A. D., and Yanagimachi, R. (1981). Only acrosome-reacted spermatozoa can bind to and penetrate zona pellucida: A study using the guinea pig. *J. Exp. Zool.* **217,** 287–290.
Huang, T. T. F., Jr., Hardy, D., Yanagimachi, H., Teuscher, C., Tung, K., Wild, G., and Yanagimachi, R. (1985). pH and protease control of acrosomal content stasis and release during the guinea pig sperm acrosome reaction. *Biol. Reprod.* **32,** 451–462.
Huttner, W. B., Gerdes, H.-H., and Rosa, P. (1991). The granin (chromogranin/secretogranin) family. *Trends Biochem. Sci.* **16,** 27–30.
Hyatt, H., and Gwatkin, R. B. (1988). Characterization of isolated acrosomal matrices from hamster spermatozoa. *J. Reprod. Fertil.* **83,** 419–429.
Ikawa, M., Wada, I., Kominami, K., Watanabe, D., Toshimori, K., Nishimune, Y., and Okabe, M. (1997). The putative chaperone calmegin is required for sperm fertility. *Nature* **387,** 607–611

Jones, R., and Williams, R. M. (1990). Identification of zona- and fucoidan-binding proteins in guinea-pig spermatozoa and mechanism of recognition. *Development* **109,** 41–50.

Karnik, N. S., Newman, S., Kopf, G. S., and Gerton, G. L. (1992). Developmental expression of G protein α subunits in mouse spermatogenic cells: Evidence that Gαi is associated with the developing acrosome. *Dev. Biol.* **152,** 393–402.

Kasahara, M., Gutknecht, J., Brew, K., Spurr, N., and Goodfellow, P. N. (1989). Cloning and mapping of a testis-specific gene with sequence similarity to a sperm-coating glycoprotein gene. *Genomics* **5,** 527–534.

Kashiwabara, S., Baba, T., Takada, M., Watanabe, K., Yano, Y., and Arai, Y. (1990). Primary structure of mouse proacrosin deduced from the cDNA sequence and its gene expression during spermatogenesis. *J. Biochem. (Tokyo)* **108,** 785–791.

Katafuchi, K., Mori, T., Toshimori, K., and Iida, H. (2000). Localization of a syntaxin isoform, syntaxin 2, to the acrosomal region of rodent spermatozoa. *Mol. Reprod. Dev.* **57,** 375–383.

Kim, K.-S., Foster, J. A., and Gerton, G. L. (2001a). Differential release of guinea pig sperm acrosomal components during exocytosis. *Biol. Reprod.* **64,** 148–156.

Kim, K.-S., Cha, M. C., and Gerton, G. L. (2001b). Mouse sperm protein sp56 is a component of the acrosomal matrix. *Biol. Reprod.* **64,** 36–43.

Kligman, I., Glassner, M., Storey, B. T., and Kopf, G. S. (1991). Zona pellucida-mediated acrosomal exocytosis in mouse spermatozoa: Characterization of an intermediate stage prior to the completion of the acrosome reaction. *Dev. Biol.* **145,** 344–355.

Klinefelter, G. R., and Hamilton, D. W. (1985). Synthesis and secretion of proteins by perfused caput epididymal tubules, and association of secreted proteins with spermatozoa. *Biol. Reprod.* **33,** 1017–1027.

Kohno, N., Yamagata, K., Yamada, S., Kashiwabara, S., Sakai, Y., and Baba, T. (1998). Two novel testicular serine proteases, TESP1 and TESP2, are present in the mouse sperm acrosome. *Biochem. Biophys. Res. Commun.* **245,** 658–665.

Kolb-Bachofen, V. (1991). A review on the biological properties of C-reactive protein. *Immunobiology* **183,** 133–145.

Konrad, R. J., Young, R. A., Record, R. D., Smith, R. M., Butkerait, P., Manning, D., Jarett, L., and Wolf, B. A. (1995). The heterotrimeric G-protein G_i is localized to the insulin secretory granules of β-cells and is involved in insulin exocytosis. *J. Biol. Chem.* **270,** 12869–12876.

Kopf, G. S., and Gerton, G. L. (1991). The mammalian sperm acrosome and the acrosome reaction. *In* "Elements of Mammalian Fertilization" (P. M. Wassarman, ed), Vol. 1, pp. 153–203. CRC Press, Boca Raton.

Krege, J. H., John, S. W. M., Langenbach, L. L., Hodgin, J. B., Hagaman, J. R., Bachman, E. S., Jennette, J. C., O'Brien, D. A., and Smithies, O. (1995). Male-female differences in fertility and blood pressure in ACE-deficient mice. *Nature* **375,** 146–148.

Lagriffoul, A., Charpentier, N., Carrette, J., Tougard, C., Bockaert, J., and Homburger, V. (1996). Secretion of protease nexin-1 by C6 glioma cells is under the control of a heterotrimeric G protein, G_{o1}. *J. Biol. Chem.* **271,** 31508–31516.

Larson, J. L., and Miller, D. J. (1999). Simple histochemical stain for acrosomes on sperm from several species. *Mol. Reprod. Dev.* **52,** 445–449.

Lathrop, W. F., Carmichael, E. P., Myles, D. G., and Primakoff, P. (1990). cDNA cloning reveals the molecular structure of a sperm surface protein, PH-20, involved in sperm–egg adhesion and the wide distribution of its gene among mammals. *J. Cell Biol.* **111,** 2939–2949.

Lee, M. A., and Storey, B. T. (1985). Evidence for plasma membrane impermeability to small ions in acrosome-intact mouse spermatozoa bound to mouse zonae pellucidae, using an aminoacridine fluorescent pH probe: Time course of the zona-induced acrosome reaction monitored by both chlortetracylcine and pH probe fluorescence. *Biol. Reprod.* **33,** 235–246.

Lewis, C. A., Talbot, C. F., and Vacquier, V. D. (1982). A protein from abalone sperm dissolves the egg vitelline layer by a nonenzymatic mechanism. *Dev. Biol.* **92,** 227–239.

Liang, L. F., and Dean, J. (1993). Oocyte development: molecular biology of the zona pellucida. *Vitam. Horm.* **47,** 115–159.

Lin, Y., Kimmel, L. H., Myles, D. G., and Primakoff, P. (1993). Molecular cloning of the human and monkey sperm surface protein PH-20. *Proc. Natl. Acad. Sci. U.S.A* **90,** 10071–10075.

Liu, D. Y., and Baker, H. W. (1993). Inhibition of acrosin activity with a trypsin inhibitor blocks human sperm penetration of the zona pellucida. *Biol. Reprod.* **48,** 340–348.

Lu, G., Villalba, M., Coscis, M. R., Hoffman, D. R., and King, T. P. (1993). Sequence analysis and antigenic cross-reactivity of a venom allergen, antigen 5, from hornets, wasps, and yellow jackets. *J. Immunol.* **150,** 2823–2830.

Maeda, T., Sakashita, M., Ohba, Y., and Nakanishi, Y. (1998). Molecular cloning of the rat Tpx-1 responsible for the interaction between spermatogenic and Sertoli cells. *Biochem. Biophys. Res. Commun.* **248,** 140–146.

Marshburn, P. B., Stovall, D. W., Hammond, M. G., Talbert, L. M., and Shabanowitz, R. B. (1991). Fertility rates in men with normal semen characteristics: spermatozoal testing by induction of the acrosome reaction and Wright-Giemsa staining for subtle abnormal forms [see comments]. *Obstet. Gynecol.* **77,** 250–255.

Michaut, M., De Blas, G., Tomes, C. N., Yunes, R., Fukuda, M., and Mayorga, L. S. (2001). Synaptotagmin VI participates in the acrosome reaction of human spermatozoa. *Dev. Biol.* **235,** 521–529.

Michaut, M., Tomes, C. N., De Blas, G., Yunes, R., and Mayorga, L. S. (2000). Calcium-triggered acrosomal exocytosis in human spermatozoa requires the coordianted activation of Rab3A and N-ethylmaleimide-sensitive factor. *Proc. Natl. Acad. Sci. U.S.A.* **97,** 9996–10001.

Mizuki, N., Sarapata, D. E., Garcia-Sanz, J. A., and Kasahara, M. (1992). The mouse male germ cell-specific gene *Tpx-1:* Molecular structure, mode of expression in spermatogenesis, and sequence similarity to two non-mammalian genes. *Mamm. Genome* **3,** 274–280.

Mochca-Morales, J., Martin, B. M., and Possani, L. D. (1990). Isolation and characterization of helothermine, a novel toxin from *Heloderma horridum horridum* (Mexican beaded lizard) venom. *Toxicon* **28,** 299–309.

Monck, J. R., and Fernandez, J. M. (1996). The fusion pore and mechanisms of biological membrane fusion. *Curr. Opin. Cell Biol.* **8,** 524–533.

Monck, J. R., Oberhauser, A. F., De Toledo, G. A., and Fernandez, J. M. (1991). Is swelling of the secretory granule matrix the force that dilates the exocytotic fusion pore? *Biophys. J.* **59,** 39–47.

Moore, A., Ensrud, K. M., White, T. W., Frethem, C. D., and Hamilton, D. W. (1994). Rat epididymis-specific sperm maturation antigens. I. Evidence that the 26 kD 4E9 antigen found on rat caudal epididymal sperm tail is derived from a protein secreted by the epididymis. *Mol. Reprod. Dev.* **37,** 181–194.

Morales, P., Cross, N. L., Overstreet, J. W., and Hanson, F. W. (1989). Acrosome-intact and acrosome-reacted human sperm can initiate binding to the zona pellucida. *Dev. Biol.* **133,** 385–392.

Mori, E., Kashiwabara, S.-I., Baba, T., Inagaki, Y. and Mori, T. (1995). Amino acid sequences of porcine sp38 and proacrosin required for binding to the zona pellucida. *Dev. Biol.* **168,** 575–583.

Mukerji, S. J., and Meizel, S. (1979). Rabbit testis proacrosin: Purification, molecular weight estimation, and amino acid and carbohydrate composition of the molecule. *J. Biol. Chem.* **254,** 11721–11728.

Mundy, A. J., Ryder, T. A., and Edmonds, D. K. (1994). A quantitative study of sperm head ultrastructure in subfertile males with excess sperm precursors. *Fertil. Steril.* **61,** 751–754.

Myles, D. G., Hyatt, H., and Primakoff, P. (1987). Binding of both acrosome-intact and acrosome-reacted guinea pig sperm to the zona pellucida during *in vitro* fertilization. *Dev. Biol.* **121,** 559–567.

NagDas, S. K., Skudlarek, M. D., Ogrebin-Crist, M.-C., and Tulsiani, D. R. P. (1992). Biochemical alterations in the proacrosin-acrosin system during epididymal maturation of the rat spermatozoa. *J. Androl.* **13,** 36–43.

NagDas, S. K., Winfrey, V. P., and Olson, G. E. (1996a). Identification of hydrolase binding activities of the acrosomal matrix of hamster spermatozoa. *Biol. Reprod.* **55,** 1405–1414.

NagDas, S. K., Winfrey, V. P., and Olson, G. E. (1996b). Proacrosin-acrosomal matrix binding interactions in ejaculated bovine spermatozoa. *Biol. Reprod.* **54,** 111–121.

Nolan, J. P., and Hammerstedt, R. H. (1997). Regulation of membrane stability and the acrosome reaction in mammalian sperm. *FASEB J.* **11,** 670–682.

Noland, T. D., Davis, L. S., and Olson, G. E. (1989). Regulation of proacrosin conversion in isolated guinea pig sperm acrosomal apical segments. *J. Biol. Chem.* **264,** 13586–13590.

Noland, T. D., Friday, B. B., Maulit, M. T., and Gerton, G. L. (1994). The sperm acrosomal matrix contains a novel member of the pentaxin family of calcium-dependent binding proteins. *J. Biol. Chem.* **269,** 32607–32614.

O'Brien, R. J., Xu, D., Petralia, R. S., Steward, O., Huganir, R. L., and Worley, P. (1999). Synaptic clustering of AMPA receptors by the extracellular immediate- early gene product Narp. *Neuron* **23,** 309–323.

Ohmura, K., Kohno, N., Kobayashi, Y., Yamagata, K., Sato, S., Kashiwabara, S., and Baba, T. (1999). A homologue of pancreatic trypsin is localized in the acrosome of mammalian sperm and is released during acrosome reaction. *J. Biol. Chem.* **274,** 29426–29432.

Ohnishi, H., Ernst, S. A., Yule, D. I., Baker, C. W., and Williams, J. A. (1997). Heterotrimeric G-protein $G_{q/11}$ localized on pancreatic zymogen granules is involved in calcium-regulated amylase secretion. *J. Biol. Chem.* **272,** 16056–16061.

Olson, G. E., and Winfrey, V. P. (1985a). Structure of membrane domains and matrix components of the bovine acrosome. *J. Ultrastruct. Res.* **90,** 9–25.

Olson, G. E., and Winfrey, V. P. (1985b). Substructure of a cytoskeletal complex associated with the hamster sperm acrosome. *J. Ultrastruct. Res.* **92,** 167–179.

Olson, G. E., and Winfrey, V. P. (1994). Structure of acrosomal matrix domains of rabbit sperm. *J. Struct. Biol.* **112,** 41–48.

Olson, G. E., Winfrey, V. P., and Davenport, G. R. (1988). Characterization of matrix domains of the hamster acrosome. *Biol. Reprod.* **39,** 1145–1158.

Olson, G. E., Winfrey, V. P., and NagDas, S. K. (1998). Acrosome biogenesis in the hamster: Ultrastructurally distinct matrix regions are assembled from a common precursor polypeptide. *Biol. Reprod.* **58,** 361–370.

Olson, G. E., Winfrey, V. P., Neff, J. C., Lukas, T. J., and NagDas, S. K. (1997). An antigenically related polypeptide family is a major structural constituent of a stable acrosomal matrix assembly in bovine spermatozoa. *Biol. Reprod.* **57,** 325–334.

Olson, G. E., Winfrey, V. P., Winer, M. A., and Davenport, G. R. (1987). Outer acrosomal membrane of guinea pig spermatozoa: Isolation and structural characterization. *Gamete Res.* **17,** 77–84.

Pelletier, R. M., and Friend, D. S. (1983). Development of membrane differentiations in the guinea pig spermatid during spermiogenesis. *Am. J. Anat.* **167,** 119–141.

Primakoff, P., Hyatt, H., and Myles, D. G. (1985). A role for the migrating sperm surface antigen PH-20 in guinea pig sperm binding to the egg zona pellucida. *J. Cell Biol.* **101,** 2239–2244.

Ramalho-Santos, J., Moreno, R. D., Sutovsky, P., Chan, A. W., Hewitson, L., Wessel, G. M., Simerly, C. R., and Schatten, G. (2000). SNAREs in mammalian sperm:possible iimplications for fertilization. *Dev. Biol.* **223,** 54–69.

Ramaraj, P., Kessler, S. P., Colmenares, C., and Sen, G. C. (1998). Selective restoration of male fertility in mice lacking angiotensin-converting enzymes by sperm specific expression of the testicular isozyme. *J. Clin. Invest.* **102,** 371–378.

Reid, M., and Blobel, C. P. (1994). Apexin, an acrosomal pentaxin. *J. Biol. Chem.* **269,** 32614–32620.

Saling, P. M. (1981). Involvement of trypsin-like activity in binding of mouse spermatozoa to zonae pellucidae. *Proc. Natl. Acad. Sci. U.S.A.* **78,** 6231–6235.

Saling, P. M., Sowinski, J., and Storey, B. T. (1979). An ultrastructural study of epididymal mouse spermatozoa binding to the zonae pellucidae *in vitro:* Sequential relationship to the acrosome reaction. *J. Exp. Zool.* **209,** 229–238.

Sandoz, D. (1979). Evolution des ultrastructures au cours de la formation de l'acrosome du spermatozoide chez la souris. *J. Microsc. (Paris)* **9,** 535–558.

Saxena, D. K., Tanii, I., Yoshinaga, K., and Toshimori, K. (1999). Role of intra-acrosomal antigenic molecules acrin 1 (MN7) and acrin 2 (MC41) in penetration of the zona pellucida in fertilization in mice. *J. Reprod. Fertil.* **117,** 17–25.

Schill, W. B. (1991). Some disturbances of acrosomal development and function in human spermatozoa. *Hum. Reprod.* **6,** 969–978.

Schroer, S. C., Yudin, A. I., Myles, D. G., and Overstreet, J. W. (2000). Acrosomal status and motility of guinea pig spermatozoa during in vitro penetration of the cumulus oophorus. *Zygote* **8,** 107–117.

Schulz, J. R., Sasaki, J. D., and Vacquier, V. D. (1998). Increased association of synaptosome-associated protein of 25 kDa with syntaxin and vesicle-associated membrane protein following acrosomal exocytosis of sea urchin sperm. *J. Biol. Chem.* **273,** 24355–24359.

Schulz, J. R., Wessel, G. M., and Vacquier, V. D. (1997). The exocytosis regulatory proteins syntaxin and VAMP are shed from sea urchin sperm during the acrosome reaction. *Dev. Biol.* **191,** 80–87.

Seethaler, G., and Huttner, W. (1991). Secretory protein sorting, processing and granule biogenesis. *Trends Cell Biol.* **1,** 35–36.

Shalgi, R., Matityahu, A., Gaunt, S. J., and Jones, R. (1990). Antigens on rat spermatozoa with a potential role in fertilization. *Mol. Reprod. Dev.* **25,** 286–296.

Sotomayor, R. E., and Handel, M. A. (1986). Failure of acrosome assembly in a male sterile mouse mutant. *Biol. Reprod.* **34,** 171–182.

Srivastava, P. N., Adams, C. E., and Hartree, E. F. (1965). Enzymatic action of acrosomal preparations on the rabbit ovum *in vitro. J. Reprod. Fertil.* **10,** 61.

Stambaugh, R., and Buckley, J. (1968). Zona pellucida dissolution enzymes of the rabbit sperm head. *Science* **161,** 585–586.

Suzuki-Toyota, F., Maekawa, M., Cheng, A., and Bleil, J. D. (1995). Immuno-colloidal gold labeled surface replica, and its application to detect sp56 the egg recognition and binding protein, on the mouse spermatozoon. *J. Electron Microsc.* **44,** 135–139.

Talbot, P., and DiCarlantonio, G. (1985). Cytochemical localization of dipeptidyl peptidase II (DPP-II) in mature guinea pig sperm. *J. Histochem. Cytochem.* **33,** 1169–1172.

Tanii, I., Yoshinaga, K., and Toshimori, K. (1999). Morphogenesis of the acrosome during the final steps of rat spermiogenesis with special reference to tubulobulbar complexes. *Anat. Rec.* **256,** 195–201.

ten Have, J., Beaton, S., and Bradley, M. P. (1998). Cloning and characterization of the cDNA encoding the PH20 protein in the European red fox *Vulpes vulpes. Reprod. Fertil. Dev.* **10,** 165–172.

Thomas, T. S., Wilson, W. L., Reynolds, A. B., and Oliphant, G. (1986). Chemical and physical characterization of rabbit sperm acrosome stabilizing factor. *Biol. Reprod.* **35,** 691–703.

Tollner, T. L., Yudin, A. I., Cherr, G. N., and Overstreet, J. W. (2000). Soybean trypsin inhibitor as a probe for the acrosome reaction in motile cynomolgus macaque sperm. *Zygote* **8,** 127–137.

Tooze, S. A. (1991). Biogenesis of secretory granules. Implications arising from the immature secretory granule in the regulated pathway of secretion. *FEBS Lett.* **285,** 220–224.

Tooze, S. A. (1992). Biogenesis of secretory granules. *Semin. Cell Biol.* **3,** 357–366.

Tooze, S. A., and Huttner, W. B. (1990). Cell-free protein sorting to the regulated and constitutive secretory pathways. *Cell* **60,** 837–847.

Tooze, S. A., Chanat, E., Tooze, J., and Huttner, W. B. (1993). Secretory granule formation. *In* "Mechanisms of Intracellular Trafficking and Processing of Proproteins" (Y. P. Loh, ed.), pp. 157–177. CRC Press, Boca Raton.

Toshimori, K., Saxena, D. K., Tanii, I., and Yoshinaga, K. (1998). An MN9 antigenic molecule, equatorin, is required for successul sperm–oocyte fusion in mice. *Biol. Reprod.* **59,** 22–29.

Toshimori, K., Tanii, I., and Araki, S. (1995). Intra-acrosomal 155,000 dalton protein increases the antigenicity during mouse sperm maturation in the epididymis: A study using a monoclonal antibody MC101. *Mol. Reprod. Dev.* **42,** 72–79.

Toshimori, K., Tanii, I., Araki, S., and Oura, C. (1992). Characterization of the antigen recognized by a monoclonal antibody MN9: Unique transport pathway to the equatorial segment of sperm head during spermiogenesis. *Cell Tissue Res.* **270,** 459–68.

Tsui, C. C., Copeland, N. G., Gilbert, D. J., Jenkins N. A., Barnes, C., and Worley, P. F. (1996). Narp, a novel member of the pentraxin family, promotes neurite outgrowth and is dynamically regulated by neuronal activity. *J. Neurosci.* **16,** 2463–2478.

Tulsiani, D. R., Abou-Haila, A., Loeser, C. R., and Pereira, B. M. (1998). The biological and functional significance of the sperm acrosome and acrosomal enzymes in mammalian fertilization. *Exp. Cell Res.* **240,** 151–164.

Urch, U. A. (1991). Biochemistry and function of acrosin. *In* "Elements of Mammalian Fertilization" (P. M. Wassarman, ed.), pp. 233–248. CRC Press, Boca Raton.

Vacquier, V. D., and Moy, G. W. (1977). Isolation of bindin: The protein responsible for adhesion of sperm to sea urchin eggs. *Proc. Natl. Acad. Sci. U.S.A.* **74,** 2456–2460.

Vacquier, V. D., and Moy, G. W. (1997). The fucose sulfate polymer of egg jelly binds to sperm REJ and is the inducer of the sea urchin sperm acrosome reaction. *Dev. Biol.* **192,** 125–135.

Valdivia, M., Sillerico, T., De Ioannes, A., and Barros, C. (1999). Proteolytic activity of rabbit perivitelline spermatozoa. *Zygone* **7,** 143–149.

VandeVoort, C. A., Yudin, A. I., and Overstreet, J. W. (1997). Interaction of acrosome-reacted macaque sperm with the macaque zona pellucida. *Biol. Reprod.* **56,** 1307–1316.

Visconti, P. E., Bailey, J. D., Moore, G. D., Pan, D., Olds-Clarke, P., and Kopf, G. S. (1995). Capacitation of mouse spermatozoa. I. Correlation between the capacitation state and protein tyrosine phosphorylation. *Development* **121,** 1129–1137.

Visconti, P. E., Ning, X., Fornes, M. W., Alvarez, J. G., Stein, P., Connors, S. A., and Kopf, G. S. (1999). Cholesterol efflux-mediated signal transduction in mammalian sperm: Cholesterol release signals an increase in protein tyrosine phosphorylation during mouse sperm capacitation. *Dev. Biol.* **214,** 429–443.

Wassarman, P. M. (1990). Profile of a mammalian sperm receptor. *Development* **108,** 1–17.

Westbrook-Case, V. A., Winfrey, V. P., and Olson, G. E. (1994). A domain-specific 50-kilodalton structural protein of the acrosomal matrix is processed and released during the acrosome reaction in the guinea pig. *Biol. Reprod.* **51,** 1–13.

Wilson, W. L., and Oliphant, G. (1987). Isolation and biochemical characterization of the subunits of the rabbit sperm acrosome stabilizing factor. *Biol. Reprod.* **37,** 159–169.

Yamagata, K., Honda, A., Kashiwabara, S. I., and Baba, T. (1999). Difference of acrosomal serine protease system between mouse and other rodent sperm. *Dev. Genet.* **25,** 115–122.

Yamagata, K., Murayama, K., Kohno, N., Kashiwabara, S., and Baba, T. (1998a). *p*-Aminobenzamidine-sensitive acrosomal protease(s) other than acrosin serve the sperm penetration of the egg zona pellucida in mouse. *Zygote* **6,** 311–319.

Yamagata, K., Murayama, K., Okabe, M., Toshimori, K., Nakanishi, T., Kashiwabara, S., and Baba, T. (1998b). Acrosin accelerates the dispersal of sperm acrosomal proteins during acrosome reaction. *J. Biol. Chem.* **273,** 10470–10474.

Yanagimachi, R. (1994). Mammalian fertilization. *In* "The Physiology of Reproduction" (E. Knobil and J. D. Neill, eds.), pp. 189–317. Raven Press, New York.

Yoo, S. H. (1993). pH-Dependent binding of chromogranin B and secretory vesicle matrix proteins to the vesicle membrane. *Biochim. Biophys. Acta Mol. Cell Res.* **1179,** 239–246.

Yoshinaga, K., Tanii, I., Saxena, D. K., and Toshimori, K. (1998). Immunocytochemical alterations in the intra-acrosomal antigen MN7 during epididymal maturation of guinea pig spermatozoa. *Cell Tissue Res.* **292,** 427–433.

Yudin, A. I., Vandevoort, C. A., Li, M. W., and Overstreet, J. W. (1999). PH-20 but not acrosin is involved in sperm penetration of the macaque zona pellucida. *Mol. Reprod. Dev.* **53,** 350–362.

9

Gamete Fusion in Mammals

Paul Primakoff* and Diana G. Myles†

*Department of Cell Biology and Anatomy, and †Section of Molecular and Cell Biology, University of California, Davis

I. Introduction

Sperm–egg fusion is the culmination of gametogenesis and all the preceding steps in fertilization. Recent work on the molecular basis of sperm–egg fusion has implicated certain key sperm and egg surface proteins as functioning in this process. However, much is still unknown. A number of outstanding questions related to the mechanism of gamete membrane fusion remain unanswered:

1. Will gametes fuse with other cell types or only with each other?
2. Is there sperm–egg plasma membrane binding that precedes membrane fusion?

3. Does sperm–egg fusion occur in a single step or are there multiple steps, i.e., is there a pathway of sperm–egg membrane interactions that leads to fusion?

4. What interaction between the membranes actually leads to fusion of the two lipid bilayers?

5. To what degree does the mechanism of gamete fusion resemble that of other membrane fusion systems, particularly somatic cell–cell fusion, virus–cell fusion, and intracellular fusion (e.g., exocytosis)?

6. How does the fusion of one sperm with an egg alter the ability of that egg to fuse with a second sperm?

II. SPECIFICITY OF GAMETE FUSION

The known topology of sperm–egg fusion suggests there is considerable specificity as to which regions of the sperm and egg plasma membrane can fuse. From a variety of studies, certain morphological features of gamete fusion among eutherian mammals are widely accepted. Sperm must acrosome react in order to be capable of fusing. The fertilizing sperm makes initial contact, via its inner acrosomal membrane (IAM), with the microvillar surface of the egg. The region of the egg plasma membrane that has a low concentration of microvilli, the cortical granule-free region, generally does not participate in sperm binding or fusion. The inner acrosomal membrane never fuses with the microvillar egg plasma membrane; instead, the sperm pivots, then lies flat on its side, and initial membrane fusion occurs between the sperm plasma membrane in the equatorial region (at least in several studied species) and the egg microvilli. Fusion continues in the sperm plasma membrane in the posterior head region and usually along the sperm tail.

It has been widely assumed that sperm and eggs will fuse only with each other. It is of importance to realize that little effort has been expended to test this assumption. Scofield and colleagues have reported that human sperm will fuse with tissue culture cells transfected to express the human major histocompatibility class II (MHC II) antigen. Our attempts to repeat this result using mouse sperm and mouse MHC II antigen indicate that the tissue culture cells expressing MHC II can phagocytose the sperm, but do not fuse with them (P. Primakoff and D. Myles, unpublished results). As mentioned above, eggs will not fuse with acrosome-intact sperm and will fuse with acrosome-reacted sperm, suggesting a substantial degree of specificity. But to our knowledge, no one has tested sperm or eggs with a large battery of different cell types to find out if there is another kind of cell with which a gamete will fuse.

III. A HYPOTHETICAL PATHWAY LEADING TO SPERM–EGG FUSION

As we discuss below, it is possible that sperm–egg fusion occurs in a single step in which a sperm adhesion protein, fertilin, binds to an egg integrin, leading to a

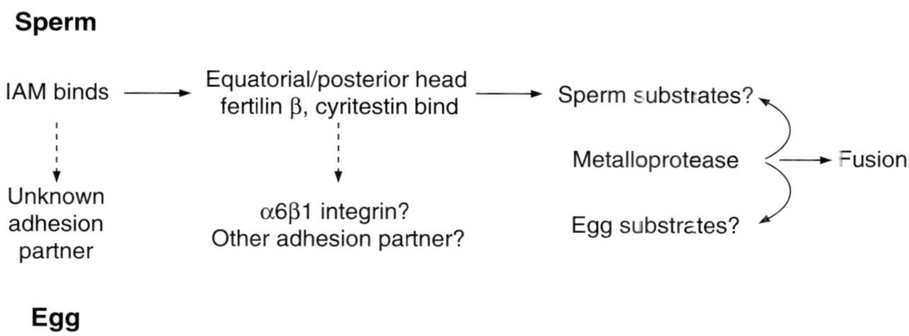

FIGURE 9.1 Proposed sperm–egg membrane interactions.

conformational change in fertilin that promotes fusion. However, there is also evidence that gamete fusion is a multistep process in which multiple adhesion molecules and a surface metalloprotease must act. To have a framework for considering this putative multistep process (pathway) to membrane fusion, we can diagram the proposed sperm–egg membrane interactions (see Figure 9.1).

This working model is based on several kinds of data. In watching the sperm first interact with the egg plasma membrane (in eggs with or without a zona pellucida), it can be observed that the sperm initially contacts the egg in the region of the anterior tip of the sperm (inner acrosomal membrane). It has been suggested that this binding may be a critical step in the sperm–egg plasma membrane interactions that lead to fusion. Thus far cyritestin (see below), a putative adhesion molecule, has been reported to be present on the inner acrosomal membrane (IAM) and the equatorial region (Forsbach and Heinlein, 1998; Yuan *et al.,* 1997). No other potential adhesion molecules on the IAM or on the egg have been identified that might play a role in IAM binding.

The initial interaction between the IAM and the egg plasma membrane potentially serves two different functions leading up to fusion. One possibility is that the interaction of IAM and egg plasma membrane serves simply to capture the sperm. The second possibility is that the IAM–egg plasma membrane interaction signals either of the two gametes to modulate the adhesion proteins that participate in subsequent binding steps. The capture or modulation may happen in a manner analogous to the initial interaction between leukocytes and endothelial cells during leukocyte extravasation. In the leukocyte–endothelial cell system the initial interaction between the two cells serves at least two purposes: it captures the moving leukocytes that are moving along the surface of the endothelium and allows them to receive cytokine signals and up-regulates the avidity of their subsequent binding steps.

As mentioned above, the initial attachment of sperm by the IAM can convert to a flattening of the sperm on the egg plasma membrane so that the sperm is binding via the equatorial or posterior head region. If binding via the equatorial/posterior head region is experimentally prevented in the golden hamster, the sperm pivot back and forth on the egg surface, remaining attached by the inner acrosomal membrane. Fusion occurs between the equatorial/posterior head plasma membrane of the sperm and the egg plasma membrane and eventually the sperm tail plasma membrane may also become incorporated into the zygote membrane.

Following sperm–egg binding but preceding fusion there may be additional steps. The best supported is the action of a sperm surface metalloprotease on unknown sperm or egg surface substrates (P. Primakoff and D. Myles, unpublished results).

The inner acrosomal membrane of mammalian sperm does not fuse with the egg plasma membrane, but is taken up into the egg cytoplasm, along with a piece of the egg plasma membrane. The engulfment of the inner acrosomal membrane has been described as a type of phagocytotic process. A major difference from classical phagocytosis in terms of the morphology is that the "phagosome" is surrounded by a hybrid membrane that includes sperm IAM and egg plasma membrane.

IV. SPERM AND EGG SURFACE PROTEINS INVOLVED IN GAMETE BINDING AND FUSION

A. DO SPERM BIND TO THE EGG BEFORE THEY FUSE?

In the early literature it was considered that sperm and egg might fuse in the absence of a preceding adhesion step, just as, under appropriate circumstances, two phospholipid vesicles in aqueous media can fuse without binding. However, current data support the idea that an adhesion step precedes fusion. Conditions have been found in which sperm bind to the egg plasma membrane, but do not fuse. For instance, with mouse gametes in the absence of glucose or presence of glucose and glucose metabolism inhibitors, sperm bind to the egg plasma membrane but do not fuse. Also gamete membrane binding without subsequent fusion is seen in the absence of calcium. When glucose or calcium is restored , fusion occurs.

During *in vivo* fertilization, one sperm reaches the egg plasma membrane and fuses. Typically, using *in vitro* fertilization assays with zona-free eggs, it is found that in 30 minutes one sperm will fuse with the egg plasma membrane and another ~10 sperm will bind. Although the correct interpretation of this result is uncertain, it may mean that the ~10 bound sperm are bound in a physiologically relevant way and could proceed to fusion, but have not yet done so. Some exogenous reagents (e.g., antibodies, peptides) that inhibit the one sperm fusing also inhibit the ~10 sperm binding (see below). This finding is consistent with the idea that sperm binding is a mandatory prerequisite to fusion. In other well-studied systems, particularly virus–cell membrane fusion and intracellular membrane fusion, an adhesion step precedes the fusion step.

B. SPERM ADHESION PROTEINS

A question not yet fully answered concerns which of the sperm surface adhesion proteins are involved in binding sperm to the egg plasma membrane. Various sperm proteins, including the secreted epididymal protein DE and antigens recognized by the antibodies M29, M37, OBF 13, and MH61, have been proposed to have a role in gamete adhesion/fusion. In the majority of cases, this putative role for the proteins has been initially proposed based on the finding that an antibody to the protein inhibits gamete binding or fusion.

Our own initial studies concerned the sperm protein fertilin, which we proposed to function in gamete fusion. The evidence for assigning this function came from monoclonal antibody (mAb) inhibition studies. Two antifertilin mAbs, PH-30 and PH-1, bind to the posterior head plasma membrane of guinea pig sperm. PH-30 (function-blocking mAb) strongly inhibits sperm fusion with zona-free guinea pig eggs whereas PH-1 (non-function-blocking mAb) has no effect on sperm fusion. Fertilin (originally called PH-30) was purified by PH-30 mAb affinity chromatography and found to be a membrane-anchored heterodimer of two noncovalently associated subunits, α and β.

Cloning of cDNAs for guinea pig fertilin-α and -β led to several important conclusions and hypotheses. Both fertilin-α and fertilin-β have the same modular organization and turned out to be the first identified members of a new gene family. Fertilin-α and fertilin-β and other family members are each about 750 residues long and each has these domains: pro-, metalloprotease, disintegrin, cysteine-rich, epidermal growth factor (EGF)-like, transmembrane, and cytoplasmic tail (Figure 9.2). This gene family of membrane-anchored proteins is called the ADAM family, because members contain a disintegrin and metalloprotease domain. The ADAM family is closely related to a family of soluble snake venom proteins that have the same N-terminal domains (modules) but lack the EGF and transmembrane domains.

Fertilin-α and fertilin-β are both processed by proteolytic cleavage during sperm differentiation. In guinea pig, fertilin-α is processed during spermatogenesis whereas fertilin-β is processed during sperm transit through the epididymis. For both types of fertilin, the cleavage is between the metalloprotease and disintegrin domains so that on cauda epididymal sperm the N-terminal domain of each subunit is the disintegrin domain.

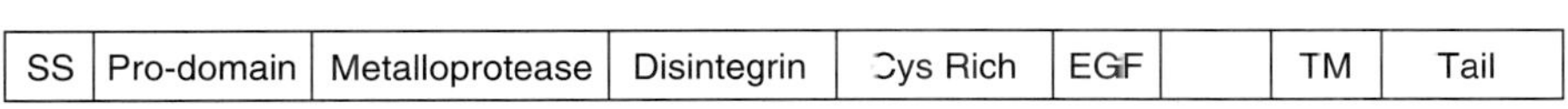

FIGURE 9.2 Domain organization of the ADAM gene family. ADAMs have large N-terminal extracellular domains and a short C-terminal cytoplasmic domain. SS, signal sequence; Cys Rich, cysteine rich; EGF, EGF-like; TM, transmembrane; tail, cytoplasmic tail.

C. ROLE OF FERTILIN-β IN SPERM BINDING

The disintegrin domain has high sequence homology with a class of soluble peptides, present in snake venom, called "disintegrins." The snake disintegrins are known integrin ligands. Thus, we hypothesized that a fertilin disintegrin domain could act to bind sperm to the egg (presumably by binding to an egg integrin) and focused initially on fertilin-β (to which the function-blocking PH-30 mAb binds).

Many of the snake disintegrins bind to the integrin αIIbβ3, thereby preventing fibrinogen binding and inhibiting blood clotting. Structural solutions of two snake disintegrins obtained using nuclear magnetic resonance show that these disintegrins have an active site region containing the signature sequence RGD at the tip of a flexible loop with two disulfide bonds at its base. Other snake disintegrins have a different sequence (not RGD) in the homologous position in the loop, and almost all ADAMs, including fertilin-β, also have a different sequence in this position. These various sequences are compared in Figure 9.3. We tested the idea that this putative binding loop region of the fertilin-β disintegrin domain functions in sperm–egg adhesion/fusion. The experiments tested whether peptides containing the loop sequence TDE (Myles *et al.,* 1981) or QDE (mouse), in the homologous position with the RGD of the snake disintegrins, would inhibit *in vitro* fertilization (IVF) assays. We found that TDE-containing peptides are potent blockers of guinea pig sperm–egg fusion, inhibiting 80–90% sperm fusion with either zona-free or zona-intact eggs. Scrambled (control) peptides, containing the same amino acids in a scrambled sequence, show little or no effect on the assay. Also, fluorescent beads conjugated with a TDE-containing peptide bind to the surface of zona-free eggs. Subsequent experiments with mouse gametes confirmed these finding. In these experiments peptides representing the hypothesized mouse fertilin-β active site loop, containing QDE, were found to inhibit sperm–egg binding and/or fusion. Also, an antibody raised against the QDE-containing loop peptide blocks (80–90%) sperm–egg binding and fusion (Yuan *et al.,* 1997) and a recombinant fertilin-β extracellular domain, made in *Escherichia coli,* inhibits sperm–egg binding and fusion (Evans *et al.,* 1997). The evidence from this group of experiments suggests that fertilin-β has an adhesion activity, its active site is the TDE/QDE-containing loop, and that it functions in sperm binding to the egg plasma

Snake disintegrin #1 (echistatin)	C	R	A	R	*R*	*G*	*D*	–	D	M	D	D	Y	C
Snake disintegrin #2 (HR1b)	C	R	A	A	*E*	*S*	*E*	C	D	I	P	E	S	C
Guinea pig fertilin β	C	R	E	S	*T*	*D*	*E*	C	D	L	P	E	Y	C
Mouse fertilin β	C	R	L	A	*Q*	*D*	*E*	C	D	V	T	E	Y	C
Cyritestin	C	R	K	S	*K*	*D*	*Q*	C	D	F	P	E	F	C

FIGURE 9.3 The sequences of the active sites of disintegrin domains. The RGD binding motif in snake disintegrin #1 is italicized as are the corresponding tripeptide sequences in the other disintegrin domains. The underlined sequences are the eight-residue peptides used to test inhibition in IVF assays.

membrane, a binding that is prerequisite for and leads to membrane fusion. *In vitro* fertilization assays using sperm from fertilin-β knockout mice confirm that fertilin-β is required for sperm–egg plasma membrane adhesion (Cho *et al.,* 1998).

Tests of a possible adhesion function of the fertilin-α subunit have produced less clear-cut results. Yuan *et al.* (1997) tested if a mouse fertilin-α peptide, representing the homologous loop sequence just discussed, affected mouse sperm adhesion to zona-free eggs. Only limited inhibition (~30%) was seen and a scrambled peptide control also inhibited (~20%). In a different approach, a recombinant fertilin-α, made as a fusion partner of maltose binding protein (MBP), was expressed in *E. coli.* This fertilin-α construct coded for the extracellular portion of fertilin-α, which is C-terminal to the disintegrin loop sequence and thus did not include the loop. The investigators found that the fusion protein bound to the egg plasma membrane and inhibited sperm–egg binding compared to MBP alone as a control. The level of inhibition seen was relatively high in experiments with low sperm binding in the control; however, the inhibition by the recombinant fusion protein was substantially less in experiments with high sperm binding in the control (Evans *et al.,* 1997). These results collectively suggest that fertilin-α could make a contribution to sperm adhesion and suggest the need for additional experiments, possibly with native fertilin and site-directed mutagenesis of the α subunit.

D. ROLE OF CYRITESTIN IN SPERM BINDING

Following the cloning and sequencing of fertilin-α and -β, many other members of the ADAM gene family have been cloned and sequenced (full-length sequences were available for ~30 ADAMs in summer, 1999). Five mouse ADAMs that are expressed in testis were examined for their presence on sperm and their possible role in sperm–egg adhesion. These five were mouse fertilin-α and -β (results discussed above), cyritestin (an ADAM, which like fertilin-β has testis-specific expression) (Heinlein *et al.,* 1994), and ADAMs 4 and 5. The experiments showed that cyritestin is present on mature sperm and examined whether its disintegrin domain functions in sperm–egg adhesion. An eight-residue peptide from the cyritestin disintegrin loop sequence inhibits (80–90%) sperm–egg adhesion and fusion. Equivalent eight-residue peptides from the disintegrin loops of ADAMs 4 and 5 have no effect on adhesion and fusion. The cyritestin peptide is about a five times more potent inhibitor of sperm–egg fusion (50% inhibition at ~70 μM) compared to the corresponding fertilin-β peptide (50% inhibition at ~400 μM) (Yuan *et al.,* 1997). A longer peptide from the cyritestin disintegrin loop was found to inhibit sperm–egg binding and fusion in another lab (Linder and Heinlein, 1997). Furthermore, an antibody raised against the active site loop peptide of cyritestin also strongly inhibits sperm–egg adhesion and fusion (Yuan *et al.,* 1997). These results suggest that the disintegrin domains of two ADAMs, fertilin-β and cyritestin, both act in adhesion of sperm to the egg plasma membrane.

The finding that (at least) two ADAM family members function in sperm adhesion to the egg plasma membrane is similar to findings in cell–cell adhesion in somatic cells. In the system of leukocyte adhesion to the walls of blood vessels, where, similar to fertilization, a moving cell (leukocyte) binds to a stationary (endothelial) cell, five different pairs of adhesion partners on the two cell types function in sequence (Springer, 1994). That two sperm surface proteins function in the adhesion/fusion process suggests that the adhesion mechanism is more complicated than previously supposed and even more complexity may await discovery.

E. EGG RECEPTORS FOR SPERM

Given that fertilin-β and cyritestin are putative sperm adhesion proteins with active sites in their disintegrin domains, it is an obvious prediction that they have egg adhesion partners and that these adhesion partners may be integrins. (Clearly, fertilin-β and cyritestin might bind to the same adhesion partner or two different adhesion partners.) Several investigators have now identified integrins on the plasma membrane of eggs from various mammalian species and there is agreement that integrins are present. Exactly which integrins are present on eggs is less clear, and eggs of different species (mouse, hamster, and human) have been tested using various assays, including enzyme-linked immunosorbent assay (ELISA), immunobead binding, RGD-coated bead binding, immunoprecipitation, IVF inhibition, and polymerase chain reaction (PCR) (Fusi *et al.,* 1992; Tarone *et al.,* 1993; Almeida *et al.,* 1995; Campbell *et al.,* 1995; Evans *et al.,* 1995; de Nadai *et al.,* 1996). From the various data and particularly the consistency with which specific integrins have been found present in different studies, we conclude that these three species' eggs probably have on their surface at least integrins α5β1, α6β1, and αVβ3.

Almeida and co-workers presented experiments suggesting that sperm bind to integrin α6β1 on the plasma membrane of zona-free mouse eggs. The previously described function of α6β1 was as a receptor for laminin. A rat monoclonal antibody, GoH3, recognizing α6 and known to inhibit laminin binding to α6β1, inhibits sperm binding to the egg plasma membrane. Half-maximal inhibition of sperm binding is observed between 50 and 100 μg/ml of GoH3, and 200 μg/ml GoH3 results in 80–90% inhibition. Sperm–egg fusion, on the other hand, is unaffected at 200 μg/ml GoH3 but is significantly inhibited by 400 μg/ml GoH3. A non-function-blocking rat monoclonal, J1B5, also recognizing α6, does not affect sperm binding or fusion. In addition to these antibody inhibition experiments, Almeida and colleagues compared tissue culture cells (particularly P388D mouse macrophages), which do not express α6β1, to P388D cells transfected so that they express α6β1. The transfected, α6β1-expressing cells bind sperm at higher levels than do mock-transfected cells. However, sperm do not fuse with transfected cells.

There is some support for the idea that sperm fertilin-β and egg integrin α6β1 are adhesion partners. A 57-kDa protein from a sperm extract, recognized by an antibody to the C terminus of fertilinβ, can be bound by this antibody to fluorescent beads. The resulting fluorescent beads will bind to eggs and the bead binding is inhibited by the GoH3 antibody or a fertilin-β active site peptide (Chen and

Sampson, 1999). In a different approach, Chen and Sampson synthesized an ^{125}I-labeled 13-residue peptide with the active site sequence of fertilin-β and an N-terminal photoactivatable cross-linker. Cross-linking this peptide to eggs results in label in only one surface protein, the integrin α6β1.

These experiments are important, but not definitive. Also other results argue that fertilin-β and egg integrin α6β are not adhesion partners. The 57-kDa sperm protein bound to fluorescent beads (Chen and Sampson, 1999) may not be fertilin-β but another cross-reactive protein in the sperm extract. In a study of a gene knockout of fertilin-β, mouse fertilin-β is found to be 45 kDa (Cho *et al.*, 1998). Although the 13-residue fertilin-β peptide used by Chen and Sampson cross-links exclusively to α6β1, native fertilin may bind to a different receptor (integrin) than does the peptide. It has been found that the recombinant extracellular domain of mature fertilin-β, expressed in *E. coli,* binds to eggs. However, this binding is not inhibited by GoH3 (Evans *et al.,* 1997). The test of genetic deletion of the α6 integrin subunit from the mature oocyte has not been reported.

F. PLASMA MEMBRANE DOMAIN LOCALIZATION OF FERTILIN, CYRITESTIN, AND INTEGRIN α6β1

Mouse fertilin is localized to the equatorial region of the plasma membrane (Yuan *et al.,* 1997), whereas guinea pig fertilin is restricted to the posterior head plasma membrane of guinea pig sperm (Primakoff *et al.,* 1987). It would be expected that a molecule involved in the initial fusion would be in the region determined morphologically to be the region where initial fusion occurs. The equatorial region localization of fertilin and the equatorial region initiation of fusion are consistent in mouse. In guinea pig the posterior head localization of fertilin, which we have observed by transmission electron microscopy (TEM), would not be consistent with an initiation of fusion in the equatorial region. This could be explained if a minor population of guinea pig sperm have fertilin on the equatorial region or if, in guinea pig, initial sperm–egg fusion occurs in the posterior head region of the sperm. Published TEM images of guinea pig sperm–egg fusion are consistent with initial fusion occurring in either the equatorial region or the adjacent anterior part of the posterior head region (Noda and Yanamigachi, 1976). Fertilin localization in both species could also be explained if fertilin has a role only in adhesion (and not directly in fusion), because in that case it could be localized to either region.

Cyritestin has been reported to be restricted to the plasma membrane equatorial region (Yuan *et al.,* 1997) or the acrosomal membrane (Linder *et al.,* 1995; Forsbach and Heinlein, 1998). Another study has traced cyritestin's appearance during spermatogenesis using electron microscopy and reported that cyritestin appears on and remains restricted to the acrosomal membrane (Forsbach and Heinlein, 1998). These conflicting reports on the localization of cyritestin have not yet been resolved.

Sperm fuse predominantly or exclusively with the microvillar surface of the egg plasma membrane. Staining of the integrin α6β1 with GoH3 or J1B5 is restricted to the microvillar surface, consistent with a role for α6β1 in sperm binding.

V. HYPOTHETICAL STEPS AFTER BINDING AND BEFORE FUSION

Because relatively little is known about sperm–egg fusion, it is possible that many steps occur between initial sperm adhesion (perhaps through cyritestin) and fusion of the gamete membranes (Figure 9.1) However, only one step has been suggested. The proposal is that after sperm–egg membrane binding, a sperm surface metalloprotease must act (on unknown substrates) before fusion occurs.

Lennarz and colleagues studied a zinc metalloprotease activity that acts after sea urchin sperm have acrosome reacted and bound to the egg plasma membrane. Inhibiting this metalloprotease activity resulted in a virtually complete block (>95%) to gamete fusion. However, sperm bound to the egg plasma membrane in normal numbers. The metalloprotease inhibitors used were phenanthrolene, a zinc chelator, and the tripeptide alanine-alanine-phenylalanine, presumed to be a metalloprotease substrate able to act as a competitive inhibitor at high concentration. Experiments indicated that the putative metalloprotease is on the sperm surface (Roe *et al.,* 1988). Using mouse gametes, we have found results essentially the same as obtained by Lennarz and colleagues in sea urchin and also found more specific metalloprotease inhibitors that block gamete fusion (P. Primakoff and D. Myles, unpublished results).

VI. SPERM TAIL STIFFENING

A dramatic event that occurs at the time of fusion has received little attention. The sperm tail, beating rapidly and forcefully as sperm bind to the egg plasma membrane, suddenly ceases beating, becomes stiff and extends straight out from the sperm head (Yanagimachi, 1994). We call this the "tail reaction" because it is a specific (presumably regulated) response of the sperm in the sequence of sperm–egg interactions. It is known that the tail reaction occurs close to the time of initial membrane fusion (in the sperm head), but it has not been determined if it occurs just before fusion or just after fusion. Also, little is known about the signaling or regulatory features governing tail movement in this situation.

In speculating on a possible biological function of the cessation of tail beating and straightening of the tail, it is easy to suppose that it has a role in initiating membrane fusion. Although powerful motility is a key to the sperm's mission up until the final moment of fusion, it is possible that continued motility would rip apart the incipient contacts and bilayer rearrangements that initiate fusion. Even if fusion in the sperm head can occur in the presence of sperm motility, one can wonder how motility would affect fusion of the sperm tail membrane with the egg plasma membrane. On the one hand, motility might impede or prevent tail membrane fusion, but on the other, the stiff tail that is extended out from the egg surface must at some point relax and again approach this surface. Thus, many interesting questions about the mechanism and function of the tail reaction have not previously been considered.

VII. FUSION IN OTHER SYSTEMS

A. CELL–CELL FUSION IN OTHER CELL TYPES

In mammals various somatic cell types also exhibit cell–cell fusion. These include fusion of cytotrophoblasts to produce a protective nutritive layer in the placenta; fusion of myoblasts to form myotubes; fusion of monocytes to form osteoclasts; and fusion of macrophages to produce "multinucleated giant cells" (Saginario *et al.,* 1995; Hernandez *et al.,* 1996). It is unknown if these cell–cell fusion processes depend in part on molecular events related to those in gamete fusion. In general, less is known about these somatic cell–cell fusion systems than about gamete fusion.

Based on our studies of a potential role of fertilin in gamete fusion, meltrin α, a protein from the ADAM family, was identified and reported to function in myoblast fusion (Yagami-Hiromasa *et al.,* 1995). However, its precise function remains unclear. A surface receptor for extracellular ATP, the P2z/P2X_7 receptor, has been implicated as having a role in macrophage fusion (Chiozzi *et al.,* 1997). P2z/P2X_7 is both necessary and in some cell lines apparently sufficient for cell–cell fusion (Chiozzi *et al.,* 1997). If this single receptor is overexpressed in J774 mouse macrophages or by transfection into HEK 293 cells, the cells spontaneously fuse as they contact each other in culture. This fusion is blocked by a receptor antagonist, oxidized ATP (Murgia *et al.,* 1993; Chiozzi *et al.,* 1997).

B. VIRUS–CELL FUSION

The mechanism of membrane fusion is best understood for certain membrane-enveloped viruses that fuse with cells, particularly influenza and human immunodeficiency virus (HIV). In the case of influenza, the coat protein hemagglutinin (HA) is cleaved by a cellular protease into two disulfide-linked fragments, HA1 and HA2. HA2 contains, at its N terminus, a hydrophobic sequence called a "fusion peptide." HA1 binds to carbohydrate on the target cell surface, leading to endocytosis of the virion. In the endosome, the low pH causes radical conformational changes in HA. One change is that the HA2 fusion peptide, buried deep within the HA structure at neutral pH, is extended at acidic pH toward the endosomal membrane and inserts into the endosomal membrane. Additional conformational changes in the HA structure act to draw the HA2 transmembrane anchor (in the viral membrane) toward the HA2 fusion peptide (in the endosomal membrane), forcing the two bilayers into very close apposition and providing a driving force for their fusion (Hernandez *et al.,* 1996).

The case of HIV is similar but has two important differences: HIV fusion occurs with the cell's plasma membrane at neutral pH and target cells for HIV have both a receptor (Doyle and Strominger, 1987) and a coreceptor (CCR5 or CXCR4) for the virus. (The two major coreceptor types, CCR5 and CXCR4, are present on distinct cell types and bind different HIV strains.)

HIV has a membrane glycoprotein, gp160, which is proteolytically processed to yield two subunits, gp120 and gp41. The gp41 subunit has a fusion peptide at its N terminus. The adhesion process begins with viral gp120 binding to cell CD4. This binding leads to conformational rearrangements in gp120 and possibly also CD4. These changes create a binding site, probably on gp120, that allows formation of a heterotrimeric complex composed of gp120, CD4, and the coreceptor. It is likely that this association of gp120 with the coreceptor triggers conformational changes in gp41 that include the exposure of its fusion peptide and its insertion into the plasma membrane (Clapham, 1997; Wu *et al.,* 1996).

C. INTRACELLULAR FUSION

At first glance, intracellular fusion (e.g., exocytosis, fusion in the Golgi, and other stages of the secretory pathway) appears different and more complicated than virus–cell fusion. The full process of calcium-regulated exocytosis is estimated to utilize ~50 gene products (Martin, 1997). However, Rothman's group (as an example) has proposed a simple model positing that the minimal fusion machinery is relatively simple and works on a principle similar to the virus–cell fusion mechanism (Weber, 1998). In their experiments, small phospholipid vesicles containing vesicle soluble NSF attachment protein receptors (v-SNAREs) are found to fuse with other small phospholipid vesicles containing target SNAREs (t-SNAREs). Both the v-SNARE and the t-SNARE are transmembrane proteins and they bind each other using membrane-proximal repeat regions. Such v-SNARE/t-SNARE complexes (termed SNAREpins) have been visualized as long, narrow rods with the two membrane anchors emerging at the same end of the rod. The formation of these complexes is proposed to drive lipid bilayer fusion either by forcing close approach of the two bilayers or by making fusion energetically favorable as the complexes release energy in transiting from being in two lipid bilayers to being in one. The proposed mechanism resembles the virus–cell fusion mechanism: the SNAREpin is a (very stable) complex of two proteins, each with transmembrane regions (in the vesicle membrane and target membrane, respectively) brought very close together in the complex. The viral fusion protein is a single protein with its transmembrane region (in the viral membrane) brought very close to its fusion peptide (inserted in the cell membrane).

Although this intracellular fusion model is of substantial interest, it has been challenged and will be refined over time (Mayer, 1999). Compared to limited work on gamete fusion, a large group of scientists work on intracellular fusion and will be able to test this model in different ways and replace it or expand it with new data.

D. RELEVANCE OF FU314SION MECHANISMS IN OTHER SYSTEMS TO GAMETE FUSION

Will the mechanism responsible for gamete membrane fusion prove to be similar to either the viral or intracellular membrane fusion mechanisms? It is hard to speculate wisely now because too little is known. The basic message of both the virus–cell fusion mechanism and some proposed intracellular fusion mechanisms is that "it's simple": the adhesion machinery (Wenhao and Hamilton, 1996) also acts as the fusion machinery (Wenhao and Hamilton, 1996). The first key step these adhesion/fusion proteins accomplish is to achieve initial adhesion. They then seamlessly progress through conformational changes to bring the two lipid bilayers into intimate contact and make their fusion energetically favorable.

Assuming that gamete fusion closely resembles these systems, one might suppose that the putative roles of cyritestin, the proposed metalloprotease, and other possible players will be refuted by further experiments or proved to be regulatory and not actors in the mechanism of membrane fusion. Then, one might suppose that fertilin binding to the integrin $\alpha 6\beta 1$ is the key step and these two are both the adhesion and the fusion proteins. Along this line, we initially postulated a specific mechanism like that of virus–cell fusion (Blobel *et al.*, 1992; Myles, 1993). In this model fertilin-β binds to $\alpha 6\beta 1$ (Almeida *et al.*, 1995), and this binding leads to a conformational change in fertilin-α. As part of the conformational change, fertilin-α would extend its "fusion peptide" into the egg plasma membrane. Thus fertilin-α would become anchored in both the sperm and egg lipid bilayers, forcing the bilayers very close together and initiating their fusion.

There are several reasons now to suggest that this specific model is incomplete or incorrect. First, crucial sequence variations in fertilin-α among mammalian species call into question the universality of the model. The initial fertilin-α sequence determined was for guinea pig fertilin-α, which has a hydropohobic 20-residue region with all the properties of a viral "fusion peptide" (Blobel *et al.*, 1992). Since 1992, fertilin-α specimens from mouse, rabbit, bull, and monkey have also been sequenced (Wolfsberg *et al.*, 1995; Perry *et al.*, 1997; Waters and White, 1997; Hardy and Holland, 1996). Bovine fertilin-α does not have a hydrophobic sequence in the same region as guinea pig fertilin-α. Another sequence, however, in a different region of bovine fertilin-α, has been suggested to serve as a fusion peptide (Bigler *et al.*, 1997). Also rabbit (Perry *et al.*, 1997) and monkey (Hardy and Holland, 1996) fertilin-α have sequences in the same region as guinea pig fertilin-α that less clearly have the features of a fusion peptide. Although it has been suggested that the rabbit and monkey fertilin-α sequences are adequate as fusion peptides (Bigler *et al.*, 1997), the investigators who reported each sequence state that the sequences could not function as fusion peptides. Thus, it is clear that there is not a fusion peptide in fertilin-α that is conserved across mammalian species.

The finding that sperm will bind to tissue culture cells expressing the integrin α6β1 but do not fuse with such cells suggests that fusion is not so simple as to require only fertilin and α6β1 (Almeida *et al.,* 1995). Also, cauda epididymal sperm from mice carrying a knockout of fertilin-β have no detectable fertilin-α or fertilin-β and yet are able to fuse [Cho *et al.* (1998) and P. Primakoff and D. Myles, unpublished results]. These mutant sperm, tested in IVF, adhere to the egg plasma membrane at a level eightfold lower compared to wild type, but can fuse at 45–50% the rate of wild-type sperm. This indicates that fertilin is required for normal fusion and apparently enhances the rate of fusion. On the other hand, substantial fusion occurs in the absence of fertilin, suggesting several possibilities. One is that another ADAM can act in fusion in fertilin's absence; another is that fertilin's role is facilitative/regulatory and the fusion machinery is composed of other surface molecules. These and related interpretations will remain possibilities until substantial further work is done.

VIII. PROSPECTUS

Mammalian gamete fusion is inherently difficult to study because of the temperamental nature of IVF assays and the miniscule amount of eggs obtainable. The strategy of beginning with the identification of key molecules on the sperm surface and making guesses about adhesion partners on eggs has carried the field a long way. Even though the system is difficult and the current, more precise questions about mechanism are difficult, progress may be possible by continuing this strategy. In addition, new strategies, particularly gene knockout and structural analysis of sperm protein–egg protein complexes, may prove to be effective in providing deeper insight into the molecular basis of gamete fusion.

REFERENCES

Almeida, E. A., Huovila, A. P., Sutherland, A. E., Stephens, L. E., Calarco, P. G., Shaw, L. M., Mercurio, A. M., Sonnenberg, A., Primakoff, P., Myles, D. G., and White, J. M. (1995). Mouse egg integrin α6β1 functions as a sperm receptor. *Cell* **81,** 1095–1104.

Bigler, D., Chen, M., Waters, S., and White, J. M. (1997). A model for sperm–egg binding and fusion based on ADAM's and integrins. *Trends Cell Biol.* **7,** 220–225.

Blobel, C. P., Wolfsberg, T. G., Turck, C. W., Myles, D. G., Primakoff, P., and White, J. M. (1992). A potential fusion peptide and an integrin ligand domain in a protein active in sperm–egg fusion [see comments]. *Nature* **356,** 248–252.

Campbell, S., Swann, H. R., Seif, M. W., Kimber, S. J., and Aplin, J. D. (1995). Cell adhesion molecules on the oocyte and preimplantation human embryo. *Hum. Reprod.* **10,** 1571–1578.

Chen, H., and Sampson, N. S. (1999). Mediation of sperm–egg fusion: Evidence that mouse egg alpha6beta1 integrin is the receptor for sperm fertilinbeta. *Chem. Biol.* **6,** 1–10.

Chiozzi, P., Sanz, J. M., Ferrari, D., Falzoni, S., Aleotti, A., Buell, G. N., Collo, G., and Di Virgilio, F. (1997). Spontaneous cell fusion in macrophage cultures expressing high levels of the P2Z/P2X7 receptor. *J. Cell Biol.* **138,** 697–706.

Cho, C., Bunch, D. O., Faure, J. E., Goulding, E. H., Eddy, E. M., Primakoff, P., and Myles, D. G. (1998). Fertilization defects in sperm from mice lacking fertilin beta. *Science* **281,** 1857–1859.

Clapham, P. R. (1997). HIV and chemokines; ligands sharing cell-surface receptors. *Trends Cell Biol.* **7,** 264–268.

de Nadai, C., Fenichel, P., Donzeau, M., Epel, D., and Ciapa, B. (1996). Characterisation and role of integrins during gametic interaction and egg activation. *Zygote* **4,** 31–40.

Doyle, C., and Strominger, J. L. (1987). Interaction between CD4 and class II MHC molecules mediates cell adhesion. *Nature* **330,** 256–259.

Evans, J. P., Kopf, G. S., and Schultz, R. M. (1997) Characterization of the binding of recombinant mouse sperm fertilin beta subunit to mouse eggs Evidence for adhesive activity via an egg beta1 integrin-mediated interaction. *Dev. Biol.* **187,** 79–93.

Evans, J. P., Schultz, R. M., and Kopf, G. (1995). Identification and localization of integrin subunits in oocytes and eggs of the mouse. *Mol. Reprod. Dev.* **40,** 211–220.

Forsbach, A., and Heinlein, U. A. (1998). Intratesticular distribution of cyritestin, a protein involved in gamete interaction. *J. Exp. Biol.* **201** (Pt. 6), 851–867.

Fusi, F., Vignali, M., Busacca, M., and Bronson, R. A. (1992). Evidence fot the presence of an integrin cell adhesion receptor on the oolemma of unfertilized human oocytes. *Mol. Reprod. Dev.* **31,** 215–222.

Hardy, C. M., and Holland, M. K. (1996). Cloning and expression of recombinant rabbit fertilin. *Mol. Reprod. Dev.* **45,** 107–116.

Heinlein, U. A. O., Wallat, S., Senftleben, A., and Lemaire, L. (1994). Male germ cell-expressed mouse gene TAZ83 encodes a putative, cysteine-rich transmembrane protein (cyritestin) sharing homologies with snake toxins and sperm–egg fusion proteins. *Dev. Growth Differ.* **36,** 49–58.

Hernandez, L. D., Hoffman, L. R., Wolfsberg, T. G., and White, J. M. (1996). Virus–cell and cell–cell fusion. *Annu. Rev. Cell Dev. Biol.* **12,** 627–661.

Linder, B., and Heinlein, U. A. (1997). Decreased *in vitro* fertilization efficiencies in the presence of specific cyritestin peptides. *Dev. Growth Differ.* **39,** 243–247.

Linder, B., Bammer, S., and Heinlein, U. A. (1995). Delayed translation and posttranslational processing of cyritestin, an integral transmembrane protein of the mouse acrosome. *Exp. Cell Res.* **221,** 66–72.

Martin, T. F. J. (1997). Stages of regulated exocytosis. *Trends Cell Biol.* **7,** 271–275.

Mayer, A. (1999). Intracellular membrane fusion: SNAREs only? *Curr. Opin. Cell Biol.* **11** (4) 447–452.

Murgia, M., Hanau, S., Pizzo, P., Rippa, M., and Di Virgilio, F. (1993). Oxidized ATP. An irreversible inhibitor of the macrophage purinergic P2Z receptor. *J. Biol. Chem.* **268,** 8199–8203.

Myles, D. G. (1993). Molecular mechanisms of sperm–egg membrane binding and fusion in mammals. *Dev. Biol.* **158,** 35–45.

Myles, D. G., Primakoff, P., and Bellvë, A. R. (1981). Surface domains of the guinea pig sperm defined with monoclonal antibodies. *Cell* **23,** 433–439

Noda, Y. D., and Yanamigachi, R. (1976). Electron microscopic observations of guinea pig spermatozoa penetrating eggs *in vitro. Dev. Growth Differ.* **18,** 15–23.

Perry, A. C., Gichuhi, P. M., Jones, R., and Hall, L. (1997). Cloning and analysis of monkey fertilin reveals novel alpha subunit isoforms. *Biochem. J.* **307,** 843–850.

Primakoff, P., Hyatt, H., and Tredick-Kline, J. (1987). Identification and purification of a sperm surface protein with a potential role in sperm-egg membrane fusion. *J. Cell Biol.* **104,** 141–149.

Roe, J. L., Farach, H. A., Strittmatter, W., and Lennarz, W. J. (1988). Evidence for the involvement of metalloendoproteases in a step in sea urchin gamete fusion. *J. Cell Biol.* **107,** 539–544.

Saginario, C., Qian, H. Y., and Vignery, A. (1995). Identification of an inducible surface molecule specific to fusing macrophages. *Proc. Natl. Acad. Sci. U.S.A.* **92,** 12210–12214.

Springer, T. A. (1994). Traffic signals for lymphocyte recirculation and leukocyte emigration; the multistep paradigm. *Cell* **76,** 301–314.

Tarone, G., Russo, M. A., Hirsch, E., Odorisio, T., Altruda, F., Silengo, L., and Siracusa, G. (1993). Expression of β1 integrin complexes on the surface of unfertilized mouse oocyte. *Development* **117,** 1369–1375.

Waters, S. I., and White, J. M. (1997). Biochemical and molecular characterization of bovine fertilin α and β (ADAM1 and ADAM2): a candidate sperm-egg binding/fusion complex. *Biol. Reprod.* **56,** 1245–1254.

Weber T., Zemelman, B. V., McNew, J. A., Westermann, B., Gmachl, M., Parlati, F., Sollner, T. H., and Rothman, J. E. (1998). SNAREpins: minimal machinery for membrane fusion. *Cell* **92**(6), 759–772.

Wenhao, X., and Hamilton, D. W. (1996). Identification of the rat epididymis-secreted 4E9 antigen as protein E: Further biochemical characterization of the highly homologous epididymal secretory proteins D and E. *Mol. Reprod. Dev.* **43,** 347–357.

Wolfsberg, T. G., Straight, P. D., Gerena, R. L., Huovila, A. J., Primakoff, P., Myles, D. G., and White, J. M. (1995). ADAM, a widely distributed and developmentally regulated gene family encoding membrane proteins with A Disintegrin And Metalloprotease domain. *Dev. Biol.* **169,** 378–383.

Wu, L., Gerard, N. P., Wyatt, R., Choe, H., Parolin, C., Ruffing, N., Borsetti, A., Cardoso, A. A., Desjardin, E., Newman, W., Gerard, C., and Sodroski, J. (1996). CD4-induced interaction of primary HIV-1 gp120 glycoproteins with the chemokine receptor CCR-5 [see comments]. *Nature* **384,** 179–183.

Yagami-Hiromasa, T., Sato, T., Kurisaki, T., Kamijo, K., Nabeshima, Y., and Fujisawa-Sehara, A. (1995). A metalloprotease-disintegrin participating in myoblast fusion [see comments]. *Nature* **377,** 652–656.

Yanagimachi, R. (1994). Mammalian fertilization. *In* "The Physiology of Reproduction" (E. Knobil and J. D. Neill, eds.), pp. 189–317. Raven Press, New York.

Yuan, R., Primakoff, P., and Myles, D. G. (1997). A role for the disintegrin domain of cyritestin, a sperm surface protein belonging to the ADAM family, in mouse sperm-egg plasma membrane adhesion and fusion. *J. Cell Biol.* **137,** 105–112.

10

MEMBRANE EVENTS OF EGG ACTIVATION

KARL SWANN* AND KEITH T. JONES†

*Department of Anatomy and Developmental Biology, University College, London, United Kingdom; and †Department of Physiological Sciences, University of Newcastle, The Medical School, Newcastle, United Kingdom

I. INTRODUCTION

There are two key membranes involved in activating the development of the egg at fertilization. The plasma membrane of the egg undergoes fusion with the sperm to initiate the process of fertilization, and the endoplasmic reticulum inside the egg undergoes a concerted release of Ca^{2+} ions that are now known to be the essential trigger for the development of the egg into an embryo. To understand how a sperm activates an egg at fertilization we need to know the sequence and logic of events that take place between sperm–egg fusion and sperm-induced Ca^{2+} release.

An increase in intracellular Ca^{2+} at fertilization was observed directly for the first time in the eggs of the medaka fish (Ridgway *et al.,* 1977). Similar increases in Ca^{2+} have been observed in all animal eggs examined thus far across widely different phyla, and interestingly even in eggs from the plant kingdom (Roberts *et al.,* 1994; Digonnet *et al.,* 1997). Therefore, a rise in the free cytoplasmic Ca^{2+} concentration within the egg appears to be a feature of fertilization in the differentiated gametes of all species. This may not be surprising considering the ubiquitous nature of Ca^{2+} as an intracellular signaling element. Its role at fertilization in potentially all eggs raises the possibility of an evolutionarily conserved mechanism for sperm-induced Ca^{2+} release at least in the vast majority of phyla.

The proposal that Ca^{2+} is the important signal for development in eggs is supported by the finding that artificially inducing a Ca^{2+} increase can trigger many, if not all, of the early events of egg activation (Whittingham, 1980; Whitaker and Steinhardt, 1982; Jaffe, 1983; Swann and Ozil, 1994). Furthermore, the introduction of Ca^{2+} chelators into the egg cytoplasm, in order to prevent a sperm-induced rise in Ca^{2+}, abolishes all events associated with activation (Whitaker and Steinhardt, 1982; Kline,1988; Kline and Kline, 1992). These data show that a Ca^{2+} increase is both sufficient and necessary to explain the central features of how a sperm initiates the development of the egg. On the basis of these findings one of the most fundamental problems of the fertilization field is to understand how sperm–egg interaction leads to the release of Ca^{2+} from intracellular membranes.

Throughout this chapter we refer to the "activation" of an egg. Activation involves a number of morphological and biochemical changes; the most obvious ones are those caused by exocytosis, such as the raising of the fertilization envelope in the sea urchin (Whitaker and Steinhardt, 1982). Probably the most important aspect of activation is the completion of meiotic stages and the initiation of mitotic cell cycles. Because eggs of different species arrest before fertilization at different stages of meiotic division, this means that activation can involve passage through different cell cycle control points (Whitaker and Patel, 1990). Some of the most commonly studied species, such as frogs and rodents, have eggs that are fertilized at metaphase of the second meiotic division (MII). Sea urchin eggs have completed meiosis and are fertilized with an intact female pronucleus. A number of marine invertebrates and worms are fertilized at metaphase I (Sagata, 1996). Whenever the female gamete is at a stage before the completion of meiosis it should be referred to as an oocyte rather than an egg. Only the sea urchin egg, of all the female gametes discussed presently, has completed both meiotic divisions before being fertilized, and is therefore correctly termed an egg. However, in this review we adopt the more lax terminology used in the literature for vertebrates such as frogs and mammals whereby the mature female gamete is called an egg. We also use the term "egg" for the generic cases.

II. CA^{2+} WAVES AND OSCILLATIONS AT FERTILIZATION

The spatial and temporal aspects of Ca^{2+} release are dependent on the species of egg being studied. In its simplest form, a single Ca^{2+} wave passes across the egg from the site of sperm fusion and leads to an elevated Ca^{2+} level that is maintained for several minutes. This sort of response is seen in medaka fish (Ridgway *et al.,* 1977), sea urchin (Steinhardt *et al.,* 1977), starfish (Eisen and Reynolds, 1984), and frog (Busa and Nuccitelli, 1985). Of note is that the Ca^{2+} wave is always initiated from the site of sperm attachment.

In mammals the Ca^{2+} signal at fertilization is more complex—it consists of a series of oscillations (Cuthbertson and Cobbold, 1985) that last for several hours, ceasing around the time of pronucleus formation (Jones *et al.,* 1995). Figure 10.1 shows a typical example of the temporal pattern of Ca^{2+} oscillations measured after fertilization in the mouse egg. As in frog, starfish, and sea urchin the first Ca^{2+} increase in hamster has also been shown to be a wave of Ca^{2+} propagating from the point of sperm fusion, with later oscillations coming from more diffuse regions in the egg (Miyazaki *et al.,* 1986). In mammals the frequency of oscillations may be species specific, but each Ca^{2+} transient lasts for about 1 minute and the series of oscillations tend to continue at regular intervals of about 10 minutes (Miyaza-

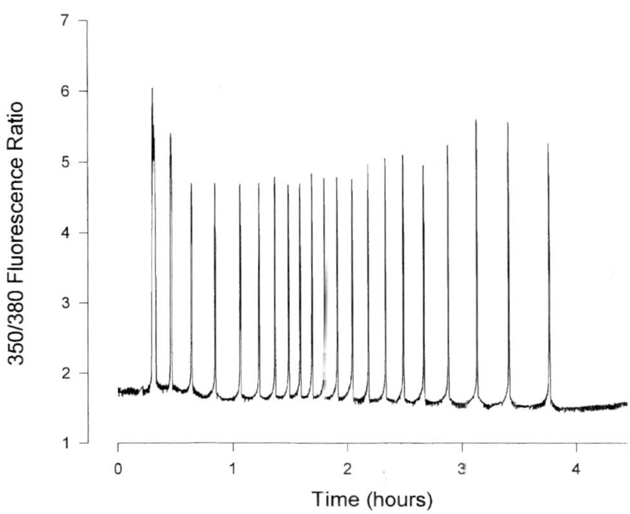

FIGURE 10.1 Ca^{2+} oscillations during *in vitro* fertilization of a mouse egg. The egg was loaded with the acetoxymethyl form of the fura-2-like dye PE3 and the fluorescence was measured with an excitation ratio of 350 and 380 nm. This fluorescence ratio indicates the Ca^{2+} levels in the egg versus time. Sperm were added at time 0.

TABLE 10.1 Responses to Ca^{2+} at Fertilization

Egg species	Stage fertilized[a]	Oscillations	Frequency	References
Plantae				
Fucus	—	—	—	Roberts *et al.* (1994)
Flowering plant (maize)	Interphase	No	N/A	Digonnet *et al.* (1997)
Animalia				
Protostomia				
Nemertina	MI	Yes	High, 1/3 minutes	Stricker (1996)
Mollusca (bivalves)	MI	Yes	High, <1/minute	Deguchi and Osanai, (1994)
Annelida (*Chaetopterus*)	MI	Yes	High, <1/minute	Eckberg and Miller (1995)
Deuterostomia				
Echinodermata				
Sea urchin	Interphase	No	N/A	Steinhardt *et al.* (1977)
Starfish	Prophase I–MI	No	N/A	Eisen and Reynolds (1984)
Chordata				
Urochordata (Ascidian)	MI	Yes	High, 1/3 minutes	Speksnijder *et al.* (1989)
Vertebrata				
Medaka fish	MII	No	N/A	Ridgway *et al.* (1977)
Frog	MII	No	N/A	Busa and Nuccitelli (1985)
Mammals	MII	Yes	Low, 1/20 minutes	Cuthbertson and Cobbold (1985)

[a] MI, MII, Meiosis I and II.

ki *et al.,* 1993; Swann and Ozil, 1994). An oscillating Ca^{2+} signal at fertilization is not unique to mammals. Oscillations in Ca^{2+} are also observed in molluscs (Deguchi and Osanai, 1994), nemertean worms (Stricker, 1996), and ascidians (Speksnijder *et al.,* 1989). These oscillations tend to be different from those of mammals: the frequency is generally higher than that of mammals and also the oscillations cease within a shorter time frame. Table 10.1 lists some of the many species of eggs that have been studied with respect to Ca^{2+} changes at fertilization and notes the different types of Ca^{2+} responses.

It is not clear why there are differences between species in the pattern of Ca^{2+} oscillations, with some showing only one transient and others showing oscillations for tens of minutes or even several hours. However, in mammalian zygotes, the time taken for completion of the cell cycle may account for the need for a particularly prolonged series of Ca^{2+} oscillations at fertilization. In MII mammalian eggs the application of a brief Ca^{2+} pulse is enough to degrade transiently maturation promoting factor (MPF), which is the factor responsible for MII arrest (Collas *et al.,* 1995). However, in such treated eggs MPF activity is rapidly restored. Therefore it is likely that the role of repetitive transients is to ensure that MPF activity is repressed for a long enough period for eggs to escape meiosis (Ozil and Swann, 1995). A similar mechanism may operate in ascidian eggs (McDougall and Levasseur, 1998). Maintaining a sustained rise in Ca^{2+} for an hour or so is not possible for mammalian eggs because prolonged rises in Ca^{2+} are generally detrimental or toxic to cells. A series of pulsatile Ca^{2+} increases may offer a way to keep a Ca^{2+} signaling process switched on for several hours while not exhausting the cell's ATP (Meyer and Stryer, 1990).

In the vast majority of species sperm–egg fusion leads to an initial wave of Ca^{2+} and this involves release from intracellular stores. In a few species, however, there are interesting variations in this theme of Ca^{2+} release at fertilization. For example, in newts, in which physiological polyspermy occurs, the Ca^{2+} release associated with a single sperm is not enough to generate a wave across the egg, but rather involves spatially restricted release from intracellular stores (Grandin and Charboneau, 1992). Each sperm fusing appears to cause its own Ca^{2+} increase, and these increases summate and lead to egg activation. In the eggs of the marine worm *Urechis* there is a predominant role for Ca^{2+} influx through plasma membrane channels, rather than internal release, in activating the egg (Jaffe, 1983; Stephano and Gould, 1997). Finally, in crabs and shrimps, egg activation occurs immediately after shedding into seawater. Although the sperm enters around the same time, it is the contact with seawater Mg^{2+} that provides the stimulus to trigger Ca^{2+} release from intracellular stores and then activation (Lindsay *et al.,* 1992; Goudeau and Goudeau, 1996). In zebrafish there is also evidence that the activating Ca^{2+} waves is triggered independently of sperm entry (Lee *et al.,* 1999). These variations suggest that there is some flexibility in the way egg or oocyte activation can be linked to a Ca^{2+} increase and this may implicate different mechanisms operating at fertilization as different species are examined. Nevertheless, it is our working hypothesis that those eggs sharing a common source of Ca^{2+} and responding specifically to the sperm with a large rise in Ca^{2+} will share a common signaling mechanism. We concentrate our discussion on the eggs that display release from intracellular stores at fertilization, so this includes the vertebrates and the most extensively studied echinoderms.

In general, when the source of Ca^{2+} activating the egg is intracellular, then the evidence clearly points to the endoplasmic reticulum as being the relevant store (Jaffe, 1983; Whitaker and Steinhardt, 1982; Swann and Ozil, 1994). In sea urchin eggs it has been shown that the mitochondria may constitute a substantial store of Ca^{2+} after fertilization, but do not appear to be able to release any significant amounts of Ca^{2+} before fertilization (Eisen and Reynolds, 1985). The membrane of the endoplasmic reticulum has been shown to contain two major classes of Ca^{2+} release channels (Berridge, 1993). The first is the inositol 1,4,5-trisphosphate receptor (IP_3 receptor), which can open in response to both IP_3 and Ca^{2+}. The second type of channel is the ryanodine receptor, which is characterized by its binding to the alkaloid ryanodine and whose endogenous agonists are Ca^{2+} and most probably the NAD^+ metabolite cyclic ADP ribose (Galione *et al.,* 1991). There is a third type of unidentified Ca^{2+} release channel in sea urchin eggs. The channel is sensitive to the NADP metabolite NAADP (Lee, 1997). There is evidence that both ryanodine and IP_3 receptors are present in sea urchin, ascidian, starfish, mouse, and bovine eggs (Parys *et al.,* 1994; Galione *et al.,* 1993a,b; Ayabe *et al.,* 1995; Swann and Ozil, 1994; Yue *et al.,* 1995; Kume *et al.,* 1997; Albrieux *et al.,* 1997; He *et al.,* 1997), whereas frog eggs appear to have only IP_3 receptors (Parys *et al.,* 1994). Whatever the proteins involved in the signaling of Ca^{2+} release at fertilization, in most eggs the Ca^{2+} release appears to be through opening of these channels. The problem of signal transduction at fertilization requires the tracing of the sequence of events from sperm–egg interaction at the plasma membrane to the opening of the Ca^{2+} release channels in the endoplasmic reticulum.

III. ELECTRICAL EVENTS AND FERTILIZATION

In many species the first sign that the sperm has begun the process of fertilizing the egg is a change in the egg plasma membrane potential, which in the sea urchin is around −70 mV. Shortly after sperm–egg contact the sea urchin sperm induces an inward current that causes a small depolarization. This depolarization leads to the eggs undergoing an action potential that takes the membrane potential to positive values (Hagiwara and Jaffe, 1979; Whitaker and Steinhardt, 1982). The action potential is generated mainly by an influx from the extracellular medium of Na^+ and Ca^{2+} ions. Action potentials are also seen at fertilization in other eggs, such as those of starfish, frogs, tunicates, annelids, and *Urechis* (Hagiwara and Jaffe, 1979). However, in mammalian eggs there is no evidence for an action potential at fertilization (Miyazaki and Igusa, 1981a; Peres, 1986). Mammalian eggs such as those of the hamster and mouse are capable of generating action potentials only if they are artificially hyperpolarized to −70 mV and then allowed to depolarize, but this appears to be nonphysiological (Miyazaki and Igusa, 1981a). In the sea urchin egg an influx of Ca^{2+} during the depolarization phase can account for all the measurable influx of Ca^{2+} into the egg at fertilization (Schmidt *et al.,* 1982). The influx of Ca^{2+} into the egg causes a measurable increase in the free Ca^{2+} concentration (Whitaker and Swann, 1993). This is also seen as a cortical flash of Ca^{2+} increase immediately beneath the plasma membrane in other eggs, such as those of the nemertean worm *Cerebratulus* (Stricker, 1996). However, in both these species the cortical increase caused by influx of Ca^{2+} is insufficient to activate the egg, as judged by cortical granule exocytosis or cell cycle events (Whitaker and Steinhardt, 1982; Stricker, 1996). Moreover, in all chordates studied, including eggs without physiological action potentials, the actual depolarization of the egg membrane does not lead to egg activation (Hagiwara and Jaffe, 1979). Consequently the electrical change in the plasma membrane potential is not directly linked to release of Ca^{2+} and the egg activation process. It is clear that other signals must link the sperm with the relevant changes in Ca^{2+}.

Although depolarization by itself does not activate most types of eggs, there are a number of exceptions. In eggs of *Urechis*, *Chaetopterus* and *Spisula* the depolarization caused by an action potential at fertilization may provide the key stimulus to activation. In these eggs the addition of high-concentration K^+ seawater to depolarize the plasma membrane potential can cause increased Ca^{2+} uptake and lead to activation events such as meiotic resumption (Eckberg and Miller, 1995; Dube, 1988). The main source of Ca^{2+} for such parthenogenetic activation is extracellular, but it is possible that at fertilization a component of Ca^{2+} release from intracellular stores may also play an important role (Bloom *et al.*, 1988). Nevertheless, the basic observation that Ca^{2+} influx through a voltage-gated channel can cause activation suggests that there may be some fundamental differences between these marine invertebrates and other animal eggs, such as those of the vertebrates, in which voltage-gated Ca^{2+} influx does not lead to egg activation.

The action potential at fertilization, where it exists, does have a clear role to play that is not related to activation. The action potential acts as a fast block to polyspermy (Hagiwara and Jaffe, 1979; Jaffe and Cross, 1986). The entry of extra sperm is prevented by the positive potentials established during the depolarizing phase of the action potential. However, once sperm entry has been initiated in the sea urchin the egg membrane requires depolarization in order for complete sperm entry (McCulloh *et al.*, 1987). Therefore voltage clamping eggs at potentials more negative than -50 mV is inhibitory for full sperm incorporation (Lynn *et al.*, 1988; McCulloh *et al.*, 1987). Thus, in sea urchins at least, the membrane potential has to be at negative potentials in order to allow a sperm to initiate the fusion process. Then, once initiated, the membrane potential has to remain positive for several minutes in order to stop extra sperm from fusing and to allow the successful sperm to further its entry into the egg cytoplasm. The full biophysical mechanism behind the electrical block to polyspermy and the electrical requirements to sperm entry are not known. Interestingly in the crab, *Maia squinodo*, there is a story similar to that in sea urchins, but the membrane potentials are reversed. Following sperm–egg interaction there is a membrane potential hyperpolarization that is responsible for the fast block to polyspermy, and preventing this hyperpolarization phase interferes with sperm incorporation (Goudeau and Goudeau, 1989).

Even though the membrane potential change may not play a direct role in egg activation in sea urchins, studies of the electrophysiology of fertilization in this species have provided some important data for theories of how the sperm activates an egg. Increases in the plasma membrane electrical capacitance can be measured at fertilization. The main capacitance increase is due to the addition of membrane surface area caused by the exocytosis of cortical granules, which is itself caused by the Ca^{2+} wave crossing the egg (Jaffe *et al.*, 1978; Whitaker and Steinhardt, 1982). The increase in membrane capacitance occurs within a minute of the first inward current in these eggs, demonstrating that cortical events and the Ca^{2+} wave follow the initial depolarization. As well as monitoring the increase in surface area brought about by exocytosis, it has also proved possible, in sea urchins, to monitor the increase in membrane surface area caused by the fusion of a single sperm with the egg plasma membrane. McCulloh and Chambers (1992) carried out a series of elegant experiments whereby the capacitance of the membrane under a loose, attached patch pipette was measured during fertilization. The sperm were only able to reach the egg by swimming through the inside of the loose patch pipette. The egg was also voltage clamped during the entire procedure so that the timing of sperm–egg membrane fusion could be measured relative to the changes in membrane current and to the larger capacitance increase that occurs during the Ca^{2+} wave. They showed that the initial depolarization at fertilization coincides with sperm–egg fusion. Furthermore, the initial inward current (which normally triggers the action potential at fertilization) was shown to be due to ions flowing into the egg via the sperm's plasma membrane (McCulloh and Chambers, 1992). These data showed that sperm–egg membrane fusion is the very first event in sea urchin egg fertilization and it is clear that the sperm and egg are fused several seconds before any detectable rise in Ca^{2+} or exocytosis. This result has implications for our understanding of the events that lead the sperm to initiate Ca^{2+} release because it indicates that the sperm and egg cytoplasms are continuous when signaling events are occurring. In fertilizing mouse eggs a small capacitance increase has also been shown to occur before the initial Ca^{2+} release at fertilization (Lee *et al.,* 2001). As with sea urchins this small capacitance increase may represent sperm–egg fusion in mammals. The continuity established by Chambers and McCulloh, strictly speaking, applies only to ions passing between sperm and egg rather than large signaling molecules. However, in the mouse, where fusion precedes Ca^{2+} release by several minutes (Lawrence *et al.,* 1997), it is possible for globular proteins of at least 250 kDa to pass from the egg to the fused sperm before any Ca^{2+} release (Jones *et al.,* 1998a). Taken together these results imply sperm and egg continuity for either seconds (sea urchin) or minutes (most mammals) before any detectable change in egg Ca^{2+}.

After the initial action potential in sea urchin and frog eggs there is a longer lasting phase of depolarization that is caused by the Ca^{2+} increase at fertilization (Whitaker and Steinhardt, 1982; Busa, 1990). The relative timing of the action potential and later depolarization can therefore give an indication of the timing of the early events in fertilization. In frog the second phase of depolarization can even be detected as a wave moving across the surface of the activating egg, and this reflects the underlying Ca^{2+} wave (Kline and Nuccitelli, 1985). Electrical changes at fertilization have also provided valuable data on events of activation in some eggs such as the hamster. During fertilization in hamster the membrane potential undergoes a series of hyperpolarizing responses from -40 to around -70 mV, each lasting about 30 seconds and occurring at intervals from once every minute to once every 10 minutes, depending on the number of sperm entries (Miyazaki and Igusa, 1981a). The hyperpolarizations are due to a Ca^{2+}-activated K^+ conductance (Miyazaki and Igusa, 1981b, 1982). Consequently the series of repetitive hyperpolarizations can be taken as an indication of underlying repetitive rises in Ca^{2+}. Even though it does not involve direct measurements of Ca^{2+} it has the advantage of being less invasive and relatively simple to carry out compared with many of the methods for measuring Ca^{2+}. Hyperpolarizations similar to those in hamster egg response also appear to exist in human eggs, in which there is also evidence for a Ca^{2+}-activated outward current (Homa and Swann, 1994). Much smaller amplitude hyperpolarizations have been reported during fertilization in mouse and rabbit eggs, where they may also reflect underlying Ca^{2+} oscillations (Jaffe *et al.*, 1983; McCulloh *et al.*, 1983).

IV. THE LATENT PERIOD OF FERTILIZATION

The latent period of fertilization has been a term used in the sea urchin field to indicate the delay between the initial sperm–egg interaction and the initiation of the cortical reaction (Allen and Griffen, 1958). The cortical reaction is triggered by the rise in cytosolic Ca^{2+}, so the latent period represents the delay between sperm–egg interaction and the initial release of Ca^{2+} from intracellular stores. Because sperm–egg membrane fusion is the first detectable change in the egg at fertilization, in sea urchins the latent period can be viewed as the delay between the sperm–egg fusion and the onset of the initial activating wave of Ca^{2+} increase at fertilization (Whitaker and Swann,1993). In this case the most precise data on the latent period are derived from the sea urchin. Shen and Steinhardt (1984) were able to control the timing of sperm–egg fusion by holding the egg at positive potentials and allowing sperm to attach to the surface of sea urchin eggs. The fusion of attached sperm could be triggered by briefly switching the membrane potential to negative values. The time from this switch to the onset of the Ca^{2+} wave indicated the duration of the latent period, which was about 15 seconds in sea urchins. The period is divided into two parts with an absolute latent period of 7 seconds when no sperm can ever induce a Ca^{2+} transient, followed by a variable second period.

The data on the timing of the latent period are not as extensive in any species outside the sea urchin. However, if the findings of McCulloh and Chambers (1992) suggesting that the action potential may be a sign of sperm–egg membrane fusion are extrapolated to other species, then in frog eggs the latent period is about 1 minute (Busa and Nuccitelli, 1985; Busa, 1990). In mammals there is no action potential as an indicator of sperm–egg fusion. However, the delay between mouse gamete fusion and Ca^{2+} release has been measured by transfer of either injected fluorescent probes, or Hoechst dye, from the egg into the sperm following fertilization (Lawrence *et al.*, 1997; Jones *et al.*, 1998a). Both of these studies suggest that the latent period in the mouse eggs is at least 1–3 minutes (Lawrence *et al.*, 1997; Jones *et al.*, 1998a).

One other indicator of sperm–egg fusion is to correlate this event with sperm motility. The cessation of sperm motility has been suggested as a sign that sperm and egg membranes have fused (Whitaker and Swann, 1993; Yanagimachi,1994), and in the sea urchin it is known that fusion has definitely occurred by the time the sperm's tail stops beating, because this is a few seconds after the action potential in the egg (Whitaker and Steinhardt,1982; Whitaker and Swann,1993). Earlier observations on the delay between sperm-tail cessation and the initial small hyperpolarization in the mouse egg also suggest a delay of several minutes, consistent with measurements of dye transfer in this species (Jaffe *et al.,* 1983; Lawrence *et al.,* 1997). If the time of sperm tail cessation is indicative of sperm–egg fusion, then the latent period in hamster eggs is likely to be less than 10 seconds because a hyperpolarizing response starts within seconds of sperm motility cessation (Miyazaki and Igusa, 1981a). The data in hamster eggs also suggest that the latent period can be quite variable, depending on the species, and that some variability may be explained by the sperm. This is borne out by experiments carried out on zona-free hamster eggs, which can be fertilized with sperm from other species (Yanagimachi, 1994). When a mouse sperm fuses with a hamster egg it takes at least 10 minutes to initiate a hyperpolarizing response (Igusa *et al.,* 1983). This delay is more characteristic of the mouse egg rather than the hamster egg (Jaffe *et al.,* 1983). Mouse sperm are similarly much less effective at causing repetitive hyperpolarizations in a hamster egg than are hamster sperm (Igusa *et al.,* 1983). This suggests basic differences in the potency and speed of action in the agents that the hamster and mouse sperm use to cause Ca^{2+} release in the egg.

Understanding the events during the latent period is critical to establishing how the sperm initiates Ca^{2+} release within an egg. There is generally little information available on what is happening in the egg during this time; however, the latent period is very temperature dependent. This was shown by monitoring the delay between sperm–egg contact and the start of the cortical reaction, which in the sea urchin egg has a Q_{10} of around 2.3, indicative of a high temperature dependence (Allen and Griffen, 1958). A marked temperature dependence ($Q_{10} \sim 2.5$) for the equivalent time delay between the initial depolarization and the Ca^{2+}-dependent activation potential in the sea urchin egg is also seen (Dale *et al.,* 1978). In contrast, the wave of envelope elevation, and thus by implication the Ca^{2+} wave at fertilization, is much less temperature dependent and has a Q_{10} of about 1.3 (Allen and Griffen, 1958). The temperature dependence of the latent period suggests that the rate-limiting step involves enzymatic reactions. The differences in temperature dependencies also suggest that some parts of the mechanism that triggers the Ca^{2+} wave in the sea urchin are different from the mechanism that propagates it (Whitaker and Swann, 1993). Our interest in finding a link between the initial interaction of a sperm with the plasma membrane and the release of Ca^{2+} from the endoplasmic reticulum really concerns the nature of the events occurring during the temperature-dependent latent period. Because mammalian eggs undergo sustained oscillations, it is possible to view the mammalian sperm as providing a continuous source of stimulation that may be analogous to that given transiently by the sperm in sea urchins. In this sense we might expect to find a connection between the molecules that initiate the wave in sea urchins and those that trigger oscillations in mammalian eggs and some invertebrate eggs.

V. SIGNALING MOLECULES AND MECHANISMS LEADING TO CA^{2+} RELEASE

There are three basic models for the mechanism of Ca^{2+} release at fertilization following sperm–egg fusion. All three have some experimental data to support them. As an *aide memoire* Lionel Jaffe (1991) has referred to these three models as (1) the sperm conduit model, (2) the sperm contact model, and (3) the sperm content model. It has been suggested that the three models are not mutually exclusive and that *in vivo* the Ca^{2+} response may be mediated via the operation of multiple mechanisms (Miyazaki *et al.,* 1993; Foltz and Shilling, 1993). This may indeed turn out to be the case, but there is still nothing to rule out the possibility that one mechanism operates exclusively and for a wide range of species. We consider it premature to invoke multiple signaling systems because some of the molecules proposed to carry the message of the sperm are relatively uncharacterized. Before each model is discussed in detail it is stressed that most of the arguments outlined here are based on the spatial and temporal aspects of Ca^{2+} release following sperm–egg fusion. This is done because in some species, such as those of mammals, the sperm induces a very characteristic pattern of Ca^{2+} oscillations over several hours (Swann and Ozil, 1994). Due to its unambiguous, readily identifiable Ca^{2+} pattern this can be regarded as the "fingerprint" of the sperm. Matching the Ca^{2+} fingerprint is the key methodology in solving how the sperm activates the egg. It allows rejection of the many agents (e.g., ethanol, ionophores) that cause egg activation by raising intracellular Ca^{2+} but do not induce spermlike Ca^{2+} oscillations. As a consequence of this argument we are not particularly concerned with listing all the events that can ensue after a given stimulus has caused a Ca^{2+} increase. For example, ethanol causes a rise in Ca^{2+}, so it induces all the events of egg activation with parthenotes developing to the blastocyst stage. However, the elegant Ca^{2+} fingerprint of the sperm is not matched by the rather crude one that is left by ethanol, a large single rise in Ca^{2+} lasting for the period of ethanol treatment (Swann and Ozil, 1994). It is therefore our view that recording events that are the inevitable consequence of a Ca^{2+} increase in the egg adds little to the arguments over whether a given activating agent is relevant to fertilization. Determining the pattern of Ca^{2+} changes for a given stimulus, especially in mammalian eggs, is the factor in deciding its physiological relevance.

VI. SPERM AS A CA^{2+} CONDUIT

The first class of hypothesis considered is the proposal that the sperm acts as a source of Ca^{2+} entry into the egg. In the original version of this hypothesis, following egg fusion the sperm delivers a "bomb" of Ca^{2+} that is sufficient by itself to set off a wave of Ca^{2+}-induced Ca^{2+} release (CICR), thereby activating the egg (Jaffe, 1983). This model seems plausible in some species because Ca^{2+} injection can set off a wave of Ca^{2+} release in frog, fish, and hamster eggs (Gilkey, 1983; Busa, 1990; Igusa and Miyazaki, 1983). However, Ca^{2+} injection or electroporation in the presence of Ca^{2+} does not appear to trigger a Ca^{2+} wave in sea urchin or mammalian eggs other than hamster eggs (Swann and Whitaker,1986; Swann, 1994; Ozil and Swann, 1995). Critically, a single pulse of Ca^{2+} increase is also unable to trigger repetitive Ca^{2+} increases in any species that displays them at fertilization (Swann and Ozil, 1994). In addition, it is not clear how the "Ca^{2+} bomb" model would be in keeping with the known latent period of fertilization, because Ca^{2+} injection leads to an immediate triggering of a Ca^{2+} wave (Igusa and Miyazaki,1983). The Ca^{2+} bomb hypothesis has since been modified to the "conduit" hypothesis. In this hypothesis the passage of Ca^{2+} through the sperm from the external medium following sperm–egg fusion acts as a conduit for Ca^{2+} entry into the egg (Jaffe, 1991; Creton and Jaffe, 1995). The sustained influx of Ca^{2+} then leads to local overloading of Ca^{2+} stores in the region next to sperm–egg fusion such that the stores eventually discharge their contents. This model is more reasonable because, in readiness for fertilization, the sperm takes up Ca^{2+}(Jaffe, 1991), and the rise in intracellular Ca^{2+} associated with the acrosome reaction in many species opens channels for Ca^{2+} influx (Florman, 1994). The temperature dependence of the latent period could also be explained by the action of the Ca^{2+}-ATPases. The experimental evidence for this hypothesis is in sea urchin studies. Eggs were inseminated in the presence of either the Ca^{2+} channel blocker La^{3+} or 1,2-bis-(2-aminophenoxy)ethane-*N,N,N′,N′*-tetraacetic acid (BAPTA) to chelate extracellular Ca^{2+}. Both of these procedures inhibited egg activation after a time that the sperm should have fused with the egg (Creton and Jaffe, 1995). This implies that Ca^{2+} influx is required for the sperm to activate the sea urchin egg. However, sperm–egg fusion was not actually monitored in these experiments, so it has yet to be shown that the sperm had actually fused and then fertilization was blocked. It is not clear how these results can be reconciled with a previous report that fertilization and normal egg activation can occur in ethyleneglycol tetraacetic acid (EGTA)-buffered Ca^{2+}-free conditions (Schmidt *et al.,* 1982).

In mammals the Ca^{2+} conduit hypothesis is far less attractive because it has not proved possible in any egg to generate Ca^{2+} oscillations by sustained injection of Ca^{2+} or by repeated electroporation in the presence of Ca^{2+} (Igusa and Miyazaki, 1983; Swann, 1994; Ozil and Swann, 1995; Fissore and Robl, 1994). At best, one regenerative Ca^{2+} transient can be generated in hamster eggs, but this is never continued. In fact, as further Ca^{2+} is microinjected the egg becomes more resistant to undergoing CICR (Igusa and Miyazaki, 1983), the opposite of what occurs following sperm addition. These results suggest that before fertilization CICR is not evident in most mammalian eggs. This makes it difficult to envisage how the Ca^{2+} conduit model could be instrumental in triggering Ca^{2+} oscillations in mammalian eggs. Furthermore, there is evidence against there being any significant Ca^{2+} increase, or influx near the point of sperm–egg fusion in mammals. Using a Mn^{2+} quench technique as a surrogate way of monitoring Ca^{2+} fluxes, it has been shown that there is no measurable Ca^{2+} influx before the first rise in Ca^{2+} at fertilization (McGuinness *et al.*, 1996). The only measurable influx is after, rather than before, the initial Ca^{2+} increase (McGuinness *et al.*, 1996). Moreover, experiments using fluorescent Ca^{2+} indicators and two-photon imaging have shown that when the mouse sperm and egg fuse at fertilization, the cytoplasm of the sperm maintains a low Ca^{2+} and that the initial Ca^{2+} increase at fertilization starts in the egg and then spreads into the sperm about 0.2 seconds later (Jones *et al.*, 1998a). These data all suggest that if the sperm does act as a conduit for entry of Ca^{2+} into the egg after fertilization, then the flux of Ca^{2+} through the sperm is rather small and of no significance for egg activation.

VII. SPERM CONTACT AS THE SIGNAL

The most widely held hypothesis for signal transduction at fertilization is that sperm acts as an honorary hormone and triggers internal Ca^{2+} release via interaction between receptors at the plasma membrane. The main proposal of the hypothesis is that sperm–egg adhesion at the level of the plasma membrane generates inositol 1,4,5,-trisphosphate, which then opens its specific Ca^{2+} channel on the endoplasmic reticulum. IP_3 appears to be a good candidate messenger for fertilization because in a number of eggs IP_3 is effective at causing Ca^{2+} release (Whitaker and Irvine, 1984; Busa, 1990; Swann and Whitaker, 1986; Miyazaki, 1988; Fissore and Robl, 1993; Fissore *et al.*, 1995). In addition, in mammalian eggs, sustained injection of IP_3 can trigger Ca^{2+} oscillations (Swann *et al.*, 1989; Swann, 1992; Jones and Whittingham, 1996). Increased turnover of phosphoinositide-containing phospholipids has also been shown to occur at fertilization in frog and sea urchin eggs (Turner *et al.*, 1984; Snow *et al.*, 1996). In analogy with signaling pathways in many somatic cells the sperm is considered to stimulate one or more isoforms of phosphatidylinositol-specific phospholipase C (PI-PLC) to generate IP_3. Either the PI-PLCβ isoforms, which couple to membrane receptors through G-proteins, or the PI-PLCγ isoforms, which are linked to tyrosine kinases, have been suggested to be involved (Foltz and Shilling, 1993). Mouse eggs have been shown to express three isoforms: PLCβ_1, PLCβ_3, and PLCγ_1 (DuPont *et al.*, 1996). As yet the main evidence has implicated PLCβ in causing Ca^{2+} oscillations. Injection of nonhydrolyzable analogs of GTP, such as GTPγS, activates sea urchin eggs (Turner *et al.*, 1985). In addition, in many mammalian eggs injection of GTPγS triggers Ca^{2+} oscillations (Miyazaki, 1988; Swann, 1992; Fissore *et al.*, 1995). Further, in hamster eggs injection of GDPβS, which inactivates G-proteins, blocks both GTPγS- and sperm-induced Ca^{2+} oscillations (Miyazaki, 1988). In sea urchin and mouse eggs the main events of egg activation at fertilization are also blocked by GDPβS (Moore *et al.*, 1994). Overexpression of a muscarinic receptor, which is G-protein coupled, in frog and mouse eggs leads to egg activation after application of acetylcholine (Kline *et al.*, 1988; Williams *et al.*, 1992). Addition of serotonin to hamster eggs, or carbachol to mouse eggs, also induces Ca^{2+} oscillations (Miyazaki *et al.*, 1990; Swann, 1992). Also in starfish, frog, and mouse eggs overexpression of tyrosine kinase-linked receptors can lead to egg activation after application of the appropriate ligand and the generation of Ca^{2+} release or, in the case of mouse eggs, Ca^{2+} oscillations (Shilling *et al.,* 1994; Yim *et al.,* 1994; Mehlmann *et al.,* 1998). Thus the signaling machinery within many types of eggs appears to be present for receptors, G-protein, and PLCβ or PLCγ isoforms.

Evidence to implicate IP_3 production at fertilization makes use of U73122, a potent inhibitor of PLC activity that can inhibit sperm-induced Ca^{2+} oscillations in mouse eggs (Dupont *et al.,* 1996). In some species high concentrations of the competitive inhibitor of the IP_3 receptor, heparin, have also been shown to block Ca^{2+} oscillations at fertilization (Fissore and Robl, 1994, Fissore *et al.,* 1995). Furthermore, an antibody to the IP_3 receptor pore region blocks Ca^{2+} oscillations induced by sperm and IP_3 in hamster and mouse eggs, suggesting it is this receptor, and by implication IP_3, that is involved at fertilization (Miyzaki *et al.,* 1992, 1993). It has also been shown that IP_3 receptors can be down-regulated in mouse eggs and oocytes by prior injection of adenophostin (Brind *et al.,* 2000). When IP_3 receptors are down-regulated, using adenophostin, in mouse eggs prior to fertilization, the sperm-induced Ca^{2+} oscillations are inhibited (Brind *et al.,* 2000). These data, along with the other inhibitor experiments, clearly implicate IP_3 receptors and IP_3 production in generating Ca^{2+} release at fertilization.

In addition to these observations implying that the sperm could use the receptor mechanism, there is now some evidence that molecules on the sperm surface can trigger Ca^{2+} changes in eggs. In *Urechis* eggs a sperm-derived basic protein is able to trigger depolarization, a Ca^{2+} increase, and activation in a manner similar to that of the sperm (Stephano and Gould, 1997). In frog eggs it has been shown that RGD-containing and disintegrin-like peptides are able to induce a Ca^{2+} increase and egg activation (Iwao and Fujimura, 1996; Shilling *et al.,* 1998). RGD peptides have also been reported to cause some Ca^{2+} release responses in bovine eggs (Campbell *et al.,* 2000). The RGD peptides are proposed to mimic the effect of the sperm by binding to potential integrin-like egg surface receptors (Foltz and Shilling, 1993). However, the egg receptors that have been identified so far in mammals appear to play roles in sperm binding and fusion rather than in Ca^{2+} release (Wasserman *et al.,* 2001).

These data collectively build a substantial case to suggest that receptors, and by implication the phosphatidylinositol signaling system, are used by the sperm at fertilization. However, there are some critical aspects of the data that question a simple story involving receptors and IP_3. First, in sea urchin eggs it has also been shown that the sperm does not appear to cause any phosphoinositide turnover in the absence of a Ca^{2+} increase, even though GTPγS does (Crossley *et al.,* 1991). This suggests fundamental differences in the way sperm and G-proteins mediate Ca^{2+} release. Given its potential role in fertilization, it is also surprising that when IP_3, or a nonhydrolyzable analog, is microinjected into mammalian eggs, the resulting pattern of Ca^{2+} oscillations does not correlate very well with those induced by sperm (Swann and Ozil, 1994). In fact, in mammalian eggs, stimulation with GTPγS and neurotransmitters induces a dynamic response consistently different from that induced by the sperm (Swann *et al.*, 1989; Miyazaki *et al.*, 1990; Swann, 1992). Most notably in the hamster egg, the response to IP_3 or GTPγS injection leads to high-frequency, critically damped Ca^{2+}oscillations (Swann *et al.*, 1989; Galione *et al.*, 1994). This is quite unlike the sperm, which can cause oscillations that vary in frequency but show little sign of a decline in amplitude (Miyazaki, 1991). The response to GTPγS can also be completely blocked by prior incubation with phorbol esters, and yet the same treatment does not stop oscillations at fertilization (Swann *et al.*, 1989). The Ca^{2+} response of mouse eggs to carbachol, which generates IP_3 through activation of the acetylcholine receptor, is oscillatory, but these oscillations are of lower amplitude and much higher frequency than those seen at fertilization (Swann, 1992). In porcine eggs, where exogenous acetylcholine receptors were expressed, the Ca^{2+} response to activating doses of acetycholine consisted of a single large Ca^{2+} rise rather than oscillations (Machaty *et al.*, 1997). Some of these reservations about G-protein signaling in mammalian eggs are not apparent in sea urchin, frog, or many other eggs, where only a single Ca^{2+} increase at fertilization is seen and very many parthenogenetic activating agents cause the same monotonic Ca^{2+} increase. This reiterates our tenet that mammalian eggs provide a more stringent test for potential sperm-borne activating molecules.

Further evidence has been presented arguing against the receptor G-protein hypothesis for mammalian fertilization. Injecting antibodies to G_q protein was found to block egg activation stimulated by the acetylcholine pathway in eggs overexpressing the muscarinic receptor (Williams *et al.*, 1998). However, the same antibodies failed to inhibit events at fertilization (Williams *et al.*, 1998). This result may mean that the sperm does not use a PLCβ to generate IP_3 at fertilization but instead uses a tyrosine kinase pathway that activates a PLCγ. The best evidence that the tyrosine kinase pathway is involved in egg activation comes from studies in eggs of echinodermata. Injection of repeat SH2 domains from bovine PLCγ, but not SH2 domains from a phosphatase, into starfish, ascidian, and sea urchin eggs blocks sperm-induced Ca^{2+} release (Carroll *et al.*, 1997, 1999; Shearer *et al.*, 1999; Runft and Jaffe, 2000). The SH2 PLCγ construct does not block sperm entry, so this result suggests that it is interfering in the mechanism leading to the generation of Ca^{2+} release during the latent period (Carroll *et al.*, 1999; Shearer *et al.*, 1999). SH2-containing protein from the Src family of kinases also inhibits Ca^{2+} release at fertilization in sea urchins and ascidians (Jaffe *et al.*, 2001; Runft and Jaffe, 2000). Because injecting excess SH2 domains is expected to bind to phosphorylated tyrosine residues, these data provide a strong case for the involvement of tyrosine-phosphorylated proteins in generating Ca^{2+} release in fertilizing sea urchin and starfish eggs. However, resolving the way the SH2 construct blocks echinoderm egg activation will require making a direct link with a receptor, and so far this has not been done. What is also unclear is how generic these data may be, because the same bovine SH2 PLCγ constructs do not block Ca^{2+} oscillations or activation during fertilization of mouse or frog eggs (Mehlmann *et al.*, 1998; Runft *et al.*, 1999). This might imply that there are somewhat different signaling mechanisms operating in sea urchin and starfish compared with vertebrate eggs.

So far we have considered that Ca^{2+} release at fertilization has to involve IP_3 receptors, which do appear to be the predominant Ca^{2+} release channels present in most eggs. There is, however, evidence for ryanodine receptor Ca^{2+} release channels in some eggs. Most notably in the sea urchin egg ryanodine receptors appear to contribute to the Ca^{2+} wave at fertilization and they can be specifically activated by the NAD metabolite cyclic ADP ribose (cADPR) (Galione *et al.,* 1991, 1993a). A role for cADPR at fertilization may be in propagating the Ca^{2+} across the egg or else in prolonging the duration of the Ca^{2+} rise. To be involved in initiating events at fertilization, cADPR synthesis would have to be stimulated by the sperm. One pathway that might activate cADPR production involves cyclic GMP (Galione *et al.,* 1993b). Cyclic GMP production might be stimulated by nitric oxide (NO), and it has been suggested that sperm-derived NO triggers egg activation in sea urchins (Kuo *et al.,* 2000). However, the role of this pathway is unclear because cyclic GMP-dependent kinase inhibitors fail to block Ca^{2+} release at fertilization in sea urchin eggs (Lee *et al.,* 1996). NO does not appear to play any role in Ca^{2+} release in mammalian eggs (Hyslop *et al.,* 2001). A role for cADPR in fertilization may exist in some other species, such as bovine eggs (Yue *et al.,* 1995), but it does not cause any Ca^{2+} increase in the eggs of other domestic animals or in any rodent eggs (Whitaker and Swann, 1993). Even though it may be possible to link cADPR to a contact hypothesis for fertilization, as yet its exact role in initiating the Ca^{2+} release at fertilization remains unclear.

VIII. THE SPERM CONTENT HYPOTHESIS

So far, well-known signaling molecules have been discussed, and although they have been good candidates some difficulties remain. Most notably it has proved difficult to match these signaling messengers with the pattern of Ca^{2+} oscillations seen at fertilization for some mammalian eggs. In the "content" theory the reverse is true. Microinjection of soluble extracts of sperm into mammalian, ascidian, and nemertean worm eggs triggers Ca^{2+} oscillations identical to those at fertilization (Swann, 1990; 1994; Homa and Swann, 1994; Wu *et al.,* 1997; Palermo *et al.,* 1997; Stricker, 1997; Kyozuka *et al.,* 1998). Ca^{2+} oscillations are not triggered by sperm extracts that have been heat or trypsin treated, suggesting a sperm protein is involved in Ca^{2+} release. Earlier in this chapter we argued that any proposed hypothesis for sperm-induced Ca^{2+} release must involve a mechanism that can match the Ca^{2+} fingerprint of the sperm, a series of low-frequency Ca^{2+} oscillations. This is the case for the relatively simple procedure in which soluble extracts from mammalian sperm are injected into the egg cytoplasm. However, although we have managed to match the signature of the sperm, the problem remains in establishing which molecules in the sperm are involved.

The sperm does not have much, if any, cytoplasm in a conventional sense. The cytoplasmic droplet of immature sperm is lost by the time sperm are ejaculated.

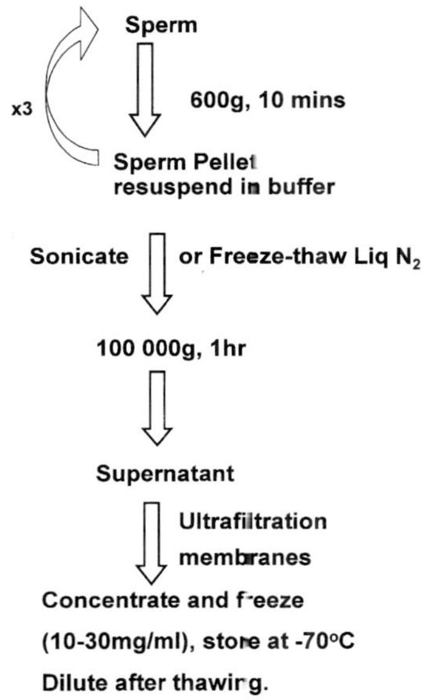

FIGURE 10.2 Preparation of sperm extracts for microinjection. The sperm is washed into a Ca^{2+}-free medium containing pH buffers and protease inhibitors [see Jones *et al.* (1998b)]. The sperm suspensions are lysed and then spun at 100,000 *g* and the supernatant "cytosolic" fraction is saved for injection. Concentration of extracts can be carried out on ultrafiltration membranes with cut-off sizes up to 30 or 100 kDa. The extracts are frozen and stored in a concentrated form and then diluted into buffer shortly before microinjection.

So how is the soluble extract made? Schematically this is shown in Figure 10.2. The preparation involves lysing the sperm in an intracellular buffer and then taking the cytosolic (soluble) fraction after a high-speed centrifuge spin. No detergents are used. The extract, once prepared, when injected into eggs gives the characteristic and prolonged oscillations normally associated with fertilization. Extracts from other tissues, such as brain and liver, do not cause any Ca^{2+} oscillations when injected into eggs, and adding the sperm extracts to the outside of eggs does not trigger oscillations (Swann, 1990; Palermo *et al.*, 1997; Stricker, 1997). Thus the factor is specific to sperm cytosol. It therefore appears that there is a factor in the sperm extract, hitherto generally referred to as a "sperm factor," which is a good candidate for triggering the Ca^{2+} oscillations at fertilization. The ability of sperm extracts to cause Ca^{2+} oscillations does not appear to be limited to one or two mammalian species, but is rather a general phenomenon. Sperm extracts prepared from human, mouse, rabbit, cow, and hamster sperm all seem to have a sperm factor activity in their own and other species of egg (Swann, 1990, 1994; Homa and Swann, 1994; Wu *et al.*, 1997; Palermo *et al.*, 1997). Sperm extracts from frog or chicken sperm have also been shown to cause Ca^{2+} oscillations in mouse eggs (Dong *et al.*, 2000), and boar sperm extracts have been shown to cause oscillations in nemertean worm eggs (Stricker *et al.,* 2000). As a consequence of its ability to raise Ca^{2+}, injection of sperm extracts is able to induce all the events of egg activation in mammals and ascidian and nemertean worms (Stice and Robl, 1990; Stricker, 1997; Palermo *et al.,* 1997; Kyozuka *et al.,* 1998). The actions of the sperm factor also appear to be fairly widespread, because injection of sperm extracts can induce Ca^{2+} oscillations in both dorsal root ganglion neurons and hepatocytes (Currie *et al.,* 1992; Berrie *et al.,* 1996). This suggests that its mechanism of action involves components of signaling pathways present in many different cell types.

The estimates of the amount of sperm factor injected suggest that there is, certainly within an order of magnitude, enough sperm factor activity in a single sperm to be able to cause the response at fertilization (Swann, 1993; Stricker, 1996; Kyozuka *et al.,* 1998). Any calculation, however, has to assume 100% recovery from sperm with no loss of activity during preparation. The existence of a relevant factor inside a sperm is also consistent with the finding that injection of intact sperm into mouse or nemertean worm eggs can trigger the same pattern of oscillations seen at fertilization in these species (Stricker, 1996; Nakano *et al.,* 1997). In mammals the clinical success of intracytoplasmic sperm injection, whereby gamete membrane interaction is bypassed by the direct injection of the sperm into the egg cytoplasm, also suggests that there is enough of the sperm factor inside a single sperm to cause the relevant pattern of oscillations seen at fertilization.

We concluded that the sperm factor is not Ca^{2+} during discussion of the "conduit" hypothesis. It has been suggested that IP_3 could be a sperm factor (Iwasa *et al.,* 1990; Tosti *et al.,* 1993). However, if the sperm factor were a small molecular weight agonist of Ca^{2+} release channels, such as IP_3 and cADPR, then this would not explain the temperature dependence of the latent period in sea urchins, or the prolonged Ca^{2+} oscillations in mammalian eggs. Importantly, it would also not account for the trypsin sensitivity of the sperm factor. In fact, the sperm factor is unlikely to be any combination of small-molecular-weight molecules because the agent that causes Ca^{2+} oscillations from nemertean worms or mammals is retained on ultrafiltration membranes of 30-kDa cut-off, suggesting it is a high-molecular-weight molecule (Swann,1990; Wu *et al.,* 1997; Stricker, 1996). The only evidence for small-molecular-weight sperm factors is based on egg activation alone, which is not a stringent criterion by which to judge any sperm factor candidate. In this respect it is worth noting that a low-molecular-weight component of sperm extract induces a single Ca^{2+} increase when injected into hamster eggs, but only the high-molecular-weight fraction triggers oscillations (Swann, 1990). It has now been shown by different groups that the sperm factor activity, as defined by the ability to induce Ca^{2+} oscillations, is lost following either trypsin incubation or heat treatment (Swann, 1990; Wu *et al.,* 1997; Stricker, 1997). All these observations point to the sperm factors being a high-molecular-weight protein or group of proteins. Because the sperm factor is defined by its ability to trigger Ca^{2+} oscillations, rather than its ability to activate eggs, it has also been referred to as an oscillogen (Swann, 1993). The important questions for the sperm factor or oscillogen concern what proteins are involved and how they work.

The first candidate for the sperm factor protein was a protein termed "oscillin." A protein of 33 kDa was correlated with oscillogen activity from hamster sperm. This was achieved by serial chromatographic purification of mammalian sperm extracts and assaying fractions for Ca^{2+} oscillation inducing activity in mouse eggs (Parrington *et al.,* 1996). An antibody raised against the purified 33-kDa protein stains most intensely in a region inside the sperm head around in the region of the equatorial segment. This is the site of sperm–egg fusion in mammalian sperm and is the prime and expected location of a signaling molecule for initiating Ca^{2+} release after sperm–egg fusion (Yanagimachi, 1994). The 33-kDa protein was N-terminal sequenced and this led to the identification of an amino acid sequence that had 53% identity to a bacterial glucosamine-6-phosphate deaminase (Parrington *et al.,* 1996). However, recombinant oscillin, which has deaminase activity, is not a Ca^{2+} releasing agent when injected into mammalian eggs (Wolosker *et al.,* 1998; Shevchenko *et al.,* 1998). More importantly, some fractionation procedures have been found that separate the Ca^{2+} oscillation-inducing activity in sperm extracts from the 33-kDa protein (Wu *et al.,* 1998; Parrington *et al.,* 1999). These data prove that the 33-kDa protein is not a component of the sperm factor and that other proteins in the sperm extracts must be responsible for triggeriing Ca^{2+} oscillations in eggs.

Another candidate sperm factor is tr-kit (truncated kit), a 23-kDa truncated form of the 150-kDa c-kit receptor important in germ cell migration (Sette *et al.,* 1997). Microinjection of extracts from cells transfected with tr-kit or with mRNA for tr-kit causes mouse egg activation (Sette *et al.,* 1997). The tr-kit factor is specifically expressed during the latter stages of spermatogenesis from a testis-specific promoter. It is presumed that tr-kit activates an egg PLC because (1) its actions are blocked by the PLC inhibitor U73122, (2) it generates IP_3 when transfected into COS cells, and (3) it is blocked by a bovine PLCγ SH3 construct (Sette *et al.,* 1997; 1998). How tr-kit specifically activates PLC is not known because no direct interaction appears to occur (Sette *et al.,* 1998) and tr-kit lacks a full kinase domain and so cannot function as a tyrosine kinase. Of particular note in assessing the relevance of tr-kit in fertilization is that SH3 constructs that block tr-kit-induced egg activation do not appear to block fertilization (Mehlmann *et al.,* 1998) In addition, tr-kit is mainly located in the cytoplasmic droplet of the immature sperm and tail midpiece of mature sperm, a location not consistent with the known activating ability of sperm heads rather than tails (Yanagimachi, 1998) Clearly, at present tr-kit is a candidate with reservations. The critical experiment still to perform is to ascertain if tr-kit can induce a spermlike series of Ca^{2+} oscillations when injected into mammalian eggs. This important experiment to test the suitability of tr-kit as a sperm factor has not been reported. Its role in fertilization, therefore, remains unclear because we have consistently argued that activation of eggs does not provide a sufficient criterion to assess the physiological relevance of a sperm factor candidate.

The only other candidate sperm factor protein has come from the authors' studies examining the mechanism by which mammalian sperm extract causes Ca^{2+} release in sea urchin egg homogenate. These egg homogenates undergo homologous desensitization to the Ca^{2+}-releasing agents IP_3, cADPR, and NAADP (Galione *et al.,* 1991). For example, following addition of repetitive maximal doses of IP_3 the homogenate becomes desensitized for prolonged periods to further additions of IP_3, but a normal Ca^{2+} rise is observed when cADPR or NAADP are added. We found that desensitization to IP_3, but not cADPR or NAADP, led to a loss of sperm extract response (Jones *et al.,* 1998b). Consistent with this we found that the sperm factor caused an increase in IP_3 in sea urchin egg homogenates and that blocking this increase by depleting phosphoinositide phospholipids inhibited the action of the sperm extract (Jones *et al.,* 1998b). The ability of sperm extracts to generate IP_3 in the homogenate was due to the action of a sperm-derived PLC, because sperm extracts generated IP_3 following incubation with phosphatidylinositol 4,5-biphosphate (PIP_2). These experiments clearly suggest the mechanism by which sperm extracts released Ca^{2+} in sea urchin egg homogenate. Moreover, when different chromatography columns were used to separate sperm proteins, on all columns the same protein fractions that caused Ca^{2+} oscillations when microinjected into mouse eggs also caused Ca^{2+} release when added to sea urchin egg homogenate (Parrington *et al.,* 1999). These fractions also contain the sperm PLC that is extremely active in free solution compared to those present in other tissues (Jones *et al.,* 2000; Rice *et al.,* 2000). The activity in a single sperm equivalent is such that it may be sufficient to account for the generation of IP_3 in an egg after gamete membrane fusion at fertilization (Rice *et al.,* 2000). It is therefore our working hypothesis that the sperm factor is some form of PLC and that a sperm-derived PLC is responsible for the Ca^{2+} oscillations at fertilization.

A schematic illustrating this suggestion is shown in Figure 10.3. PLC activity has been previously identified in mammalian sperm extracts (Ribbes *et al.,* 1987; Vanha-Perttula and Kasurinen, 1989), although the full inventory of isozymes has not been reported. The identity of the sperm PLC that may explain Ca^{2+} release in eggs, therefore, remains to be established. Studies suggest that the common γ or β isoforms, by themselves, could not be the sperm factor PLC (Jones *et al.,* 2000; Wu *et al.,* 2001). PLCδ4 is also excluded as a sperm factor because mice lacking the PLCδ4 gene still induce Ca^{2+} oscillations in eggs at fertilization (Fukami *et al.,* 2001). As well as identifying the sperm PLC, it has also to be established how IP_3 production by a sperm factor PLC could explain the distinctive pattern of Ca^{2+} oscillations seen after its injection. A number of differences may exist between the sperm PLC and the egg-derived PLCs. For example, the source and location of phosphatidylinositol 4,5-bisphosphate, which is the PLC substrate, may be different for the sperm PLC (Rice *et al.,* 2000).

A sperm factor model in which a sperm PLC generates IP_3 in the egg would be a good way of reconciling some aspects of the "contact" and "content" models. For a long time the "contact" model has provided a clear signal transuction pathway based on an egg PLC. It has, however, lacked a definitive egg receptor. On the other hand the "content" model has had strong experimental support for a sperm-borne factor triggering Ca^{2+} oscillation, but it has lacked a definitive signal transduction pathway for Ca^{2+} release. A sperm PLC involved in triggering

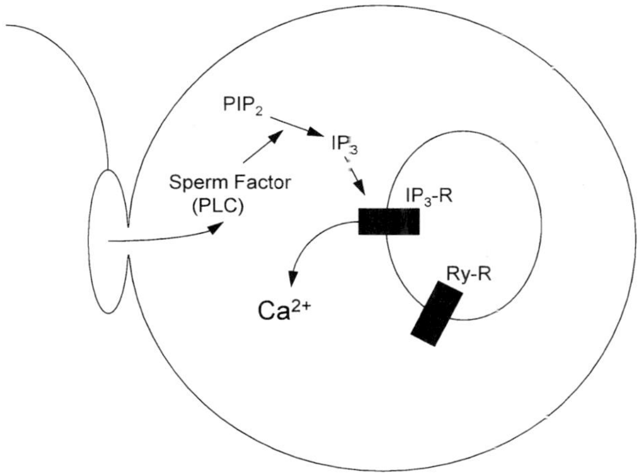

FIGURE 10.3 Schematic of the hypothesis for signaling at fertilization in vertebrate eggs. The sperm fuses with the egg plasma membrane and as a result introduces a soluble protein into the egg cytoplasm. The sperm factor is a protein complex that contains an essential phospholipase C (PLC) activity. This generates inositol 1,4,5-triphosphate (IP_3) from phosphatidylinositol bisphosphate (PIP_2) within the egg. This leads to the generation of Ca^{2+} release via the IP_3 receptor (IP_3-R), which is probably within a Ca^{2+} store that also contains ryanodine-sensitive Ca^{2+} release channels. Ry-R, Ryanodine receptor.

Ca^{2+} oscillations would therefore be in agreement with a "content" model but makes use of the "contact" model's signal transduction process.

IX. CONCLUSIONS

Understanding how a sperm activates an egg at fertilization has been a major and as yet unresolved problem. The process of answering this question helps focus attention on the way that the sperm triggers Ca^{2+} release from the intracellular stores. Despite many years of research effort there have been few sperm-derived molecules that can actually be shown to trigger the correct pattern of Ca^{2+} changes in appropriate eggs. In some species, sperm-derived molecules have been applied to the surface and shown to activate eggs. These examples tend to be eggs that are readily activated, with the same Ca^{2+} changes induced by many non-physiological stimuli. In contrast, there is good evidence that the sperm of some species, most notably mammals, contain an intracellular protein that triggers a pattern of Ca^{2+} changes very similar to that seen at fertilization. This effect cannot be readily mimicked by parthenogenetic stimuli. A model in which the sperm activates the egg via the introduction of such a protein sperm factor, or "oscillogen," is compatible with the known sequence of events at fertilization. We consider the identification of the soluble sperm factor as the key issue in advancing the understanding of fertilization. One good sperm factor candidate is a PLC that would be responsible for generating IP_3 and thus Ca^{2+} release seen at fertilization. We hope that the hypothesis that the sperm factor is a PLC helps unite some aspects of the various different models put forward that hitherto have appeared contradictory.

REFERENCES

Albrieux, M., Sardet, C., and Villaz, M. (1997). The two intracellular Ca^{2+} release channels, ryanodine receptors and inositol 1,4,5-trisphosphate receptor, play different roles during fertilization in ascidians. *Dev. Biol.* **189,** 174–185.

Allen, R. D., and Griffen, J. L. (1958). The time sequence of early events in the fertilization of sea urchin eggs. 1. The latent period and the cortical reaction *Exp. Cell Res.* **15,** 163–173.

Ayabe, T., Kopf, G. S., and Schultz, R. M. (1995). Regulation of mouse egg activation: Presence of ryanodine receptors and effects of microinjected ryanodine cyclic ADP ribose on uninseminated and inseminated eggs. *Development* **121,** 2233–2244.

Berridge, M. J. (1993). Inositol trisphophate and calcium signalling. *Nature* **361,** 315–325.

Berrie, C. P., Cuthbertson, K. S. R., Parrington, J., Lai, F. A., and Swann, K. (1996). A cytosolic sperm factor triggers calcium oscillations in rat hepatocytes. *Biochem. J.* **313,** 369–372.

Bloom, T. L., Szuts, E. Z., and Eckberg, W. R. (1988). Inositol trisphosphate, inositol trisphosphate metabolism and germinal vesicle breakdown in surf clam oocytes. *Dev. Biol.* **129,** 532–540.

Brind, S., Swann, K., and Carroll, J. (2000). Inositol 1,4,5-trisphosphate receptors are downregulated in mouse oocytes in response to sperm and adenophostin A but not to increase in intracellular Ca^{2+} or egg activation. *Dev. Biol.* **223,** 251–265.

Busa, W. B. (1990). Involvement of calcium and inositol phosphates in amphibian egg activation. *J. Reprod. Fertil. Suppl.* **42,** 155–161.

Busa, W. B., and Nuccitelli, R. (1985). An elevated free cytosolic Ca^{2+} wave follows fertilization in the frog, *Xenopus laevis. J. Cell Biol.* **100,** 1325–1329.

Campbell, K. D., Reed, W. A., and White, K. L. (2000). Ability of integrins to mediate fertilization, intracellular calcium release, and parthenogenetic activation in bovine oocytes. *Biol. Reprod.* **62,** 1702–1709.

Carroll, D. J., Albay, D. T., Terasaki, M., Jaffe, L. A., and Foltz. K. R. (1999). Identification of PLCγ-dependent and -independent events during fertilization of sea urchin eggs. *Dev. Biol.* **206,** 232–247.

Carroll, D. J., Ramarao, C. S., Mehlmann, L. M., Roche, S., Terasaki, M., and Jaffe, L. A. (1997). Calcium release at fertilization in starfish eggs is mediated by phospholipase Cγ. *J. Cell Biol.* **138,** 1303–1311.

Collas, P., Chang, T., Long, C., and Robl, J. M. (1995). Inactivation of histone H1 kinase by Ca^{2+} in rabbit oocytes. *Mol. Reprod. Dev.* **40,** 253–258.

Creton, R., and Jaffe, L. F. (1995). Role of calcium influx during the latent period in sea urchin eggs. *Dev. Growth Differ.* **37,** 703–709.

Crossley, I., Whalley, T., and Whitaker, M. J. (1991). Guanosine 5′-thiotriphosphate may stimulate phosphoinositide messenger production in sea urchin eggs by a different route than the fertilizing sperm. *Cell Reg.* **2,** 121–133.

Currie, K. P. M., Swann, K., Galione, A., and Scott, R. H. (1992). Activation of Ca^{2+}-dependent currents in cultured dorsal root ganglion neurons by a sperm factor and cyclic-ADP ribose. *Mol. Biol. Cell* **3,** 1415–1425.

Cuthbertson, K. S. R., and Cobbold, P. H. (1985). Phorbol ester and sperm activate mouse oocytes by inducing sustained oscillations in cell Ca^{2+}. *Nature* **316,** 541–542.

Dale, B., De Felice, L. J., and Taglietti, V. (1978). Membrane noise and conductance increase during single spermatozoa–egg interaction. *Nature* **275,** 217–219.

Deguchi, R., and Osanai, K. (1994). Repetitive intracellular Ca^{2+} increases at fertilization and the role of Ca^{2+} in meiosis reinitiation from the first metaphase in oocytes of marine bivalves. *Dev. Biol.* **163,** 162–174.

Digonnet, C., Aldon, D., Leduc, N., Dumas, C., and Rougier, M. (1997). First evidence of a calcium transient in flowering plants at fertilization. *Development* **124,** 2867–2874.

Dong, J. B., Tang, T. S., and Sun, F. Z. (2000). *Xenopus* and chicken sperm contain a cytosolic soluble protein factor which can trigger calcium oscillations in mouse eggs. *Biochem. Biophys. Res. Commun.* **268,** 947–951.

Dube, F. (1988) The relationship between early ionic events, the pattern of protein synthesis and oocyte activation in the surf clam, *Spisula solidissima. Dev. Biol.* **126,** 233–241.

Dupont, G., McGuiness, O., Johnson, M. H., Berridge, M. J., and Borgese, F. (1996). Phospholipase C in mouse oocytes: Characterization of isoforms and their possible involvement in sperm-induced Ca^{2+} spiking. *Biochem. J.* **316,** 583–591.

Eckberg, W. R., and Miller, A. L. (1995). Propagated and nonpropagated calcium transients during egg activation in the annelid, *Chaetopterus. Dev. Biol.* **172,** 654–664.

Eisen, A., and Reynolds, G. T. (1984). Calcium transients during early development in single starfish (*Asterias forbesi*) oocytes. *J. Cell Biol.* **99,** 1878–1882.

Eisen, A., and Reynolds, G. T. (1985). Source and sinks for calcium release during fertilization of single sea urchin eggs. *J. Cell Biol.* **100,** 1522–1527.

Fissore, R. A., and Robl, J. M. (1993). Sperm, inositol trisphosphate and thimerosal-induced Ca^{2+} elevations in rabbit eggs. *Dev. Biol.* **159,** 122–130.

Fissore, R. A., and Robl, J. M. (1994). Mechanism of calcium oscillations in fertilized rabbit eggs. *Dev. Biol.* **166,** 634–642.

Fissore, R. A., Pinto-Correla, C., and Robl, J. M. (1995). Inositol trisphosphate-induced calcium release in the generation of calcium oscillations in bovine eggs. *Biol. Reprod.* **53,** 766–774.

Florman, H. M. (1994). Sequential focal and global elevations of sperm intracellular Ca^{2+} are initiated by zona pellucida during acrosomal exocytosis. *Dev. Biol.* **152,** 152–164.

Foltz, K. R., and Schilling, F. M. (1993). Receptor-mediated signal transduction and egg activation. *Zygote* **1** 276–279.

Fukami, K., Nakao, K., Inoue, T., Kataoka, Y., Kurokawa, M., Fissore, R. A., Nakamura, K., Katsuki, M., Mikoshiba, K., Yoshida, N., and Takenawa, T. (2001). Requirement of phsopholipase Cδ4 for the zona pellucida induced acrosome reaction. *Science* **292,** 920–923.

Galione, A., Jones, K. T., Lai, F. A., and Swann, K. (1997). A cytosolic sperm protein factor mobilizes Ca^{2+} from intracellular stores by activating mutiple Ca^{2+} release mechanisms independently of low molecular weight messengers. *J. Biol. Chem.* **272,** 28901–28905.

Galione, A., Lee, H. C., and Busa, W. B. (1991). Ca^{2+}-induced Ca^{2+} release in sea urchin egg homogenates: Modulation by cyclic ADP-ribose. *Science* **253,** 1143–1146.

Galione, A., McDoughall, A., Busa, W. B., Willmot, N., Gillot, I., and Whitaker, M. (1993a). Redundant mechanisms of calcium-induced calcium release underlying calcium waves during fertilization of sea urchin eggs. *Science* **261,** 348–352.

Galione, A., Swann, K., Georgiou, P., and Whitaker, M. J. (1994). Regenerative and non-regenerative calcium transients in hamster eggs triggered by inositol 1,4,5-trisphosphate. *J. Physiol.* **480,**(3), 465–474.

Galione, A., White, A., Willmot, N., Turner, M., Potter, B. V., and Watson, S. P. (1993b). cGMP mobilizes intracellular Ca^{2+} in sea urchin eggs by stimulating cyclic ADP-ribose synthesis. *Nature* **365,** 456–459.

Gilkey, J. C. (1983). Roles of calcium and pH in activation of eggs of the medaka fish, *Oryzias latipes. J. Cell Biol.* **97,** 669–678.

Goudeau, M., and Goudeau, H. (1989). A long lasting electrically mediated block, due to the egg membrane hyperpolarization at fertilization, ensures physiology monospermy in eggs of the crab *Maia squinado. Dev. Biol.* **133,** 348–360.

Goudeau, M., and Goudeau, H. (1996). External Mg2+ triggers oscillations and a subsequent sustained level of intracellular free Ca^{2+}, correlated with changes in membrane conductance in the oocyte of the prawn *Palaemon serratus. Dev. Biol.* **177,**178–189.

Grandin, N., and Charbonneau, M. (1992). Intracellular free Ca^{2+} changes during physiological polyspermy in amphibian eggs. *Development* **114,** 617–624.

Hagiwara, S., and Jaffe, L. A. (1979). Electrical properties of egg cell membranes. *Annu. Rev. Biophys. Bioeng.* **8,** 385–416.

He, C. L., Damiani, P., Parys, J. B., and Fissore, R. A. (1997). Calcium, calcium release receptors, and meiotic matuation in bovine oocytes. *Biol. Reprod.* **57,** 1245–1255.

Homa, S. T., and Swann, K. (1994). A cytosolic sperm factor triggers calcium oscillations and membrane hyperpolarizations in human oocytes. *Hum. Reprod.* **9,** 2356–2361.

Hyslop, L. A., Carroll, M., Nixon, V. L., McDougall, A., and Jones, K. T. (2001). Simultaneous measurement of intracellular nitric oxide and free calcium levels in chordate eggs demonstrates that nitric oxide has no role at fertilization. *Dev. Biol.* **234,** 216–230.

Igusa, Y., and Miyazaki, S. (1983). Effects of altered extracelluar and intracellular calcium concentration on hyperpolarizing responses of hamster egg. *J. Physiol.* **340,** 611–632.

Igusa, Y., Miyazaki, S., and Yamashita, N. (1983). Periodic hyperpolarizing responses in hamster and mouse eggs fertilized with mouse sperm. *J. Physiol.* **340,** 643–647.

Iwao, Y., and Fujimura, T. (1996). Actvation of *Xenopus* eggs by RGD-containing peptides accompanied by intracellular Ca^{2+} release. *Dev. Biol.* **177,** 558–567.

Iwasa K. H., Ehrenstein G., DeFelice L. J., and Russell J. T. (1990). High concentrations of inositol 1,4,5-trisphosphate in sea urchin sperm. *Biochem. Biophys. Res. Commun.* **172,** 932–938.

Jaffe, L. A., and Cross, N. L. (1986). Electrical regulation of sperm–egg fusion. *Annu. Rev. Physiol.* **48,** 191–200.

Jaffe, L. A., Giusti, A. F., Carroll, D. J., and Foltz, K. R. (2001). Ca^{2+} signalling during fertilization of echinoderm eggs. *Semin. Cell Dev. Biol.* **12,** 45–51.

Jaffe, L. A., Hagiwara, S., and Kado, R. T. (1978). The time course of cortical vesicle fusion in sea urchin eggs observed as membrane capacitance changes. *Dev. Biol.* **67,** 243–248.

Jaffe, L. A., Sharp, A. P., and Wolf, D. P. (1983). Abscence of electrical polyspermy block in the mouse. *Dev. Biol.* **96,** 317–323.

Jaffe, L. F. (1983). Sources of calcium in egg activation; a review and hypothesis. *Dev. Biol.* **99,** 265–276.

Jaffe, L. F. (1991). The path of calcium in cytosolic calcium oscillations: A unifying hypothesis. *Proc. Natl. Acad. Sci. U.S.A.* **88,** 9883–9887.

Jones, K. T., and Whittingham, D. G. (1996). A comparison of sperm- and $InsP_3$-induced Ca^{2+} release in activated and aging mouse oocytes. *Dev. Biol.* **178,** 229–237.

Jones, K. T., Carroll, J., and Whittingham, D. G. (1995). Ionomycin, thapsigargin, ryanodine and sperm induced Ca^{2+} release increase during meiotic maturation of mouse oocytes. *J. Biol. Chem.* **270,** 6671–6677

Jones, K. T., Cruttwell, C., Parrington, J., and Swann, K. (1998b). A mammalian sperm cytosolic phospholipase C activity generates inositol trisphosphate and causes Ca^{2+} release in sea urchin egg homogenates. *FEBS Lett.* **437,** 297–300.

Jones, K. T., Matsuda, M., Parrington, J., Katan, M., and Swann, K. (2000). Different Ca^{2+} releasing abilities of sperm extracts compared with tissue extracts and phospholipase C isoforms in sea urchin egg homogenate and mouse eggs. *Biochem. J.* **346,** 743–749.

Jones, K. T., Soeller, C., and Cannell, M. B. (1998a). The passage of Ca^{2+} and fluorescent markers between the sperm and egg after fusion in the mouse. *Development* **125,** 4627–4635.

Kline, D. (1988). Calcium dependent events at fertilization of the frog egg: Injection of a calcium buffer blockes ion channel opening, exocytosis and formation of pronuclei. *Dev. Biol.* **126,** 346–361

Kline, D., and Kline, J. T. (1992). Repetitive calcium transients and the role of calcium in exocytosis and cell cycle activation in the mouse egg. *Dev. Biol.* **149,** 80–89.

Kline, D., and Nuccitelli, R. (1985). The wave of activation current in the Xenopus egg. *Dev. Biol.* **111,** 471–487.

Kline, D., Simoncini, L., Mandel, G., Mau, R. A, Kado, R. T., and Jaffe, L. A. (1988). Fertilization events induced by neurotransmitters after injection of mRNA in *Xenopus* eggs. *Science* **241,** 464–467.

Kume, S., Yamamoto, A., Inoue, T., Muto, A., and Mikoshiba, K. (1997). Developmental expression of the inositol 1,4,5-trisphosphate receptor. *Dev. Biol.* **182,** 228–239.

Kuo, R. C., Baxter, G. T., Thompson, S. H., Stricker, S. A., Patton, C., Bonaventura, J., and Epel, D. (2000). NO is necessary and sufficient for egg activation at fertilization. *Nature* **406,** 633–636.

Kyozuka, K., Deguchi, R., Mohri, T., and Miyazaki, S (1998). Injection of sperm extract mimics spatiotemporal dynamics of Ca^{2+} responses and progression of meiosis at fertilization of ascidian oocytes. *Development* **125,** 4099–4105.

Lawrence, Y., Whitaker, M., and Swann, K. (1997). Sperm–egg fusion is the prelude to the initial Ca^{2+} increase at fertilization in the mouse. *Development* **124,** 223–241.

Lee, H. C. (1997). Mechanisms of calcium signaling by cyclic ADP-ribose and NAADP. *Physiol. Rev.* **77,** 1133–1164.

Lee, K. W., Webb, S. E., and Miller, A. L. (1999). A wave of free cytosolic calcium traverses zebrafish eggs on activation. *Dev. Biol.* **214,** 68–80.

Lee, S. C., Fissore, R. A., and Niccitelli, R. (2001). Sperm factor initiated capacitated and conductance changes in mouse eggs that are more similar to fertilization than IP_3- or Ca^{2+}-induced changes. *Dev. Biol.* **232,** 127–148.

Lee, S. J., Christenson, L., Martin, T., and Shen, S. S. (1996). The cyclic GMP-mediated calcium release pathway in sea urchin eggs is not required for the rise in calcium during fertilization. *Dev. Biol.* **180,** 324–335.

Lindsay, L. L., Hertzler, P. L., and Clark, W. H. (1992). Extracellular Mg^{2-} induces an intracellular Ca^{2+} wave during oocyte activation in the marine shrimp *Sicyonia ingentis. Dev. Biol.* **152,** 94–102.

Lynn, J. W., MCculloh, D. H., and Chambers, E. L. (1988). Voltage clamp studies of fertilization in sea urchin eggs. II. Current patterns in relation to sperm entry, nonentry and activation. *Dev. Biol.* **128,** 305–323.

Machaty, Z., Mayes, M. A., Kovacs, L., Balatti, P. A., Kim, J. H., and Prather, R. S. (1997). Activation of porcine oocytes via exogenously introduced rat muscarinic M1 receptor. *Biol. Reprod.* **57,** 85–91.

McCulloh, D. H., and Chambers, E. L. (1992). Fusion of membranes during fertilization: Increases of sea urchin egg's membrane capacitance and membrane conductance at the site of contact with the sperm. *J. Gen. Physiol.* **99,** 137–175.

McCulloh, D. H., Lynn, J. W., and Chambers, E. L. (1987). Membrane delopalization facilitates sperm entry, large fertilization cone formation, and prolonged current responses in sea urchin oocytes. *Dev. Biol.* **124,** 177–190.

McCulloh, D. H., Rexroad, C. E., and Levitan, H. (1983). Insemination of rabbit eggs is assosciated with slow depolarization and repetitive membrane potentials. *Dev. Biol.* **95,** 372–377.

McDougall, A., and Levasseur, M. (1998). Sperm-triggered calcium oscillations during meiosis in ascidian oocytes first pause, restart, then stop: Correlations with cell cycle kinase activity. *Development* **125,** 4451–4459.

McGuinness, O., Moreton, R. B., Johnson, M. H., and Berridge, M. J. (1996). A direct measurement of increased divalent cation influx in fertilized mouse oocytes. *Development* **122,** 2199–2206.

Mehlmann, L. M., Carpenter, G., Rhee, S. G., and Jaffe, L. A. (1998). SH2 domain-mediated activation of phospholipase C is not required to initiate Ca^{2+} release at fertilization of mouse eggs. *Dev. Biol.* **203,** 221–232.

Meyer, T., and Stryer, L. (1992). Calcium spiking. *Annu. Rev. Biophys. Biophys. Chem.* **20,** 153–174.

Miyazaki, S. (1988). Inositol 1,4,5-trisphosphate-induced calcium release and guanine nucleotide-binding protein-mediated period calcium rises in golden hamster eggs. *J. Cell Biol.* **106,** 345–353.

Miyazaki, S. (1991). Repetitive calcium transients in hamster oocytes *Cell Calcium* **12,** 205–216.

Miyazaki, S., and Igusa, Y. (1981a). Fertilization potential in golden hamster eggs consists of recurring hyperpolarizations. *Nature* **290,** 706–707.

Miyazaki, S., and Igusa, Y. (1981b). Ca-dependent action potential and Ca-induced fertilization potential in golden hamster eggs. *In* "The Mechanism of Gaited Calcium Transport across Biological Membranes" (S. T. Ohnishi and M. Endo, eds.), pp. 305–311. Academic Press, New York.

Miyazaki, S., and Igusa, Y. (1982). Calcium mediated activation of a K current at fertilization of golden hamster eggs. *Proc. Natl. Acad. Sci. U.S.A.* **79,** 931–935.

Miyazaki, S., Hashimoto, N., Yoshimoto, Y., Kishimoto, T., Igusa, Y., and Hiramoto, Y. (1986). Temporal and spatial dynamics of the period incease in intracellular free calcium at fertilization of golden hamster eggs. *Dev. Biol.* **118,** 259–267.

Miyazaki, S., Katayama, Y., and Swann, K. (1990). Synergistic activation by serotonin and GTP analogue and inhibition by phorbol ester of cyclic Ca^{2+} rises in hamster eggs. *J. Physiol.* **426,** 209–227.

Miyazaki, S., Shirakawa, H., Nakada, K., and Honda, Y. (1993). Essential role of the inositol 1,4,5-trisphosphate/Ca^{2+} release channel in Ca^{2+} waves and Ca^{2+} oscillations at fertilization of mammalian eggs. *Dev. Biol.* **158,** 62–78.

Miyazaki, S., Yazuki, M., Nakada, K., Shirakawa, H., Nakanishi, S., Nakade, S., and Mikoshiba, K. (1992). Block of Ca^{2+} wave and Ca^{2+} oscillation by antibody to the inositol 1,4,5-trisphosphate receptor in fertilized hamster eggs. *Science* **257,** 251–255.

Nakano, Y., Shirakawa, H., Mitsuhasho, N., Kuwabara, Y., and Miyazaki, S. (1997). Spatiotemporal dynamics of intracellular calcium in the mouse egg injected with a spermatozoa. *Mol. Hum. Reprod.* **3,** 1087–1093.

Oda, S., Deguchi, R., Mohri, T., Shikano, T., Nakanishi, S., and Miyazaki, S. (1999). Spatiotemporal dynamics of the $[Ca^{2+}]_i$ rise induced by microinjection of sperm extract into mouse eggs: Preferential induction of a Ca^{2+} wave from the cortex mediated by the inositol 1,4,5-trisphosphate receptor. *Dev. Biol.* **209,** 172–185.

Ozil, J. P., and Swann, K. (1995). Stimulation of repetitive calcium transients in mouse eggs. *J. Physiol.* **483**(2), 331–346.

Palermo, G. D., Avrech, O. M., Colombero, L. T., Wu, H., Wolny, Y. M., Fissore, R. A., and Rosenwaks, Z. (1997). Human sperm cytosolic factor triggers Ca^{2+} oscillations and overcomes activation failure of mammalian oocytes. *Mol. Hum. Reprod.* **3,** 367–374.

Parrington, J., Jones, K. T., Lai, F. A., and Swann, K. (1999). The soluble sperm factor that causes Ca^{2+} release from sea-urchin (*Lytechinus pictus*) egg homogenates also triggers Ca^{2+} oscillations after injection into mouse eggs. *Biochem. J.* **341,** 1–4.

Parrington, J., Swann, K., Shevchenko, V. I., Sesay, A. K., and Lai, F. A. (1996). A soluble sperm protein that triggers calcium oscillations in mammalian eggs. *Nature* **379,** 364–368.

Parys, J. D., McPherson, S. M., Mathews, L., Campbell, K. P., and Longo, F. J. (1994). Prescence of inositol 1,4,5-trisphosphate receptor, calreticulin, and calsequestrin in eggs of sea urchins and *Xenopus laevis. Dev. Biol.* **161,** 466–476.

Peres, A. (1986). Resting membrane potential and inward current properties of mouse ovarian oocytes and eggs. *Pflugers Arch.* **407,** 534–540.

Ribbes, H., Plantavid, M., Bennet, J. P., Chap, H., and Douste-Blazy, L. (1987). Phospholipase C from human sperm specific for phosphoinositides. *Biochim. Biophys. Acta* **919,** 245–244.

Rice, A., Parrington, J., Jones, K. T., and Swann, K. (2000). Mammalian sperm contain a Ca^{2} sensitive phospholipase C activity that can generate $InsP_3$ from PIP_2 associated with intracellular organelles. *Dev. Biol.* **227,** 125–135.

Ridgway, E. B., Gilkey, J. C., and Jaffe, L. F. (1977). Free calcium increases explosively in activating medaka eggs. *Proc. Natl. Acad. Sci. U.S.A.* **74,** 623–627.

Roberts, S. K., Gillot, I., and Brownlee, C. (1994). Cytoplasmic calcium and *Fucus* egg activation. *Development* **120,** 155–163.

Runft, L. L., and Jaffe, L. A. (2000). Sperm extract injection into ascidian eggs signals Ca^{2+} release by the same pathway as fertilization. *Development* **127,** 3227–3236.

Runft, L. L., Watras, J., and Jaffe, L. A. (1999). Calcium release at fertilization of *Xenopus* eggs requires type I IP(3) receptors, but not SH2 domain-mediated activation of PLCγ or G(q)-mediated activation of PLCβ. *Dev. Biol.* **214,** 399–411.

Sagata, N. (1996). Meiotic metaphase arrest in animal oocytes: Its implication *Trends Cell Biol.* **6,** 22–28.

Schmidt, T., Paton, C., and Epel, D. (1982). Is there a role for Ca^{2+} influx during fertilization of the sea urchin egg? *Dev. Biol.* **90,** 284–290.

Sette, C., Bevilacqua, A., Bianchini, A., Mangia, F., and Geremia, R. (1997). Parthenogenetic activation of mouse eggs by microinjection of a truncated c-kit tyrosine kinase present in spermatozoa. *Development* **124,** 2267–2274.

Sette. C., Bevilacqua, A., Geremia. R., and Rossi, P. (1998). Involvement of phospholipase Cγ1 in mouse egg activation induced by a truncated form of the C-kit tyrosine kinase present in spermatozoa. *J. Cell Biol.* **142,** 1063–1074.

Shearer, J., De Nadai, C., Emily-Fenouil, F., Gache, C., Whitaker, M., and Ciapa, B. (1999). Role of phospholipase Cγ at fertilization and during mitosis in sea urchin eggs and embryos. *Development* 126, 2273–2284.

Shen, S. S., and Steinhardt, S. A. (1984). Time and voltage windows for reversing the electrical block to fertilization. *Proc. Natl. Acad. Sci. U.S.A.* **81,** 1436–1439.

Shevchenko, V., Hogben, M., Ekong, R., Parrington, J., and Lai, F. A. (1998). The human glucosamine-6-phosphate deaminase gene: cDNA cloning and expression, genomic organisation and chromosomal localisation. *Gene* **216,** 31–8.

Shilling, F. M., Carroll, D. J., Muslin, A. J., Escobedo, J. A., Williams, L. T., and Jaffe, L. A. (1994). Evidence for both tyosine kinase and G-protein coupled pathways leading to starfish egg activation. *Dev. Biol.* **162,** 590–599.

Shilling, F. M., Magie, C. R., and Nucittelli, R. (1998). Voltage-dependent activation of frog eggs by a sperm surface disintegrin peptide. *Dev. Biol.* **202,** 113–124.

Snow, P., Yim, D. L., Leibow, J. D., Saini, S., and Nuccitelli, R. (1996). Fertilization stimulates an increase inositol trisphosphate and inositol phospholipid levels in *Xenopus* eggs. *Dev. Biol.* **180,** 108–118.

Speksnijder, J. E., Corson, D. W., Sardet, C., and Jaffe, L. F. (1989). Free calcium pulses following fertilization in the ascidian egg. *Dev. Biol.* **135,** 182–190.

Steinhardt, R. A., Zucker, R., and Schatten, G. (1977). Intracellular calcium release at fertilization in the sea urchin egg. *Dev. Biol.* **58,** 185–196.

Stephano, J. K., and Gould, M. C. (1997). The intracellular calcium increase at fertilization in *Urechis caupo* oocytes: Actvation without waves. *Dev. Biol.* **191,** 53–68.

Stice., S. L., and Robl, J. M. (1990). Activation of mammalian oocytes by a factor obtained from rabbit sperm. *Mol. Reprod. Dev.* **25,** 272–280.

Stricker, S. A. (1996). Repetitive calcium waves induced by fertilization in the nemertean worm *Cerebratulus lactues. Dev. Biol.* **176,** 243–263.

Stricker, S. A. (1997). Intracellular injections of a soluble sperm factor trigger calcium oscillations and meiotic maturation in unfertilized oocytes of a marine worm. *Dev. Biol.* **186,** 185–201.

Stricker, S. A., Swann, K., Jones, K. T., and Fissore, R. A. (2000). Injection of porcine sperm extracts trigger fertilization-like calcium oscillations in oocytes of a marine worm. *Exp. Cell Res.* **257,** 341–347.

Swann, K. (1990). A cytosolic sperm factor stimulates repetitive calcium increases and mimics fertilization in hamster eggs. *Development* **110,** 1295–1302.

Swann, K. (1992). Different triggers for calcium oscillations in mouse eggs involve a ryanodine-sensitive store. *Biochem. J.* **287,** 79–84.

Swann, K. (1993). The soluble sperm oscillogen hypothesis. *Zygote* **1,** 273–276.

Swann, K. (1994). Ca^{2+} oscillations and sensitization of Ca^{2+} release in unfertilized mouse eggs injected with a sperm factor. *Cell Calcium* **15,** 331–339.

Swann, K., and Ozil, J. P. (1994). Dynamics of the calcium signal that triggers mammalian egg activation. *Int. Rev. Cytol.* **152,** 183–222.

Swann, K., and Whitaker, M. J. (1986). The part played by inositol trisphosphate and calcium in the propagation of the fertilization wave in sea urchin eggs. *J. Cell Biol.* **103,** 2333–2342.

Swann, K., Igusa, I., and Miyazaki, S. (1989). Evidence for an inhibitory effect of protein kinase C on G-protein-mediated repetitve calcium transients in hamster eggs. *EMBO J.* **8,** 3711–3718.

Tosti, E., Palumbo, A., and Dale, B. (1993). Inositol tri-phosphate in human and ascidian spermatozoa. *Mol. Reprod. Dev.* **35,** 52–56.

Turner, P. R., Jaffe, L. A., and Fein, A. (1985). Regulation of cortical granule exocytosis in sea urchin eggs by inositol 1,4,5-trisphophate and GTP-binding protein. *J. Cell Biol.* **102,** 70–76.

Turner, P. R., Scheetz, M. P., and Jaffe, L. A. (1984). Fertilization increases the polyphosphoinositide turnover of sea urchin eggs. *Nature* **310,** 414–415.

Vanha-Perttula, T., and Kasurinen, J. (1989). Purification and characterization of phosphatidylinositol-specific phospholipase C from bovine spermatozoa. *Int. J. Biochem.* **21,** 997–1007.

Wasserman, P., Jovine, L., and Litscher, E. S. (2001). A profile of fertilization in mammals. *Nature Cell Biol.* **3,** 359–E64.

Whitaker, M., and Irvine, R. F. (1984). Insoitol 1,4,5-trisphophate microinjection activates sea urchin eggs. *Nature* **312,** 636–639.

Whitaker, M., and Patel, R. (1990). Calcium and cell cycle control. *Development* **108,** 525–542.

Whitaker, M. J., and Steinhardt, R. A. (1982). Ionic regulation of egg activation. *Q. Rev. Biophys.* **15,** 593–666.

Whitaker, M., and Swann, K. (1993). Lighting the fuse at fertilization. *Development* **117,** 1–12.

Whittingham, D. G. (1980). Parthenogenesis in mammals. *In* "Oxford Reviews of Reproductive Biology," pp. 205–231. Oxford University Press, New York.

Williams, C. J., Mehlmann, L. M., Jaffe, L. A., Kopf, G. S., and Schultz, R. M. (1998). Evidence that G_q family G proteins do not function in mouse egg activation at fertilization. *Dev. Biol.* **198,** 116–127.

Williams, C. J., Schultz, R. M., and Kopf, G. S. (1992). Role of G-proteins in mouse egg activation: Stimulatory effects of acetylcholine on ZP2 to ZP2f conversion and pronuclear formation in eggs expressing a functional m1 muscarinic receptor. *Dev. Biol.* **151,** 288–296.

Wolosker, H., Kline, D., Bian, Y., Blackshaw, S., Cameron, A. M., Fralich, T. J., Schnaar, R. L., and Snyder, S. H. (1998). Molecularly cloned mammalian glucosamine-6-phosphate deaminase localizes to transporting epithelium and lacks oscillin activity. *FASEB J.* **12,** 91–99.

Wu, H., He, C. L., and Fissore, R. A. (1997). Injection of a porcine sperm factor triggers calcium oscillations in mouse oocytes and bovine eggs. *Mol. Reprod. Dev.* **46,** 176–189.

Wu, H., He, C.-L., Jehn, B., Black, S. J., and Fissore, R. A. (1998). Partial characterization of the calcium-releasing activity of porcine sperm cytosolic extracts. *Dev. Biol.* **203,** 369–381.

Wu, H., Smyth, J., Luzzi, V., Kukami, K., Takenawa, T., Black, S. L., Allbritton, N. L., and Fissore, R. A. (2001). Sperm factor induces intracellular free calcium oscillations by stimulating the phosphoinositide pathway. *Biol. Reprod.* **64,** 1338–1349.

Yanagimachi, R. (1994). Mammalian fertilization. *In* "The Physiology of Reproduction" (E. Knobil and D. Niell, eds.), 2nd Ed., pp. 189–317. Raven Press, New York.

Yanagimachi, R. (1998). Intracytoplasmic sperm injection experiments using the mouse as a model. *Hum. Reprod.* (Suppl.) **1,** 87–98.

Yim, D. L., Lee, K., Opresko, H., Wiley, S., and Nuccitelli, R. (1994). Highly polarized EGF receptor tysoine kinase activity initiates egg activation in *Xenopus. Dev. Biol.* **164,** 41–55.

Yue, C., White, K. L., Reed, W. A., and Bunch, T. D. (1995). The existence of inositol trisphosphate and rynodine receptors in mature bovine oocytes. *Development* **121,** 2645–2654.

SECTION

II

Unique Problems and Applications

11

Molecular Genetics of Fertilization

Patricia Olds-Clarke and Stephen H. Pilder

Department of Anatomy and Cell Biology, Temple University School of Medicine, Philadelphia, Pennsylvania

I. Introduction

This review describes unique gamete characteristics that determine which approaches can be used to study them, and discusses systems that are particularly good for the genetic analysis of fertilization. In the past, genetic approaches to a molecular understanding of fertilization have often been overlooked. However, advances in genome sequencing, linkage maps, and targeted gene disruption methods have opened many more species to serious genetic dissection (Miklos and Rubin, 1996). This chapter is written with newcomers to the field in mind, but should provide some food for thought for the oldtimers as well. It is not comprehensive, because it does not include multicellular plants. There is also a bias toward mammalian gametes, because of advances in targeted mutagenesis and the authors' special interests.

II. GAMETES HAVE UNUSUAL CHARACTERISTICS

Before studying gametes, it is important to understand how they differ from somatic cells. These differences have implications for their manipulation as well as for the mechanisms gametes employ to carry out their functions. Thus, specific characteristics of gametes dictate the approaches that can be used to study them. Although some classical approaches are not practical with gametes, it is possible to take advantage of the unusual features of these cells to develop unique approaches to their study. Some examples of this are discussed below.

A. OOCYTES

1. Translation, but Not Transcription, Occurs in Mature Oocytes

At the time of fertilization, the oocytes of almost all vertebrates are arrested in some stage of meiosis. The relationship of meiosis to fertilization in invertebrates' oocytes, however, is much more variable, with oocytes of various species being fertilized in the germinal vesicle stage, during meiosis, of after completion of meiosis. For those species that have been studied, as a rule there is no transcription in the mature oocyte after germinal vesicle breakdown. However, oocytes have a very robust translational system that has been used to determine whether specific signal transduction pathways are functional. A mRNA specifying a receptor of interest can be injected into oocytes. After the oocytes are given time to synthesize the protein and incorporate it into the oolemma, they can be challenged with the appropriate ligand, and observed for signs of oocyte activation (e.g., exocytosis, Ca release, resumption of the cell cycle). This approach has been used in oocytes of many species (see Chapter 10, this volume).

2. Other Specializations

An advantage of working with oocytes is their size, ranging from large to gigantic, relative to somatic cells. Their size facilitates electrophysiological studies, as well as injection of mRNAs, antibodies, inhibitors, and ion indicators (see Chapter 10).

Oocytes of all species are surrounded by an extracellular matrix (Correa and Carroll, 1997; Hedrick, 1996). Various components of this matrix function to activate sperm motility (see Chapter 2), to induce an acrosome reaction (see Chapter 6), or to prevent polyspermy (see Chapter 4). In mammals, the extracellular matrix is also important for transport to the uterus and later development (Rankin *et al.,* 2001). The composition of the matrix is not similar to that of somatic cells or basement membranes; instead, it is composed of a small number of glycoproteins. Homology is clearly evident among glycoproteins of the extracellular matrix of vertebrate oocytes (see Chapter 4).

Mammalian oocytes have unusual mitochondria. Unlike most somatic cells, oocyte mitochondria have an electron-dense matrix and peripherally located cristae (Stern *et al.*, 1971). Also, oocytes at the time of fertilization cannot utilize glucose directly as an energy source; only pyruvate or oxaloacetate can support resumption of meiosis (Biggers *et al.*, 1967). *In vivo,* cumulus cells metabolize glucose and provide pyruvate to the oocyte mitochondria (Donahue and Stern, 1968). These specialized mitochondria make it imperative to include pyruvate (but not >3 m*M* lactate) in any fertilization medium used with cumulus-free oocytes.

B. SPERMATOZOA

1. Neither Transcription nor Translation Occurs in Sperm Nuclei

The mature sperm nucleus is highly condensed and inactive in transcription, but there may be some mitochondrial transcription (Bartoov *et al.*, 1981). Furthermore, because the polysomes and rough endoplasmic reticulum are shed in the residual body, mature spermatozoa have no means of translating any cytoplasmic messages that may remain. However, mitochondrial RNA transcripts are present (Alcivar *et al.*, 1989), and there may be some protein synthesis in the mitochondrial sheath (Bragg and Handel, 1979).

2. Other Specializations

Like mammalian oocytes, mitochondria in mammalian spermatozoa have an unusual morphology and molecular composition. Their matrices are highly condensed (DeMartino *et al.*, 1979) and they have a keratinous capsule containing phospholipid hydroperoxide glutathione peroxidase, which may act as both a structural protein and a deterrent to oxidative damage (Ursini *et al.*, 1999). Invertebrate sperm mitochondria also have an unusual composition (Mann and Lutwak-Mann, 1981). In mammalian fertilization, sperm mitochondria are incorporated into the oocyte, yet only maternal mitochondria are inherited by the zygote. Specific mitochondrial proteins are ubiquinated in the spermatid. Sperm mitochondria are destroyed in the oocyte by ubiquitin-stimulated proteolysis (Sutovsky *et al.*, 2000; Shitara *et al.*, 2000). Diffusion of ATP from mitochondria to the distal end of the sperm tail is a major problem. Many invertebrate spermatozoa have solved this problem with a phosphocreatine shuttle (Tombes and Shapiro, 1989). However, mammalian spermatozoa have apparently solved this problem in another way. In these cells, mitochondria are probably not the source of much of the ATP that drives motility, because they have a glycolytic pathway tethered within the principal piece of the flagellum (Travis *et al.*, 2001).

Flagellar movement can be studied by partially demembranating spermatozoa by treatment with Triton X-100 and reactivating them by incubation with Mg-ATP (Moritz *et al.*, 2001). This procedure allows the introduction of antibodies and other reagents that are not membrane permeable, and thus facilitates studies of the molecular mechanism of axonemal function. A drawback to using this approach with mammalian sperm cells is that there is as yet no automated method for measuring flagellar characteristics, such as bend amplitude and beat frequency. Instead, flagellar characteristics have to be inferred from head movement parameters (e.g., Si and Olds-Clarke, 2000).

Spermatozoa have little if any cytosol, making microinjection and patch-clamping very difficult. Study of sperm ion channels is critical, because ion fluxes regulate many sperm functions such as motility, chemotaxis, and the acrosome reaction (See Chapter 2). However, some alternative strategies have been developed, such as swollen sea urchin spermatozoa (Babcock *et al.,* 1992) and immature mammalian germ cells (O'Toole *et al.,* 2000) (see Chapter 7).

Is sperm morphology important in fertilization? Although abnormal morphology can serve as a sign of abnormalities at the molecular level, there are several indications that morphological abnormalities in and of themselves might not be detrimental to fertilization. For example, most rodent spermatozoa have a large asymmetrical "hook," which has led to speculation about the potential function of this hook in penetrating the oocyte investments. In two closely related species of vesper mice, one produces spermatozoa that are hooked, the other produces spatulate-shaped spermatozoa. Yet spermatozoa of one species can penetrate the zonae pellucidae of the other, suggesting that the shape of the sperm head may have no obvious function in penetration of this extracellular matrix (Roldan *et al.,* 1985). Many studies of abnormally shaped spermatozoa from house mice have claimed that morphological abnormalities were responsible for the inability of the cells to penetrate the oocyte investments or to complete fertilization, yet these studies have not always included assays of motility, capacitation, or other functions that could have been the underlying cause of the inability to fertilize. For example, spermatozoa from *azh/azh* mice are all abnormally shaped, and had difficulty penetrating the cumulus oophorus and zona pellucida. However, the reduced zona penetration could have been due to the significantly fewer motile sperm cells and their reduced swimming speeds (Meistrich *et al.,* 1994), because good motility is essential for penetration of the ocyte investments (Olds-Clarke, 1996). Spermatozoa from *azh/azh* mice, however, were able to penetrate the zona-free oocyte (Meistrich *et al.,* 1994), a step in fertilization that does not require motility. Furthermore, when grossly abnormal spermatozoa from mice of the BALB/c strain were selected for intracellular sperm injection, some were successful in fertilization and supported normal development (Burruel *et al.,* 1996). These examples suggest that an abnormally shaped sperm head per se is not always associated with defects in the ability of that spermatozoan to complete fertilization.

III. GENETIC MODEL SYSTEMS

Any species to be used as a genetic model system should have a reasonable number of well-spaced genetic markers that can be used to locate a gene of interest. Natural mutations provide useful information about the function of the gene product, and can lead via positional cloning to identification of previously unknown genes. Induced mutations can be useful in sorting out the essential functions of a gene from those that are ancillary. Sometimes temperature-sensitive mutations are available, so that each mutant can be its own control. Each of the models discussed below has characteristics that allow the use of approaches impossible with other species. Species that are more difficult to study, e.g., mammals, have benefited from the study of homologous gene products active in gametes from other vertebrate and invertebrate phyla.

How are mutations causing sterility propagated? For species requiring two sexes, if a recessive mutation causes infertility in both sexes, it is necessary to propagate the mutation by mating heterozygotes. Because most mutations affecting fertility are recessive, one-fourth of the offspring will be homozygous for the mutation and thus will display the infertile phenotype. If the mutation alters fertility in only one sex, then homozygotes of the opposite sex can be mated to heterozygotes of the affected sex. This will result in one-fourth of the offspring being homozygous and of the appropriate sex to express the mutation. Some species may be hermaphroditic, others may propagate by parthenogenesis. If so, this is a bonus that can be exploited in propagating mutations. Mutations that do not cause sterility but alter gamete function can also be useful if the phenotype is consistent among affected animals, and characteristic of virtually all spermatozoa in the population.

Whenever possible, mutant animals should be compared to wild-type animals that are genetically identical to the mutants except at the locus of interest. Otherwise, phenotypic differences between wild type and mutant could be due at least in part to genetic differences at loci other than the one under study. In other words, a mutation can result in different phenotypes when it is placed on different genetic backgrounds (Miklos and Rubin, 1996; Redkar *et al.*, 2000). If it is not possible to compare mutant and wild-type controls in a coisogenic background (i.e., where all individuals are genetically identical), it is best to compare large numbers of mutants to sibling controls or some other group that shares as much as possible of the same genetic background.

A. YEAST

Although yeast species are undoubtedly the best understood eukaryotic genetic systems, yeast mutants have not as yet been very helpful in understanding processes important for fertilization in multicellular animal systems, because most mutations that do not allow fertilization are also involved in other functions important for cell metabolism and growth (Rose, 1996; Ng and Walter, 1996). However, a mutation specifically affecting mating cell fusion has been identified (Heiman and Walter, 2000). Thus yeast may yet become a genetic model system for plasma membrane fusion and nuclear fusion.

B. *CHLAMYDOMONAS*

Chlamydomonas is a single-cell alga that moves by means of two flagella. It is the model system for axonemal-based motility, and also a genetic system for mechanisms important for gamete binding and fusion. Its advantages as a genetic model are manifold: a haploid/diploid life cycle, numerous genetic markers, simple screens for mutations, tetrad analysis, and complementation tests. In addition, insertional mutagenesis can be performed with *Chlamydomonas* genomes (Smith and Lefebvre, 1996). This makes this system a very powerful tool for genetic studies of motility and cell–cell interactions.

Chlamydomonas flagella are the *sine qua non* of our understanding of the mechanism of axonemal function. Over 200 polypeptides have been isolated from *Chlamydomonas* flagella; most have been associated with a specific axonemal structure, and more than 80 are phosphorylated (Dutcher, 1995; King, 2000; Yang and Sale, 2000). Because of the existence of many mutations affecting flagellar motility, the function of many of these proteins is known or suspected. *Chlamydomonas* mutants have been instrumental in determining the functions of inner and outer dynein arms (Witman, 1992), and understanding the outer arm dynein complex (Wakabayashi *et al.,* 2001). Furthermore, many genes specifying proteins of mammalian sperm axonemes are homologous to those of *Chlamydomonas,* so that information from *Chlamydomonas* has been an invaluable help in uncovering mechanisms important in mammalian axonemes. For example, mutations in the mouse genome that cause defective sperm motility are within a locus containing genes homologous to members of the axonemal dynein complex in *Chlamydomonas* (see Section III,F,1). *Chlamydomonas* axonemal genes that produce relevant mutant phentypes have also been used to isolate and map homologous counterparts in the human genome (Kastury *et al.,* 1997; Pennarun *et al.,* 1999; Neilson *et al.,* 1999).

Chlamydomonas is also a good model for the cell–cell adhesion and fusion that occurs in fertilization. Adhesion of the flagella of opposite mating types occurs via a cAMP-mediated signal transduction pathway (Pan and Snell, 2000). This stimulates cell wall lysis and formation of discrete "mating structures" in the region between the bases of the two flagella: a 3-μm-long, microvillus-like fertilization tubule (containing actin) on the plus cell and a dome-shaped mating structure on the minus cell. Binding and subsequent fusion of the two cells first occur between the membranes at the tip of the fertilization tubule and the top of the domed mating structure [see references in Wilson *et al.* (1997)]. A method has been developed to isolate functional fertilization tubules (Wilson *et al.,* 1997). This approach, together with the use of mutants in genes that alter development of the fertilization tubule (Ohara *et al.,* 1998), will make analysis of this fertilization system much easier.

C. *CAENORHABDITIS ELEGANS*

The nematode worm *Caenorhabditis elegans* is a useful model of a simple multicellular eukaryotic system. This species has a relatively small genome (100 Mb) that has been completely sequenced and is highly amenable to genetic manipulation (Stein *et al.,* 2001). *Caenorhabditis elegans* is normally a self-fertilizing hermaphrodite, but a male sex is also present. As hermaphroditic worms mature, they first produce spermatozoa, then oocytes. The male gametes are unusual in that they lack a conventional acrosome and are nonflagellated. Instead, these cells use something akin to ameboid movement (Roberts and Stewart, 2000). The oocyte has no obvious cellular or acellular investments, but a "shell" is formed around the oocyte after fertilization. Oocytes are fertilized as they pass through the spermatheca. Every spermatozoan will fertilize an oocyte. Studies of mutations that inhibit fertilization have revealed several genes whose products affect sperm–oocyte interactions and egg activation (Singson, 2001). Another mutation produces spermatozoa with no nuclei. Such cells can crawl, penetrate and activate oocytes, and direct polarization of the zygote, demonstrating that the nucleus is not involved in these activities (Sadler and Shakes, 2000).

D. *DROSOPHILA*

The fruit fly is the epitome of the multicellular genetic models. This system has many well-known advantages, including a relatively small genome that has been sequenced (Adams *et al.,* 2000), polytene chromosomes to facilitate coordination of physical with genetic mapping, and a short generation time. It has contributed extensively to an understanding of developmental mechanism, but it has not been much utilized for studies of fertilization.

An interesting feature of *Drosophila* spermatozoa is their long tail, which is about the same length (1.8 mm) as the fly that produces them. This is true of most insect species, with the record length being 17 mm (Karr, 1996). The entire spermatozoan enters the egg (about 500 μm long). Once in the ooplasm, the tail coils into a stereotypical structure and persists throughout much of embryogenesis (Pitnik and Karr, 1998). The purpose of this enormous tail is not known, but it could act as a nutritive source for the egg or provide a substance important in egg activation or early development. Although fruit flies have been used extensively to study sperm competition (Prout and Clark, 2000; Civetta and Clark, 2000), the cellular and molecular mechanisms used appear to be unique to insects. Furthermore, an ultrastructural study suggested that sperm penetration does not involve sperm–egg membrane fusion (Perotti, 1975). Furthermore, relatively few gametes are produced, fertilization is internal, and an *in vitro* fertilization system has not yet been developed. Thus, it is not clear that *Drosophila* fertilization is a useful model system for mammalian systems.

E. ZEBRAFISH (*DANIO RERIO*)

Few vertebrate models are as yet suitable for genetic analysis, other than the mouse (see Section III,F). However, the zebrafish has become very popular for vertebrate developmental studies, because it allows simultaneous application of classical experimental embryological techniques and extensive genetic analysis (Haffter *et al.,* 1996). Improvements in the genetic linkage map have made positional cloning technically feasible (Woods *et al.,* 2000), and the complete genome sequence should be available in the next several years.

Cellular and biochemical studies are feasible because gametes are produced in relatively large numbers, the eggs are optically clear, and they can be fertilized *in vitro*. Furthermore, it is possible to create viable diploid parthenogenetic or androgenetic fish (Sakai *et al.,* 1997). The gametes have many characteristics in common with other vertebrates, including an egg envelope protein with homology to human zona pellucida proteins (Mold *et al.,* 2001), an oocyte calcium activation wave (Lee *et al.,* 1999), and activation of a oocyte Fyn protein kinase in response to fertilization (Wu and Kinsey, 2000). Although a chemical mutagenesis screen found 11 gonadal mutations (Bauer and Goetz, 2001), as yet no screen for mutations affecting fertilization has been performed. Because of the extensive genetic infrastructure now available for zebrafish, this system appears to be fertile ground for a genetic approach to understanding vertebrate fertilization.

F. MOUSE

This species is the genetic standard for mammalian organisms. There are many reasons for this: extensive information available about mouse genetics (see the Mouse Genome Informatics web page, www.informatics.jax.org/), development of techniques for targeted mutagenesis, broad homology to the human genome, and availability of inbred strains. Mice of the same inbred strain are genetically identical and homozyous at all loci, i.e., clones. This is particularly important for assessing gamete functions when the effects of a mutation are quantitative in nature (e.g., reduced efficiency of fertilization, fewer acrosome reactions, decreased cAMP levels).

1. Natural Mutations

Few natural mutaitons that primarily interfere with steps in fertilization have been identified, but several mutations altering sperm function in fertilization have been mapped within the *t* haplotype. The *t* haplotype is a variant form of the proximal third of chromosome 17. Separate genes within this region interact to alter sperm function in fertilization without greatly affecting sperm production. Genes in different parts of a *t* haplotype remain linked from generation to generation because of the presence of four large inversions (Figure 11.1), relative to the homologous wild-type region (referred to as the *t* complex). Within the most proximal, *In(17)1,* the second to most distal *In(17)3,* and the most distal inversion, *In(17)4,* are interacting genetic factors causing or contributing to male sterility (Lyon, 1986; Olds-Clarke, 1997). Males carrying two *t* haplotypes produce spermatozoa that are morphologically normal, but have very poor motility and are unable to penetrate either zona pellucida-intact or zona pellucida-free oocytes (Redkar *et al.,* 2000). Spermatozoa from males heterozygous for a *t* haplotype are fertile, but have observable motility defects and are retarded in their penetration of the zona-free oocyte (Olds-Clarke and Johnson, 1993; Johnson *et al.,* 1995).

The only gene of interest that resides between two of the inversions (which encodes the *t* complex responder known as *Smok*) has been cloned and partially characterized (Herrmann *et al.,* 1999). In the past it was difficult to localize to high

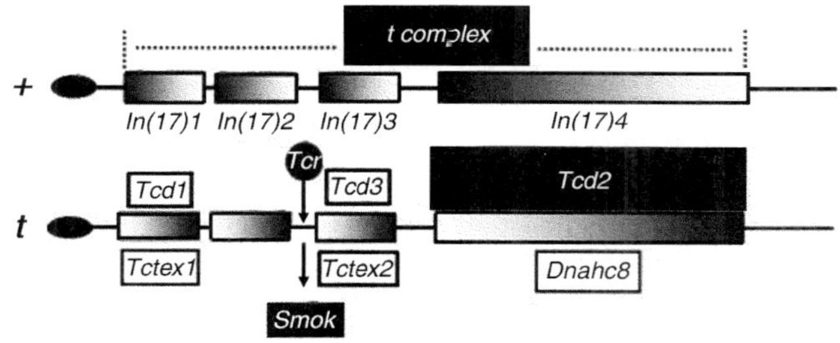

FIGURE 11.1 The wild-type (+) and *t* haplotype (*t*) homologs of mouse chromosome 17. Differentially shaded boxes represent relative inversions, *In(17)1–4*. Tcd1, Tcd2, and Tcd3 represent members of one class of factor, known as distorters, that contribute to *t* haplotype sperm function defects, with Tcd2 having the largest effect. *Tcr* (the *t* complex responder) represents a single gene not mapping to an inversion. *Tcr* is thought to interact with the distorters to produce male nonmendelian inheritance of chromosome 17 in +/*t* heterozygotes. *Tcr* encodes a kinase named *Smok* (sperm motility kinase), with unproved function. Candidates for the three distorter factors, Tctex1, Tctex2, and Dnahc8, are all components of testis-expressed axonemal dynein complexes.

resolution a majority of the genes of interest that were in or near the large inversions. However, the finding that male laboratory mice heterozygous for a *t* haplotype and a *Mus spretus* (a species of mouse distantly related to the lab mouse, *Mus musculus*) chromosome 17 homolog are sterile (Pilder *et al.,* 1991) has allowed investigators to map the genes responsible for this unique sterility phenotype to a high resolution by producing a series of *M. musculus* wild-type and *M. spretus* wild-type recombinant chromosome 17 homologs. Male lab mice heterozygous for a *t* haplotype and various *M. musculus–M. spretus* recombinant homologs have been tested for sterility and effects on sperm functions. Using this approach, several loci affecting sperm motility have been mapped within *In(17)1* (Pilder, 1997; Pilder *et al.,* 1997) and within *In(17)4* (Pilder *et al.,* 1991, 1993). A locus affecting capacitation and sperm penetration of the zona-free oocyte has been mapped within *In(17)4* (Redkar *et al.,* 1998, 2000). More importantly, a candidate gene for the *t* haplotype male sterility-diagnostic sperm motility phenotype, "curlicue," has been localized centrally within *In(17)4* (Samant *et al.,* 1999; Fossella *et al.,* 2000). This candidate is an axonemal dynein heavy chain gene known as *Dnahc8* (Vaughan *et al.,* 1996), which by virtue of alternative splicing, encodes two radically different isoforms with strikingly different functions. Moreover, the *t* and wild-type alleles are substantially different (Pilder and Samant, 2001).

Random cloning experiments have identified other testis-expressed genes from the *t* complex region with potential effects on spermatozoa/sperm function [see discussion in Olds-Clarke (1997)]. One interesting gene mapping to *In(17)1* is *Tctex1*. It codes for a protein highly similar to an inner dynein arm light chain of *Chlamydomonas reinhardtii* (Harrison *et al.,* 1998). The wild-type and *t* haplotype alleles of *Tctex1* have been sequenced, and the *t* haplotype allele contains a mutation that could affect dynein function and sperm motility (O'Neill and Artzt, 1995). Additionally, the gene for an outer dynein arm light chain, *Tctex2,* whose *t* allele differs from the wild-type allele, has been mapped to *In(17)3,* the apparent site of a factor that makes a weak contribution to *t* haplotype homozygous male sterility (Lyon, 1986). In view of the fact that the three regions of the *t* haplotype that contribute to male sterility carry genes that encode altered dynein components, it is tempting to speculate that defective axonemal dynein complexes are the primary cause of *t* haplotype-specific male sterility.

Another approach to identifying, within the *t* haplotype, genes that affect sperm function in fertilization is the creation in embryonic stem cells of a series of overlapping deletions in the region of *In(17)1* (Planchart *et al.,* 2000). This provides an important new tool for the analysis of loss-of-function phenotypes in this region. Although the *t* haplotype has been a difficult system in which to work, it is well worth the effort to isolate and identify within the *t* haplotype the many genes that alter sperm motility or egg penetration.

2. Targeted Mutagenesis

A great advantage of the mouse as a genetic system is the ability to perform targeted mutagenesis (knockouts). Because many loss-of-function mutants with some effect on reproduction have been generated, only knockouts of genes whose products are known or suspected to have a function in mature gametes will be considered. As more and more genes are specifically targeted for mutagenesis, it is becoming clear that most genes do not have a unique function, nor do they function independently. Thus, targeted mutagenesis does not always confirm the gene's suggested function.

The creation of c-*mos* knockout mice confirmed the hypothesis that the Mos protein kinase (encoded by the c-*mos* protooncogene) is an essential component of the metaphase II arrest in vertebrate ocytes. These mice produced oocytes that progressed through MII without arrest, subsequently undergoing parthenogenetic activation (Sagata, 1996). Oocytes from these mice were used to understand the relationship of the Mos protein kinase to mitogen-activated protein kinase (MAPK) (Araki *et al.,* 1996; Choi *et al.,* 1996). Another study suggests that Mos activates MAPK in maturing oocytes by two pathways (Verlhac *et al.,* 2000).

The mouse zona pellucida is composed of three glycoproteins. Targeted disruption of the gene *mZP3,* which encodes the (glyco)protein to which the spermatozoan binds and which triggers the acrosome reaction (McLeskey *et al.,* 1998), causes female infertility. The oocytes of these females lack a zona pellucida (Rankin *et al.,* 1996; Liu *et al.,* 1996), demonstrating that ZP3 is essential for assembly of the zona pellucida matrix, and that a zona is required for fertilization *in vivo.* *Zp2*-null mice are also sterile, producing only a few mature oocytes, all without a zona pellucida (Rankin *et al.,* 2001). Targeted mutagenesis of genes specifying zona pellucida proteins demonstrate that a normal zona pellucida is also required for transport through the oviduct (see Chapter 4).

Targeted mutagenesis can reveal unexpected interactions at the protein level.

Fertilin-β is a transmembrane sperm protein containing disintegrin and metalloprotease domains. Antibodies against fertilin-β inhibited sperm–oolemma adhesion and fusion (see Chapter 9). Spermatozoa from sterile mice lacking fertilin-β were deficient in oolemma adhesion and fusion, as well as transport to the oviduct and adhesion to the zona pellucida (Cho *et al.,* 1998). Because fertilin-β has a disintegrin domain, it was not surprising that the integrin α6 (as a complex with β1) bound specifically to fertilin-β (Almeida, 1995). However, oocytes of mice homozygous for a null mutation of α6 were fully functional in sperm–oocyte adhesion and fusion (Miller *et al.,* 2000). Nevertheless, gamete fusion did not occur with oocytes from mice lacking CD9, a membrane protein functioning in somatic cell adhesion and often associated with a subset of β1 integrins (Miyado *et al.,* 2000; Kaji *et al.,* 2000). These latter studies indicate that for sperm–egg fusion, the disintegrin–integrin interaction involves other factors.

Acrosin, an endoprotease with a trypsinlike cleavage specificity, is thought to be the major protease involved in digesting a path for the sperm cell through the zona pellucida (see Chapter 8). Mice homozygous for a targeted mutation in the acrosin gene were fertile (Baba *et al.,* 1994; Adham *et al.,* 1997), suggesting that acrosin is not essential for fertilization. However, spermatozoa from these males penetrated the zona at a rate slower than controls (Baba *et al.,* 1994), implying that acrosin could contribute to zona adhesion/penetration. The mouse is unusual among mammals in that acrosin is only a minor component of the acrosome. Thus it is likely that in the mouse, other sperm proteases also are involved in zona penetration (Ohmura *et al.,* 1999) (see Chapter 8).

The discovery of β1,4-galactosyltransferase, a protein that has long been thought to play an important role in adhesion to the zona pellucida, was entirely fortuitous (see Chapters 4 and 5). In a study of various enzyme activities in spermatozoa from sterile mice homozygous for *t* haplotypes (see Section III,F,1), galactosyltransferase activity was two to four times that of controls (Shur, 1981), suggesting that activity of this enzyme could be important in fertilization. Since then Shur's group has studied β1,4-galactosyltransferase exclusively (Lu and Shur, 1997). Because the β1,4-galactosyltransferase structural gene is located on chromosome 4 (Shaper *et al.,* 1990), the elevated galactosyltransferase activity of *t* haplotype spermatozoa cannot be due to mutations in this enzyme, although there could be a gene in the *t* complex that regulates the activity of β1,4-galactosyltransferase in spermatozoa. Another galactosyltransferase, β1,3-galactosyltransferase, is located in the *t* complex but is not expressed in male germ cells (Johnston *et al.,* 1995). Thus, it remains uncertain whether the elevated galactosyltransferase activity in spermatozoa from *t* haplotype mice plays a role in sperm dysfunction (see Sectoin III,F,1). Mice homozygous for a targeted β1,4-galactosyltransferase are fertile (Lu *et al.,* 1997), but their spermatozoa have difficulties in the acrosome reaction and zona pellucida penetration *in vivo* (Lu and Shur, 1997). This is consistent with the hypothesis that β1,4-galactosyltransferase is one of the proteins involved in sperm–zona pellucida interaction (Nixon *et al.,* 2001).

Targeted mutagenesis of the RIIα form of the regulatory subunit of protein kinase A (PKA) is another example of how knockouts that do not confirm the supposed function of the gene can indicate novel protein interactions. RIIα is found within the flagellum, and has long been suspected to be a key player in sperm motility (Vijayaraghavan *et al.,* 1997). However, male mice homozygous for a targeted mutation of the gene encoding the RIIα subunit are fertile and the sperm from these mice have normal motility. In the absence of RIIα, up-regulation of RIα occurs, but PKAI is not anchored to the flagellum. It is possible that unlocalized RIα substitutes satisfactorily for RIIα in PKA activity. However, because these mutants discard most of the PKA in the cytoplasm, it is also possible that PKA tethered to the flagellum is not necessary for the motility of mature spermatozoa (Burton *et al.,* 1999).

Considering how vital fertilization is to the continuation of the species, it is reasonable to assume that there might be significant evolutionary pressure for gene redundancy or multiple unrelated genes with a similar function. However, it is then necessary to ask why some genes, e.g., c-*mos, ZP3,* and *fertilin*-β, do not have a "backup" system, so that their loss results in sterility. Clearly there is much to understand about how genes interact in gametes.

REFERENCES

Adams, M. D., Celniker, S. E., Holt, R. A., Evens, C. A., Gocayne, J. D., Amanatides, P. G., Scherer, S. E., Li, P. W., Hoskins, R. A., Galle, R. F., George, R. A., Lewis, S. E., Richards, S., Ashburner, M., Henderson, S. N., Sutton, G. G., Wortman, J. R., Yandell, M. D., Zhang, Q., Chen, L. X., *et al.* (2000). The genome sequence of *Drosophila melanogaster. Science* **287,** 2185–2196.

Adham, I. M., Nayernia, K., and Enel, W. (1997). Spermatozoa lacking acrosin protein show delayed fertilization. *Mol. Reprod. Dev.* **46,** 370–376.

Alcivar, A. A., Hake, L. E., Millette, C. F., Trasler, J. M., and Hecht, N. B. (1989). Mitochondrial gene expression in male germ cells of the mouse. *Dev. Biol.* **135,** 263–271.

Almeida, E. A. C., Huovila, A.-P. J., Sutherland, A. E., Stephens, L. E., Calarco, P. G., Shaw, L. M., Mercurio, A. M., Sonnenberg, A., Primakoff, P., Myles, D. G., and White, J. M. (1995). Mouse egg integrin α6β1 funcitons as a sperm receptor. *Cell* **81,** 1095–1104.

Araki, K., Naito, K., Haraguchi, S., Suzuki, R., Yokoyama, M., Inoue, M., Aizawa, S., Toyoda, Y., and Sato, E. (1996). Meiotic abnormalities of c-*mos* knockout mouse oocytes: Activation after first meiosis or entrance into third meiotic metaphase. *Biol. Reprod.* **55,** 1315–1324.

Baba, T., Azuma, S., Kashiwabara, S., and Toyoda, Y. (1994). Sperm from mice carrying a targeted mutation of the acrosin gene can penetrate the oocyte pellucida and effect fertilization. *J. Biol. Chem.* **269,** 31845–31849.

Babcock, D. F., Bosma, M. M., Battaglia, D. F., and Darszon, A. (1992). Early persistent activation of sperm K^+ channels by the egg peptide speract. *Proc. Natl. Acad. Sci. U.S.A.* **89,** 6001–6005.

Bartoov, B., Reis, I., and Fisher, J. (1981). Incorporation of [^{3}H]thymidine into mtDNA of ram spermatozoa. *J. Reprod. Fertil.* **61,** 295–301.

Bauer, M. P., and Goetz, F. W. (2001). Isolation of gonadal mutations in adult zerbrafish from a chemical mutagenesis screen. *Biol. Reprod.* **64,** 548–554.

Biggers, J. D., Whittingham, D. G., and Donahue, R. P. (1967). The pattern of energy metabolism in the mouse oocyte and zygote. *Proc. Natl. Acad. Sci. U.S.A.* **58,** 560–567.

Bragg, P. W., and Handel, M. A. (1979). Protein synthesis in mouse spermatozoa. *Biol. Reprod.* **20,** 333–337.

Burruel, V. R., Yanagimachi, R., and Whitten, W. K. (1996). Normal mice develop from oocytes injected with spermatozoa with grossly misshapen heads. *Biol. Reprod.* **55,** 709–714.

Burton, K. A., Treash-Osio, B., Muller, C. H., Dunphy, E. L., and McKnight, G. S. (1999). Deletion of type IIα regulatory subunit delocalizes protein kinase A in mouse sperm without affecting motility or fertilization. *J. Biol. Chem.* **274,** 24131–24136.

Cho, C., O'Dell-Bunch, D., Faure, J.-E., Goulding, E. H., Eddy, E. M., Primakoff, P., and Myles, D. G. (1998). Fertilization defects in sperm from mice lacking fertilinβ. *Science* **281,** 1857–1859.

Choi, T., Fukasawa, K., Zhou, R., Tessarollo, L., Borror, K., Resau, J., and Vande Woude, G. F. (1996). The Mos/mitogen-activated protein kinase (MAPK) pathway regulates the size and degradation of the first polar body in maturing mouse oocytes. *Proc. Natl. Acad. Sci. U.S.A.* **93,** 7032–7035.

Civetta, A., and Clark, A. G. (2000). Correlated effects of sperm competition and postmating female mortality. *Proc. Natl. Acad. Sci. U.S.A.* **97,** 13162–13165.

Correa, L. M., and Carroll, E. J. (1997). Characterization of the vitelline envelope of the sea urchin *Strongylocentrotous purpuratus. Dev. Growth Differ.* **39,** 69–85.

DeMartino, C., Floridi, A., Marcante, M. L., Malorni, W., Scorza, B. P., Bellocci, M., and Silvestrin, B. (1979). Morphological, histochemical, and biochemical studies on germ cell mitochondria of normal rats. *Cell Tissue Res.* **196,** 1–22.

Donahue, R. P., and Stern, S. (1968). Follicular cell support of oocyte maturation: Production of pyruvate *in vitro. J. Reprod. Fertil.* **17,** 395–398.

Dutcher, S. K. (1995). Flagellar assembly in two hundred and fifty easy-to-follow steps. *Trends Genet.* **11,** 398–404.

Fossella, J., Samant, S. A., Silver, L. M., King, S. M., Vaughan, K. T., Olds-Clarke, P., Johnson, K. A., Mikami, A., Vallee, R. B., and Pilder, S. H. (2000). An axonemal dynein at the *Hybrid Sterility 6* locus: Implications for *t* haplotype-specific male sterility and the evolution of species barriers. *Mamm. Genome* **11,** 8–15.

Haffter, P., Granato, M., Brand, M., Mullins, M. C., Hammerschmidt, M., Kane, D. A., Odenthal, J., van Eeden, J. M., Jiang, Y.-J., Heisenberg, C.-P., Kelsh, R. N., Furutani-Seiki, M., Vogelsang, E., Beuchie, D., Schach, U., Fabian, C., and Nüsslein-Volhard, C. (1996). The identification of genes with unique and essential functions in the development of the zebrafish, *Danio rerio. Development* **123,** 1–36.

Harrison, A., Olds-Clarke, P., and King, S. M. (1998). Identification of the *t* complex-encoded cytoplasmic dynein light chain *Tctex1* in inner arm I1 supports the involvement of flagellar dyneins in meiotic drive. *J. Cell Biol.* **140,** 1137–1147.

Hedrick, J. L. (1996). Comparative structural and antigenic properties of zona pellucida glycoproteins. *J. Reprod. Fertil.* (Suppl.) **50,** 9–17.

Heiman, M. G., and Walter, P. (2000). Prm1p, a pheromone-regulated multispanning membrane protein, facilitates plasma membrane fusion during yeast mating. *J. Cell Biol.* **151,** 719–730.

Herrmann, B. G., Koschorz, B., Wertz, K., McLaughlin, J., and Kispert, A. (1999). A protein kinase encoded by the *t complex responder* gene causes non-Mendelian inheritance. *Nature* **402,** 141–146.

Johnson, L. R., Pilder, S. H., Bailey, J. L., and Olds-Clarke, P. (1995). Sperm from mice carrying one or two *t* haplotypes are deficient in investment and oocyte penetration. *Dev. Biol.* **168,** 138–149.

Johnston, D. S., Shaper, J. H., Shaper, N. L., Joziasse, K. H., and Wright, W. W. (1995). The gene encoding the murine α1,3-galactosyltransferase is expressed in female germ cells but not in male germ cells. *Dev. Biol.* **171,** 224–232.

Kaji, K., Oda, S., Shikano, T., Ohnuki, T., Yematsu, Y., Sakagami, J., Tada, N., Miyazaki, S., and Kudo, A. (2000). The gamete fusion process is defective in eggs of CD9-deficient mice. *Nature Genet.* **24,** 279–282.

Karr, T. L. (1996). Paternal investment and intracellular sperm-egg interactions during and following fertilization in *Drosophila. Curr. Top. Dev. Biol.* **34,** 89–115.

Kastury, K., Taylor, W. E., Shen, R., Arver, S., Gutierrez, M., Fisher, C. E., Coucke, P. J., Van Hauwe, P., Van Camp, G., and Bhasin, S. (1997). Complementary DNA cloning and characterization of the human axonemal dynein light chain gene: a candidate gene for the immobile cilia syndrome. *J. Clin. Endocrinol. Metab.* **82,** 3047–3053.

King, S. J. (2000). The dynein microtubule motor. *Biochim. Biophys. Acta* **1496,** 60–75.

Lee, K. W., Webb, S. E., and Miller, A. L. (1999). A wave of free cytosolic calcium traverses zebrafish eggs on activation. *Dev. Biol.* **214,** 168–180.

Liu, C., Litscher, E. S., Mortillo, S., Sakai, Y., Kinloch, R. A., Stewart, C. L., and Wassarman, P. M. (1996). Targeted disruption of the *mZP3* gene results in production of eggs lacking a zona pellucida and infertility in female mice. *Proc. Natl. Acad. Sci. U.S.A.* **93,** 5431–5436.

Lu, Q., and Shur, B. D. (1997). Sperm from β1,4-galactosyltransferase-null mice are refractory to ZP3-induced acrosome reactions and penetrate the zona pellucida poorly. *Development* **124,** 4121–4131.

Lu, Q., Hasty, P., and Shur, B. D. (1997). Targeted muation in β1,4-galactosyltransferase leads to pituitary insufficiency and neonatal lethality. *Dev. Biol.* **181,** 257–267.

Lyon, M. F. (1986). Male sterility of the mouse *t* complex is due to homozygosity of the distorter genes. *Cell* **44,** 357–363.

Mann, T., and Lutwak-Mann, C. (1981). "Male Reproductive Function and Semen." Springer-Verlag, Berlin.

McLeskey, S. B., Dowds, C., Carballada, R., White, R. R., and Saling, P. M. (1998). Molecules involved in mammalian sperm-egg interaction. *Int. Rev. Cytol.* **177,** 57–113.

Meistrich, M. L., Kasai, K., Olds-Clarke, P., MacGregor, G. R., Berkowitz, A. D., and Tung, K. S. K. (1994). Deficiency in fertilization by morphologically abnormal sperm produced by *azh* mutant mice. *Mol. Reprod. Dev.* **37,** 69–77.

Miklos, G. L. G., and Rubin, G. M. (1996). The role of the genome project in determining gene function: Insights from model organisms. *Cell* **86,** 521–529.

Miller, B. J., Georges-Labouesse, E., Primikoff, P., and Myles, D. G. (2000). Normal fertilization occurs with eggs lacking the integrin α6β1 and is CD9-dependent. *J. Cell Biol.* **149,** 1289–1295.

Miyado, K., Yamada, G., Yamada, S., Hasuwa, H., Nakamura, Y., Ruy, F., Suzuki, K., Kosai, K., Inoue, K., Ogura, A., Okabe, M., and Mekada, E. (2000). Requirement of CD9 on the egg plasma membrane for fertilization. *Science* **287,** 321–324.

Mold, D. E., Kim, I. F., Tsai, C. M., Lee, D., Chang, C. Y., and Huang, R. C. C. (2001). Cluster of genes encoding the major egg envelope protein of zebrafish. *Mol. Reprod. Dev.* **58,** 4–14.

Moritz, M. J., Schmitz, K. A., and Lindemann, C. B. (2001). Measurement of the force and torque produced in the calcium response of reactivated rat sperm flagella. *Cell Mot. Cytoskeleton* **49,** 33–40.

Neilson, L. I., Schneider, P. A., Van Deerlin, P. G., Kiriakidou, M., Driscoll, D. A., Pellegrini, M. C., Millinder, S., Yamamoto, K. K., French, C. K., and Strauss, J. F. (1999). cDNA cloning and characterization of a human sperm antigen (SPAG6) with homology to the product of the *Chlamydomonas PF16* locus. *Genomics* **60,** 272–280.

Ng, D. T., and Walter, P. (1996). ER membrane protein complex required for nuclear fusion. *J. Cell Biol.* **132,** 499–509.

Nixon, B., Lu, Q. X., Wassler, M. J., Foote, C. I., Ensslin, M. A., and Shur, B. D. (2001). Galactosyltransferase function during mammalian fertilization. *Cells Tissues Organs* **168,** 46–57.

Ohara, A., Kata-Minoura, T., Kamiya, R., and Hirono, M. (1998). Recovery of the flagellar inner-arm dynein and the fertilization tubule in *Chlamydomonas ida5* mutant by transformation with actin genes. *Cell Struct. Funct.* **23,** 273–281.

Ohmura, K., Kohno, N., Kobayashi, Y., Yamagata, K., Sato, S., Kashiwabara, S., and Baba, T. (1999). A homologue of pancreatic trypsin is localized in the acrosome of mammalian sperm and is released during acrosome reaction. *J. Biol. Chem.* **274,** 29426–29432.

Olds-Clarke, P. (1996). How does poor motility alter sperm fertilizing ability? *J. Androl.* **17,** 183–186.

Olds-Clarke, P. (1997). Models for male infertility: The *t* haplotypes. *Rev. Reprod.* **2,** 157–164.

Olds-Clarke, P., and Johnson, L. (1993). *t* Haplotypes in the mouse compromise sperm flagellar function. *Dev. Biol.* **155,** 14–25.

O'Neill, M. J., and Artzt, K. (1995). Identification of a germ-cell-specific transcriptional repressor in the promoter of *Tctex-1*. *Development* **121,** 561–568.

O'Toole, C. M. B., Arnoult, C., Darszon, A., Steinhardt, R., and Florman, H. M. (2000). Ca^{2+} entry through store-operated channels in mouse sperm is initiated by egg CP3 and drives the acrosome reaction. *Mol. Biol. Cell* **11,** 1571–1584.

Pan, J. M., and Snell, W. J. (2000). Signal transduction during fertilization in the unicellular green alga, *Chlamydomonas. Curr. Opin. Microbiol.* **3,** 596–602.

Pennarun, G., Escudier, E., Chapelin, C., Bridoux, A.-M., Cacheux, V., Roger, G., Clément, A., Goossens, M., Amselem, S., and Duriez, B. (1999). Loss-of-function mutations in aa human gene related to *Chlamydomonas reinhardtii* Dynein IC78 result in primar ciliary dyskinesia. *Am. J. Hum. Genet.* **65,** 1508–1519.

Perotti, M. E. (1975). Ultrastructural aspects of fertilization in *Drosophila. In* "The Functional Anatomy of the Spermatozoan" (B. Afzelius, ed.), pp. 57–68. Pergamon, New York.

Pilder, S. H. (1997). Identification and linkage mapping of *Hst7,* a new *M. spretius/M. m. domesticus* chromosome 17 hybrid sterility locus. *Mamm. Genome* **8,** 290–303.

Pilder, S., and Samant, S. (2001). The *Hybrid Sterility 6* locus: A model system for studying sperm tail-related infertility in humans. *In* "Proceedings of the 7th International Congress of Andrology," pp. 317–326. Medimond Medical Publishing, Englewood, NJ.

Pilder, S. H., Hammer, M. F., and Silver, L. M. (1991). A novel mouse chromosome 17 hybrid sterility locus: Implications for the origin of *t* haplotypes. *Genetics* **129,** 237–246.

Pilder, S. H., Olds-Clarke, P., Orth, J. M., Jester, W. F., and Dugan, L. M. (1997). *Hst7:* A male sterility mutation perturbing sperm motility, flagellar assembly, and mitochondrial sheath differentiation. *J. Androl.* **18,** 663–671.

Pilder, S. H., Olds-Clarke, P., Phillips, D. M., and Silver, L. M. (1993). *Hybrid sterility-6:* A mouse *t* complex locus controlling sperm flagellar assembly and movement. *Dev. Biol.* **159,** 631–642.

Pitnick, S., and Karr, T. L. (1998). Paternal products and by-products in *Drosophila development. Proc. Royal Soc. Lond. (B) Biol. Sci.* **265,** 821–826.

Planchart, A., You, Y., and Schimenti, J. (2000). Physical mapping of male fertility and meiotic drive quantitative trait loci in the mouse *t* complex using chromosome deficiencies. *Genetics* **155,** 803–812.

Prout, T., and Clark, A. G. (2000). Seminal fluid causes temporarily reduced egg hatch in previously mated females. *Proc. Natl. Acad. Sci. U.S.A.* **267,** 201–203.

Rankin, T., Familari, M., Lee, E., Ginsberg, A., Dwyer, N., Blanchette-Mackie, J., Drago, J , Westphal, H., and Dean, J. (1996). Mice homozygous for an insertional mutation in the ZP3 gene lack a zona pellucida and are infertile. *Development* **122,** 2903–2910.

Rankin, T. L., O'Brien, M., Lee, E., Wigglesworth, K., Eppig, J., and Dean, J. (2001). Defective zonae pellucidae in *Zp2*-null mice disrupt folliculogenesis, fertility and development. *Development* **128,** 1119–1126.

Redkar, A. A., Olds-Clarke, P., Dugan, L. M., and Pilder, S. H. (1998). High-resolution mapping of sperm function defects in the *t* complex fourth inversion. *Mamm. Genome* **9,** 825–830.

Redkar, A. A., Si, Y., Twine, S. N., Pilder, S. H., and Olds-Clarke, P. (2000). Genes in the first and fourth inversions of the mouse *t* complex synergistically mediate sperm capacitation and interactions with the oocyte. *Dev. Biol.* **226,** 267–280.

Roberts, T. M., and Stewart, M. (2000). Acting like actin: The dynamics of the nematode major sperm protein (MSP) cytoskeleton indicate a push–pull mechanism for amoeboid cell motility. *J. Cell Biol.* **149,** 7–12.

Roldan, E. R. S., Vitullo, A. D., Merani, M. S., and von Lawzewitsch, I. (1985). Cross fertilization *in vivo* and *in vitro* between three species of vesper mice, *Calomys* (Rodentia, Cricetidae). *J. Exp. Zool.* **233,** 433–442.

Rose, M. D. (1996). Nuclear fusion in the yeast *Saccharomyces cerevisiae. Annu. Rev. Cell Dev. Biol.* **12,** 663–695.

Sadler, P. L., and Shakes, D. C. (2000). Anucleate *Caenorhabditis elegans* sperm can crawl, fertilize oocytes and direct anterior-posterior polarization of the 1-cell embryo. *Development* **127,** 355–366.

Sagata, N. (1996). Meiotic metaphase arrest in animal oocytes: its mechanisms and biological significance. *Trends Cell Biol.* **6,** 22–28.

Sakai, N., Burgess, S., and Hopkins, N. (1997). Delayed *in vitro* fertilization of zebrafish eggs in Hank's saline containing bovine serum albumin. *Mol. Marine Biol. Biotechnol.* **6,** 84–87.

Samant, S. A., Fossella, J., Silver, L. M., and Pilder, S. H. (1999). Mapping and cloning recombinant breakpoints demarcating the *Hybrid Sterility 6*-specific sperm tail assembly defect. *Mamm. Genome* **10,** 88–94.

Shaper, N. L., Shaper, J. H., Peyser, M., and Kozak, C. A. (1990). Localization of the gene for β1,4-galactosyltransferase to a position in the centromeric region of mouse chromosome 4. *Cytogenet. Cell Genet.* **54,** 172–174.

Shitara, H., Kaneda, H., Sato, A., Inoue, K., Ogura, A., Yonekawa, H., and Hayashi, J.-I. (2000). Selective and continuous elimination of mitochondria micro-injected into mouse eggs from spermatids, but not from liver cells, occurs throughout embryogenesis. *Genetics* **156,** 1277–1284.

Shur, B. (1981). Galactosyltransferase activities on mouse sperm bearing multkple t^{lethal} and t^{viable} haplotypes of the *T/t* complex. *Genet. Res.* **38,** 225–236.

Si, Y., and Olds-Clarke, P. (2000). Evidence for the involvement of calmodulin in mouse sperm capacitation. *Biol. Reprod.* **62,** 1231–1239.

Singson, A. (2001). Every sperm is sacred: Fertilization in *Caenorhabditis elegans. Dev. Biol.* **230,** 101–109.

Smith, E. F., and Lefebvre, P. A. (1996). *PF16* encodes a protein with armadillo repeats and localizes to a single microtubule of the central apparatus in *Chlamydomonas* flagella. *J. Cell Biol.* **132,** 359–370.

Stein, L., Sternberg, P., Durbin, R., Thierry-Mieg, J., and Spieth, J. (2001). WormBase: Network access to the genome and biology of *Caenorhabditis elegans. Nucleic Acids Res.* **29,** 82–86.

Stern, S., Biggers, J. D., and Anderson, E. (1971). Mitochondria and early development of the mouse. *J. Exp. Zool.* **176,** 179–192.

Sutovsky, P., Moreno, R. G., Ramalho-Santos, J., Dominko, T., Simerly, C., and Schatten, G. (2000). Ubiquinated sperm mitochondria, selective proteolysis, and the regulation of mitochondrial inheritance in mammalian embryos. *Biol. Reprod.* **63,** 582–590.

Tombes, R. M., and Shapiro, B. M. (1989). Energy transport and cell polarity; relationship of phosphagen kinase activity to sperm function. *J. Exp. Zool.* **251,** 82–90.

Travis, A. J., Jorgez, C. J., Merdiushev, T., Jones, B. H., Dess, D. M., Diaz-Cueto, L, Storey, B. T., Kopf, G. S., and Moss, S. B. (2001). Functional relationships between capacitation-dependent cell signaling and compartmentalized metabolic pathways in murine spermatozoa. *J. Biol. Chem.* **276,** 7630–7636.

Ursini, F., Heim, S., Kiess, M., Maiorino, M., Roveri, A., Wissing, J., and Flohé, L. (1999). Dual function of the selenoprotein PHGPx during sperm maturation. *Science* **285,** 1393–1396.

Vaughan, K. T., Mikami, A., Paschal, B. M., Holzbaur, E. L. F., Hughes, S. M., Echeverri, C. J., Moore, K. J., Gilbert, D. J., Copeland, N. G., Jenkins, N. A., and Vallee, R. B. (1996). Multiple mouse chromosomal loci for dynein-based motility. *Genomics* **36,** 29–38.

Verlhac, M. H., Lefebvre, C., Kubiak, J. Z., Umbhauer, M., Rassinier, P., Colledge, W., and Maro, B. (2000). Mos activates MAP kinase in mouse oocytes through two opposite pathways. *EMBO J.* **19,** 6065–6074.

Vijayaraghavan, S., Olson, G. E., NagDas, S., Winfrey, V. P., and Carr, D. W. (1997). Subcellular localization of the regulatory subunits of cyclic adenosine 3′,5′-monophosphate-dependent protein kinase in bovine spermatozoa. *Biol. Reprod.* **57,** 1517–1523.

Wakabayashi, K., Takada, S., Witman, G. B., and Kamiya, R. (2001). Transport and arrangement of the outer-dynein-arm docking complex in the flagella of *Chalamydomonas* mutants that lack outer dynein arms. *Cell Motil. Cytoskel.* **48,** 277–286.

Wilson, N. F., Foglesong, M. J., and Snell, W. J. (1997). The *Chlamydomonas* mating type plus fertilization tubule, a prototypic cell fusion organelle: Isolation, characterization, and *in vitro* adhesion to mating type micus gametes. *J. Cell Biol.* **137,** 1537–1553.

Witman, G. B. (1992). Axonemal dyneins. *Curr. Opin. Cell Biol.* **4,** 74–79.

Woods, I. G., Kelly, P. D., Chu, F., Ngo-Hazelett, P., Yan, Y. L., Huang, H., Postlethwait, J. H., and Talbot, W. S. (2000). A comparative map of the zebrafish genome. *Genome Res.* **10,** 1903–1914.
Wu, W. J., and Kinsey, W. H. (2000). Fertilization triggers activation of Fyn kinase in the zebrafish egg. *Int. J. Dev. Biol.* **44,** 837–841.
Yang, P., and Sale, W. S. (2000). Casein kinase I is anchored on axonemal doublet microtubules and regulates flagellar dynein phosphorylation and activity. *J. Biol. Chem.* **275,** 18905–18912.

12

GAMETE IMMUNOBIOLOGY

M. G. O'RAND AND I. A. LEA

Department of Cell and Developmental Biology
University of North Carolina, Chapel Hill

I. INTRODUCTION

Mammalian reproductive systems have a unique and precarious relationship with the immune system. This relationship in the male appears highly protective of the developing gametes, whereas in the female it appears maddeningly paradoxical, potentially exposing the gametes to the full force of immune destruction. Different mechanisms have arisen in males and females to provide protection for the gametes and for the single most important event in biology, namely, the reproduction of the species. What are these different mechanisms and why do they seem so precarious at times? This chapter explores the relationship between the immune system and the gametes, examining both the morphological and immunological bases for protection and the enormous potential for destruction.

II. FETAL AND NEONATAL GERM CELLS

Early in fetal life primordial germ cells migrate from the endodermal walls of the fetal yolk sac along the dorsal mesentery into the gonadal ridges to become incorporated into the primary sex cords (Figure 12.1). Thus begins the differentiation of all gametes. If a Y chromosome is present the cords condense, extend into the medulla and begin the formation of seminiferous tubules. If no Y chromosome is present development is somewhat slower and eventually (week 16 in humans) secondary sex cords containing the primordial germ cells form primordial follicles beneath the germinal epithelium. Long before the formation of primordial follicles begins, however, the 10,000 germ cells present during weeks 6–7 of gestation divide by mitosis until some 6 or 7 million are present at 20 weeks in the human female (Adashi, 1991). Mitotic proliferation in the fetal testis usually ends by 18 weeks [in humans (Pelliniemi *et al.,* 1993)]. In the ovary the initiation of meiosis begins during week 8, at which time there are perhaps 60,000 oogonia present. Concomitant with the onset of meiosis is the onset of atresia such that now the increasing number of oogonia from mitosis is offset by the loss of oogonia to meiosis and atresia.

Eventually two-thirds of the 6–7 million oogonia will have entered S phase of the first meiotic division, proceeding to the diplotene stage of meiotic prophase I, where they remain arrested until shortly before ovulation, which may be many years later in primates. In this arrested state the chromosomes decondense and RNA synthesis occurs, providing material to be stored for later use by the growing oocyte. As the oogonia become primary oocytes they are surrounded by a single layer of squamous follicular cells and appear in the ovarian cortex as primordial follicles. Oogonia that do not enter meiosis are eventually lost by atresia and by the sixth month of gestation in humans, follicular atresia has also begun. The

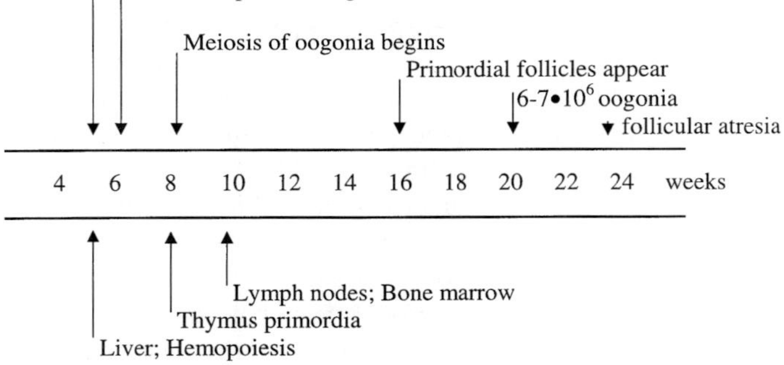

FIGURE 12.1 Developmental time line between 4 and 24 weeks of human gestation. The arrows indicate the approximate start of each event in the immune and reproductive systems.

consequence of oogonial and follicular atresia is a dramatic drop in the number of oocytes present in the ovary. Only 1–2 million primary oocytes in primordial follicles and no oogonia remain at birth. In spite of the fact that meiosis is arrested, follicular atresia continues, and by puberty approximately 300,000 primordial follicles remain (Adashi, 1991). With the onset of both oogonial and follicular atresia and, significantly, the continued follicular atresia in the neonate, circulating leukocytes and the resident ovarian macrophages (Gaytan *et al.,* 1998), which express class II major histocompatibility complex (MHC) molecules (Rhim *et al.,* 1992), are set in motion for their lifelong duties of removing ovarian debris. Importantly in the present context, this would include the oocyte and granulosa cell constituent protein antigens of potential immunological significance, not the least of which is the zona pellucida.

Immunocytochemical evidence indicates that the three major glycoprotein components of the zona pellucida, namely ZP1, ZP2, and ZP3, are present in primordial follicles of humans, rhesus monkeys, marmosets, and rabbits (Grootenhuis *et al.,* 1991). In mice ZP2 has been reported to be present in the primordial follicle stage (Castle and Dean, 1996).

Returning to the development of the testis during fetal life, the seminiferous tubules form from the primary sex cords, as do the tubuli recti and the rete testis. During this time all connections with the overlying germinal epithelium are lost and a characteristically thick tunica albuginea is present. Meanwhile the spermatogonia, derived from the primordial germ cells, are nestled between supporting Sertoli cells, and both Leydig cells and macrophages are present in the interstitial connective tissue. The presence of a meiosis-inhibiting substance, probably produced by Sertoli cells, ensures that spermatogonia do not enter meiosis (Pelliniemi *et al.,* 1993). Shortly before puberty the Sertoli–Sertoli cell junctions will appear (Vitale *et al.,* 1973) such that with the onset of puberty the spermatogonia resume their mitotic cycle, but this time, under the influence of follicle-stimulating hormone (FSH) from the pituitary and testosterone from Leydig cells, some will enter meiosis. Proceeding to the primary spermatocyte stage they will cross the basal occluding junctions of the Sertoli cells at the preleptotene stage and enter the protected adluminal compartment of the seminiferous tubule. Unlike anything found in the ovary, this morphological and physiological barrier is the "blood–testis barrier" (Setchell *et al.,* 1969; Dym and Fawcett, 1970), which is thought to provide a special environment to nourish the completion of spermatogenesis and protect against immunological insult (O'Rand and Romrell, 1977).

III. THE DEVELOPING IMMUNE SYSTEM

The development of the thymus begins earlier than the beginning of primordial follicle and primary oocyte development in the female and somewhat later than the onset of seminiferous tubule development in the male. Indeed, by week 8 of human fetal life the two primordia of the thymus derived from the third pharyngeal pouch meet and fuse to form a dense mass of endothelial cells. Quickly invaded by stem cells (possibly prethymocytes) originating from the fetal liver and yolk sac, and somewhat later by stem cells from the bone marrow, this mass organizes into the thymic cortex and medulla and begins the differentiation of thymocytes into mature, competent T cells. During fetal life most circulating T cells are antigen naive and probably cannot respond as fully as adult cells can to stimulatory signals. Differentiation of stem cells into pre-B cells begins in the fetal liver, continues in the bone marrow after its appearance somewhat later (week 10 in humans), and finally results in the migration of B cells into the developing lymph nodes (weeks 10–12 in humans; see Figure 12.1). The developmental timing of these differentiation processes for both B and T cells would indicate that they occur simultaneously with germ cell development. Hence by the second trimester of human pregnancy, mature B and T cells are present in the fetus and surveillance for foreign antigens as well as the individual's ability to distinguish between self and non-self are well underway. Given these developmental realities, what are the consequences for gametes?

IV. IMMUNE RESPONSE TO GAMETES IN THE FETAL, NEONATAL, AND PREPUBERTAL STAGES

A. THE FEMALE

To understand the relationship between gametes and the immune system in the fetus and neonate it is necessary to review several important observations regarding the immune response. Although a genetic predisposition exists for an individual's response to any particular antigen, in those species in which it has been tested, immunization of a female with homologous zona pellucidae (ZP) or ovary, and by implication with oocytes, does not produce circulating anti-ZP antibodies (Tsunoda and Chang, 1976; Gwatkin *et al.,* 1977; Dunbar *et al.,* 1994). Presumably the self-reacting ZP-specific lymphocytes were prevented from responding to the endogenous ZP because these lymphocytes had been eliminated by clonal deletion, down-regulated (causing clonal anergy), or suppressed through cytokine or idiotype networks. Any or all of these mechanisms develop during fetal and neonatal life. Indeed, Taguchi and colleagues (Taguchi and Nishizuka, 1980; Taguchi *et al.,* 1980) demonstrated that neonatal thymectomy in female mice results in infertile adults with circulating antibodies that recognize both oocyte cytoplasm and extracellular zona pellucida. They also reported that some sera from day 3 thymectomized mice recognized ovarian theca, interstitial, luteal, and granulosa cells. This result would imply that there was a generalized antiovarian antigen response. More extensive studies of the effects of neonatal thymectomy (Smith *et al.,* 1989, 1991) indicate that the oophoritis effect can only be transferred to syngeneic mice and can be reversed by adult $CD4^+$ T cells if they are given before the thymectomized mice reach age 10–12 days (Tung, 1995). These experiments are interpreted to mean that a set of effector and regulatory T cells develops in the thymus, maintaining a balance that in normal individuals would inhibit self-reacting lymphocytes. Hence, ZP-specific lymphocytes have not been eliminated by clonal deletion. A day 3 thymectomy (D3TX) in the female neonatal mouse would upset this balance specifically resulting in the depletion of $CD25^+$ and $CD4^+$ cells (Asano *et al.,* 1996). Indeed, when $CD25^+$ and $CD4^+$ enriched T cell populations were transferred to D3TX mice, autoimmune oophritis was inhibited (Alard *et al.,* 2001). $CD4^+CD25^+$ cells play a critical role in the control of autoimmunity because the pathogenic potential of $CD25^-$ T cells is restricted by the presence of $CD25^+$ T cells (Asano *et al.,* 1996). Autoimmune oophritis as well as other autoimmune diseases (Salomon *et al.,* 2000; Takahashi *et al.,* 2000; Seddon and Mason, 1999) are prevented by the presence of $CD4^+CD25^+$ cells, which are also important for tolerance to alloantigen (Taylor *et al.,* 2001). The loss of $CD25^+$ cells by D3TX would result in the later dominance of aggressive T cells in the adult ovary. It should be recognized, however, that transfer of spleen cells from normal adult male mice to D3TX females also suppresses oophoritis, as would the transfer of cells from mice ovarectomized at or immediately before birth (Tung, 1995), the implication being that antigen is not required for the development of the appropriate suppressor population. Moreover, transfer of T cells from D3TX male mice to adult females caused only mild ovarian lesions and only if large numbers of cells were transferred (Tung, 1995), and under these conditions it is unlikely that sterility would be induced. However, recent evidence from studies of autoimmune thyroiditis (Seddon and Mason, 1999) and tolerance to zona pellucida protein 3 (Garza *et al.,* 2000) has shown that the presence of the specific autoantigen is required. Thus it would appear that the normal T-cell response to endogenous ovarian proteins is being suppressed in mice through at least two mechanisms: regulatory T cell populations independent of cytokine production but probably requiring autoantigen specificity and inhibitory cytokine networks from normal T helper cell populations (Alard *et al.,* 2001; Abbas *et al.,* 1996). These experiments using D3TX mice demonstrate the general concept that the immune system exists in a dynamic balance between regulation and destruction, and those specific endogenous antigens heighten its vigilance. This concept is consistent with the observation that the occurrence of anti-ZP antibody in the serum of both normal and infertile women is extremely low (Dunbar, 1995; Van Voorhis and Stovall, 1997). Clearly the time-dependent development of tolerance to gamete self-antigens in the female is critical for reproductive success.

If suppression of autoimmunity in normal female mice is brought about through a network of regulatory T cells and inhibitory cytokines, then it remains to be seen exactly how this network is thrown off balance when oophoritis is induced. With regard to such cytokine networks, it may be appropriate and helpful to the reader to point out here that mouse $CD4^+$ T cell clones can be divided into different groups based on their cytokine production (Mosmann *et al*, 1986, 1991) and that these different groups have distinct functional attributes that are responsible for many of the heterogeneous responses of the immune system [for review, see Abbas *et al.* (1996) and Constant and Bottomly (1997)]. Similar but probably not identical functionally distinct populations are also thought to exist in humans (Abbas *et al.,* 1996). The significance of these populations of T cells lies in their ability to direct the immune response, and consequently the factors that influence these cells in one direction or the other become critical for understanding the kind of immune response observed. For example, it has recently been reported (Maity *et al.,* 1997) that neonatal thymectomy in female mice may disrupt the T helper (Th) cell balance between Th1-type and Th2-type $CD4^+$ cells. The result of such a disruption is the production of a predominant Th2-type response [increased production of interleukin-4 (IL-4) and decreased production of interferon-γ (INF-γ)] that appears to be correlated with the appearance of autoimmune oophoritis. The observed autoimmune oophoritis, and particularly the complete lack of ovarian follicular development, could be reversed by treatment with IL-12 (restoring a Th1/Th2 balance) either before or after the day 3 neonatal thymectomy. If a Th2-type response is involved in the development of oophoritis, then perhaps this favors the development of a strong humoral response with IgG_1 circulating anti-ZP or other antiovarian antigen antibodies. Such non-complement-fixing antibodies might lead to the inability of oocytes and granulosa cells to maintain their gap junction connections (Dunbar, 1995) and therefore to the failure of ovarian follicular development. Alternatively, Th2 cytokines could influence the development and maturation of ovarian follicles either directly by activating granulocytes (eosinophils, mast cells) or indirectly by blocking Th1 effector functions such as the production of IFN-γ (Abbas *et al.,* 1996). Despite this, we cannot exclude the possibility that induction of a Th1 dominant response will also give rise to autoimmune oophritis. In order to understand autoimmune oophritis in detail, the roles of NK T cells and $CD4^+CD25^+$ regulatory T cells await further study. Nonetheless, it is obvious that ovarian follicular development is dependent on maintaining a balance between various T cell responses.

B. THE MALE

In contrast to the female gametes within ovarian follicles, the male gametes within their seminiferous tubules (peritubular and Sertoli cells) seem relatively protected and perhaps more quiescent until puberty, because no particular prepubertal atresia has been noted. Additionally, immune surveillance of the testis and spermatogonial cells present in the tubules would not detect any meiotic prophase antigenic products until the onset of puberty. Some orchitis has been demonstrated in adult male mice after neonatal thymectomy (Tung, 1995), but the incidence is low and does not seem to provoke the same aggressive-cell response seen in the ovary. Consequently, although regulator and effector T cells are present in males for the immune surveillance of the testis, the absence of significant amounts of specific antigens to drive a cellular immune response may prevent strong autoimmune reactions. Of course such a precarious balance of regulatory factors in the absence of antigen can easily be overridden by immunizations with testis antigens in complete Freund's adjuvant and *Bordetella pertussis* toxin or with large numbers of viable syngeneic testis cells, either of which will result in experimental autoimmune orchitis (Tung, 1995; Itoh *et al.,* 1991).

The male environment changes dramatically, however, with the onset of puberty. By day 19 after birth in the rat, for example, the occluding junctions between Sertoli cells have differentiated and matured into the "blood–testis barrier" (Setchell *et al.,* 1969; Dym and Fawcett, 1970; Vitale *et al.,* 1973). Moreover, the lamina propria of the seminiferous tubule develops into a layer (single in rodents, multiple in primates) of peritubular contractile cells (myoid cells) interspersed with basal lamina and collagen fibrils, forming an epitheloid structure capable of rhythmic contractions that is not easily penetrated by cellular elements. During this time of awakened differentiation in the male, specific mRNAs for sperm antigens begin to appear. For example, the mRNA for nuclear autoantigenic sperm protein (NASP), a sperm and testis differentiation antigen (Welch and O'Rand, 1990), is upregulated in the rat as early as 21 days and persists throughout spermatogenesis. The appearance of several other mRNAs for testis and sperm-specific antigens has also been detected at this time (O'Rand and Romrell, 1977; O'Brien and Milette, 1984; Kurpisz and Janitz, 1995), including the testis-specific histone H1t (Kremer and Kistler, 1992). The appearance of these antigens is consistent with the appearance of meiotic prophase cells and their movement into the adluminal compartment of the seminiferous tubules as primary spermatocytes. Protein synthesis of a number of proteins, including NASP, is also upregulated at this time in pachytene spermatocytes (Welch and O'Rand, 1990). Significant levels of serum antisperm antibody can be detected between 56 and 91 days of development in the male rat (Flickinger *et al.,* 1997). This is entirely consistent with the development of a systemic cellular and humoral immune response to newly synthesized male gamete antigens at the initiation of meiosis some 38 days previously. There is no indication that these newly circulating antisperm antibodies affect fertility in any way. The presence of antisperm antibodies in normal males has been known for many years (Edwards, 1960; Johnson, 1968; Tung *et al.,* 1976; Bronson *et al.,* 1992), and although there are numerous reports in the literature of the adverse effects of such antibodies on spermatozoa and fertilization *in vitro,* their adverse effects *in vivo* remain to be proved. Given our current knowledge of $CD4^+$ T cell clones, it would be of interest to determine the dominant T cell epitopes that emerge during meiotic prophase in the male.

Unlike the nonresponse of females immunized with homologous ZP discussed above, males immunized with whole sperm do respond by the production of antibodies. For example, when B6AF1 male mice were immunized with mouse sperm in complete Freund's adjuvant either with or without *B. pertussis* toxin they developed antisperm antibodies and specifically developed antibodies to epitopes in the sperm protein Sp17 (Kong *et al.,* 1995). However, these males were completely fertile in two successive breeding experiments with normal females. Considering all of the evidence to date, it may be reasonably concluded that there is a fundamental difference between the male and female immune response to gametes. This difference, however, is not inherent in the immune response—rather it is a result of the developmental history of each separate population of gametes, the female gametes quickly reaching meiotic prophase in the fetus and the male gametes delaying entry into meiotic prophase until puberty. Such developmental differences in the presentation of meiotic prophase antigenic constituents to the immune system have profound effects on the adult's immune response to normal reproductive processes as well as to experimental manipulation.

V. IMMUNE RESPONSE TO GAMETES IN THE ADULT

A. THE MALE

The rise of serum antisperm antibodies with the onset of puberty and the presence of these antibodies in normal males does not preclude the adverse effects that high-titer antisperm antibodies can have on the male reproductive system. Why should this be the case? Two important aspects need to be considered. The first is the physical barrier or lack of barriers between the male reproductive system and the immune system, and the second is the characteristics of the antigens (immunogens), including their molecular properties and species of origin. Although the seminiferous tubules have a well-developed lamina propria and a system of Sertoli–Sertoli cell occluding junctions that function as the "blood–testis barrier," the junctions end rather quickly within the short straight tubules (tubuli recti), giving way to the simple columnar epithelium of the rete testis. Throughout the male excurrent duct system, lymphocytes ($CD4^+$ and $CD8^+$) can be observed within the epithelium that lines the duct, and macrophages are present in the underlying connective tissue, where they are known to scavenge within the duct (Nashan *et al.,* 1989, 1990; Anderson, 1994). Moreover, antisperm antibodies can enter the rete testis to bind to testicular spermatozoa (Tung, 1980). It is therefore reasonable to expect that testis-specific antigens be presented to antigen-presenting cells (APCs) of the immune system. These antigens may be present on spermatozoa leaving the testis, on cytoplasmic droplets or remnants of cells leaving the testis, or even in the seminiferous tubular fluid that is reabsorbed after leaving the testis. Professional APCs such as monocytes and B cells, which have or can be induced to have both class II MHC and costimulatory molecules on their surfaces, would be available to interact with testicular antigens along the excurrent duct system. The resulting presentation of testicular peptides to T helper cells would initiate the effector arm of the immune response, although the presence of a large number of suppressor T cells helps preserve a bias against an autoimmune response to spermatozoa (El Demiry and James, 1988). As discussed previously, some testicular tolerance mechanisms do exist in the male early in development, but these are presumably nonantigen specific and therefore may not be directly relevant to the regulation of the production of specific $CD4^+$ Th1- or Th2-type T cells after the onset of puberty.

Immunization with testis antigen homogenates in the appropriate adjuvants may result in the development of experimental autoimmune orchitis (EAO) in genetically susceptible species (Teuscher, 1985; Teuscher *et al.*, 1985; Tung, 1995). Although this result should not be unexpected given the heroic nature of the immunization protocols necessary to achieve orchitis in mice (Yule *et al.*, 1988), the study of EAO has been helpful in furthering our understanding of the relationship between male gametes and the immune response. One particularly relevant finding is that activated $CD4^+$ T cells that are capable of transferring disease from EAO mice to naive mice seem to create the first detectable signs of pathology in the area around the straight tubules (Mahi-Brown *et al.*, 1988). This is the area, as described above, where the protected environment of the seminiferous tubule lumen ends. Consequently it is quite clear that the "blood–testis barrier" is incomplete in the sense that it does not protect the excurrent duct system, it cannot protect against retrograde transmission in the lumen, and, as discussed above, the initial meiotic prophase transition occurs before the barrier forms around the preleptotene spermatocyte and incorporates it into the adluminal compartment. Indeed, all the spermatogonia committed to meiosis (some type A forms, intermediate, and type B forms) occur outside the barrier. These committed spermatogonia and preleptotene cells external to the barrier have stage-specific antigens, many of which are also present in somatic cells (Kurpisz and Janitz, 1995). Not unexpectedly, immunizations with testis preparations containing such spermatogonia and preleptotene spermatocytes result in the deposit of specific circulating antibodies on preleptotene cells. And, in fact, antibody deposits of both IgG_1 and IgG_3 isotypes have been observed on preleptotene spermatocytes (Mahi-Brown *et al.*, 1988). However, such antibodies may be present normally after the initiation of meiosis in the male, as discussed above, and given a balanced Th1/Th2 cytokine environment there is no reason to assume that they would interact with later stages of spermatogenesis. Additionally, there is no current evidence to demonstrate that antipreleptotene antibodies interfere with sperm function or fertility. In fact, early studies (O'Rand and Romrell, 1977; Romrell and O'Rand, 1978) conclusively demonstrated that sperm antigens, present on ejaculated spermatozoa, were present only on the surface of primary spermatocytes *after* they cross the "blood–testis barrier." Hence, reports (Yule *et al.*, 1988; Mahi-Brown *et al.*, 1988) of antibodies to preleptotene spermatocytes being present on preleptotene spermatocytes after immunization with testis are of no particular significance.

The presence of circulating antisperm antibodies directed toward essential sperm antigens on epididymal and ejaculated spermatozoa can, however, be of significance for fertility. The characteristics of the antigen (immunogen) become of primary importance in determining its interaction with the immune system. The APC that first encounters the immunogen (B cell or macrophage, for example), the processing to peptide fragments, and the presentation of the peptide in the context of the individual's MHC are all important factors in the immune system's response.

Subsequently, the characteristics of the effector immune response will determine what, if any, physiological consequences result. Is there a significant cytotoxic T cell ($CD4^-CD8^+$) response that could result in sperm-immobilizing antibodies, or is there a significant Th1-type or Th2-type $CD4^+$ T cell response indicative of a particular cytokine response? Sperm surface antigens that bind complement-fixing antibodies can immobilize spermatozoa in the presence of complement (O'Rand and Metz, 1976) and have been shown in some patients to be carbohydrate epitopes (O'Rand, 1980; Kameda *et al.,* 1992; Tsuji, 1995; Diekman *et al.,* 1997).

A consideration of results from the study of vasectomy patients points out some of the important aspects of the formation of antisperm antibodies in the male. When such antibodies are present in the serum and reproductive tract of patients following vasectomy, they may cause secondary infertility (Hendry, 1992; Hjort and Meinertz, 1988; Bronson *et al.,* 1984) should a reversal of the vasectomy (vasovasostomy) be desired. It has been pointed out (Clarke, 1988; Barratt *et al.,* 1992) that the localization and nature of the sperm antigen are more important than the actual titer. Our study (Lea *et al.,* 1997) of the immune response after vasectomy points out an interesting aspect of the immune response to sperm antigens. The sperm protein Sp17 was clearly demonstrated to be autoantigenic in humans, because sera from 87% of men tested either pre- or postvasovasostomy exhibited a statistically significant increase in anti-Sp17 reactivity when compared to control sera (Lea *et al.,* 1997). The levels of reactivity to Sp17 did not correlate with the titer of antisperm antibodies in the serum. Although the anti-Sp17 reactivity varied considerably between individuals, mimotope analysis of the sera's reactivity to recombinant human Sp17 showed that the dominant linear B cell epitopes were constant. The individuals' mimotope profiles varied in the magnitude of their response and in the actual number of epitopes recognized (the less dominant epitopes were recognized by fewer individuals). However, in this presumed heterogeneous population of patients, the most immunodominant epitopes elicited a consistent response (Lea *et al.,* 1997). This pattern of restricted reactivity was also seen in antisperm autoimmune mouse and rabbit serum (Kong *et al.,* 1995; O'Rand and Widgren, 1994). Consequently, using recombinant human Sp17 as a model sperm immunogen, we have demonstrated that the immunodominant linear B cell epitopes of sperm antigens recognized by the immune system are largely independent of genetic background, even across species.

The significance of this observation on immunodominant B cell epitopes is that the effector T cell immune response becomes the first important variable in determining whether gamete function is affected (see below). The second important variable is whether the immunodominant epitope recognized by the immune system is located in an exposed position on the sperm surface. If this is the case, then circulating antibodies in the male reproductive tract or in the seminal fluid at the time of ejaculation should be able to bind to it. The observation that antisperm antibody agglutinates or immobilizes spermatozoa present in the ejaculates of some males supports this premise. It has been clear for many years that sperm surface autoantigens, which can be detected by the sperm immobilization reaction of autoantisera, are intrinsic to the sperm plasma membrane, and of obvious concern to infertile patients (O'Rand, 1980). However, short of the complete absence of functional spermatozoa in the ejaculate, the presence of such antibodies in the ejaculate may not be the primary cause of male infertility.

B. The Female

The presence of antibodies to homologous gametes in the female was discussed in Section III,A. The delicate balance between regulation and destruction can easily be tipped by the simple presentation of a homologous ovarian or ZP antigen with a foreign T cell epitope to the female immune system. This was clearly demonstrated by the inhibition of fertility and the abnormal nature of the ovulated ZP in mice immunized with hamster zonae (Gwatkin *et al.,* 1977) and the abnormal development of ovarian follicles in rabbits after immunization with heterologous ZP (Wood *et al.,* 1981). Variations of this initial observation in numerous species (Dunbar, 1995) have substantiated the general case that heterologous immunization does tip the balance against the ovary, with the production of both serum antibodies and disease. The production of both antibody and disease requires the presentation of a foreign T cell epitope, as demonstrated in the case of a single $ZP3^{328-342}$ peptide attached to the carrier keyhole limpet hemocyanin (KLH) (Millar *et al.,* 1989). Immunized female mice produced anti-ZP antibodies that recognized the native ZP and inhibited fertilization, yet some of these mice recovered their fertility after several months, indicating that oophoritis was not of the same severity in each individual (Millar *et al.,* 1989; Epifano and Dean, 1994). More extensive studies using ZP3 peptides demonstrated that the production of autoantibodies to a specific ZP peptide and the appearance of autoimmune oophoritis are H2 haplotype dependent (Rhim *et al.,* 1992; Lou *et al.,* 1995). Immunization of female mice with a chimeric peptide containing a mouse ZP3 peptide and a promiscuous T cell epitope, seen as foreign by the mouse, clearly demonstrated that the production of antibodies to zona pellucida and the onset of disease in the ovary were separate immunological events (Lou *et al.* 1995; Sun *et al.,* 1999).

In a series of experiments with female B6AF1 mice immunized with a zona pellucida peptide ($ZP3^{330-340}$), Bagavant *et al.* (1999) reported that the mice developed a predominantly Th1-type T cell response (increased IL-2 and IFN-γ production). The reduced fertility in these mice correlated with anti-ZP antibody titer, but not with the severity of oophoritis. Moreover, cloned Th1 cells from these mice passively transferred into syngeneic mice targeted atretic follicles, resulting in oophoritis, but without significant effects on follicular development or fertility. Consequently, it would appear that a humoral immune response to specific ovarian targets is required for the loss of follicular development and subsequent infertility.

Monoclonal antibodies to both ZP3 and ZP2 given passively to female mice inhibit fertility by coating the ovulated oocytes (East *et al.,* 1984). Consequently, antibodies to peptides from mouse ZP2 would be expected to have effects on the ovary similar to those of ZP3 peptides. Peptides from the ZP proteins in other species have also been shown to affect fertility in the female. For example, peptides from pig ZP3α and ZP3β (Sacco and Yurewicz, 1994; Kaul *et al.,* 1996) as well as ZP1 (ZP2 and ZP4) (Koyama *et al.,* 1994; Hasegawa *et al.,* 1995) also inhibit fertility.

VI. IMMUNE RESPONSE TO MALE GAMETES IN THE ADULT FEMALE

The fertilization process begins when spermatozoa released from the male enter into the environment of the oocyte (external fertilization) or are deposited into the female reproductive tract (internal fertilization), and it ends when the zygote is formed (O'Rand, 1986). In the normal course of events, mammalian spermatozoa are therefore deposited into the female reproductive tract to initiate the process of fertilization. The mucosal surface of the tract—the lining epithelia and the underlying basal lamina—like that of the gastrointestinal tract, is continuous with the outside environment. Unlike the gastrointestinal tract, however, the female reproductive tract opens into the peritoneal cavity at the distal end of the infundibulum. In spite of this anatomical fact, it is well adapted to prevent bacteria and other foreign organisms from reaching both uterus and oviduct. Nevertheless, the reproductive tract rather quickly transports spermatozoa to the site of fertilization, usually in the upper regions of the oviduct, through peristaltic smooth muscle contractions in the wall and with some help from the sperm's own motility. In many species this initial phase may not contain the fertilizing spermatozoa and may be followed by a second wave that more slowly makes its way into the oviduct to await the oocyte (O'Rand and Nikolajczyk, 1991). Because the millions of spermatozoa deposited into the female tract in a single ejaculate contain numerous antigens that are certainly foreign to the female immune system, the relationship between the female immune system and the need for the survival of the spermatozoa (while minimizing the risk of infection) remains a balance. It is vital that there is some level of cellular and humoral control of the immune system if the spermatozoa are to fertilize the oocyte successfully. In humans, the necessary protection of the spermatozoa may be afforded by the lack of human leukocyte antigen (HLA) markers on the surface of the spermatozoa (Anderson *et al.,* 1982), although this is a controversial finding, with other reports suggesting the expression of HLA genes in spermatozoa (Chiang *et al.,* 1994) [see, however, Kurpisz *et al.* (1995)]. In addition, seminal plasma contains several immunosuppressive factors that assist in minimizing the immune response to the spermatozoa. These include prostaglandins (E series) that are known to modulate the immune response, in particular by inhibiting the production of Th1-inducing cytokines, hence favoring a Th2-type response (Betz and Fox, 1991). Other seminal plasma factors that modulate the immune response to ejaculated spermatozoa (human), at least initially, include cytokines (transforming growth factor-α and -β) IL-1, IL-6, IL-8, Fc receptor and Fc binding molecules, and complement (James and Skibinski, 1995). Thus components of the seminal plasma have the potential to block antigen binding either by masking sperm surface proteins or by binding to the immune cell receptors. Following this, the sequence of events leading to an effective immune response can also be blocked or modified, allowing spermatozoa to proceed with the process of fertilization. Once the spermatozoa have begun to move toward the oviduct, it should not be surprising that an immune response to sperm antigens would occur with sufficient exposure to spermatozoa. Macrophages in the peritoneal cavity, within the female reproductive tract, and those in the underlying connective tissue might all be exposed to sperm antigens at one time or another and function as APCs. They would also be able to carry out receptor-mediated phagocytosis of left-over spermatozoa if the appropriate opsonizing antibodies were present. Not surprisingly, numerous studies have shown the existence of antisperm antibodies in females, even in prepubertal girls (Tung *et al.*, 1976; O'Rand, 1995, and references therein). Of course, many of these antibodies that exist in normal serum were raised against nonsperm antigens—for example, bacterial carbohydrates (Sarkar, 1974)—and are simply cross-reactive with spermatozoa (O'Rand, 1980).

Our studies (Lea *et al.*, 1998a) on the immunization of female monkeys with sperm antigens point out the pervasive nature of circulating antisperm antibodies. Immunization (intramuscular) with the recombinant human sperm protein Sp17 or with Sp17 peptides elicited an immune response that was detected in serum and oviduct fluid. Each monkey had an oviduct fluid antibody titer no more than 10-fold lower than the serum titer, although the titer was variable over the course of successive menstrual cycles. In two out of the three Sp17-immunized monkeys, a sudden drop in the level of oviduct fluid antibody reactivity occurred after the midpoint of some menstrual cycles, concomitant with a reported drop in the concentration of protein present in the oviduct after ovulation. Analysis of the oviduct fluid antibody showed that the Sp17 and Sp17 peptide antibodies were solely IgG, presumably derived from serum, and that no specific anti-Sp17 IgA was detectable. Specific IgA class antibodies to some sperm proteins do occur in the oviduct fluid, as demonstrated by studies on the sperm protein SP-10 (Kurth *et al.*, 1997). Comparisons of the Sp17-specific antibodies from the oviduct fluid and serum showed that they have a reactivity identical to that of linear B cell epitopes of Sp17 (Lea *et al.*, 1998a). These studies lead to the conclusion that if there is an immune response to sperm antigens, then those antibodies are almost certainly present in the female reproductive tract, and therein have the potential to bind to sperm.

Returning to a consideration of the T cell immune response as the first important variable in determining whether gamete function is affected, we have studied the difference between two inbred strains of mice using the same synthetic immunogen construct (Lea *et al.*, 1998b). This construct consisted of an immunodominant linear B cell epitope from the human sperm protein Sp17 and a promiscuous T cell epitope from RNase. In this situation both B6AF1 and BALB/c strains recognized the same immunodominant linear B cell epitope, both had similar antibody titers, and both had antibodies that recognized spermatozoa. Significantly, however, the BALB/c strain became infertile and the B6AF1 strain was unaffected (Lea *et al.,* 1998b). It was determined that the two strains had a substantially different T cell response to the immunogen. The B6AF1 strain responded with a predominant Th1-type response with high levels of INF-γ, and its T cells recognized the RNase promiscuous T cell epitope. In contrast, the BALB/c strain responded with a predominant Th2-type response with high levels of IL-4. Its T cells recognized a bridging T cell epitope consisting of the amino acids between the end of the promiscuous T cell epitope and the beginning of the Sp17 B cell epitope. Thus the strain-specific infertility depended on the T cell epitope recognized in the context of the major histocompatibility complex and the cytokine response.

VII. CONCLUDING REMARKS

Study of the immunobiology of gametes gives us an understanding of the mechanisms that operate to protect the gametes and prevent destruction in both males and females. These mechanisms are rooted in the morphology of the testis and ovary and in the significant developmental difference in the time of meiotic prophase between the sexes. The immune system has evolved to deal with this reality by prioritizing the T cell response in a network of cytokines that most often directs the response away from destruction. The presence of antigamete antibodies in either males or females is not inherently dangerous and may in fact indicate in many cases an adaptation to deal with the late-arriving antigens of the reproductive system. It would seem that it is only when an excess of antigen stimulation occurs, particularly when this stimulation is driven by foreign B and T cell epitopes, that immune destruction of the gametes ensues, leading to sterility and ultimately to the inability of the species to reproduce.

ACKNOWLEDGMENTS

The work reported in this chapter was supported in part by NIH grants HD14232 and U54HD29099 and CONRAD (CIG-96–06). The authors thank Dr. Iglika Batova for critical reading of the manuscript.

REFERENCES

Abbas, A. K., Murphy, K. M., and Sher, A. (1996). Functional diversity of helper T lymphocytes. *Nature* **383,** 787–793.

Adashi, E. (1991). The ovarian life cycle. *In* "Reproductive Endocrinology" (S. Yen and R. B. Jaffe, eds.), pp. 181–237. W. B. Saunders, Philadelphia.

Alard, P., Thompson, C., Agersborg, S. S., Thatte, J., Setiady, Y., Samy, E., and Tung, K. S. K. (2001). Endogenous oocyte antigens are required for rapid induction and progression of autoimmune ovarian disease following day-3 thymectomy. *J. Immunol.* **166,** 4363–4369

Anderson, D. J. (1994). Mucosal immunity of the male urogenital tract and the development of HIV-1 vaccines. *In* "Recombinant and Synthetic Vaccines" (G. P. Talwar, K. V. S. Rao, and V. S. Chauhan, eds.), pp. 502–512. Narosa Publishing House, New Delhi.

Anderson, D. J., Bach, D. L., Yunis, E. L., and De Wolf, W. C. (1982). Major histocompatibility antigens are not expressed on human epididymal sperm. *J. Immunol.* **129,** 452–461.

Asano, M., Toda, M., Sakaguchi, N., and Sakaguchi, S. (1996). Autoimmune disease as a consequence of developmental abnormality of a T cell population. *J. Exp. Med.* **184,** 387–396.

Bagavant, H., Adams, S., Terranova, P., Chang, A., Kraemer, F. W., Lou, Y, Kasai, K., Luo, A. M., and Tung, K. S. K. (1999). Autoimmune ovarian inflammation triggered by proinflammatory (Th1) cells is compatible with normal ovarian function in mice. *Biol. Reprod.* **61,** 635–642.

Barratt, C. L. R., Dunphy, B. C., McLead, I., and Cooke, I. D. (1992). The poor prognostic value of low to moderate levels of sperm surface-bound antibodies. *Hum. Reprod.* **7,** 95–98.

Betz, M., and Fox, B. S. (1991). Prostaglandin E_2 inhibits production of Th1 lymphokines but not Th2 lymphokines. *J. Immunol.* **146,** 108–113.

Bronson R., Cooper, G., and Rosenfeld, D. (1984). Sperm antibodies: Their role in infertility. *Fertil. Steril.* **42,** 171–183.

Bronson, R. A., O'Connor, W. J., Wilson, T. A., Bronson, S. K., Chasalow, F I., and Droesch, K. (1992). Correlation between puberty and the development of autoimmunity to spermatozoa in men with cystic fibrosis. *Fertil. Steril.* **58,** 1199–1204.

Castle, P. E., and Dean, J. (1996). Molecular genetics of the zona pellucida: Implications for immunocontraceptive strategies. *J. Reprod. Fertil. Suppl.* **50,** 1–8.

Chiang, M. H., Steuewald, N., Main, E. K., Lambert, H., Main, E. K , and Steinleitner, A. (1994). Detection of human leukocyte antigen class I messenger ribonucleic acid transcripts in human spermatozoa via reverse transcription–polymerase chain reaction. *Fertil. Steril.* **61,** 276–280.

Clarke, G. N. (1988). Immunoglobulin class and regional specificity of antispermatozoal autoantibodies blocking cervical mucus penetration. *Am. J. Reprod. Immunol. Microbiol.* **16,** 135–138.

Constant, S. L., and Bottomly, K. (1997). Induction of Th1 and Th2 CD4+ T cell responses: The alternative approaches. *Annu. Rev. Immunol.* **15,** 297–322.

Diekman, A. B., Westbrook-Case, A., Naaby-Hansen, S., Klotz, K. L., Flickinger, C. J., and Herr, J. C. (1997). Biochemical characterization of sperm agglutination antigen-1, a human sperm surface antigen implicated in gamete interaction. *Biol. Reprod.* **57,** 1136–1144.

Dunbar, B. S. (1995). Role of ovarian antigens in fertility and infertility. *In* "Immunology of Human Reproduction" (M. Kurpisz and N. Fernandez, eds.), pp. 115–132. Bios Scientific Publishers, Oxford, UK.

Dunbar, B. S., Avery, S., Lee, V., Prasad, S., Schwahn, D., Schwoebel, E., Skinner, S., and Wilkins, B. (1994). The mammalian zona pellucida: Its biochemistry, immunochemistry, molecular biology, and developmental expression. *Reprod. Fertil. Dev.* **6,** 331–347.

Dym, M., and Fawcett, D. W. (1970). The blood–testis barrier in the rat and the physiological compartmentation of the seminiferous epithelium. *Biol. Reprod.* **3,** 308–326.

East, I. J., Mattison, D. R., and Dean, J. (1984). Monoclonal antibodies to the major protein of the murine zona pellucida: Effects on fertilization and early development. *Dev. Biol.* **104,** 49–56.

Edwards, R. G. (1960). Complement-fixing activity of normal rabbit serum with rabbit spermatozoa and seminal plasma. *J. Reprod. Fertil.* **1,** 268–282.

El-Demiry, M., and James, K. (1988). Lymphocyte subsets and macrophages in the male genital tract in health and disease. *Eur. Urol.* **14,** 226–235.

Epifano, O., and Dean, J. (1994). Biology and structure of the zona pellucida: A target for immunocontraception. *Reprod. Fertil. Dev.* **6,** 319–330.

Flickinger, C. J., Howards, S. S., Baran, M. L., Pressoa, N., and Herr, J. C. (1997). Appearance of 'natural' antisperm autoantibodies after sexual maturation of normal Lewis rats. *J. Reprod. Immunol.* **33,** 127–145.

Garza, K. M., Agersborg, S. S., Baker, E., and Tung, K. S. K. (2000). Persistence of physiological self antigen is required for the regulation of self tolerance. *J. Immunol.* **164,** 3982–9.

Gaytan, F., Morales, C., Bellido, C., Aguilar, E., and Sanchez-Criado, J. E. (1998). Ovarian follicle macrophages: Is follicular atresia in the immature rat a macrophage-mediated event? *Biol. Reprod.* **58,** 52–59.

Grootenhuis, A. J., Philipsen, H. L. A., de Breet-Grijsbach, J. T. M., and van Duin, M. (1991). Immunocytochemical localization of ZP3 in primordial follicles of rabbit, marmoset, rhesus monkey and human ovaries using antibodies against human ZP3. *J. Reprod. Fertil.* Suppl. **50,** 43–54.

Gwatkin, R. B. L., Williams, D. T., and Carlo, D. J. (1977). Immunization of mice with heat-solubilized hamster zonae: Production of anti-zona antibody and inhibition of fertility. *Fertil. Steril.* **28,** 871–877.

Hasegawa, A., Yamasaki, N., Inoue, M., Koyama, K., and Isojima, S. (1995). Analysis of an epitope sequence recognized by a monoclonal antibody MAb-5H4 against a porcine zona pellucida glycoprotein (pZP4) that blocks fertilization. *J. Reprod. Fertil.* **105,** 295–302.

Hendry W. F. (1992). The significance of antisperm antibodies measurement and management. *Clin. Endocrinol.* **36,** 219–21.

Hjort, T., and Meinertz, H. (1988). Anti-sperm antibodies and immune subfertility. *Hum. Reprod.* **3,** 59–62.

Itoh, M., Hiramine, C., and Hojo, K. (1991). A new murine model of autoimmune orchitis induced by immunization with viable germ cells alone. 1. Immunological and histological studies. *Clin. Exp. Immunol.* **83,** 137–142.

James, K., and Skibinski, G. (1995). Immunosuppressive factors in human seminal plasma: Their effects, characterization and possible mode of action. *In* "Immunology of Human Reproduction" (M. Kurpisz and N. Fernandez, eds.), pp. 267–283. Bios Scientific Publishers, Oxford, UK.

Johnson, M. H. (1968). Characterization of a natural antibody in normal guinea-pig serum reacting with homologous spermatozoa. *J. Reprod. Fertil.* **16,** 503–506.

Kameda, K., Tsuji, Y., Koyama, K., and Isojima, S. (1992). Comparative studies of the antigens recognized by sperm-immobilizing monoclonal antibodies. *Biol. Reprod.* **46,** 349–357.

Kaul, R., Afzalpurkar, A., and Gupta, S. K. (1996). Strategies for designing an immunocontraceptive vaccine based on zona pellucida synthetic peptides and recombinant antigen. *J. Reprod. Fertil.* (Suppl.) **50,** 127–134.

Kong, M., Richardson, R. T., Widgren, E. E., and O'Rand, M. G. (1995). Sequence and localization of the mouse autoantigenic protein, Sp17. *Biol. Reprod.* **53,** 579–590.

Koyama, K., Hasegawa, A., and Isojima, S. (1994). Porcine zona pellucida antigen 4 (PZP-4) as a candidate vaccine. *In* "Recombinant and Synthetic Vaccine" (G. P. Talwar, K. V. S. Rao, and V. S. Chauhan, eds.), pp. 502–512. Narosa Publishing House, New Delhi.

Kremer, E. J, and Kistler, W. S. (1992). Analysis of the promoter for the gene encoding the testis-specific histone H1t in a somatic cell line: Evidence for cell-cycle regulation and modulation by distant upstream sequences. *Gene* **110,** 167–173.

Kurpisz, M., and Janitz, J. (1995). Antigens and immunomodulatory factors in spermatogenesis. *In* "Immunology of Human Reproduction" (M. Kurpisz and N. Fernandez, eds.), pp. 33–52. Bios Scientific Publishers, Oxford, UK.

Kurpisz, M., Fernandez, N., and Fiszer, D. (1995). The expression of MHC genes in reproductive organs and their role in reproduction. *In* "Immunology of Human Reproduction" (M. Kurpisz and N. Fernandez, eds.), pp. 163–184. Bios Scientific Publishers, Oxford, UK.

Kurth, B. E., Weston, C., Reddi, P. P., Bryant, D., Bhattacharya, R., Flickinger, C. J., and Herr, J. C. (1997). Oviductal antibody response to a defined recombinant sperm antigen in Macaques. *Biol. Reprod.* **57,** 981–989.

Lea, I. A., Adoyo, P., and O'Rand, M. G. (1997). Autoimmunogenicity of the human sperm protein Sp17 in vasectomized men and identification of linear B cell epitopes. *Fertil. Steril.* **67,** 355–361.

Lea, I. A., Kurth, B., and O'Rand, M. G. (1998a). The immune response to immunization with sperm antigens in the macaque oviduct. *Biol. Reprod.* **58,** 794–800.

Lea, I. A., van Lierop, M. J. C., Widgren, E. E., Grootenhuis, A., Wen, Y., van Duin, M., and O'Rand,

M. G. (1998b). A chimeric sperm peptide induced antibodies and strain-specific reversible infertility. *Biol. Reprod.* 59, 527–536.

Lou, Y. H., Ang, J., Thai, H., Mcelveen, F., and Tung, K. S. K. (1995). A zona pellucida 3 peptide vaccine induces antibodies and reversible infertility without ovarian pathology. *J. Immunol.* **155,** 2715–2720.

Mahi-Brown, C. A., Yule, T. D., and Tung, K. S. K (1988). Evidence for active immunological regulation in prevention of testicular autoimmune disease independent of the blood–testis barrier. *Am. J. Reprod. Immunol.* **16,** 165–170.

Maity, R., Caspi, R. R., Nair, S., Rizzo, L. V., and Nelson, L. M. (1997). Murine postthymectomy autoimmune oophoritis develops in association with a persistent neonatal-like Th2 response. *Clin. Immunol. Immunopathol.* **83,** 230–236.

Millar, S. E., Chamow, S. M., Baur, A. W., Oliver, C., Robey, F., and Dean, J. (1989). Vaccination with a synthetic zona pellucida peptide produces long-term contraception in female mice. *Science* **246,** 935–938.

Mosmann, T. R., Cherwinski, H., Bond, M. W., Giedlin, M. A., and Coffman, R. L. (1986). Two types of murine helper T cell clone. I. Definition according to profiles of lymphokine activities and secreted proteins. *J. Immunol.* **136,** 2348–2357.

Mosmann, T. R., Schumacher, J. H., Street, N. F., Budd, R., O'Garra, A., Fong, T. A., Bond, M. W., Moore, K. W., Sher, A., and Fiorentino, D. F. (1991). Diversity of cytokine synthesis and function of mouse CD4+ T-cells. *Immunol. Rev.* **123,** 209–229.

Nashan, D., Cooper, T. C., Knuth, U. A., Schubeis, P., Sorg, C., and Nieschlag, E. (1990). Presence and distribution of leucocyte subsets in the murine epididymis after vasectomy. *Int. J. Androl.* **13,** 39–49.

Nashan, D., Malorney, U., Sorg, C., Cooper, T. G., and Nieschlag, E. (1989). Immuno-competent cells in the murine epididymis. *Int. J. Androl.* **12,** 85–94.

O'Brien, D. A., and Millette, C. F. (1984). Identification and immunochemical characterization of spermatogenic cell surface antigens that appear during early meiotic prophase. *Dev. Biol.* **101,** 307–317.

O'Rand, M. G. (1980). Antigens of spermatozoa and their environment. *In* "Immunological Aspects of Infertility and Fertility Regulation" (D. H. Dhindsa and G. F. B. Schumaker, eds.), pp. 155–171. Elsevier, New York.

O'Rand, M. G. (1986). Steps in the fertilization process. *In* "The Molecular and Cellular Biology of Fertilization" (J. L. Hedrick, ed.), pp. 383–393. Plenum Press, New York.

O'Rand, M. G. (1995). Antigenic targets on sperm for immunological interception. *In* "Immunology of Human Reproduction" (M. Kurpisz and N. Fernandez, eds.), pp. 473–484. Bios Scientific Publishers, Oxford, UK.

O'Rand, M. G., and Metz, C. B. (1976). Isolation of an "immobilizing antigen" from rabbit sperm membranes. *Biol. Reprod.* **14,** 586–598.

O'Rand, M. G., and Nikolajczyk, B. S. (1991). Fertilization in the rabbit. *In* "A Comparative Overview of Mammalian Fertilization" (B. S. Dunbar and M. G. O'Rand, eds.), pp. 271–279. Plenum Press, New York.

O'Rand, M. G., and Romrell, L. J. (1977). Appearance of cell surface auto- and isoantigens during spermatogenesis in the rabbit. *Dev. Biol.* **55,** 347–358.

O'Rand, M. G., and Widgren, E. E. (1994). Identification of sperm antigen targets for immunocontraception: B-cell epitope analysis of Sp17. *Reprod. Fertil. Dev.* **6,** 289–296.

Pelliniemi, L. J., Frojdman, K., and Paranko, J. (1993). Embryological and prenatal development and function of Sertoli cells. *In* "The Sertoli Cell" (L. D. Russell and M. D. Griswold, eds.), pp. 88–113. Cache River Press, Clearwater, FL.

Rhim, S. H., Millar, S. E., Robey, F., Luo, A., Lou, Y., Yule, T., Allen, P. Dean, J., and Tung, K. S. K. (1992). Autoimmune disease of the ovary induced by a ZP3 peptide from the mouse zona pellucida. *J. Clin. Invest.* **89,** 28–35.

Romrell, L. J., and O'Rand, M. G. (1978). Capping and ultrastructural localization of sperm surface isoantigens during spermatogenesis. *Dev. Biol.* **63,** 76–93.

Sacco, A. G., and Yurewicz, E. C. (1994). Assessment of synthetic peptides from porcine zona glycoproteins ZP3α and ZP3β for contraceptive vaccine development. *In* "Recombinant and Synthetic Vaccines" (G. P. Talwar, K. V. S. Rao, and V. S. Chauhan, eds.), pp. 502–512. Narosa Publishing House, New Delhi.

Salomon, B., Lenschow, D. J., Rhee, L., Ashourian, N., Singh, B., Sharpe, A., and Bluestone, J. A. (2000). B7/CD28 costimulation is essential for the homeostasis of the CD4+CD25+ immunoregulatory T cells that control autoimmune diabetes. *Immunity* **12,** 431–40.

Sarkar, S. (1974). Carbohydrate antigens of human sperm and autoimmune induction of infertility. *J. Reprod. Med.* **13,** 93–99.

Seddon, B., and Mason, D. (1999). Peripheral autoantigen induces regulatory T cells that prevent autoimmunity. *J. Exp. Med.* **189,** 877–882.

Setchell, B. P., Scott, T. W., Voglmayr, J. K., and Waites, G. M. (1969). Characteristics of testicular spermatozoa and the fluid which transports them into the epididymis. *Biol. Reprod.* (Suppl.) **1,** 46–66.

Smith, H., Chen, I. M., Kubo, R., and Tung, K. S. K. (1989). Neonatal thymectomy results in a repertoire enriched in T cells deleted in adult thymus. *Science* **245,** 749–752.

Smith, H, Sakamoto, Y., Kasai, K., and Tung, K. S. K. (1991). Effector and regulatory cells in autoimmune oophoritis elicited by neonatal thymectomy. *J. Immunol.* **147,** 2928–2933.

Sun, W., Lou, Y. H., Dean, J., and Tung, K. S. K. (1999). A contraceptive peptide vaccine targeting sulfated glycoprotein ZP2 of the mouse zona pellucida. *Biol. Reprod.* **60,** 900–907.

Taguchi, O., and Nishizuka, Y. (1980). Autoimmune oophoritis in thymectomized mice: T cell requirement in adoptive cell transfer. *Clin. Exp. Immunol.* **42,** 324–331.

Taguchi, O., Nishizuka, Y., Sakakura, T., and Kojima, A. (1980). Autoimmune oophoritis in thymectomized mice: Detection of circulating antibodies against oocytes. *Clin. Exp. Immunol.* **40,** 540–553.

Takahashi, T., Tagami, T., Yamazaki, S., Uede, T., Shimizu, J., Sakaguchi, N., Mak, T. W., and Sakaguchi, S. (2000). Immunologic self-tolerance maintained by CD25+CD4+ regulatory T cells constitutively expressing cytotoxic T lymphocyte-associated antigen 4. *J. Exp. Med.* **192,** 303–310.

Taylor, P. A., Noelle, R. J., and Blazar, B. R. (2001). CD4+CD25+ immune regulatory cells are required for induction of tolerance to alloantigen via costimulatory blockade. *J. Exp. Med.* **193,** 1311–1318.

Teuscher, C. (1985). Experimental allergic orchitis in mice, II: Association of disease susceptibility with the locus controlling *Bordetella pertussis*-induced sensitivity to histamine. *Immunogenetics* **22,** 417–425.

Teuscher, C., Smith, S. M., Goldberg, E. H., Shearer, G. M., and Tung, K. S. K. (1985). Experimental allergic orchitis in mice, I: Genetic control of susceptibility and resistance to induction of autoimmune orchitis. *Immunogenetics* **22,** 323–333.

Tsuji, Y. (1995). Carbohydrate antigens recognized by anti-sperm antibodies. *In* "Immunology of Human Reproduction" (M. Kurpisz and N. Fernandez, eds.), pp. 101–114. Bios Scientific Publishers, Oxford, UK.

Tsunoda, Y., and Chang, M. C. (1976). Reproduction in rats and mice isoimmunized with homogenates of ovary or testes with epididymis, or sperm suspensions. *J. Reprod. Fertil.* **46,** 379–382.

Tung, K. S. K. (1980). Autoimmunity of the testis. *In* "Immunological Aspects of Infertility and Fertility Regulation" (D. H. Dhindsa and G. F. B. Schumaker, eds.), pp. 33–91. Elsevier, New York.

Tung, K. S. K. (1995). Experimental autoimmune orchitis in the mouse: Immunoregulation and pathogenesis. *In* "Immunology of Human Reproduction" (M. Kurpisz and N. Fernandez, eds.), pp. 101–114. Bios Scientific Publishers, Oxford, UK.

Tung, K. S. K., Cooke, W. D., McCarthy, T. A., and Robitaille, P. (1976). Human sperm antigens and antisperm antibodies II. Age related incidence of antisperm antibodies. *Clin. Exp. Immunol.* **25,** 73–79.

Van Voorhis, B. J., and Stovall, D. W. (1997). Autoantibodies and infertility: A review of the literature. *J. Reprod. Immunol.* **33,** 239–256.

Vitale, R., Fawcett, D. W., and Dym, M. (1973). The normal development of the blood–testis barrier and the effects of clomiphene and estrogen treatment. *Anat. Rec.* **176,** 333–344.
Welch, J. E., and O'Rand, M. G. (1990). Characterization of a sperm-specific nuclear autoantigenic protein. II. Expression and localization in the testis. *Biol. Reprod.* **43,** 569–578.
Wood, D., Liu, C., and Dunbar, B. (1981). Effect of alloimmunization and heteroimmunization with zonae pellucidae on fertility in rabbits. *Biol. Reprod.* **25,** 439–450.
Yule, T. D., Montoya, G. D., Russell, L. D., Williams, T. M., and Tung, K. S. K. (1988). Autoantigenic germ cells exist outside the blood testis barrier. *J. Immunol.* **141,** 1161–1167.

13

FERTILIZATION BIOPHYSICS

D. P. L. GREEN

Department of Anatomy and Structural Biology, University of Otago Medical School, Dunedin, New Zealand

I. INTRODUCTION

Fertilization biophysics is not a subject with a well-defined meaning. Broadly speaking, however, it can be construed as encompassing those processes in fertilization that lend themselves to the application of physical principles. In general terms, this means those processes associated with sperm motility, sperm as force generators, the collision of gametes, the mechanics of sperm adhesion to surfaces, and the mechanics of sperm movement through barriers such as jelly coats and zonae pellucidae. Much of the research on animal fertilization has focused on a few experimentally tractable systems, notably mammals and the marine invertebrates. These systems are anisogamous, systems in which a small motile sperm fuses with a large egg. The physics of these fertilizations is focused strongly on the flagellar movement of the sperm. Within the past decade or so, some of the physical problems associated with this kind of anisogamous fertilization have yielded to theoretical analysis and experiment. It is on these developments that this review largely focuses. The review excludes fertilization in plants. Nor does the review include collisions and attachments such as are seen in *Chlamydomonas* or the movement of crawling sperm seen, for example, in nematodes.

II. SPERM AS FORCE-GENERATING MACHINES

A typical sperm from a mammal or a marine invertebrate is a single cell with a small head driven by a flagellum. Flagellar lengths vary considerably between species (Cummins and Woodall, 1985). In mammals, there is a common but not universal inverse relationship between the body weight of the animal and the length of the flagellum. Many large mammals (>10 kg) have sperm that are <60 μm in length, whereas the noolbenger, for example, a small Australian marsupial that has an adult body weight of about 10 g, has sperm that are ~350 μm long. Posterior to the head is a mitochondrial sheath with a length related to the overall length of the tail (Cardullo and Baltz, 1991). The mitochondria are responsible for generating ATP. The relationship between flagellar length and mitochondrial length reflects the need to generate a concentration of ATP that, on diffusion through the flagellum, will power the production of bending waves along its length. Unlike the movement of cilia, which have a clear power stroke in one direction and a recovery stroke in the reverse direction, sperm flagella are propelled by successive waves of bending traveling to their distal tips. In some of the marine invertebrates, these bending waves approximate sine waves fairly closely, but the picture is more complex in mammalian sperm (Green, 1988).

Sperm swim at low Reynold's number, where inertial forces can be ignored. Rigorous analysis of flagellar movement in this environment is due to Lighthill (1976). Lighthill analyzed flagellar hydrodynamics in general, however, and did not focus specifically on sperm. The first calculation of the forces generated by sperm using resistive force theory gave a calculated thrust for bull sperm of 8 pN (Green and Purves, 1984). An alternative calculation of thrust undertaken at the same time and based on a simple Stokes' law argument gave a crude, order-of-magnitude value for the thrust exerted by bull sperm of 5 pN. With hindsight, an approximate estimate of the thrust exerted by sperm could have been derived at any time in the past 40 years or so. Further calculations assessed the effects on sperm thrusts of proximity to a surface such as a zona pellucida (Katz and deMestre, 1985). Thrusts were calculated to increase by 20% when sperm are orientated at right angles, and up to 85% when sperm are orientated obliquely to the zona pellucida. Thrusts (maximal thrusts, presumably) were calculated as 38 pN for human, 80 pN for rabbit, and 357 pN for hamster (3.8, 8, and 35.7 μdynes, respectively).

An experimental examination of sperm mechanics undertaken subsequently showed that forces generated by sperm attached to surfaces were more complex than simple thrusts (Baltz *et al.*, 1988). In a demanding set of experiments, human sperm were sucked onto the tips of micropipettes and allowed to become free by their own movements as the suction pressure was reduced. Becase $P = F/A$, where P is the pressure difference across the sperm head that just holds sperm onto the pipette, F is the force just holding the sperm onto the pipette, and A is the cross-sectional area of the pipette orifice, a knowledge of P and A gives a value for F. It is this force that sperm must overcome if they are to detach themselves. Prior to these experiments, it was not known how sperm would come free, but essentially there were only two possibilities. Sperm would either slide across the pipette orifice by forward or sideway thrust (thrust-derived force) or pry themselves off by using the edge of the pipette as a fulcrum and the flagellum as a lever (torque-derived force). All these forces, irrespective of their direction or method of generation, were described by Baltz *et al.* (1988) as pulling forces. Comparison of the measured force that sperm need to pull themselves off the pipette with calculations of thrust-derived and torque-derived forces showed that the pulling force was always similar to the torque-derived force and an order of magnitude larger than any thrust-derived force. (Because time-dependent, three-dimensional waveforms were not available, calculations were undertaken on an idealized waveform represented by a helix of elliptical cross-section contained in a conical envelope with its point at the junction of head and flagellum.) For human sperm, the *measured* pulling force was 200 pN (20 μdynes) against a calculated torque-derived force of 190 pN (19 μdynes). The calculated thrust-derived force was 16 pN (1.6 μdynes) for human sperm held on a pipette and 19 pN (1.9 μdynes) for sperm against the zona pellucida. Hence the conclusion that the pulling force is torque derived, not thrust derived. Figures calculated for *thrusts* for hamster, guinea pig and rabbit against zonae pellucidae were 99, 43, and 28 pN (9.9, 4.3, and 2.8 μdynes), respectively. By contrast, torque-derived pulling forces were calculated for human, hamster, guinea pig, and rabbit sperm against zonae pellucidae as 100, 2150, 140, and 90 pN, respectively. An altogether different approach to measurement of sperm forces using optical tweezers to hold sperm by their heads gave values of 10–60 pN for sea urchin and human sperm thrusts (Bonder *et al.*, 1990) and a mean force of 44 ± 22 pN for human sperm (Konig *et al.*, 1996). There can be little doubt that human sperm exert thrusts of the order of 10–50 pN, depending on their immediate situation, and that calculation using resistive force theory reflects this. It follows that sperm thrusts for other species can be calculated using relevant experimental data.

The values for sperm thrusts and pulling forces are important in three distinct contexts. On the one hand, sperm need, in many cases, to become tethered to eggs during fertilization, either to jelly coats or to zonae pellucidae. The tethering has to be strong enough to resist pulling by sperm. On the other hand, sperm have to avoid becoming tethered accidentally on surfaces from which they cannot pull away. This applies particularly to animals with internal fertilization where surfaces abound. Third, sperm have to drive through egg vestments. Confidence in the magnitude of sperm thrusts and pulling forces allows some useful statements to be made about sperm in these situations.

III. TETHERING SPERM

Sperm are captured on the surface of jelly coats in fertilization of marine invertebrates and can be captured by the surface of the zona pellucida of cumulus-free mammalian oocytes. The forces that bind sperm under these circumstances have to be larger than those that sperm exert to pull away, otherwise fertilization would seldom take place. To make progress in assessing the tethering requirements of sperm, one needs to know the strengths both of covalent bonds and of reversible (or affinity) interactions such as those between receptor and ligand. There has been considerable progress in the past two decades or so in both theoretical analysis and experimental measurement of forces needed to break bonds.

The force needed to break bonds can be calculated using the kinetic theory of fracture (Zhurkov, 1965). This was adapted by Bell (1978) to accommodate rupture of noncovalent interactions such as those between receptors and ligands. Any tensile force applied to a bond or interaction has the effect of shortening bond lifetime. If sufficient, the force ruptures the bond, reducing its lifetime to a small fraction of its normal value. Covalent bonds have long lifetimes for the most part, and a rupturing force will reduce bond lifetime to the order of seconds. Noncovalent (or affinity) interactions are marked by lower bond energies, rapid reversibility, and short bond or interaction lifetimes. The force needed to rupture such bonds or interactions is that needed to cause rupture or unbinding in the absence of spontaneous dissociation.

The tensile strength of bonds and interactions is, to a first approximation, E/d, where, for a noncovalent interaction, E is the free energy of binding ligand to receptor and d is the effective length of the bond (Zhurkov, 1965; Bell, 1978). The free energy of binding is directly related to the binding constant. Using this approach, the main uncertainty in calculating the tensile strengths of noncovalent interactions is the effective bond length, because the calculated force needed to rupture the noncovalent bond is inversely proportional to its length. By contrast, the effect of the binding constant is much smaller. For example, a four-order-of-magnitude shift in the value of the binding constant, from 10^6 to 10^{10} M^{-1} reflects an increase in the free energy of binding of only 60%. Forces needed to rupture noncovalent interactions were calculated, on this approach, to lie between 60 and 2.5 nN (Baltz *et al.,* 1988; Baltz and Cone, 1990).

In practice, the situation is more complex. Experimental values for a number of related interactions based on the avidin–biotin system show no relationship between the force needed to rupture the interaction (the unbinding force) and the free energy change on binding (Moy *et al.,* 1994). Instead, a simple relationship, $F_u = \Delta H/r_{eff}$, exists in this system between rupturing force and enthalpy change (where F_u is the rupturing or unbinding force, r_{eff} is the effective width of the binding potential (or effective rupture length), and ΔH is the enthalpy change associated with bond rupture or unbinding). Even this relationship is not valid for the general case of ligand–receptor systems. A more appropriate expression would be $F_u = \Delta H^*/r_{eff}$, where ΔH^* is the enthalpy of activation. Measured rupturing forces (unbinding forces) in the avidin–biotin system ranged from 94 to 257 pN (Moy *et al.,* 1994). The rupturing force for a single actin filament has also been measured and found to be 110 pN (Kishino and Yanagida, 1988).

Of some interest, given the influence of the effective bond length on calculated rupturing forces, are the respective effective bond lengths for actin monomer–monomer binding and avidin–biotin binding. The effective bond length, r_{eff}, for actin monomer–monomer binding is 0.1–0.3 nm, whereas that for avidin–biotin binding is about 0.95 nm (Moy *et al.,* 1994). The difference reflects the fact that recognition between actin monomers is over flat surfaces and is effectively the range of van der Waals contact, whereas biotin binds to a pocket in avidin with an interaction potential that extends inward from the entrance to the pocket. Taken together, there is broad agreement between theory and experiment that forces of the order of 100–300 pN can be needed to break specific ligand–receptor interactions.

Forces that are below those needed to rupture bonds also have effects on bond lifetime, whether the bond is covalent or noncovalent. Examples of the reduction in bond lifetime in two noncovalent interactions following application of a tensile load are the approximate twofold increase in detachment rate in CD2–CD48 interactions following an increase in the force exerted on the bond from about 11 to 22 pN (Pierres *et al.,* 1996), and the approximately 3.5-fold increase in dissociation rate of selectin–ligand interactions following application of a force of 110 pN (Alon *et al.,* 1995).

Although the identities of the receptors and ligands mediating adhesion between sperm and jelly coats or zonae pellucidae are still not established with complete certainty, individual sperm–egg receptor bonds are likely to require forces of the order of 100 pN or more for rupture (reasons are given in Section IV). These forces are larger than the purely thrusting forces of sperm discussed earlier. They are also of the same order of magnitude as the pulling forces of many sperm, although hamster sperm can apparently exert a larger pulling force. It was concluded a decade ago that sperm can be tethered by a few noncovalent interactions, possibly as few as one, between themselves and a surface (Baltz *et al.,* 1988; Baltz and Cone, 1990). This conclusion remains sound. Even where sperm pry themselves from an interaction by pitching or rolling against a surface, there must remain a distinct possibility that the point of contact needed to generate torque simply allows another noncovalent interaction to form, which requires a further prying movement to break. If that happened, sperm would never break free.

Other important consequences flow from these conclusions. First, because sperm capture by the egg plays an important role in normal fertilization, it seems likely that only a few noncovalent bonds are required to tether sperm for this initial capture. Second, when sperm traverse egg vestments, they must either avoid establishing noncovalent interactions in any number, or have some means of severing the interaction by means other than force. Third, the magnitudes of the forces generated by sperm and those required to rupture bonds become an important consideration when deciding how sperm penetrate egg vestments. Fourth, sperm must avoid binding to cells and viscoelastic secretions that would provide unbreakable attachments and prevent them from reaching the egg.

IV. SPERM CAPTURE BY EGGS

Sperm capture by eggs is an important feature of many fertilizations, for without it sperm are not always in a position to proceed to penetration of the egg coat. It is a potentially complex problem. Clearly, capture has to be preceded by prior collision between sperm and egg. This needs to be followed by the interaction of receptors and ligands or counterreceptors on sperm and eggs to produce a tether. This tether, if it survives, allows sperm and egg to consolidate their adhesion by establishing multiple interactions.

Possibly the best characterized system in biophysical terms is that between mouse sperm and the zona pellucida (ZP). The receptor on the zona pellucida for mouse sperm is the zona pellucida protein ZP3 [see Thaler and Cardullo (1996a) for references]. Like all zona pellucida proteins, ZP3 is a glycoprotein. The sperm binding capacity of ZP3 is associated with O-linked oligosaccharides, not the polypeptide. The identity of the specific saccharide residues responsible for binding remains controversial, however (Thaler and Cardullo, 1996a,b). There is evidence that the terminal monosaccharide involved in binding is α-galactose but transgenic mice lacking all α-1,3-galactose residues are fertile (Thall *et al.,* 1995). There is also some evidence to suggest that the terminal monosaccharide, β-*N*-acetylglucosamine (β-GlcNAc), acts as a ligand (rather than substrate) for β-1,4-galactosyltransferase on the sperm surface. However, transgenic mice lacking β-1,4-galactosyltransferase are fertile (Lu and Shur, 1997). It is, perhaps, dangerous to conclude from these results that the respective terminal saccharides play no role whatsoever in fertilization, although that is the simplest conclusion. It is clearly possible that a dual (or multiple) system of receptors and ligands operates to bind sperm to the zona pellucida. It would be of interest to know the phenotype of the Galα1,3Gal/β‴1,4-galactosyltransferase double-null mutant. Other candidate receptors have been identified on sperm, such as zonadhesin, but the ligands have yet to be identified (Hardy and Garbers, 1995).

Whatever the identity of the ZP3 ligand, there is evidence that the corresponding sperm receptors are proteins. Calmegin is a testis-specific protein that is present in the endoplasmic reticulum (ER) and homologous to the ER chaperone calnexin. Its role in the testis is likely to be in protein folding. Knockout mice that are homozygously null for calmegin are nearly infertile (Ikawa *et al.,* 1997). The near sterility of calmegin knockouts is apparently due to an inability of sperm to adhere to the zona pellucida. This suggests that a protein folding failure on the sperm surface is responsible for infertility, indirectly identifying the sperm receptors for ZP3 as proteins.

Data on the kinetics of ZP3 binding to glutaraldehyde-fixed mouse sperm using whole solubilized ^{125}I-labeled ZP suggest a complex interaction for sperm/ZP binding (Thaler and Cardullo, 1996a). An association rate of $3.2 \times 10^6\ M^{-1}\ min^{-1}$ is matched by two dissociation rates, indicating the presence of high- and low-affinity components of sperm/ZP binding, with K_d values of 0.72 and 50 nM respectively. The sizes of the dissociation constants suggest specific interactions. Although the effective bond lengths of ZP/sperm receptors are unknown, the rupturing (or unbinding) forces required to break these individual sperm/ZP bonds are, on the basis of previous arguments, likely to be of the order of 100 pN or more (see Section III). Other evidence suggests that the interaction between sperm and ZP reflects a multivalent ligand in ZP3 as well as the possible presence of multiple receptors on sperm (Thaler and Cardullo, 1996a). If this is the case, the rupturing force could potentially be >100 pN, because all components of the tether would need to be broken at once if sperm were to break free.

Whether a sperm becomes tethered on the surface of the zona pellucida depends on the surface density of ZP3 ligands, the surface density of ZP3 receptors on the sperm surface (and, to some extent, their position), the contact area between sperm and zona pellucida, and the diffusion coefficient of the membrane-bound receptor on the sperm surface (Baltz and Cardullo, 1989). The ZP3 ligand density may be as high as 300 molecules/μm^2 on the surface of the zona pellucida and the contact area somewhere between 0.1 and 5 μm^2. The total number of ZP3 binding sites per mouse sperm is 30,000. If this is the case, the density of these receptors must be high, because they are all present in the periacrosomal plasma membrane (see Section VI). If the sperm ZP3 receptor is treated as a cylinder of 5 nm diameter projecting outward from the membrane, 30,000 receptors would occupy close to 2 μm^2 of membrane. This is of the same general order as the area of periacrosomal plasma membrane available. The density of ZP3 ligands on the zona pellucida and ZP3 receptors on sperm suggests that tethering and rapid consolidation of adhesion is inevitable once contact between the two is made. It also suggests that, as indicated earlier, torque-derived thrusts would simply result in the establishment of new tethers as sperm attempted to break free. It seems inevitable that sperm movement after tethering promotes further contact and consolidates binding through formation of additional sperm/ZP3 bonds.

Sperm adhesion to the zona pellucida establishes the general principles that must apply to any sperm capture by a surface. These principles would apply equally, for example, to sperm capture by jelly coats or accidental sperm capture, if it occurs, by the uterine wall. Receptors and ligands have been identified for abalone (Moy *et al.,* 1996), but kinetic data are currently unavailable.

V. SPERM PENETRATION OF EGG COATS

Eggs are protected by egg coats. These can be in single layers, as in mammals, or in multiple layers, as, for example, in marine invertebrates. They consists of filaments of varying degrees of density, cross-linking, etc. giving rise to concentric layers with a wide range of thicknesses and physical properties. The primary function of these coats appears to be the protection of the egg and early-stage embryo from parasites and phagocytic cells. They also provide substrates for establishing blocks to polyspermy and, in some cases, determine the approach path of sperm toward the egg surface.

Coevolution of sperm and eggs has allowed the development of mutual specializations to allow sperm passage. One simple example is the chorion of fish eggs. Chorions are thick and mechanically tough. They possess a single small hole, the micropyle, which runs through the thickness of the chorion. Sperm swim through the micropyle to reach the egg surface. By contrast, mammals and many marine invertebrate eggs are surrounded by jellies, gels, and envelopes without preformed slits or holes.

The challenges faced by sperm in penetrating these layers depend critically on the physical properties of the material being crossed. In no cases have these properties been studied in detail although there are some data for the zona pellucida. This discussion deals first with the zona pellucida before briefly discussing other egg coats. A review of the mammalian zona pellucida structure has been published by Green (1997) and only the relevant points, mainly derived from the mouse, are summarized here.

The zona pellucida is a spherical shell of elastic gel. In the mouse, it is composed of three proteins, ZP1, ZP2, and ZP3. These are probably globular in form. Although not proved, the most likely arrangement is of filaments of repeating ZP2/ZP3 units joined with intermittent cross-linking by ZP1. Any zona pellucida is difficult to study in a physically rigorous manner: they are only available in small pieces and attempts to reconstitute them have been unsuccessful. Other, more tractable, biological gels can, however, be used as points of reference.

The biological gel that has been investigated in most detail is probably that of actin, cross-linked either with α-actinins or with avidin–biotin (Wachsstock *et al.,* 1994). Amoeba α-actinin cross-links actin with highly reversible interactions, and the cross-linked gel behaves as a viscoelastic liquid. Actin cross-linked with biotin–avidin behaves like a solid and is elastic. These differences can be attributed, to a first approximation, to differences in the affinities of cross-linking interactions. (This assumes that association rates are similar and that differences in affinity reflect dissociation rates—the smaller the dissociation constant, the longer the interaction lifetime.) Cross-linking of actin with α-actinin from chicken smooth muscle, which has a higher affinity constant than amoeba actinin but a smaller one than biotin–avidin, gives a gel with physical properties intermediate between the other two (Wachsstock *et al.,* 1994). The limited data on stress relaxation in the mammalian zona pellucida indicate that it is elastic, with slow or nonexistent stress relaxation for deformations lasting minutes (Green, 1987). Stress concentrations significantly greater than those expected of sperm do not cause stress relaxation or any other evidence of viscoelastic liquid behavior. The physical properties of zonae pellucidae therefore most closely resemble that of actin cross-linked by streptavidin–biotin. It suggests that the zona pellucida is an elastic solid on any time-scale relevant to sperm penetration. This means that, for penetration, sperm must either push obstructing filaments to one side elastically, or sever them.

There are at least two lines of evidence suggesting that sperm do not penetrate the zona pellucida by purely elastic displacement of zona pellucida filaments. First, sperm leave behind an empty slit that marks the site of their passage. The existence of this slit is inconsistent with elastic displacement of zona pellucida filaments as the only means of penetration. Second, small probes with cross-sectional areas similar to sperm fail to enter the zona pellucida by generating holes elastically (Green, 1987), suggesting that it is beyond the capacity of sperm to do likewise. Calculation suggests that the density of zona pellucida filaments is simply too high. If sperm do not penetrate the zona pellucida by elastic displacement of zona pellucida filaments, then they must sever filaments, either by breaking them mechanically or cleaving them enzymatically.

Mechanical breakage of bonds is not a realistic explanation for sperm penetration of the zona pellucida, for two reasons. The first is the argument from sperm forces and the strength of the zona pellucida. If the zona pellucida behaves as an elastic solid, it is because it is held together by high-affinity noncovalent interactions (Green, 1997). These are likely to require forces >100 pN to rupture, even if severed singly. However, the filament density of the zona pellucida is such that filaments will be recruited in parallel, with a significant increase in the force needed to tension individual filaments to the breaking point. It follows that sperm must sever zona pellucida filaments enzymatically. A second, equally compelling, reason why the zona pellucida cannot be penetrated solely by mechanical means is the residual slit following sperm penetration. Even if filaments could be broken initially by sperm imposing tensile loads, a permanent slit requires further filament rupture. There is no obvious mechanism by which the free ends of filaments, created by rupture, could be held while a further tensile load was imposed. Instead, there would be elastic opening of the slit while sperm passed through, followed by its subsequent closure.

The inevitable conclusion is that mammalian sperm penetrate the zona pellucida using a combination of their own forward movement together with an enzyme to cleave zona pellucida filaments. For many years, the principal candidate for this enzyme was considered to be acrosin. However, an acrosin knockout mouse is fertile, with little detectable phenotype (Baba *et al.,* 1994). The enzyme used by sperm to create the penetration slit may therefore be a variant of acrosin or an entirely novel enzyme.

Another animal in which features associated with egg coat penetration are increasingly well-characterized is the abalone. Abalone eggs are surrounded by three concentric layers, an outer jelly layer, the vitelline envelope, and an egg surface coat (Mozingo *et al.,* 1995). The sperm receptor responsible for binding to the jelly coat has been identified (Moy *et al.,* 1996). Abalone sperm penetrate the outer jelly coat without undergoing an acrosome reaction. Morphological evidence suggests that the jelly coat may have pores of up to 1–2 μm (Bonnell *et al.,* 1994), although their length is uncertain. It is unclear whether these pores can be stretched by sperm to allow purely mechanical penetration or whether sperm carry a surface enzyme that is used to sever filaments to clear a path. It is also unclear how sperm make the transition from initial capture and tethering on the outer surface of the jelly coat, to movement through it (a problem that is discussed in more detail for mammalian sperm in Section VI). Once through the jelly coat, sperm attach to the vitelline envelope and release a lysin that titrates sites in the envelope, destabilizing a small portion of it in the region of the sperm head. This produces a hole through which sperm swim (Swanson and Vacquier, 1997).

Other examples of egg coat penetration could include *Saccoglossus* and the sea urchins, wherein the egg is surrounded by a vitelline envelope and an outer jelly coat. Sperm are captured by the jelly coat and undergo an acrosome reaction. This reaction generates an acrosomal extension that drives through the jelly coat and vitelline envelope until its tip makes contact with the egg plasma membrane (Tilney, 1985). The sperm head, meanwhile, remains on the outer surface of the egg coat. The acrosomal extension will meet physical resistance as it drives through the jelly, but the precise mechanism of penetration remains unknown. The physical properties of the jelly coat may allow penetration by purely physical means, or an enzyme may be required (Green and Summers, 1980). The force required to drive the acrosomal extension through the jelly coat will impose a detaching force on the sperm head, which will have to be opposed by the binding interactions between the sperm head and the jelly coat, and within the jelly coat. These interactions may be numerous enough to withstand detaching forces of many nanoNewtons. The force being used by the acrosomal extension to penetrate the jelly coat may therefore be much larger than that generated by flagellar movement alone. In *Hydroides,* the acrosome reaction precipitates the elevation of a number of small processes on the anterior region of the head. Because the head retains its original cross-sectional area during penetration of the jelly coat, it appears that penetration is due primarily to enzymatic digestion of a hole, with sperm being pushed forward using their flagella. This, of course, is a mechanism reminiscent of sperm penetration of the zona pellucida.

In all these cases, the broad physical principles involved in penetrating the egg coats can be identified. Detailed models cannot be built, however, because of the dearth of experimental data.

VI. THE TRANSITION FROM SPERM ADHESION TO PENETRATION OF THE ZONA PELLUCIDA

The sperm of many mammalian sperm adhere to the zona pellucida while their acrosome is intact. From the previous discussion it is clear that a small number of noncovalent bonds will tether sperm here indefinitely unless sperm possess a mechanism for releasing themselves for penetration of the zona pellucida. The mechanism deployed appears to be that of restricting the zona pellucida binding molecules on sperm to the periacrosomal plasma membrane. This detaches from the sperm head following the acrosome reaction. Initially, a direct physical connection remains through the intermediacy of the acrosomal matrix, but as this is broken down by proteolysis, sperm become free, leaving residual vesicles of plasma and outer acrosomal membrane on the surface of the zona pellucida.

An apparent paradox emerges at this point. If sperm are released by the acrosome reaction, they will swim free and not enter the zona pellucida. (This risk may be less in oocytes with an intact cumulus oophorus, but cumulus-free oocytes can still be fertilized readily. There must be some mechanism, therefore, that operates to ensure that sperm enter the zona pellucida without any need for the cumulus oophorus.) Any fresh sperm capture would involve retethering sperm, with the attendant risk that they could not subsequently move forward. Two solutions to this problem have been suggested (Baltz *et al.*, 1988). The first is that sperm adhere obliquely to the surface of zona pellucida, undergo the acrosome reaction, and move forward while loosely bound to the zona pellucida, possibly in the posterior region of the head. In doing so, they generate a wide, oblique slit in the zona pellucida (due to the horizontal oscillation of the head about the point of attachment). This produces a flap of zona pellucida that subsequently prevents them from pulling off directly. The loose contacts with the zona pellucida surface are progressively lost, and sperm move forward into the zona pellucida. This mechanism requires a highly organized and regulated sequence of events involving adhesion molecules on sperm. The suggested alternative is that the acrosome reaction generates a collar of vestigial acrosomal and periacrosomal membranes that allows forward movement of sperm but constrains their movement away from the zona pellucida. There is ample morphological evidence for vestigial collars of acrosomal and periacrosomal membranes that remain on the zona pellucida surface after the acrosome reaction, but whether all fertilizing sperm produce these is unclear. Equally, there is also no direct evidence for low-affinity sperm receptors for the zona pellucida.

VII. SUMMARY

Fertilization biophysics is potentially a large and amorphous subject. However, it includes topics of real importance. Those discussed in this chapter reflect topics that have proved tractable and that have increased our understanding of fertilization. The first of these is the forces that sperm exert. These are now known with reasonable accuracy from experimental measurements. The measured forces agree with those calculated theoretically, providing confidence that calculation can be used in other circumstances that are less accessible experimentally.

As our knowledge of sperm forces has developed, so also has a better understanding of the mechanical strength of single bonds, including noncovalent bonds. The forces needed to break noncovalent bonds range from 10 to 250 pN. Sperm could possibly break free from a single noncovalent bond, but a small number of noncovalent bonds holding a sperm simultaneously would collectively resist rupture and be unlikely to dissociate spontaneously in a synchronous manner. They would tether a sperm indefinitely. This has important consequences. It means that sperm in reproductive tracts must avoid premature attachments of any strength if they are to reach the egg.

Knowing the magnitude of the forces that sperm exert also allows some conclusions to be drawn about sperm penetration of egg coats. Those, such as the zona pellucida, are elastic and cannot be penetrated by purely mechanical means. This conclusion should encourage the search for an enzyme that enables sperm to penetrate the zona pellucida. Inhibitors of this enzyme could prove effective contraceptives.

There are any number of biophysical problems in fertilization that could be considered in the future. The challenge is to find important problems that are experimentally tractable. Two areas that may be worth exploring are molecules involved in physiological sperm capture, either at the surface of the egg coat or the egg plasma membrane, and the effects of small mechanical forces on the egg surface. Despite the small forces they exert, sperm attached to the surface of a mammalian egg are able to produce mechanical deformation. It is possible that, following sperm attachment to the egg, there is an important interplay between receptor stimulation and activation of signaling pathways on the one hand and local deformation of membrane and cortical architecture on the other.

REFERENCES

Alon, R., Hammer, D. A., and Springer, T. (1995). Lifetime of the P-selectin-carbohydrate bond and its response to tensile force in hydrodynamic flow. *Nature (London)* **374,** 539–542.

Baba, T., Azuma, S., Kashiwabara, S., and Toyoda, Y. (1994). Sperm from mice carrying a targeted mutation of the acrosin gene can penetrate the oocyte zona pellucida and effect fertilization. *J. Biol. Chem.* **269,** 31845–31849.

Baltz, J. M., and Cardullo, R. A. (1989). On the number and rate of formation of sperm–zona bonds in the mouse. *Gamete Res.* **24,** 1–8.

Baltz, J. M. and Cone, R. A. (1990). The strength of non-covalent biological bonds and adhesions by multiple independent bonds. *J. Theoret. Biol.* **142,** 163–178.

Baltz, J. M., Katz, D. F., and Cone, R. A. (1988). Mechanics of sperm–egg interaction at the zona pellucida. *Biophys. J.* **54,** 643–654.

Bell, G. I. (1978). Models for the specific adhesion of cells to cells. *Science (Washington D.C.).* **200,** 618–627.

Bonder, E. M., Colon, J. M., Dziedzic, J. M., and Ashkin, A. (1990). Force production by swimming sperm: Analysis using optical tweezers. *J. Cell Biol.* **111,** 421a.

Bonnell, B.S., Keller, S.H., Vacquier, V.D., and Chandler, D.E. (1994). The sea urchin egg jelly coat consists of globular glycoproteins bound to a fibrous fucan superstructure. *Dev. Biol.* **162,** 313–324.

Cardullo, R. A. and Baltz, J. M. (1991). Metabolic regulation in mammalian sperm: Mitochondrial volume determines sperm length and flagellar beat frequency. *Cell Motil. Cytoskel.* **19,** 180–188.

Cummins, J. M., and Woodall, P. F. (1985). On mammalian sperm dimensions. *J. Reprod. Fertil.* **75,** 153–175.

Green, D. P. L. (1987). Mammalian sperm cannot penetrate the zona pellucida solely by force. *Exp. Cell Res.* **169,** 31–38.

Green, D. P. L. (1988). Sperm thrusts and the problem of penetration. *Biol. Rev.* **63,** 79–105.

Green, D. P. L. (1997). Three-dimensional structure of the zona pellucida. *Rev. Reprod.* **2,** 147–156.

Green, D. P. L., and Purves, R. D. (1984). Mechanical hypothesis of sperm penetration. *Biophys. J.* **45,** 659–662.

Green, J. D., and Summers, R. G. (1980). Ultrastructural demonstration of trypsin-like protease in acrosomes of sea urchin sperm. *Science (Washington D.C.)* **209,** 398–400.

Hardy, D. M., and Garbers, D. L. (1995). A sperm membrane protein that binds in a species-specific manner to the egg extracellular matrix is homologous to von Willebrand factor. *J. Biol. Chem.* **270,** 26025–26028.

Ikawa, M., Wada, I., Kominami, K., Watanabe, D., Toshimori, K., Nishimune, Y., and Okabe, M. (1997). The putative chaperone calmegin is required for sperm fertility. *Nature (London)* **387,** 607–611.

Katz, D. F., and deMestre, N. J. (1985). Thrust generation by mammalian spermatozoa against the zona pellucida. *Biophys. J.* **47** (2, Pt. 2), 123a.

Kishino, A., and Yanagida, T. (1988). Force measurement by micromanipulation of a single actin filament by glass needles. *Nature (London)* **334,** 74–76.

Konig, K., Svaasand, L., Liu, Y., Sonek, G., Patrizio, P., Tadir, Y., Berns, M. W., and Tromberg, B. J. (1996). Determination of motility forces of human spermatozoa using an 800 nm optical trap. *Cell. Mol. Biol.* **42,** 501–509.

Lighthill, M. J. (1976). Flagellar hydrodynamics. *SIAM (Soc. Ind. Appl. Math.) Rev.* **18,** 161–230.

Lu, Q., and Shur, B. D. (1997). Sperm from β1,4-galactosyltransferase-null mice are refractory to ZP3-induced acrosome reactions and penetrate the zona pellucida poorly. *Development* **124,** 4121–4131.

Moy, G. W., Mendoza, L. M., Schulz, J. R., Swanson, W. J., Glabe, C. G., and Vacquier, V. D. (1996). The sea urchin sperm receptor for egg jelly is a modular protein with extensive homology to the human polycystic kidney disease protein, PKD1. *J. Cell Biol.* **133,** 809–817.

Moy, V. T., Florin, E.-L., and Gaub, H. E. (1994). Intermolecular forces and energies between ligands and receptors. *Science (Washington D.C.)* **266,** 257–259.

Mozingo, N. M., Vacquier, V. D., and Chandler, D. E. (1995). Structural features of the abalone egg extracellular matrix and its role in gamete interaction during fertilization. *Mol. Reprod. Dev.* **41,** 493–502.

Pierres, A., Benoliel, A. M., Bongrand, P., and Van der Merwe, P. (1996). Determination of the lifetime and force dependence of interactions of single bonds between surface-attached CD2 and CD48 adhesion molecules. *Proc. Natl. Acad. Sci. U.S.A.* **93,** 15114–15118.

Swanson, W. J., and Vacquier, V. D. (1997). The abalone egg vitelline envelope receptor for sperm lysin is a giant multivalent molecule. *Proc. Natl. Acad. Sci. U.S.A.* **94,** 6724–6729.

Thaler, C. D., and Cardullo, R. A. (1996a). The initial molecular interaction between mouse sperm and the zona pellucida is a complex binding event. *J. Biol. Chem.* **271,** 23289–23297.

Thaler, C. D., and Cardullo, R. A. (1996b). Defining oligosaccharide specificity for initial sperm–zona pellucida adhesion in the mouse. *Mol. Reprod. Dev.* **45,** 535–546.

Thall, A. D., Maly, P., and Lowe, J. B. (1995). Oocyte Galα1,3Gal epitopes implicated in sperm adhesion to the zona pellucida glycoprotein ZP3 are not required for fertilization in the mouse. *J. Biol. Chem.* **270,** 21437–21440.

Tilney, L. G. (1985). The acrosomal reaction. *In* "The Biology of Fertilization" (C. Metz and A. Monroy, eds.), Vol. 2, pp. 157–213. Academic Press, New York.

Wachsstock, D. H., Schwarz, W. H. and Pollard, T. D. (1994). Cross-linker dynamics determine the mechanical properties of actin gels. *Biophys. J* **66,** 801–809.

Zhurkov, S. N. (1965). Kinetic concept of the strength of solids. *Int. J. Frac. Mech.* **1,** 311–322.

14

Applications of Fertility Regulation for the Management of Wild and Domestic Species

Mark P. Bradley* and Peter Bird†

*Xcelerator Ltd., North Ryde, New South Wales, Australia,
†Therapeutics Goods Administration (Bacterial Vaccine Stream),
Woden, Australian Capital Territory, Australia

I. Introduction

Worldwide, the effective and humane management of wild and domestic animal populations is becoming a major issue (Sinclair, 1997; Artois, 1997; Cowan and Tyndale-Biscoe, 1997). This problem can be viewed from two completely different perspectives—the problem of controlling desirable animals in their native environments, and the problem of controlling in foreign environments introduced, exotic animals that are impacting negatively on the environment.

Some examples serve to illustrate the dichotomy of this issue. Elephant populations in Africa are increasing at a significant rate. As a consequence of increased urbanization, a conflict now exists between overlapping elephant habitats and the use of land for human habitation. In recent times suggestions have been made that effective control of elephant populations could be achieved through selective culling. This has led to an international outcry and the call for more humane control methods for a species that not long ago was considered to be under the threat of extinction. In the United States, numbers of white-tailed deer have increased and are making an impact into urban areas, where they become a problem, albeit more of a nuisance than a threat. Similar problems are being experienced with the increasing coyote populations in the United States national parks that border agricultural areas. As a consequence, coyote predation on sheep is becoming an issue in the agricultural communities. In Australia, the recent overpopulation by koalas on Kangaroo Island off the South Australian coast is leading to massive native forest destruction, and the proposal to manage the koala population by selective culling has led to international condemnation of the idea.

Such wildlife management issues have arisen for a variety of complex reasons, ranging from increased land use and urbanization to the introduction of exotic animals into foreign environments with unforeseen impacts on the survival of indigenous native species.

In Australia, these problems are clearly demonstrated by the vast tracts of land that were cleared for agricultural use in the nineteenth and twentieth centuries, thus reducing habitat for native species. The problem was further exacerbated by the introduction onto the continent of a large number of exotic animals on to the continent during the same period. These animals readily adapted to their new environment and became major feral pests. Some examples include the European rabbit, the European red fox, the camel, horses, pigs, rodents, and cats.

As a result of these perturbations it is estimated that up to 20 native Australian species of mammals have become extinct. Australia is now committed to spending large amounts of money to resolve this problem. Of all the exotic species introduced into Australia, three probably have had the greatest impact: the European rabbit (*Oryctolagus cuniculus*), the European red fox (*Vulpes vulpes*), and the feral domestic cat (*Felis catus*). The rabbit has been responsible for massive land degradation, which occurs easily in an already fragile environment. The fox and the cat, both predators, have found a ready food source in the numerous marsupial and native rodent species, thus leading to an extensive list of endangered and vulnerable species in Australia.

Often in the face of intense predation and the rapid decline of a particular species, the only means of allowing the recovery of these endangered populations is to establish captive breeding programs to increase numbers for eventual release into their natural habitats. An excellent example can be found in Western Australia, which because of its geographic isolation from the rest of the continent still provides a unique refuge for many native species that were once widely represented over the entire continent. To allow both *in situ* and reintroduced populations to recover, the Western Australia Government Department of Conservation and Land

Management (CALM) has instituted a program to develop fox predator-free areas primarily through the use of the poison "1080" delivered in suitable baits. This approach has certainly seen the dramatic recovery of several species from the brink of extinction (Kinnear *et al.,* 1988), but requires continuous and ongoing commitment to maintain its desirable outcome.

Unfortunately, any reduction in the level of activity in these control measures will inevitably lead to a rise in predator numbers, thus again threatening the continued survival of the very species being protected. The eventual long-term solution being proposed is to impose methods whereby the feral pests are maintained at low levels through the imposition of fertility control (Tyndale-Biscoe and Bradley, 1997). If successful, feral pest numbers will be reduced to a level where the impact of predation and habitat destruction is reduced, thus allowing the long-term recovery and survival of a range of the native species.

II. CASE STUDIES ON WILDLIFE MANAGEMENT

In a number of countries around the world, wild animals that act as vectors or reservoirs for disease pose both a major risk to human health and agriculture. Wildlife managers are looking for effective means of controlling such diseases, and although wild animal management has often routinely invoked the use of lethal means as the primary controlling measure, growing public concerns over animal welfare issues increasingly make such approaches unacceptable. Immunocontraception for wildlife management is therefore taking on a new perspective, and many agencies charged with such management decisions are turning to fertility control as a potential new means of dealing with such issues.

A. FOXES

In Europe the red fox transmits parasitic diseases to pet dogs and people, but most importantly is the main vector for the spread of rabies (Artois *et al.,* 1993). Foxes can transmit rabies to both people and domestic animals. Culling and vaccination are the current methods for fox rabies control. Over 14 million vaccine baits are distributed annually in Europe (Aubert, 1995). This achieves a 50% immunization rate of adult foxes, which has resulted in rabies being eliminated from large areas of western Europe. Most importantly, this work has demonstrated that immunization of large populations of wild animals is feasible and effective. Preliminary modeling studies indicate that fertility control may have a role in the management of fox rabies by reducing the numbers of foxes in a particular region and thus reducing the transmission rate of the disease. Another consequence of fertility control is that a reduction in the annual birth rate would decrease the numbers of juvenile foxes, which are difficult to reach by conventional vaccination. Juveniles act as an immunologically naive reservoir, each year reducing the effectiveness of baiting campaigns and hence maintaining a rabies reservoir and ensuring its continued spread. Immunocontraception therefore has the potential to reduce the annual birth rate in a permanent way, providing a powerful tool to assist with overcoming this problem.

B. POSSUMS IN NEW ZEALAND

In New Zealand, the brush tail possum (*Trichosurus vulpecula*), an introduced species, causes untold damage to native forests and is the principal vector for the transmission of bovine tuberculosis. Since 1993 possums have been the target of key research programs in New Zealand aimed at exploring new means for their control, including fertility control (Cowan, 1996). In 1995, the Cooperative Research Centre for the Conservation and Management of Marsupials was established in Australia and New Zealand, and one of its major research programs is concerned with the use of immunocontraception for the control and management of possums in New Zealand. The primary target antigens being considered for the possum are the zona pellucida antigens ZP2 and ZP3. Female possums immunized with whole porcine zona pellucida show reduced fertility, and as such this provides supporting evidence for the potential applicability of anti-zona-pellucida-based vaccines for possum fertility control.

C. CATS

The feral cat has been identified in Australia and New Zealand as a major threat to wildlife. Cats are also problems on a number of islands throughout the world and as a result have severely impacted on native wildlife of those islands. Cats may be difficult to control by antifertility means because they are long-lived and are not easily attracted to taking baits containing an antifertility vaccine. Nevertheless, a combination of poison baits to knock down a resident population, combined with the use of fertility control baits, may work to decrease the population in critical areas to levels at which predatory impacts are significantly reduced. Cats would be amenable to delivery of an antifertility vaccine by a recombinant, freely disseminating, cat-specific virus, but this raises difficult questions as to containment of such a control measure and the high risk of cross-infection of domestic cats. Thus such methods are not considered feasible at this time.

III. FERTILITY CONTROL: TARGETS AND IMMUNOLOGICAL INTERVENTION

A. IMMUNOCONTRACEPTION: GENERAL CONSIDERATIONS

A successful contraceptive vaccine will depend on several factors: (1) blocking either fertilization or early embryonic development, (2) primarily targeting the female reproductive tract, (3) being species specific, (4) provoking a prolonged and sustained immune response, and (5) not interfering with the normal social function of the animal. In particular, an effective mechanism for transmitting the vaccine throughout the target population must be found that is both cost effective to manufacture and administer, and that does not impose hazards to the environment.

These caveats make the development of an immunocontraceptive vaccine for wild animals highly challenging. An immunocontraceptive vaccine aimed at preventing fertilization in a species can be targeted at two main sites: proteins important for fertilization on the surface of spermatozoa, or those proteins involved in sperm adhesion to the outer coat of the egg, the zona pellucida. In many species both sites have been studied and candidate antigens have been identified and assessed for their ability to induce contraception.

The Cooperative Research Centre for the Biological Control of Vertebrate Pest Populations (The Vertebrate Biocontrol Control Cooperative Research Centre, or VBCCRC) was established in 1993 to investigate the use of immunocontraception for the control of feral pests in Australia (Tyndale-Biscoe, 1994). The two primary species under investigation have been the European rabbit and the European red fox. Recently work has extended to the control of feral mice, which in Australia are a major problem in the agricultural wheat belt regions.

For fox control the initial aim has been to develop a bait-delivered contraceptive vaccine (Bradley, 1994; Bradley *et al.*, 1997). This approach is broadly based on the success of vaccine delivery in baits for the control of rabies in Europe (Artois *et al.*, 1993) and has the potential to provide a very practical method for long-term fox management. For the rabbit, the proposed delivery would be by a recombinant myxoma virus, a rabbit-specific virus that has been in Australia for at least 40 years (Tyndale-Biscoe, 1994). A recombinant viral delivery system is also being considered for contraceptive vaccine delivery to the mouse. Currently the virus under study is murine cytomegalovirus, which appears specific for *Mus musculus* and apparently is unable to infect native Australian rodent species (Shellam, 1994).

B. GAMETE ANTIGENS: TARGETS OF IMMUNOCONTRACEPTION

1. Sperm Antigens

It has long been demonstrated in several species that the immunization of females with whole sperm preparations can induce immunological infertility (Baskin, 1932; O'Rand, 1977; Bradley, 1994). In the human, it has been recognized that a significant number of cases of female infertility can result due to the presence of antisperm antibodies (Isojima *et al.*, 1968; Menge, 1980). Thus spermatozoa clearly contain antigens that can evoke immune responses capable of interfering either with sperm function or their ability to interact with the egg. Another possibility that emerges from such studies is that some sperm antigens may continue to be expressed on the developing or early implantation blastocysts, and are thus potentially susceptible to the presence of antisperm antibodies, resulting in early embryonic loss (Menge and Naz, 1988). Thus immunization against sperm antigens would be a highly desirable approach because of the potential for antibodies to interfere with reproduction at multiple sites in the reproductive process (D'Occhio, 1994).

Considerable effort has been expended by many different groups to identify sperm antigens that may be useful as contraceptive vaccine targets. Several candidates in this class include FA-1 (Naz, 1987), SP-10 (Beaton *et al.,* 1995; Herr, 1996), PH-20 (Primakoff *et al.,* 1988) and LDH-C4 (Goldberg, 1973, 1974; Goldberg and Shelton, 1986). Although antifertility effects have been reported with LDH-C4 in primates (O'Hern *et al.,* 1995), these results have not always been duplicated when applied to experiments in other species.

There are a number of ways in which candidate sperm antigens can be identified for subsequent study and testing. The most common approach has been to generate libraries of monoclonal antibodies against sperm surface antigens. Selection of target is based on criteria such as antigen cellular location or the ability of these monoclonal antibodies to agglutinate spermatozoa or block fertilization *in vitro.* The antigens recognized by these monoclonal antibodies have either been purified or their cognate genes cloned from appropriate testis or epididymal cDNA libraries (Beaton *et al.,* 1994, 1995; Bradley, 1994). Once completed, recombinant proteins are produced in selected expression systems, purified, and then tested for effects on fertility *in vivo.*

An alternative approach has been to take advantage of the naturally occurring antisperm antibodies present in some infertile women, and to use these sera to screen human testis cDNA libraries to clone and identify potential antifertility antigens. One such study identified a number of unique sperm antigens that had not previously been identified by other means (Liang *et al.,* 1994; Diekman and Goldberg, 1994). The antifertility effects of these antigens produced by recombinant or synthetic means are yet to be fully determined, but the study serves to demonstrate the potential utility of such an approach and may have application for species other than humans.

One sperm antigen, PH-20 has been shown to cause 100% effective infertility in both male and female guinea pigs (Primakoff *et al.,* 1988, 1997; Tung *et al.,* 1997). Unfortunately efforts to date to replicate this effect in other animals have not proved easy. In both the fox and the rabbit (Holland *et al.,* 1997) the PH-20 cDNAs have been cloned; recombinant forms of these antigens produced in bacterial expression systems are used to immunize females of both species. In both foxes and rabbits high-titer antibodies specific to the PH-20 protein were detected in both serum and reproductive tract secretions, but no consistent antifertility effects were observed. This highlights the difficulty of translating the results of one species in which sperm antigens prove to be effective in inducing infertility, to another species, in which the functional aspects of the homolog are not well defined. Whereas purified sperm PH-20 protein was used in the successful immunocontraception of guinea pigs, recombinant PH-20 produced in bacterial expression systems was utilized in the trials with foxes and rabbits. Therefore, further consideration needs to be given to the source of the potential immunocontraceptive antigen.

The sperm antigen SP-10 has previously been identified as a potential candidate for immunocontraceptive vaccine development (Kerr, 1995). SP-10 is located within the inner acrosomal membrane and the gene encoding this protein has been cloned from several species, including primates, humans, and the fox (Beaton *et al.,* 1995). The function of SP-10 is unknown. Although *in vitro* experiments with monoclonal antibodies to SP-10 have previously demonstrated an inhibition in fertilization (Liu *et al.,* 1990), there have been no reports of the effects of the antigen *in vivo.* Fox SP-10 (fSP-10) has been expressed as a maltose-binding protein fusion protein in *Escherichia coli* (Beaton *et al.,* 1995) and these preparations have been used to immunize female foxes. Although specific antibodies were produced in both serum and reproductive tract secretions to fSP-10, no effects on fertility were observed. It is possible that the intracellular location of the antigen makes immunological intervention against this protein difficult. This highlights a particular problem with sperm antigens as targets—unless the proteins are directly accessible to the female reproductive tract secretions, and for a prolonged period of time, it may be difficult to target such proteins immunologically.

Lactate dehydrogenase C4 (LDH-C4) (Bradley *et al.,* 1996) has been studied intensively because its homolog in other species has been shown to induce infertility (O'Hern *et al.,* 1995). For example, the fox LDH-C4 cDNA has been successfully cloned and sequenced, and an antigenic peptide (ENLIEEKISQK) corresponding to the same region previously identified in other species as both antigenic and capable of conferring infertility when used as an immunogen has been synthesized (Bradley *et al.,* 1996; O'Hern *et al.,* 1995). A fertility trial conducted with this chimeric peptide indicated that antibodies to the LDH-C4 peptide are capable of reducing fertility in foxes, a result consistent with those obtained previously in baboons (Bradley *et al.,* 1997; O'Hern *et al.,* 1995) and indicating that further work with this peptide as an immunocontraceptive antigen in other species may be warranted.

Overall, the lack of reproducible results of antifertility effects with sperm antigens in a variety of species serves to highlight the difficulty of using spermatozoa as a target for contraception. Despite the large number of investigations undertaken in this area, little conclusive evidence has been produced of effective antigens for this type of application. It may well be time to reassess carefully the approaches used for the identification of such antigens. One approach that may in the future prove useful would be the conclusive identification of the ZP3 sperm receptor. For example, in the mouse a protein, sp56, has been identified as a possible primary sperm adhesion molecule that interacts specifically with ZP3 protein (Cheng *et al.,* 1994) (see Chapter 5, this volume). Whether such a protein has applicability as a contraceptive target awaits conclusive evidence from *in vivo* immunological trials.

An alternative approach recently adopted by some groups has been to start developing maps of sperm surface proteins. This involves using the relatively new "proteome technology," which combines high-resolution two-dimensional gel electrophoresis of sperm membrane protein preparations followed by excision and rapid mass spectrometric analysis of these proteins to determine either partial amino acid sequence or composition (Wilkens *et al.,* 1997; Herr, 1996). These data can then be used to search proteome analyses databases such as those found at the web site http://expasy.org. The searches aid in either the identification of the protein of interest or the functional class to which a protein might belong based on amino acid composition and chemical characteristics. Such basic research will ultimately provide us with a comprehensive picture of the proteins that comprise the sperm surface. Combined with high-resolution microscopic imaging this approach may well yield the best information for the selection of sperm-based immunocontraceptive candidates.

2. Zona Pellucida Antigens

The zona pellucida of the egg offers another major target for immunological intervention of fertility. The egg zona pellucida of most species examined contains only three proteins, ZP1, ZP2, and ZP3 (Mahi-Brown and Tung, 1995). Although these proteins all share regions of homology between species, the cloning of the genes encoding these proteins has revealed some differences in their amino acid sequences that reflect the species-specific functions of these proteins (Epifano and Dean, 1994). ZP3 has been identified as the primary protein to which sperm adhere (Rosiere and Wassarman, 1992). Blocking the function of this protein with antibodies results in infertility (Epifano and Dean, 1994), and this has been experimentally demonstrated in a number of species (Prasad *et al.,* 1996). Therefore, ZP3 can be considered an important protein target for immunocontraception in a number of species.

In the male, gametes are protected from the immune system, whereas in the female, zona pellucida antigens are exposed to the immune system. Immune protection is afforded probably through the development of innate tolerance, although when this breaks down, immune disorders such as autoimmune oophoritis can develop (Mahi-Brown and Tung, 1995). Immunization of a range of species, such as horses, cats, hamsters, rats, or rabbits, with whole preparations of porcine zona pellucida lead to the induction of antizona pellucida antibodies and often the induction of infertility. However, the degree of both of these sequelae depends on the routes of immunization, choice of adjuvant, and the immune responses of the recipient species.

Experiments with porcine zona pellucida (PZP) demonstrate the applicability of targeting its antigens for immunocontraception, although it remains to be determined whether autologous versus heterologous delivery of zona antigens will be required for the generation of high levels of anti-ZP3 antibodies. Regardless, isolated porcine zonae pellucidae have already been used in very practical ways in the application of fertility control and wildlife management. Examples include the immunization of horses on Assateague Island National Seashore with whole PZP (Kirkpatrick *et al.,* 1996). In one such trial started in 1988, 26 female horses were administered two to three doses of 65 μg of PZP by remote dart delivery. The horses showed no alterations in social behavior, and fertility was compromised by 83% compared to nonimmunized control mares. Contraception was effective for up to 1 year following such a regime.

Similar results have been obtained with the control of fertility in white-tailed deer with porcine zona pellucida antigens. In this instance, 10 females were immunized twice with 65 μg of PZP, resulting in 100% contraception in this group; nonimmunized females showed 82% fertility (Kirkpatrick *et al.*, 1996). There was no evidence of autoimmune damage to the ovary as evidenced by normal endocrine profiles in all of the immunized females. Thus, PZP is a highly effective antigenic complex that already has practical application for the management of wild animal populations through fertility control.

In the examples cited here there appeared to be minimal effects on social behavior and an apparent lack of side effects in the animals in question. The remote delivery provided containment of the vaccine and posed no environmental impacts. As a result of the success with these small groups of animals, interest in the use of PZP for the management of fertility in captive animals has grown. For example, in zoological institutions contraception of animals has been traditionally confined to the use of steroids or of surgical sterilization (Cohnen, 1995). Both procedures can lead to pathology and unwanted stress and side effects. However, there is an increased interest in the application of PZP for fertility control, and to date 68 different species have been administered PZP, with successful immunocontraception being reported in 27 of these species.

Zona pellucida-induced infertility can be the result of two actions: immunological or antibody-mediated blocking of ZP sperm receptor sites, and the cytotoxic T cell-mediated destruction of developing ovarian follicles. Although this latter mode of interference is totally undesirable for humans and some target animal species, it is in some instances highly desirable for inducing permanent sterility in some feral pest populations.

3. Refining the Zona Pellucida Vaccine

Considerable effort has focused on the epitopes within the ZP3 protein that are important in the induction of the appropriate immune responses (Lou *et al.*, 1996; Afzalpurkar *et al.*, 1997). Miller *et al.* (1989) demonstrated that for mouse ZP3, an epitope spanning amino acids 330–340 was one of the major sequences that induced both B and T cell immune responses, and when administered to mice as a keyhole limpet hemocyanin (KLH) conjugate, it caused 75% infertility. Careful dissection of this epitope (Lou *et al.*, 1996) subsequently demonstrated that amino acids 330–335 contained the B cell epitope and 335–340 contained the T cell epitope. This region of mouse ZP3 is of considerable interest not only for its potential for use in certain immunocontraceptive vaccines but also because careful study of this region of ZP3 proteins in other species shows it has considerable variation, and may be a key species-restricted epitope with important functional sperm adhesion activity.

The availability of the cDNAs that encode both the fox and possum ZP3 proteins has allowed the production of recombinant proteins in a baculovirus expression system in order to obtain glycosylated preparations of the antigens for *in vivo* testing. Although it is recognized that the pattern of ZP glycosylation by baculovirus may not be identical to that of the native protein, other studies indicate that ZP proteins produced in this system are antigenic and capable of producing antisera that block sperm–egg adhesion (Prasad *et al.,* 1996). The generation of recombinant proteins in such an expression system, in combination with immunization of the species of interest with the recombinant protein, allows the generation of polyclonal antisera to these antigens; these antisera can in turn be used in epitope mapping studies. Ultimately certain zona pellucida-based vaccines may be based on selective epitopes chosen in order to manipulate selectively the immune responses for the particular target species.

IV. REPRODUCTIVE TRACT IMMUNE RESPONSES

A. THE TARGET SITE

An understanding of the induction, modulation, and duration of reproductive tract immune responses is central to the successful development of immunocontraceptive vaccines (Quayle and Anderson, 1995). The precise reproductive tract target site will depend to a major extent on the vaccine antigens employed for this process. For example, if the vaccine is primarily composed of sperm-derived antigens, the target sites within the female reproductive tract will primarily focus on inducing significant sperm antibody responses within the vaginal, cervical, uterine, and oviductal secretions. It has previously been shown that the reproductive tract is a mucosal effector site linked to the common mucosal system (McGhee *et al.,* 1994). This means that oral route of immunization could probably be employed as the route for vaccine delivery, provided sufficiently high antibody levels can be generated at these sites by this administration route.

In considering the case of zona pellucida antigens as contraceptive targets, immune intervention can occur at two potential sites—within the ovary prior to ovulation, or at a later stage within the oviduct, where secreted antibody would prevent sperm zona interaction. It is known that follicular fluid contains IgG that transudates from serum. Hence ZP3 antibodies have the potential of coating preovulatory oocytes. Oral immunization can also evoke serum IgG responses, thus, provided they are of a high sustainable duration, this form of delivery should also prove useful for immunization against oocyte antigens.

Immunization against zona pellucida antigens can also result in damage to ovarian follicles and follicular growth, which in turn results in disruptions to normal hormonal activity and abnormal estrous cycles in some immunized females. These additional effects on the ovary are not seen in all animals immunized with ZP antigens and, indeed, the degree of ovarian pathology that results is often highly variable (Mahi-Brown and Tung, 1995). However, for domestic species or wild animal populations, the longer term perturbations of ovarian function that result from these immune responses may be beneficial in sustaining a long-term infertility in the target species.

A number of studies have demonstrated that the sex hormones, estrogen and progesterone, can influence IgG and IgA antibody production within the female reproductive tract (Wira and Sandoe, 1987; de Jersey *et al.*, 1997). If the immunocontraceptive vaccine has been constructed in such a way that it is dependent on high levels of reproductive tract antibodies, then changes in the localized IgG and IgA concentrations during the estrous period maybe an important factor to consider. Another factor that may need to be addressed in the administration of an immunocontraceptive vaccine to an outbred population is the variability of the immune response between individuals. Effective application of a vaccine for fertility control requires that a high level of immunity be achieved among individuals exposed to the vaccine. It may therefore be necessary to include multiple antigenic determinants within a vaccine to stimulate a broad range of immune responses. In addition, the antigens may need to be presented in conjunction with other highly immunogenic carrier proteins to maintain a contraceptive level of immunity.

V. BAIT DELIVERY OF AN IMMUNOCONTRACEPTIVE VACCINE

The delivery of an antifertility vaccine to wild animal populations over large areas raises a number of unique problems that require careful consideration. Parameters to consider include the distribution of the species under study, whether large-scale or localized control is desired, and the issue of directed specificity with regard to other species. For each species a unique set of conditions will be imposed.

A number of different vaccine delivery systems are being assessed to determine which will be the most effective for immunocontraception of animal populations. These include dart delivery, disseminating viral vectors, and oral bait systems. For many species under consideration bait delivery will be the best method of choice, and this discussion will focus on bait delivery. Such a bait will need to be designed to be acceptable for the target species, and each bait will contain within it the contraceptive vaccine, either as the recombinant protein or in the form of a microbial replicating vector that will infect the host, produce the recombinant antigen, and stimulate the immune response.

The selection of the most appropriate delivery system for inclusion in a bait, for example, will be dependent on a number of factors, such as the nature of the antigen to be included in the vector (i.e., glycosylated versus nonglycosylated), the safety of the delivery system, and the type of immune response the vector assists in inducing (Cryz, 1996). Some of the potential vectors are considered here.

A. VIRAL VECTORS

There are several different viral vectors currently being investigated for their potential as baited-vaccine delivery systems. These include poxviruses, adenoviruses, and herpesviruses, with the poxviruses being the most thoroughly investigated to date. Vaccinia virus is the live poxvirus vaccine that has been used to eradicate smallpox. Although it is neither an attenuated form of the variola virus that causes smallpox, nor is it the cowpox virus, it is an orthopox virus whose origin and host are unknown (Smith, 1996). There have been many heterologous antigens expressed in vaccinia virus and its potential to induce both an antibody- and a cell-mediated immune response has been well documented. Because vaccinia virus caused vaccination complications in some individuals, there has been a concerted effort to produce an attenuated form of this virus for use as a vaccine delivery system for heterologous antigens (Smith, 1996).

The use of recombinant vaccinia viral vectors expressing genes encoding selected gamete antigens may offer an excellent delivery system for an immunocontraceptive vaccine, particularly for those antigens that are highly glycosylated, and in which posttranslational modification is important for the generation of immune responses to functionally important domains. Such a system would also allow further studies on the enhancement of mucosal immunity, possibly by constructing vectors that coexpress IgA-specific stimulating cytokines (Ramsay *et al.,* 1994).

B. BACTERIAL VECTORS

There have been many bacterial species targeted for their potential as either vaccines or delivery systems for heterologous antigens (Cryz, 1996, Lintermans and De Greve, 1995, Hodgson, 1994). These include both gram-positive and gram-negative strains, with *Salmonella typhimurium* being the most extensively studied for its potential as a delivery vector. Considerable effort has resulted in the development of attenuated strains of *Salmonella,* with the most common type in current usage being those strains that are deficient in one or more genes essential for the synthesis of aromatic amino acids and the regulation of cAMP. For example, the AroA mutants are auxotrophic for *p*-aminobenzoic acid and dihydroxybenzoate, nutrients that are not available in eukaryotes, which means that the *Salmonella* are not able to grow in the host. Although immunization of animals with Aro mutant strains has demonstrated no pathological consequences, it has resulted in a local secretory, humoral, and cell-mediated response (Cryz, 1996). This suggests that *Salmonella* would be a useful delivery system for antigens targeting either fertility or the ovary.

Oral immunization experiments with *S. typhimurium* in foxes have demonstrated that immune responses to *Salmonella* are of sufficient intensity and duration to warrant continuing work with these bacteria (de Jersey *et al.,* 1997). A *S. typhimurium* recombinant expressing the fox LDH-C4 antigen and delivered oral-ly to foxes was capable of inducing a mucosal immune responses to the LDH-C4 protein (Bird *et al.*, 1996). Similarly, *S. typhimurium* expressing the human sperm SP-10 antigen has induced antibodies in mice (Srinivasan *et al.*, 1995). Recombinant *Salmonella* bacteria expressing murine ZP3 have been constructed, and oral immunization of BALB/c mice induced significant anti-ZP3 IgG antibodies in serum and IgA antibodies in vaginal secretions, with three out of the six females immunized being infertile (Zhang *et al.*, 1997).

Salmonella has the advantage in that it can be produced cheaply in large quantities using standard fermentation technology. Nevertheless, if *Salmonella* is used in a wildlife immunocontraceptive product it will be important that the bacteria are in a usable form. Freeze-dried *Salmonella* would meet the criteria for field delivery. Initial studies of oral delivery of freeze-dried *S. typhimurium* to mice and foxes indicate that both systemic and mucosal immune responses to the *Salmonella* (Bradley *et al.*, 1997) can be generated.

The requirements for environmental safety of a bait-delivered bacterial immunocontraceptive vaccine need to be addressed. One feature that would prevent any chance of foreign DNA getting into the environment would be a vaccine that does not contain any gamete DNA. One approach is to construct suicide plasmid systems that will destroy any gamete DNA before the vaccine is released into the environment. Initial *in vitro* tests show that such systems are feasible.

C. BACTERIAL GHOSTS

Cloning and subsequent expression of the PhiX174 *geneE* in bacteria can result in the lysis of a variety of gram-negative bacteria, including *E. coli, S. typhimurium, Vibrio cholerae, Klebsiella pneumoniae,* and *Actinobacillus*. This phage lysis results in a transmembrane tunnel through which the cytoplasmic contents of the bacteria are extruded, yielding a nonliving candidate vaccine delivery system: ghost bacteria that contain only membrane-associated recombinant antigen. Recombinant bacterial ghosts are cheap to produce, can be stored for long periods, and can contain multiple antigenic determinants that are present in a highly immunostimulatory environment (Szostak *et al.*, 1996). These features make bacterial ghosts an attractive delivery system for immunocontraceptive antigens. It remains to be determined whether these preparations produce immunity after oral delivery.

D. PLANTS

Plants are potential bioreactors for recombinant biopharmaceuticals, including vaccines (Mason and Arntzen, 1995). Transgenic plants have application for the production of inexpensive vaccines in developing nations and would be suited to delivery of immunocontraceptive vaccines to many herbivore species. A contraceptive epitope of the mouse ZP3 (spanning amino acid residues 336–342) was inserted into the tobacco mosaic virus to produce a hybrid coat protein. When the virus was used to inoculate plants the ZP3 epitope was expressed as part of the hybrid virus protein. Parenteral immunization of mice with purified recombinant viral coat particles produced a serum antibody response to the ZP3 epitope (Fitchen *et al.,* 1995). A transgenic tobacco has been developed that expresses the fox LDH-C4 protein (I. Polkinghorne, 1997, personal communication). Clearly, the technology now exists to begin constructing plant-based vaccines, and this is an area of research that will rapidly evolve as more effective plant transformation systems are developed. Smith *et al.* (1997) review the current progress in the use of plant-derived vaccines and especially how this technology may be useful for the delivery of immunocontraceptive antigens to free-ranging wildlife.

E. SYNTHETIC DELIVERY SYSTEMS

There has been much research directed at particulate delivery systems for parenteral vaccines, but oral administration of vaccines is becoming increasingly desirable throughout the world and consequently there is growing research on oral particulate delivery systems. These systems include immunostimulating complexes (iscoms) (Quil-A, cholesterol, phospholipid constructs), microspheres (polylactide–coglycolide, polyphosphazenes), and liposome emulsions (Davis, 1996), which all have potential as delivery systems for immunocontraceptive antigens. A study to evaluate the efficacy of microspheres containing a recombinant sperm antigen to stimulate a mucosal immune response in rats (Muir *et al.,* 1994) demonstrates the potential utility of these agents. Microspheres were synthesized using the poly(DL-lactide)–coglycolide copolymer incorporating a recombinant source of the fox sperm protein FSA-1r (Beaton *et al.,* 1995). The oral administration of FSA-1r-loaded microspheres to rats resulted in a significant production of cells within the jejunum that were secreting IgA antibodies specific for the FSA-lr antigen. The level of stimulation was comparable to that obtained by direct immunization of the Peyer's patches with microspheres containing either antigen or unencapsulated antigen. However, the current cost of production would make these systems more suited to human and companion animal vaccination rather than broadscale application to a wildlife population. Nevertheless, the per unit production cost will decrease as these systems become more popular and production technology improves.

VI. CONCLUDING REMARKS

The application of fertility control technologies for the management of animal populations is still a relatively new concept. Although there have been demonstrations of its utility using crude antigenic preparations of porcine zona pellucida, refined, cheap, and easily delivered vaccines are still in the early phases of development and testing, especially those required for remote delivery to free-ranging wild animal populations. The relatively slow emergence of such technologies for practical application clearly reflects the complexity of the problem and highlights the requirement for a multidisciplinary research approach when attempting to develop such technologies for wild animal management. In many cases, one of the most difficult areas requiring further information is the immunobiology of such target species.

Effective immunocontraceptive vaccines will also be critically dependent on the design of the delivery systems. These will be unique to each species under study. A major challenge for free-ranging species will be to ensure that the vaccine can induce a long-lasting immune response to a high percentage of the target population, thereby reducing the frequency and hence the cost of vaccine application.

Many of the delivery systems being considered will utilize recombinant organisms. Hence considerations on the use of such organisms need to take into account the political, ethical, and safety constraints prior to any environmental release. Also of prime importance is the issue of species specificity. In some instances this will be a major challenge and careful thought will need to be given to ensure that specificity can be built into the vaccine at several levels. These might include the target antigen or epitopes, the microbial or other delivery vector, and, where applicable, the bait matrix and any target-specific attractive properties it may require.

If these caveats can be satisfied and the public is able to accept the use of such vaccines, the potential benefits for using them to manage animal populations and their impact are most certainly assured. If proved a successful technology in the field, fertility control will also provide an excellent example of the application of basic scientific knowledge to the solution of a complex biological problem.

ACKNOWLEDGMENTS

This work has been supported by the Perth Zoological Gardens, the Vertebrate Biocontrol Cooperative Research Centre, CSIRO Australia, and the Marsupial Cooperative Research Centre.

REFERENCES

Afzalpurkar, A., Sacco, A. G., Yurewicz, E. C., and Gupta, S. K. (1997). Induction of native protein reactive antibodies by immunization with peptides containing linear B-cell epitopes defined by antiporcine ZP3B monoclonal antibodies. *J. Reprod. Immunol.* **33**, 113–125.

Artois, M. (1997). Managing a problem of wildlife in the "Old World": A veterinary perspective. *Reprod. Fertil. Dev.* **9**, 17–25.

Artois, M., Mason, E., Barrat, J., and Aubert, M. (1993). Efficacy of three oral rabies vaccine baits in the red fox: A comparison. *Vet. Microbiol.* **38**, 167–172.

Aubert, M. (1995). Oral vaccination of wildlife against rabies in Europe. *In* "Proceedings from the XXVth World Veterinary Congress, Yokohama," pp. 193–200. Intervet Publisher, Bangalore.

Baskin, M. J. (1932). Temporary sterilization by the injection of human spermatozoa. A preliminary report. *Am. J. Obstet. Gynecol.* **24**, 892–897.

Beaton, S., Cleary, A., ten Have, J., and Bradley, M. P. (1994). Cloning and characterisation of a fox sperm protein FSA-1. *Reprod. Fertil. Dev.* **6**, 761–770.

Beaton, S., ten Have, J., Cleary, A., and Bradley, M. P. (1995). Cloning and partial characterisation of the cDNA encoding the fox sperm antigen FSA-Acr. 1 with similarities to the SP-10 antigen. *Mol. Reprod. Dev.* **40,** 242–252.

Bird, P. H., Hayes, C. M., and Bradley, M. P. (1996). The use of *Salmonella typhimurium* to deliver a recombinant vaccine expressing fox LDH-C4. *In* "Proceedings of the Vaccines Beyond 2000," p. 48. Australia.

Bradley, M. P. (1994). Experimental strategies for the development of an immunocontraceptive vaccine for the european red fox *Vulpes vulpes. Reprod. Fertil. Dev.* **6,** 307–317.

Bradley, M. P., Geelan, A., Leitch, V., and Goldberg, E. (1996). Cloning, sequencing and characterization of LDH-C4 from a fox testis cDNA library. *Mol. Reprod. Dev.* **44,** 452–459.

Bradley, M. P., Hinds, L. A., and Bird, P. H. (1997). A bait-delivered vaccine for the European red fox *(Vulpes vulpes)* by the year 2002? *Reprod. Fertil. Dev.* **9,** 111–116.

Cheng, A., Le, T., Palacios, M., Bookbinder, L. H., Wassarman, P. M., Suzuki, F., and Bleil, J. D. (1994). Sperm–egg recognition in the mouse: Characterization of sp56, a sperm protein having specific affinity for ZP3. *J. Cell Biol.* **125,** 867–878.

Cohnen, K. (1995). Review of reversible and irreversible contraceptive methods in captive, exotic animals. *In* "Research and Captive Propagation" (U. GansloBer and J. K. Hodges, eds.), pp. 81–89. Filander-Verlag, Furth.

Cowan, P. E. (1996). Possum biocontrol: Prospects for fertility regulation. *Reprod. Fertil. Dev.* **8,** 655–660.

Cowan, P. E., and Tyndale-Biscoe, C. H. (1997). Australia and New Zealand mammal species considered to be pests or problems. *Reprod. Fertil. Dev.* **9,** 27–36.

Cryz, S. J. (1996). Vaccine delivery systems: Live attenuated bacterial vaccine vector systems. *Vaccine* **14,** 676–80.

Davis, S. S. (1996). Vaccine delivery systems: Particulate delivery systems. *Vaccine* **14,** 672–675.

de Jersey, J., Hinds, L. A., and Bradley, M. P. (1997). Regulation of reproductive tract immunoglobulins by oestrodial-17B in the European red fox (*Vulpes vulpes*). *Reprod. Fertil. Dev.* **9,** 531–538.

Diekman, A. B., and Goldberg, E. (1994). Characterization of a human antigen with sera from infertile patients. *Biol. Reprod.* **50,** 1087–1093.

D'Occhio, M. J. (1994). Endocrine, gamete and conceptus antigens as targets for immunocontraceptive vaccines: Applications in farm animals. *In* "Vaccines in Agriculture" (P. R. Wood, P. Willadsen, J. E. Vercoe, R. M. Hoskinson, and D. Demeyer, eds.), pp. 71–80. CSIRO Publications, Australia.

Epifano, O., and Dean, J. (1994). Biology and structure of the zona pellucida: A target for immunocontraception. *Reprod. Fertil. Dev.* **6,** 319–330.

Fitchen, J., Beachy, R. N., and Hein, M. B. (1995). Plant virus expressing hybrid coat protein with added murine epitope elicits autoantibody response. *Vaccine* **13,** 1051–1057.

Goldberg, E. (1973). Infertility in female rabbits immunized with lactate dehydrogenase X. *Science* **181,** 458–459.

Goldberg, E. (1974). Effects of immunization with LDH-X on fertility. *Acta Endocrinol.* **7,** 202–222.

Goldberg, E., and Shelton, J. (1986). Immunosuppression of fertility by LDH-C4. *In* "Immunological Approaches to Contraception and Promotion of Fertility" (G. P. Talwar, ed.), p. 219. Plenum Publishing, New York.

Herr, J. (1996). Update on the center for recombinant gamete contraceptive vaccinogens. *Am. J. Reprod. Immunol.* **35,** 184–189.

Hodgson, A. L. M. (1994). Bacterial vaccine vectors. *In* "Immunological Applications to Animal Health and Production" (P. R. Wood, ed.), pp. 177–186. CSIRO Publications, Australia.

Holland, M. K., Andrews, J., Clarke, H., Walton, C., and Hinds, L. A. (1997). Selection of antigens for use in a virus-vectored immunocontraceptive vaccine: PH-20 as a case study. *Reprod. Fertil. Dev.* **9,** 117–124.

Isojima, S., Li, T. S., and Ashitaka, Y. (1968). Immunologic analysis of sperm-immobilizing factor found in sera of women with unexplained sterility. *Am. J. Obstet. Gynecol.* **101,** 677–683.

Kerr, L. E. (1995). Sperm antigens and immunocontraception. *Reprod. Fertil. Dev.* **7,** 825–830.

Kinnear, J. E., Onus, M. L., and Bromilow, R. N. (1988). Fox control and rock wallaby dynamics. *Austral. Wildlife Res.* **15,** 435–450.

Kirkpatrick, J. P., Turner, Jr., J. W, Liu, I. K. M., and Fayrer-Hosken, R. (1996). Applications of pig zona pellucida immunocontraception to wildlife fertility control. *J. Reprod. Fertil.* (Suppl.) **50,** 183–189.

Liang, Z.-G., O'Hern. P. A., Yavetz, B., and Yavertz, H., and Goldberg, E. (1994). Human testis cDNAs identified by sera from infertile patients: A molecular biological approach to immunocontraceptive vaccine development. *Reprod. Fertil. Dev.* **6,** 297–305.

Lintermans, P., and De Greve, H. (1995). Live bacterial vectors for mucosal immunization. *Adv. Drug Deliv. Rev.* **18,** 73–89.

Liu, M., Chan, K., Lau, Y., and Lee, C. G. (1990). Molecular cloning of an acrosomal sperm antigen gene and the production of its recombinant protein for immunocontraceptive vaccine. *Mol. Reprod. Dev.* **25,** 302–308.

Lou, Y.-H., Bagavant, H., Ang, J., McElveen, Thai, H., and Tung, K. S. K. (1996). Influence of autoimmune ovarian disease pathogenesis on ZP3 contraceptive vaccine design. *J. Reprod. Fertil.* (Suppl.) **50,** 159–163.

Mahi-Brown, C. A., and Tung K. S. K. (1995). Vaccines against egg antigens and pathogenesis of ovarian autoimmune disease. *In* "Birth Control Vaccines" (G. P. Talwar and R. Raghupathy, eds.), pp. 41–61. R.G. Landes Company, Austin Texas.

Mason, H. S., and Arntzen, C. J. (1995). Transgenic plants as vaccine production systems. *Trends Biotechnol.* **13,** 399–402.

McGhee, J. R., Xu-Amano, J., Millar, C. J., Jackson R. J., Fujihashi, K., Staats, H. F., and Kiyono, H. (1994). The common mucosal immune system: From basic principles to enteric vaccines with relevance for the female reproductive tract. *Reprod. Fertil. Dev.* **6,** 369–379.

Menge, A. C. (1980). Clinical immunologic infertility: Diagnostic measures. Incidence of antisperm antibodies fertility and mechanisms. *In* "Immunological Aspects of Infertility and Fertility Regulation" (D. S. Dhindsa and G. F. B. Schumacher eds.), pp. 205–224. Elsevier, North Holland.

Menge, A. C., and Naz, R. K. (1988). Immunologic reactions involving sperm cells at and preimplantation embryos. *Am. J. Reprod. Immunol.* **3,** 492–495.

Miller, S. E., Chamow, S. M., Baur, A. W., Olive, C., Robey, F., and Dean, J. (1989). Vaccination with a synthetic ZP peptide produces long term contraception in female mice. *Science* **246:** 935–938.

Muir, W., Gipps, E., Husband, A., and Bradley, M. P. (1994). Intestinal vaccination using biodegradable microspheres. *Immunol. Lett.* **42,** 203–207.

Naz, R. K. (1987). The fertilization antigen (FA-1) causes reduction of fertility in actively immunized female rabbits. *J. Reprod. Immunol.* **11,** 117–133.

O'Hern, P. A., Bambra, C. S., Isahakia, M., and Goldberg, E. (1995). Reversible contraception in female baboons immunized with a synthetic epitope of sperm-specific lactate dehydrogenase. *Biol. Reprod.* **52,** 331–339.

O'Rand, M. G. (1977). The presence of sperm-specific isoantigen on the egg following fertilization. *J. Exp. Zool.* **202,** 267–293.

Prasad, S. V., Wilkins, B., and Dunbar, B. S. (1996). Molecular biology approaches to evaluate species variation in immunogenicity and antigenicity of zona pellucida proteins. *J. Reprod. Fertil.* (Suppl.) **50,** 143–149.

Primakoff, P., Lathrop, W., Woolman, L., Cowan, A , and Myles, D. (1988). Fully effective contraception in male and female guinea pigs immunized with the sperm protein PH20. *Nature* **335,** 543–546.

Primakoff, P., Woolman-Gamer, L., Tung, K. S., and Myles, D. G. (1997). Reversible contraceptive effect of PH-20 immunization in male guinea pigs. *Biol. Reprod.* **56,** 1142–1146.

Quayle, A. J., and Anderson, D. J. (1995). Induction of mucosal immunity in the genital tract and prospects for oral vaccines. *In* "Birth Control Vaccines" (G. P. Talwar and R. Raghupathy, eds.), pp. 149–159. R.G. Landes Company, Austin Texas.

Ramsay, A. J., Leong, K.-H., Boyle, D., Ruby, J., and Ramshaw, I. A. (1994). Enhancement of mucosal IgA responses by interleukins 5 and 6 encoding recombinant vaccine vectors. Reprod. Fertil. Dev. **6,** 389–392.

Rosiere, T. K., and Wassarman, P. M. (1992). Identification of a region of mouse zona pellucida glycoprotein mZP3 that possesses sperm receptor activity. *Dev. Biol.* **154,** 309–317.

Shellam, G. R. (1994). The potential of murine cytomegalovirus as a viral vector for immunocontraception. *Reprod. Fertil. Dev.* **6,** 401–409.

Sinclair, A. R. E. (1997). Fertility control of mammal pests and the conservation of endangered marsupials. *Reprod. Fertil. Dev.* **9,** 1–16.

Smith, G. L. (1996). Vaccine delivery systems: Genetically engineered viruses as candidate vaccines. *Vaccine* **14,** 681–83.

Smith, G. L., Walmsley, A., and Polkinghome, I. (1997). Plant-derived immunocontraceptive vaccines. *Reprod. Fertil Dev.* **9,** 85–89.

Srinivasan, J., Tinge, S., Wright, R., Herr, J. C., and Curtiss III, R. (1995). Oral immunization with attenuated *Salmonella* expressing human antigen induces antibodies in serum and the reproductive tract. *Biol. Reprod.* **53,** 462–471.

Szostak, M. P., Hensel, A., Eko, P. O., Klei, R., Auer, T., Mader, H., Haslberger, A., Bunka, S., Wanner, G., and Lubitz, W. (1996). Bacterial ghosts: Non-living candidate vaccines. *J. Biotechnol.* **44,** 161–170.

Tung, K. S. K., Primakoff, P., Woolman-Gamer, L., and Myles, D. G. (1997). Mechanism of infertility in male guinea pigs immunized with sperm PH-20. *Biol. Reprod.* **56,** 1133–1141.

Tyndale-Biscoe, C. H. (1994). Virus-vectored immunocontraception of feral mammals. *Reprod. Fertil. Dev.* **6,** 281–287.

Tyndale-Biscoe, C. H., and Bradley, M. P. (1997). Vaccination for wildlife contraception. *In* "Veterinary Vaccinology" (P.-P. Pastoret, J. Blancou, P. Vannier, and C. Verschueren, eds.), pp. 492–498. Elsevier Science, Amsterdam.

Wilkins, M. R., Williams, K. L., Appel, R. D., and Hochstrasser, D. F. (1997). "Proteome Research: New Frontiers in Functional Genomics." Springer-Verlag, Berlin, Heidelberg, and New York.

Wira, C. R., and Sandoe, C. P. (1987). Specific IgA and IgG antibodies in secretions of the female reproductive tract: Effects of immunization and estradiol on expression of this response *in vivo. J. Immunol.* **138,** 4159–4164.

Zhang, X., Lou, Y., Koopman, M., Doggett, T., Tung, K. S. K., and Curtiss III, R. (1997). Antibody responses and infertility in mice following oral immunization with attenuated *salmonella typhimurium* expressing recombinant murine ZP3. *Biol. Reprod.* **56,** 33–41.

INDEX